21世纪生物学规划教材

遗传学（第10版）

（双语版）

Essentials of Genetics

10th Edition

【美】威廉·S. 克卢格（William S. Klug）
【美】迈克尔·R. 卡明斯（Michael R. Cummings）
【加】夏洛特·A. 斯潘塞（Charlotte A. Spencer） 著
【美】迈克尔·S. 帕拉迪诺（Michael A. Palladino）
【美】达雷尔·J. 基利安（Darrell J. Killian）

宣劲松 编译

著作权合同登记号 图字：01-2022-1540
图书在版编目（CIP）数据

遗传学：第10版：双语版：汉、英 / [美]威廉 · S. 克卢格（William S. Klug）等著；宣劲松编译. —北京：北京大学出版社，2023.6
ISBN 978-7-301-34106-3

Ⅰ. ①遗…　Ⅱ. ①威… ②宣…　Ⅲ. ①遗传学－教材－汉、英　Ⅳ. ①Q3

中国国家版本馆CIP数据核字(2023)第110203号

Title: Essentials of genetics / William S. Klug, Michael R. Cummings,Charlotte A. Spencer, Michael A. Palladino, Darrell J. Killian.
Description: Tenth edition. | Hoboken : Pearson, 2019.
Identifiers: LCCN 2018051574 | ISBN 9780134898414 (pbk.)

书　　名　**遗传学（第10版）（双语版）**
YICHUANXUE（DI-SHI BAN）（SHUANGYU BAN）
著作责任者　[美]威廉 · S. 克卢格（William S. Klug）等著　宣劲松　编译
责任编辑　黄　炜　刘　洋
标准书号　ISBN 978-7-301-34106-3
出版发行　北京大学出版社
地　　址　北京市海淀区成府路205号　100871
网　　址　http://www. pup. cn　新浪微博：@北京大学出版社
电子信箱　zpup@ pup. cn
电　　话　邮购部 010-62752015　发行部 010-62750672　编辑部 010-62764976
印 刷 者　北京飞达印刷有限责任公司
经 销 者　新华书店
787毫米×1092毫米　16开本　34.25印张　817千字
2023年6月第1版　2023年6月第1次印刷
定　　价　98.00元

序

遗传学是生命科学领域中的一门核心学科，它发展十分迅猛，为现代生物学的研究提供了一个大的框架，并把所有的生物学分支学科连接在了一起。遗传学一般是国内外综合性大学以及师范、农林、医药等院校相关专业开设的学科基础课程。为了让学生的知识结构能够与国际本科教育更好地接轨，我们有必要引进优秀的国外原版教材，以便于国内遗传学教学国际化的开展。*Essentials of Genetics* 是包括华盛顿大学在内的 40 多所国外高校采用的主流教材，内容全面、注重更新，目前国外最新版本为第 10 版（2019 年出版）。

当前，教学目标和教学内容需要不断融合创新，时代精神、学科前沿、能力培养等均要融合到教学之中，教材作为教学的基本工具是强化本科教育的重要载体。我们依托 *Essentials of Genetics*（第 10 版）编译了《遗传学（第 10 版）（双语版）》，将教材内容向“少而精”的方向进行了改编，以帮助学生提升阅读和使用原版教材的能力。在编译过程中，为了强化对学生遗传分析能力的培养，我们保留了原版教材中的经典遗传学部分，精减了分子遗传学相关内容；为了提高学生对遗传学的学习兴趣，拓宽学生的新知发现视野，我们同时保留了原版教材中的现代遗传学进展部分；为了锻炼学生解决实际问题的能力，方便学生根据课程进度及时进行知识巩固，我们还新增了电子版学习题库。

在本书的编译过程中，我们得到了北京大学出版社黄炜、刘洋编辑的大力支持和帮助，得到了北京科技大学教材建设经费的资助和北京科技大学教务处的全程支持，对他们付出的辛勤劳动表示衷心的感谢和敬意。

遗传学发展迅速、涉及的内容较新，但是限于我们水平的局限性，疏漏和不妥之处在所难免，恳请读者批评指正。

编译者

2023 年 5 月

Contents

目　录

1 Introduction to Genetics

1　绪论

CHAPTER CONCEPTS

■ Genetics in the twenty-first century is built on a rich tradition of discovery and experimentation stretching from the ancient world through the nineteenth century to the present day.

■ Transmission genetics is the general process by which traits controlled by genes are transmitted through gametes from generation to generation.

■ Mutant strains can be used in genetic crosses to map the location and distance between genes on chromosomes.

■ The Watson-Crick model of DNA structure explains how genetic information is stored and expressed. This discovery is the foundation of molecular genetics.

■ Recombinant DNA technology revolutionized genetics, was the foundation for the Human Genome Project, and has generated new fields that combine genetics with information technology.

■ Biotechnology provides genetically modified organisms and their products that are used across a wide range of fields including agriculture, medicine, and industry.

■ Model organisms used in genetics

本章速览

■ 21 世纪的遗传学是建立在自远古时代经由 19 世纪直至当今的大量发现和科学实验基础之上的一门学科。

■传递遗传学研究由基因决定的性状在世代交替中通过配子进行传递的过程。

■在遗传杂交实验中，突变体可用于确定基因在染色体上的位置及基因间的距离。

■ DNA 结构的沃森 - 克里克模型阐明了遗传信息如何进行储存和表达，这一发现为分子遗传学奠定了基础。

■ 重组 DNA 技术为遗传学带来了巨大的革新，为人类基因组计划奠定了基础，并由此衍生出一批集遗传学、信息学技术于一体的新型学科领域。

■ 通过生物技术获得的遗传修饰生物体及其产品已被广泛用于农业、医药和工业等各领域。

■ 应用于遗传学研究的模式生物现如

research are now utilized in combination with recombinant DNA technology and genomics to study human diseases.

今已与重组 DNA 技术、基因组学相结合，共同用于研究人类疾病。

■ Genetic technology is developing faster than the policies, laws, and conventions that govern its use.

■ 遗传学相关技术的发展速度远远超过与之相关的政策、法律法规及国际惯例的发展速度。

One of the small pleasures of writing a genetics textbook is being able to occasionally introduce in the very first paragraph of the initial chapter a truly significant breakthrough in the discipline that has started to have a major, diverse impact on human lives. In this edition, we are fortunate to be able to discuss the discovery of CRISPR-Cas, a molecular mechanism found in bacteria that has the potential to revolutionize our ability to rewrite the DNA sequence of genes from any organism. As such, it represents the ultimate tool in genetic technology, whereby the genome of organisms, including humans, may be precisely edited. Such gene modification represents the ultimate application of the many advances in biotechnology made in the last 35 years, including the sequencing of the human genome.

撰写遗传学教科书的小小乐趣之一便是间或能够在第一章的第一段介绍该学科巨大、非凡的科学突破，而该学科已经开始对人类生活产生重大、多元的影响。在这个版本的教材中，我们很幸运地能够介绍 CRISPR-Cas 系统的发现。这是一种在细菌中发现的分子机制，它具有进一步革新人类编辑源自任何生物的基因 DNA 序列的潜力。因此，它代表了遗传技术的最新工具。借助该工具，人们可以精确编辑包括人类在内的生物体基因组。该基因修饰技术代表了过去 35 年以来，含人类基因组测序在内的生物技术诸多进展的终极应用。

Although gene editing was first made possible with other methods, the CRISPR-Cas system is now the method of choice for gene modification because it is more accurate, more efficient, more versatile, and easier to use. CRISPR-Cas was initially discovered as a "seek and destroy" mechanism that bacteria use to fight off viral infection. CRISPR (clustered regularly interspersed short palindromic repeats)

尽管基因编辑最初是使用其他方法来实现的，但是现在 CRISPR-Cas 系统已经成为基因修饰的首选方法，因为该系统更加准确、有效、通用，并且也更易于使用。CRISPR-Cas 最初是作为细菌用于抵抗病毒侵染的“寻毁”机制而被发现的。CRISPR（规律成簇间隔短回文重复）是细菌基因组中合成 RNA 分子的部分，而 Cas（CRISPR 相关蛋白）指核酸酶或 DNA 剪切酶。

refers to part of the bacterial genome that produces RNA molecules, and Cas (CRISPR-associated) refers to a nuclease, or DNA cutting enzyme. The CRISPR RNA binds to a matching sequence in the viral DNA (seek) and recruits the Cas nuclease to cut it (destroy). Researchers have harnessed this technology by synthesizing CRISPR RNAs that direct Cas nucleases to any chosen DNA sequence. In laboratory experiments, CRISPR-Cas has already been used to repair mutations in cells derived from individuals with genetic disorders, such as cystic fibrosis, Huntington disease, sickle-cell disease, and muscular dystrophy. In the United States a clinical trial using CRISPR-Cas for genome editing in cancer therapy is recruiting participants, while proposals for treating a genetic form of blindness and genetic blood disorders are in preparation. In China, at least 86 patients have already started receiving treatments in CRISPR-Cas clinicaltrials for cancer.

CRISPR RNA 与病毒 DNA 中的匹配序列相结合（“寻”），然后募集 Cas 核酸酶将其进行剪切（“毁”）。研究人员已经可以通过合成能够将 Cas 核酸酶定向至任意指定 DNA 序列的 CRISPR RNA 来驾驭这项技术。在实验室中，CRISPR-Cas 系统已被用于修复源自遗传疾病患病个体的细胞突变，这些遗传疾病包括囊性纤维化、亨廷顿病、镰状细胞病和肌营养不良。在美国，一项利用 CRISPR-Cas 系统通过基因组编辑进行癌症治疗的临床试验正在招募志愿者，此外治疗遗传性失明和遗传性血液疾病的提案也正在准备中。在中国，至少有 86 名患者已经开始接受 CRISPR-Cas 癌症临床治疗试验。

The application of this remarkable system goes far beyond developing treatments for human genetic disorders. In organisms of all kinds, wherever genetic modification may benefit human existence and our planet, the use of CRISPR-Cas will find many targets. For example, one research group edited a gene in mosquitoes, which prevents them from carrying the parasite that causes malaria in humans. Other researchers have edited the genome of algae to double their output for biofuel production. The method has also been used to create disease-resistant strains of wheat and rice.

这一非凡系统的应用远不仅限于开发人类遗传疾病的治疗方法。只要遗传修饰有益于人类生存，可以造福我们的星球，那么 CRISPR-Cas 系统的应用将在所有生物物种中发现更多可改造的基因靶标。例如，某研究小组针对蚊子的一种基因进行基因编辑，从而阻止它们携带能导致人类感染疟疾的寄生虫。还有一些研究人员已经对藻类基因组进行基因编辑，从而使其生物量加倍，更有利于生物燃料的生产。该方法也已被用于创造小麦和水稻的抗病品种。

The power of this system, like any major

和任何一项重大技术的进步一样，该

technological advance, has already raised ethical concerns. For example, genetic modification of human embryos would change the genetic information carried by future generations. These modifications may have unintended and significant negative consequences for our species. In 2017, an international panel of experts discussed the science, ethics, and governance of human genome editing. The panel recommended caution, but not a ban, stating that human embryo modification should "only be permitted for compelling reasons and under strict oversight."

系统的强大同样已经引起伦理方面的关注。例如，人类胚胎的遗传修饰将改变后代所携带的遗传信息，这些修饰可能无意中将对人类物种产生意想不到的重大负面影响。2017年，国际专家小组对人类基因组编辑的科学、伦理和监管等展开讨论。该专家组建议谨慎使用而不是禁止使用基因组编辑，并且指出应该"仅在具有令人信服的理由并在严格监管下"，才可以进行人类胚胎的遗传修饰。

CRISPR-Cas may turn out to be one of the most exciting genetic advances in decades. We will return later in the text to discuss its discovery in bacteria, its development as a gene-editing tool, its potential for gene therapy (Chapter 11—Gene Therapy), and its uses in genetically edited foods (Chapter 14—Genetically Modified Foods).

CRISPR-Cas 可能是近几十年来最令人兴奋的遗传学进展之一。我们将在本书的一些章节中讨论其在细菌中的发现、作为基因编辑工具的进展、基因治疗中的应用潜力（第11章——基因治疗）以及在转基因食品中的应用(第14章——转基因食品)。

For now, we hope that this short introduction has stimulated your curiosity, interest, and enthusiasm for the study of genetics. The remainder of this chapter provides an overview of many important concepts of genetics and a survey of the major turning points in the history of the discipline.

我们希望这段简短的介绍能够激发你对遗传学研究的好奇、兴趣和热情。接下来本章将概述许多重要的遗传学概念，并将纵览该学科历史中的一些重要转折点。

1.1 Genetics Has an Interesting Early History

1.1 有趣的遗传学早期历史

While as early as 350 B.C., Aristotle proposed that active "humors" served as bearers of hereditary traits, it was not until the 1600s that initial strides were made to understand

尽管早在公元前350年，亚里士多德就提出活性"体液"是遗传性状的载体，但直到17世纪，人们才开始逐步了解生命的生物学基础。在17世纪，医生和解剖学家

the biological basis of life. In that century, the physician and anatomist William Harvey proposed the theory of **epigenesis**, which states that an organism develops from the fertilized egg by a succession of developmental events that eventually transform the egg into an adult. The theory of epigenesis directly conflicted with the theory of **preformationism**, which stated that the fertilized egg contains a complete miniature adult, called a **homunculus**. Around 1830, Matthias Schleiden and Theodor Schwann proposed the **cell theory**, stating that all organisms are composed of basic structural units called cells, which are derived from preexisting cells. The idea of **spontaneous generation**, the creation of living organisms from nonliving components, was disproved by Louis Pasteur later in the century, and living organisms were then considered to be derived from preexisting organisms and to consist of cells.

威廉·哈维提出了**后成说**理论，该理论指出生物体是由受精卵经由一系列发育事件而最终发育成成年个体的。后成说理论与**先成说**理论直接冲突，后者则认为受精卵中含有一个完整的被称为“**微型人**”的微缩成年个体。1830年左右，马蒂亚斯·施莱登和特奥多尔·施万提出**细胞学说**，指出所有生物体都是由细胞这一基本结构单元组成的，而这些基本单元又是由先前存在的细胞衍生而来的。19世纪后期，路易·巴斯德推翻了认为“生物由非活体组分产生”的**自然发生说**，此后人们开始认可生物体是由原先存在的生物衍生而来，并由细胞组成这一观点。

In the mid-1800s the work of Charles Darwin and Gregor Mendel set the stage for the rapid development of genetics in the twentieth and twenty-first centuries.

19世纪中叶，查尔斯·达尔文和格雷戈尔·孟德尔的杰出工作为20世纪和21世纪遗传学的快速发展奠定了基础。

Darwin and Mendel

In 1859, Darwin published *On the Origin of Species*, describing his ideas about evolution. Darwin's geological, geographical, and biological observations convinced him that existing species arose by descent with modification from ancestral species. Greatly influenced by his voyage on the HMS *Beagle* (1831—1836), Darwin's thinking led him to formulate the theory of **natural selection**, which presented an explanation of the

达尔文与孟德尔

1859年，达尔文出版了《物种起源》，并阐述了他关于进化的思想。达尔文在地质、地理和生物学方面的观察与发现使他确信现有物种由其祖先物种衍生而来，同时也伴随着变化。达尔文的思想受到了“贝格尔号”军舰环球考察航行经历（1831—1836）的极大影响，并促使他提出了**自然选择理论**，从而对进化变化的机理做出解释。此外，艾尔弗雷德·拉塞尔·华莱士独立

mechanism of evolutionary change. Formulated and proposed independently by Alfred Russel Wallace, natural selection is based on the observation that populations tend to produce more offspring than the environment can support, leading to a struggle for survival among individuals. Those individuals with heritable traits that allow them to adapt to their environment are better able to survive and reproduce than those with less adaptive traits. Over time, advantageous variations, even very slight ones, will accumulate. If a population carrying these inherited variations becomes reproductively isolated, a new species may result.

创立并提出的自然选择则是基于其所观察的现象：种群往往产生超出环境承受能力的后代数量，从而导致该物种的个体之间为了生存而产生竞争。那些携带能更好适应所在环境的可遗传性状的个体比那些适应性较差的个体有更好的生存能力和繁殖能力。随着时间的推移，有益变异即使非常微小也将被累积。如果携带这些遗传变异的种群被生殖隔离，那么将可能产生一种新物种。

Darwin, however, lacked an understanding of the genetic basis of variation and inheritance, a gap that left his theory open to reasonable criticism well into the twentieth century. Shortly after Darwin published his book, Gregor Johann Mendel published a paper in 1866 showing how traits were passed from generation to generation in pea plants and offered a general model of how traits are inherited. His research was little known until it was partially duplicated and brought to light by Carl Correns, Hugo de Vries, and Erich Tschermak around 1900.

然而，达尔文缺乏对遗传与变异的遗传基础的理解，这一空白使得他的理论在20世纪前受到了很多理性批判。在达尔文出版其著作不久后，格雷戈尔·约翰·孟德尔于1866年发表了一篇论文，文中展示了性状如何在豌豆的世代中进行传递，由此为性状遗传提供了模型。孟德尔的研究工作几乎不被人所知，直到1900年左右，卡尔·科伦斯、雨果·弗里斯和埃里克·切尔马克才对其工作进行了部分重复并使其研究成果重见天日。

By the early part of the twentieth century, it became clear that heredity and development were dependent on genetic information residing in genes contained in chromosomes, which were then contributed to each individual by gametes—the so-called *chromosome theory of inheritance*. The gap in Darwin's theory was closed, and Mendel's research now serves as the foundation of genetics.

20世纪初期，人们已经很清楚地知道遗传和发育取决于染色体上所含基因的遗传信息，这些遗传信息随后通过配子传递至后代的每个个体——即遗传的染色体学说。达尔文理论的鸿沟由此被弥补，孟德尔的研究工作成为现代遗传学的奠基。

1.2 Genetics Progressed from Mendel to DNA in Less Than a Century

Because genetic processes are fundamental to life itself, the science of genetics unifies biology and serves as its core. The starting point for this branch of science was a monastery garden in central Europe in the late 1850s.

Mendel's Work on Transmission of Traits

Gregor Mendel, an Augustinian monk, conducted a decade-long series of experiments using pea plants. He applied quantitative data analysis to his results and showed that traits are passed from parents to offspring in predictable ways. He further concluded that each trait in pea plants is controlled by a pair of factors (which we now call genes) and that members of a gene pair separate from each other during gamete formation (the formation of egg cells and sperm). Mendel's findings explained the transmission of traits in pea plants and all other higher organisms. His work forms the foundation for **genetics**, the branch of biology concerned with the study of heredity and variation. Mendelian genetics will be discussed later in the text (see Chapters 3 and 4).

The Chromosome Theory of Inheritance: Uniting Mendel and Meiosis

Mendel did his experiments before the structure and role of chromosomes were known. About 20 years after his work was published, advances in microscopy allowed researchers

1.2 从孟德尔到 DNA ——遗传学在一个世纪内的飞速进展

由于遗传过程是生命的基础，因此遗传学将生物学的所有学科统一起来并成为生物学的核心学科。这一学科分支的发展起点就是 19 世纪 50 年代后期位于中欧的一座修道院的花园。

孟德尔对性状传递规律的研究

格雷戈尔 · 孟德尔是一位奥古斯丁修道士，他使用豌豆进行了长达 10 年的系列实验，并对结果进行了定量数据分析，发现性状以可预测的方式从亲本个体传给了子代个体。他进一步推断在豌豆中，每种性状由一对因子（我们现在称为“基因”）控制，并且在配子（卵细胞和精子）形成过程中，每对基因的两个成员将彼此分开。孟德尔的发现阐明了性状在豌豆及所有其他高等生物的世代交替中是如何进行传递的。他的工作成为**遗传学**的基础，该学科分支关注遗传和变异的相关研究。孟德尔遗传学将在随后章节中进行讨论（请参阅第 3 章和第 4 章）。

遗传的染色体学说：孟德尔遗传学与减数分裂的统一

孟德尔的实验工作开展于染色体的结构和功能被解析之前。他的研究成果发表后约 20 年，显微镜技术的进步使得研究人员能够鉴定染色体，并确定在大多数真核

to identify chromosomes and establish that, in most eukaryotes, members of each species have a characteristic number of chromosomes called the **diploid number** ($2n$) in most of their cells. For example, humans have a diploid number of 46. Chromosomes in diploid cells exist in pairs, called **homologous chromosomes**.

生物中，每个物种成员在其大多数细胞中均含有特征数目的染色体数量，即**二倍染色体数目**（$2n$）。例如，人类的二倍染色体数目为 46。二倍体细胞中的染色体成对存在，被称为**同源染色体**。

Researchers in the last decades of the nineteenth century also described chromosome behavior during two forms of cell division, **mitosis** and **meiosis**. In mitosis, chromosomes are copied and distributed so that each daughter cell receives a diploid set of chromosomes identical to those in the parental cell. Meiosis is associated with gamete formation. Cells produced by meiosis receive only one chromosome from each chromosome pair, and the resulting number of chromosomes is called the **haploid number** (n). This reduction in chromosome number is essential if the offspring arising from the fusion of egg and sperm are to maintain the constant number of chromosomes characteristic of their parents and other members of their species.

在 19 世纪的最后几十年中，研究人员还描述了在**有丝分裂**和**减数分裂**两种细胞分裂方式中的染色体行为。在有丝分裂中，染色体被复制和分配，从而使得每个子细胞均能获得与亲代细胞数量完全相同的二倍染色体组成。减数分裂则与配子形成有关，由减数分裂产生的细胞仅获得每对染色体中的一条，因此子代细胞最终所得的染色体数目被称为**单倍染色体数目**（n）。如果想要卵细胞和精子融合产生的后代与其父母和同一物种其他成员所具有的恒定特征染色体数目维持一致，那么这种染色体数目的减半就十分必要。

Early in the twentieth century, Walter Sutton and Theodor Boveri independently noted that the behavior of chromosomes during meiosis is identical to the behavior of genes during gamete formation described by Mendel. For example, genes and chromosomes exist in pairs, and members of a gene pair and members of a chromosome pair separate from each other during gamete formation. Based on these and other parallels, Sutton and Boveri each proposed that genes are carried on chromosomes. They independently formulated the **chromosomal**

20世纪初，沃尔特·萨顿和特奥多尔·博韦里分别指出减数分裂过程中的染色体行为与孟德尔所描述的配子形成过程中的基因行为完全一致。例如，基因和染色体均成对存在，且在配子形成过程中，一对基因的两个成员和一对染色体的两个成员均彼此分开。基于这些和其他的相似之处，萨顿和博韦里分别提出基因位于染色体上。他们分别独立地阐述了**遗传的染色体学说**，该理论指出遗传性状由位于染色体上的基因决定，这些基因通过配子准确地进行世代传递，从而保证了世代之间的遗传连续

theory of inheritance, which states that inherited traits are controlled by genes residing on chromosomes faithfully transmitted through gametes, maintaining genetic continuity from generation to generation.

性。

ESSENTIAL POINT

The chromosome theory of inheritance explains how genetic information is transmitted from generation to generation.

基本要点

遗传的染色体学说解释了遗传信息如何进行世代传递。

Genetic Variation

About the same time that the chromosome theory of inheritance was proposed, scientists began studying the inheritance of traits in the fruit fly, *Drosophila melanogaster*. Early in this work, a among normal (wild-type) red-eyed flies. This variation was produced by a **mutation** in one white-eyed fly was discovered of the genes controlling eye color. Mutations are defined as any heritable change in the DNA sequence and are the source of all genetic variation.

遗传变异

几乎在遗传的染色体学说提出的同时，科学家开始了黑腹果蝇（*Drosophila melanogaster*）性状的遗传研究。在这项工作的早期研究中，人们在正常红眼果蝇（野生型）中发现了白眼果蝇。这种变异是由控制眼睛颜色的基因发生**突变**引起的。突变指 DNA 序列中任何可遗传的变化，并且是所有遗传变异的源泉。

The white-eye variant discovered in *Drosophila* is an **allele** of a gene controlling eye color. Alleles are defined as alternative forms of a gene. Different alleles may produce differences in the observable features, or **phenotype**, of an organism. The set of alleles for a given trait carried by an organism is called the **genotype**. Using mutant genes as markers, geneticists can map the location of genes on chromosomes .

果蝇中发现的白眼基因变体是控制果蝇眼睛颜色基因的一种**等位基因**。等位基因是一个基因的多种可相互替代的不同存在形式。不同的等位基因可以产生可观察的有差异的生物体特征或**表型**。生物体携带的指定性状所对应的等位基因组合被称为**基因型**。遗传学家利用突变基因作为标记可以定位基因在染色体上的位置。

The Search for the Chemical Nature of Genes: DNA or Protein?

Work on white-eyed *Drosophila* showed

探索基因的化学本质：DNA 还是蛋白质?

对白眼果蝇的研究表明该突变体性状

that the mutant trait could be traced to a single chromosome, confirming the idea that genes are carried on chromosomes. Once this relationship was established, investigators turned their attention to identifying which chemical component of chromosomes carries genetic information. By the 1920s, scientists knew that proteins and DNA were the major chemical components of chromosomes. There are a large number of different proteins, present in both the nucleus and cytoplasm, and many researchers thought proteins carried genetic information.

可以追溯到单一染色体，从而证实了基因是位于染色体上的猜想。一旦建立了这种对应关系，研究人员便开始将注意力转向鉴定染色体的哪些化学组成携带遗传信息上。直至20世纪20年代，科学家才知道蛋白质和DNA是染色体的主要化学成分。在细胞核和细胞质中均存在大量不同的蛋白质，许多研究人员因此认为遗传信息是由蛋白质携带的。

In 1944, Oswald Avery, Colin MacLeod, and Maclyn McCarty, researchers at the Rockefeller Institute in New York, published experiments showing that DNA was the carrier of genetic information in bacteria. This evidence, though clear-cut, failed to convince many influential scientists. Additional evidence for the role of DNA as a carrier of genetic information came from Alfred Hershey and Martha Chase who worked with viruses. This evidence that DNA carries genetic information, along with other research over the next few years, provided solid proof that DNA, not protein, is the genetic material, setting the stage for work to establish the structure of DNA.

1944年，纽约洛克菲勒研究所的研究人员奥斯瓦尔德·埃弗里、科林·麦克劳德和麦克林恩·麦卡蒂发表了证实细菌DNA是遗传信息载体的相关实验。这一实验的证据尽管十分明确，但却未能让许多当时有影响力的科学家信服。艾尔弗雷德·赫尔希和玛莎·蔡斯以病毒为研究材料提供了证明DNA是遗传信息载体的进一步实验证据。该证据与随后几年中的其他研究一起为DNA是遗传物质而蛋白质不是遗传物质提供了有力支持，也为DNA结构的研究工作奠定了基础。

1.3 Discovery of the Double Helix Launched the Era of Molecular Genetics

1.3 DNA双螺旋结构的发现开启了分子遗传学新纪元

Once it was accepted that DNA carries genetic information, efforts were focused on deciphering the structure of the DNA molecule

人们一旦接受了携带遗传信息的物质是DNA后，就开始将精力集中在破译DNA的分子结构以及阐明其存储信息如何

and the mechanisms by which information stored in it produce a phenotype.

产生表型的分子机制上。

The Structure of DNA and RNA

One of the great discoveries of the twentieth century was made in 1953 by James Watson and Francis Crick, who described the structure of DNA. DNA is a long, ladder like macromolecule that twists to form a double helix (Figure1.1). Each linear strand of the helix is made up of subunits called **nucleotides**. In DNA, there are four different nucleotides, each of which contains a nitrogenous base, abbreviated A (adenine), G (guanine), T (thymine), or C (cytosine). These four bases, in various sequence combinations, ultimately encode genetic information. The two strands of DNA are exact complements of one another, so that the rungs of the ladder in the double helix always consist of A═T and G ≡ C base pairs. Along with Maurice Wilkins, Watson and Crick were awarded a Nobel Prize in 1962 for their work on the structure of DNA.

DNA 和 RNA 的分子结构

20 世纪最伟大的发现之一是詹姆斯 · 沃森和弗朗西斯 · 克里克在 1953 年阐明了 DNA 的分子结构。DNA 是一个呈长梯状的生物大分子，并扭曲形成双螺旋结构（图 1.1）。双螺旋的每一条线性长链均由被称为**核苷酸**的亚单元所组成。在 DNA 中有四种不同的核苷酸组分，每种核苷酸包含一个含氮碱基，分别缩写为 A（腺嘌呤）、G（鸟嘌呤）、T（胸腺嘧啶）或 C（胞嘧啶）。这四种碱基以多种顺序进行组合，最终编码遗传信息。构成 DNA 的两条长链彼此间精确互补，因此梯形双螺旋中的梯级始终由 A ═ T 和 G ≡ C 碱基对所组成。沃森、克里克与莫里斯 · 威尔金斯一起揭开了 DNA 的分子结构，并且由于该项工作于 1962 年分享了诺贝尔奖。

Another nucleic acid, RNA, is chemically similar to DNA but contains a different sugar (ribose rather than deoxyribose) in its

另一类核酸分子 RNA 在化学组成上与 DNA 相似，但在其核苷酸中含有不同的糖分子（核糖而不是脱氧核糖），并且含氮

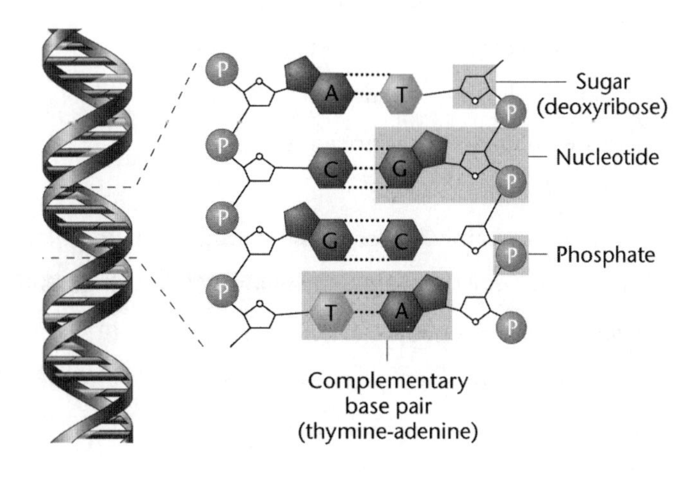

FIGURE 1.1 The structure of DNA showing the arrangement of the double helix (on the left) and the chemical components making up each strand (on the right). The dotted lines on the right represent weak chemical bonds, called hydrogen bonds, which hold together the two strands of the DNA helix.

图 1.1 DNA 双螺旋结构（左）和每条链的化学组分（右）。右侧图中的点线表示氢键，这是一种弱化学键，可以将 DNA 双螺旋中的两条链相互联系。

nucleotides and contains the nitrogenous base uracil in place of thymine. RNA, however, is generally a single-stranded molecule.

碱基尿嘧啶取代了胸腺嘧啶。此外，RNA通常是单链分子。

Gene Expression: From DNA to Phenotype

The genetic information encoded in the order of nucleotides in DNA is expressed in a series of steps that results in the formation of a functional gene product. In the majority of cases, this product is a protein. In eukaryotic cells, the process leading to protein production begins in the nucleus with **transcription**, in which the nucleotide sequence in one strand of DNA is used to construct a complementary RNA sequence (top part of Figure 1.2). Once an RNA molecule is produced, it moves to the cytoplasm, where the RNA—called **messenger RNA**, or **mRNA** for short—binds to a **ribosome**. The synthesis of proteins under the direction of mRNA is called **translation** (center part of Figure 1.2). The information encoded in mRNA (called the **genetic code**) consists of a linear series of nucleotide triplets. Each triplet, called a **codon**, is complementary to the information stored in DNA and specifies the insertion of a specific amino acid into a protein. Proteins (lower part of Figure 1.2) are polymers made up of amino acid monomers. There are 20 different amino acids commonly found in proteins.

基因表达：从DNA到表型

编码在DNA分子核苷酸序列中的遗传信息通过一系列步骤进行表达，最终形成有功能的基因产物。在大多数情况下，基因产物是蛋白质。在真核细胞中，引导蛋白质合成的过程起始于细胞核内的**转录**，其间，以DNA双链中的一条核苷酸序列为模板合成互补的RNA序列（图1.2上部）。RNA分子——即**信使RNA**（简称为**mRNA**）一旦合成，就会转移至细胞质中并与**核糖体**相结合。在mRNA指导下进行蛋白质合成的过程称为**翻译**（图1.2中部）。mRNA所编码的信息（称为**遗传密码**）由系列线性排列的核苷酸三联体组成。每个三联体称为一个**密码子**，与DNA中所存储的信息互补，并明确指定了插入蛋白质中的特定氨基酸种类。蛋白质（图1.2底部）是由氨基酸单体组成的聚合物，组成蛋白质的常见氨基酸有20种。

Protein assembly is accomplished with the aid of adapter molecules called **transfer RNA (tRNA)**. Within the ribosome, tRNAs recognize the information encoded in the mRNA codons and carry the proper amino acids for

蛋白质组装的完成需要借助接头分子，即**转移RNA（tRNA）**的帮助。在核糖体内，tRNA识别mRNA密码子中的编码信息，并在翻译过程中携带正确的氨基酸进行蛋白质组装。

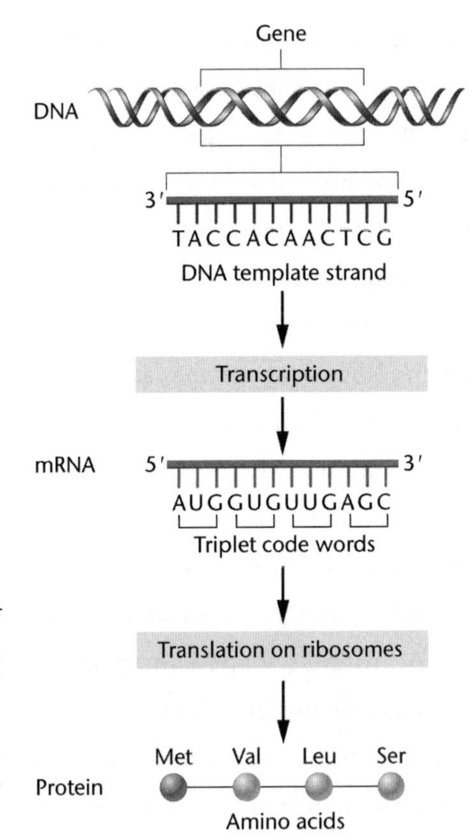

FIGURE 1.2 Gene expression consists of transcription of DNA into mRNA (top) and the translation (center) of mRNA (with the help of a ribosome) into a protein (bottom).

图 1.2 基因表达包括：DNA 转录成 mRNA（上部）、mRNA 在核糖体的协助下翻译（中部）成蛋白质（底部）。

construction of the protein during translation.

We now know that gene expression can be more complex than outlined here.

现在我们知道基因表达可能远比此处所概述的要复杂得多。

Proteins and Biological Function

In most cases, proteins are the end products of gene expression. The diversity of proteins and the biological functions they perform—the diversity of life itself—arises from the fact that proteins are made from combinations of 20 different amino acids. Consider that a protein chain containing 100 amino acids can have at each position any one of 20 amino acids; the number of possible different 100-amino-acid proteins, each with a unique sequence, is therefore equal to 20^{100}.

蛋白质与生物功能

在大多数情况下，蛋白质是基因表达的最终产物。蛋白质的多样性及其所发挥的生物学功能，即生命本身的多元化，均基于蛋白质由 20 种不同氨基酸所组成这一事实。假设有一条含有 100 个氨基酸的蛋白质长链，每个氨基酸位置上可以是 20 种氨基酸中的任意一种，那么该蛋白质可能的氨基酸序列种类数为 20^{100}，并且这些序列彼此不同。

Obviously, proteins are molecules with the

显然，蛋白质是具有巨大结构多样性

potential for enormous structural diversity and serve as a mainstay of biological systems.

潜能的分子，并且构成了生物系统的支柱。

Enzymes form the largest category of proteins. These molecules serve as biological catalysts, lowering the energy of activation in reactions and allowing cellular metabolism to proceed at body temperature.

酶是蛋白质中最大的一个类别。这些分子作为生物催化剂，降低了反应中的活化能，使得细胞代谢能够在体温下顺利进行。

Proteins other than enzymes are critical components of cells and organisms. These include hemoglobin, the oxygen binding molecule in red blood cells; insulin, a pancreatic hormone; collagen, a connective tissue molecule; and actin and myosin, the contractile muscle proteins. A protein's shape and chemical behavior are determined by its linear sequence of amino acids, which in turn is dictated by the stored information in the DNA of a gene that is transferred to RNA, which then directs the protein's synthesis.

除酶以外的其他蛋白质是细胞和生物体中至关重要的组成部分。这些蛋白质包括红细胞中与氧分子相结合的血红蛋白；胰岛分泌的激素分子胰岛素；结缔组织分子胶原蛋白；参与肌肉收缩的肌动蛋白和肌球蛋白。蛋白质的形状及其化学性质由氨基酸的线性排列顺序所决定，而氨基酸的线性排列顺序又由基因中 DNA 所存储的遗传信息所决定，该存储信息被传递至 RNA，进而指导蛋白质的合成。

Linking Genotype to Phenotype: Sickle-Cell Anemia

基因型与表型的关系：镰状细胞贫血

Once a protein is made, its biochemical or structural properties play a role in producing a phenotype. When mutation alters a gene, it may modify or even eliminate the encoded protein's usual function and cause an altered phenotype. To trace this chain of events, we will examine sickle-cell anemia, a human genetic disorder.

蛋白质一旦合成，其生化特性或结构特性便对表型的产生发挥作用。当基因发生突变时，所编码蛋白的正常功能可能被修饰甚至消失，从而导致表型改变。我们以镰状细胞贫血这一人类遗传疾病为例来讨论其中发生的系列事件。

Sickle-cell anemia is caused by a mutant form of hemoglobin, the protein that transports oxygen from the lungs to cells in the body. Hemoglobin is a composite molecule made up of two different proteins, α-globin and β-globin, each encoded by a different gene. In sickle-

血红蛋白是一种将氧气从肺部运输到人体细胞中的蛋白质，镰状细胞贫血就是由血红蛋白的一种突变体形式引起的。血红蛋白是由两种不同蛋白质亚基组成的复合分子，即 α- 珠蛋白和 β- 珠蛋白，它们分别由不同的基因所编码。在镰状细胞贫血

cell anemia, a mutation in the gene encoding β-globin causes an amino acid substitution in 1 of the 146 amino acids in the protein. Figure 1.3 shows the DNA sequence, the corresponding mRNA codons, and the amino acids occupying positions 4—7 for the normal and mutant forms of β-globin. Notice that the mutation in sickle-cell anemia consists of a change in one DNA nucleotide, which leads to a change in codon 6 in mRNA from GAG to GUG, which in turn changes amino acid number6 in β-globin from glutamic acid to valine. The other 145 amino acids in the protein are not changed by this mutation.

患者中，编码β-珠蛋白的基因有一处发生突变，从而导致蛋白质所含的146个氨基酸中的1个被取代。图1.3显示了DNA序列、相应的mRNA密码子以及β-珠蛋白中第4—7位氨基酸的正常形式和突变体形式。值得注意的是，镰状细胞贫血的突变仅含一个DNA核苷酸的变化，而这导致mRNA的第6个密码子由GAG变为GUG，进而使得β-珠蛋白中的第6位氨基酸由谷氨酸变为缬氨酸，蛋白质中的其他145个氨基酸并未发生改变。

Individuals with two mutant copies of the β-globin gene have sickle-cell anemia. Their mutant β-globin proteins cause hemoglobin molecules in red blood cells to polymerize when the blood's oxygen concentration is low, forming long chains of hemoglobin that distort the shape of red blood cells. Deformed cells are fragile and break easily, reducing the number of circulating red blood cells (anemia is an insufficiency of red blood cells). Sickle-shaped cells block blood flow in capillaries and small blood vessels, causing severe pain and damage to the heart,

同时携带β-珠蛋白基因两个突变体拷贝的个体会患镰状细胞贫血。当血液中氧气浓度低时，突变体β-珠蛋白将引起红细胞中血红蛋白分子聚合，形成长链状血红蛋白，从而使红细胞的形状发生扭曲。变形的红细胞脆弱易碎，进而减少了参与循环的红细胞数量（贫血即为红细胞数量不足）。镰状细胞阻塞毛细血管和微血管的血液流动，会引发严重疼痛并损害心脏、大脑、肌肉和肾脏。这种疾病的所有症状都由基因中单一核苷酸的改变引起，从而使得β-珠蛋白分子的146个氨基酸中的一

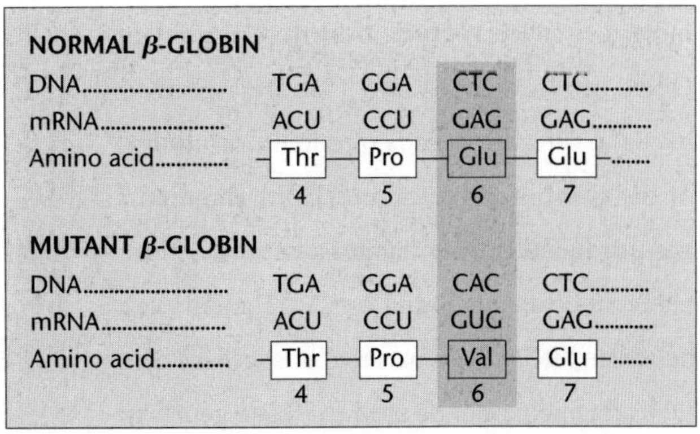

NORMAL β-GLOBIN

DNA........................	TGA	GGA	CTC	CTC..........
mRNA......................	ACU	CCU	GAG	GAG..........
Amino acid............	Thr	Pro	Glu	Glu ....
	4	5	6	7

MUTANT β-GLOBIN

DNA........................	TGA	GGA	CAC	CTC..........
mRNA......................	ACU	CCU	GUG	GAG..........
Amino acid............	Thr	Pro	Val	Glu ....
	4	5	6	7

FIGURE 1.3 A single-nucleotide change in the DNA encoding β-globin (CTC → CAC) leads to an altered mRNA codon (GAG → GUG) and the insertion of a different amino acid (Glu → Val), producing the altered version of the β-globin protein that is responsible for sickle-cell anemia.

图1.3 编码β-珠蛋白的DNA发生单核苷酸改变（CTC → CAC）导致mRNA密码子变化（GAG → GUG）以及一个氨基酸的改变（Glu → Val），由此产生的β-珠蛋白突变引发了镰状细胞贫血。

brain, muscles, and kidneys. All the symptoms of this disorder are caused by a change in a single nucleotide in a gene that changes one amino acid out of 146 in the β-globin molecule, demonstrating the close relationship between genotype and phenotype.

个随之发生变化，由此可见基因型和表型之间的密切关系。

ESSENTIAL POINT

The central dogma of molecular biology — that DNA is a template for making RNA, which in turn directs the synthesis of proteins — explains how genes control phenotype.

基本要点

分子生物学中心法则是以DNA为模板合成RNA，再在RNA的指导下合成蛋白质，这一法则解释了基因是如何控制表型的。

1.4 Development of Recombinant DNA Technology Began the Era of DNA Cloning

1.4 重组DNA技术的发展开启了DNA克隆时代

The era of recombinant DNA began in the early 1970s, when researchers discovered that **restriction enzymes**, used by bacteria to cut and inactivate the DNA of invading viruses, could be used to cut any organism's DNA at specific nucleotide sequences, producing a reproducible set of fragments.

重组DNA时代始于20世纪70年代初，研究人员发现细菌中剪切并灭活入侵病毒DNA的**限制性内切酶**，也可用于剪切任何生物体DNA中特定的核苷酸序列，产生可再生的DNA片段。

Soon after, researchers discovered ways to insert the DNA fragments produced by the action of restriction enzymes into carrier DNA molecules called **vectors** to form recombinant DNA molecules. When transferred into bacterial cells, thousands of copies, or **clones**, of the combined vector and DNA fragments are produced during bacterial reproduction. Large amounts of cloned DNA fragments can be isolated from these bacterial host cells. These DNA fragments can be used to isolate genes, to

不久，研究人员发现可以将限制性内切酶剪切产生的DNA片段插入被称为**载体**的DNA分子中，形成重组DNA分子的方法。当重组DNA分子转化入细菌细胞中后，在细菌繁殖过程中，将产生数以千计的整合有载体和DNA片段的拷贝或**克隆**。大量克隆的DNA片段可以从这些细菌宿主细胞中分离出来。这些DNA片段可用于分离基因、研究基因的组织和表达，以及研究核苷酸序列和进化。

study their organization and expression, and to study their nucleotide sequence and evolution.

Collections of clones that represent an organism's **genome**, defined as the complete haploid DNA content of a specific organism, are called genomic libraries. Genomic libraries are now available for hundreds of species.

代表生物体**基因组**所有克隆的集合被称为基因组文库，它包含了特定生物体完整单倍体 DNA 的组成。人们已经建立了数百种物种的基因组文库。

Recombinant DNA technology has not only accelerated the pace of research but also given rise to the biotechnology industry, which has grown to become a major contributor to the U.S. economy.

重组DNA技术不仅加快了研究的步伐，而且催生了生物技术产业，该产业已经成为美国的主要经济支柱之一。

1.5 The Impact of Biotechnology Is Continually Expanding

1.5 生物技术的影响正持续扩大

The use of recombinant DNA technology and other molecular techniques to make products is called **biotechnology**. In the United States, biotechnology has quietly revolutionized many aspects of everyday life; products made by biotechnology are now found in the supermarket, in health care, in agriculture, and in the court system. Now, let's look at some everyday examples of biotechnology's impact.

应用重组 DNA 技术和其他分子技术制造产品的过程被称为**生物技术**。在美国，生物技术已经悄悄地改变了日常生活的方方面面。现在，在超市、医疗卫生行业、农业和执法系统中都可以找到生物技术相关产品。这里，让我们先来看一些受到生物技术影响的日常示例。

Plants, Animals, and the Food Supply

The use of recombinant DNA technology to genetically modify crop plants has revolutionized agriculture. Genes for traits including resistance to herbicides, insects, and genes for nutritional enhancement have been introduced into crop plants. The transfer of heritable traits across species using recombinant DNA technology creates **transgenic organisms**.

植物、动物和食品

使用重组 DNA 技术对农作物进行遗传改造已经彻底改变了现代农业。一些性状基因，包括抗除草剂基因、抗虫基因、营养改良基因，均已被引入作物。使用重组 DNA 技术实现可遗传性状的跨物种转移，产生了**转基因生物**。20 世纪 90 年代中期，抗除草剂玉米和抗除草剂大豆最早进行了田间种植，目前美国约 88%的玉米作物和

Herbicide-resistant corn and soybeans were first planted in the mid-1990s, and transgenic strains now represent about 88 percent of the U.S. corn crop and 93 percent of the U.S. soybean crop. It is estimated that more than 70 percent of the processed food in the United States contains ingredients from transgenic crops.

约 93％的大豆作物均为转基因植株。据估计，美国超过 70％的加工食品中含有转基因作物成分。

We will discuss the most recent findings involving genetically modified organisms later in the text (Chapter 14—Genetically Modified Foods).

我们将在随后的章节部分继续讨论与遗传修饰生物体相关的最新发现（第 14 章—— 转基因食品）。

New methods of cloning livestock such as sheep and cattle have changed the way we use these animals. In 1996, Dolly the sheep was cloned by nuclear transfer, a method in which the nucleus of an adult cell is transferred into an egg that has had its nucleus removed. This makes it possible to produce dozens or hundreds of genetically identical offspring with desirable traits with many applications in agriculture and medicine.

对绵羊和牛等牲畜进行克隆的新方法也已经改变了我们利用这些动物的方式。1996 年，研究人员通过核转移技术克隆得到多莉羊，使用的方法是将成年细胞的细胞核转移至一个已经去除细胞核的卵细胞中。这使得生产数十甚至数百具有在农业和医学上有广泛应用前景的理想性状、遗传上完全相同的后代动物成为可能。

Biotechnology has also changed the way human proteins for medical use are produced. Through use of gene transfer, transgenic animals now synthesize these therapeutic proteins. In 2009, an anticlotting protein derived from the milk of transgenic goats was approved by the U.S. Food and Drug Administration for use in the United States. Other human proteins from transgenic animals are now being used in clinical trials to treat several diseases. The biotechnology revolution will continue to expand as gene editing by CRISPR-Cas and other new methods are used to develop an increasing array of products.

生物技术还改变了医用人源蛋白质的生产方式。通过应用基因转移技术，现在转基因动物可以合成这些药用蛋白质。2009 年，美国食品药品监督管理局批准了在美国使用一种来自转基因山羊乳汁的抗凝蛋白，其他来自转基因动物的人源蛋白目前也正在进行治疗多种疾病的临床试验。随着 CRISPR-Cas 基因编辑技术以及其他新型方法被用于开发越来越多的产品，生物技术革命将继续深入开展。

Biotechnology in Genetics and Medicine

More than 10 million children or adults in the United States suffer from some form of genetic disorder, and every child-bearing couple faces an approximately 3 percent risk of having a child with a genetic anomaly. The molecular basis for hundreds of genetic disorders is now known, and most of these genes have been mapped, isolated, and cloned. Biotechnology-derived genetic testing is now available to perform prenatal diagnosis of heritable disorders and to test parents for their status as heterozygous carriers of more than 100 inherited disorders. Newer methods now offer the possibility of scanning an entire genome to establish an individual's risk of developing a genetic disorder or having an affected child. The use of genetic testing and related technologies raises ethical concerns that have yet to be resolved.

用于遗传与医学的生物技术

在美国，超过一千万的儿童或成人患有不同类型的遗传疾病，并且每对育龄夫妇大约有3%的风险会孕育含遗传缺陷的孩子。目前数百种遗传疾病的分子机制已被破解，并且其中大多数基因已得到定位、分离和克隆。生物技术衍生的遗传检测现在已被用于遗传疾病的产前诊断，还可针对100多种遗传疾病检测父母的杂合基因携带状态。现在，更新的方法则是通过扫描整个基因组，评估个体罹患某种遗传疾病或生育患病儿的风险。遗传检测及相关技术的使用也引发了一些尚未解决的社会伦理问题。

ESSENTIAL POINT

Biotechnology has revolutionized agriculture and the pharmaceutical industry, while genetic testing has had a profound impact on the diagnosis of genetic diseases.

基本要点

生物技术已经彻底改变了农业和制药业，同时遗传检测对遗传疾病诊断已经产生了深远影响。

1.6 Genomics, Proteomics, and Bioinformatics Are New and Expanding Fields

1.6 基因组学、蛋白质组学和生物信息学是日益发展的新兴领域

The ability to create genomic libraries prompted scientists to consider sequencing all the clones in a library to derive the nucleotide sequence of an organism's genome. This

基因组文库的构建促使科学家考虑对文库中所有克隆进行测序，从而获得生物体的全基因组核苷酸序列。该序列信息将用于鉴定基因组中的每个基因并最终确定

sequence information would be used to identify each gene in the genome and establish its function.

其功能。

One such project, the Human Genome Project (HGP), began in 1990 as an international effort to sequence the human genome. By 2003, the publicly funded HGP and a private, industry-funded genome project completed sequencing of the gene-containing portion of the genome.

起始于 1990 年的人类基因组计划（HGP）即为这样一项集聚全球科研力量，旨在对人类基因组进行测序的项目。直至 2003 年，由政府资助的人类基因组计划和由行业资助的基因组项目完成了基因组中蕴含基因部分的序列测定。

As more genome sequences were acquired, several new biological disciplines arose. One, called **genomics** (the study of genomes), studies the structure, function, and evolution of genes and genomes. A second field, **proteomics**, identifies the set of proteins present in a cell under a given set of conditions, and studies their functions and interactions. To store, retrieve, and analyze the massive amount of data generated by genomics and proteomics, a specialized subfield of information technology called **bioinformatics** was created to develop hardware and software for processing nucleotide and protein data.

随着更多基因组序列的获得，一些新兴生物学领域随之产生。其一，**基因组学**，即研究基因组的科学，主要研究基因和基因组的结构、功能及进化。其二，**蛋白质组学**，用于鉴定特定条件下细胞中存在的所有蛋白质的集合，并研究它们的功能和相互作用。为了存储、检索和分析由基因组学和蛋白质组学产生的海量数据，**生物信息学**这一信息技术的特殊分支领域也应运而生，用以开发处理核苷酸和蛋白质相关数据的硬件和软件。

Geneticists and other biologists now use information in databases containing nucleic acid sequences, protein sequences, and gene-interaction networks to answer experimental questions in a matter of minutes instead of months and years.

当前，遗传学家和其他生物学家使用包含核酸序列、蛋白质序列和基因相互作用等信息的网络数据库在几分钟之内即可回答实验相关问题，而无须再像以前那样花费数月甚至数年的时间。

Modern Approaches to Understanding Gene Function

理解基因功能的现代方法

Historically, an approach referred to as **classical** or **forward genetics** was essential for studying and understanding gene function. In this approach geneticists relied on the use of

历史上，**经典遗传学**或**正向遗传学**的研究方法对于基因功能的研究和理解至关重要。在这种方法中，遗传学家依靠自发突变或使用化学试剂、X 射线、紫外线等诱

naturally occurring mutations or intentionally induced mutations (using chemicals, X-rays, or UV light as examples) to cause altered phenotypes in model organisms, and then worked through the labor-intensive and time-consuming process of identifying the genes that caused these new phenotypes. Such characterization often led to the identification of the gene or genes of interest, and once the technology advanced, the gene sequence could be determined.

发突变，从而在模式生物中引发表型变化，然后通过劳动强度大、时间耗费长的过程来鉴定产生这些新表型的基因。这一过程常常可鉴定出一种或多种目的基因，并且一旦技术发展，就可以进一步获悉基因序列。

Classical genetics approaches are still used, but as whole genome sequencing has become routine, molecular approaches to understanding gene function have changed considerably in genetic research. These modern approaches are what we will highlight in this section.

经典遗传学方法目前仍被科研工作者所使用，但是随着全基因组测序成为常规技术，研究基因功能的分子生物学方法已在遗传研究中相应地发生了巨大变化。在本节中，我们将重点介绍这些现代方法。

For the past two decades or so, geneticists have relied on the use of molecular techniques incorporating an approach referred to as **reverse genetics**. In reverse genetics, the DNA sequence for a particular gene of interest is known, but the role and function of the gene are typically not well understood. For example, molecular biology techniques such as **gene knockout** render targeted genes nonfunctional in a model organism or in cultured cells, allowing scientists to investigate the fundamental question of "what happens if this gene is disrupted?" After making a knockout organism, scientists look for both apparent phenotype changes, as well as those at the cellular and molecular level. The ultimate goal is to determine the function of the gene being studied.

在过去的二十多年里，遗传学家一直依赖于使用整合了**反向遗传学**方法的分子生物学技术。在反向遗传学中，特定目的基因的 DNA 序列已知，但是通常不了解该基因的作用和功能。例如，分子生物学技术中的**基因敲除**技术可以使目的基因在模式生物体内或培养细胞内丧失生理功能，从而使科学家能够研究“如果该基因被破坏将发生什么？”这一基本问题。创造基因敲除生物后，科学家既要寻找明显的表型变化，同时也要寻找细胞水平和分子水平的表型变化，最终目标是确定所研究基因的功能。

ESSENTIAL POINT

Recombinant DNA technology gave rise to several new fields, including genomics, proteomics, and bioinformatics, which allow scientists to explore the structure and evolution of genomes and the proteins they encode.

基本要点

重组 DNA 技术衍生出多个新生学科领域，包括基因组学、蛋白质组学和生物信息学，并使科学家能够探索基因组及其编码蛋白质的结构及进化。

1.7 Genetic Studies Rely on the Use of Model Organisms

After the rediscovery of Mendel's work in 1900, research using a wide range of organisms confirmed that the principles of inheritance he described were of universal significance among plants and animals. Geneticists gradually came to focus attention on a small number of organisms, including the fruit fly (*Drosophila melanogaster*) and the mouse (*Mus musculus*). This trend developed for two main reasons: First, it was clear that genetic mechanisms were the same in most organisms, and second, these organisms had characteristics that made them especially suitable for genetic research. They were easy to grow, had relatively short life cycles, produced many offspring, and their genetic analysis was fairly straightforward. Over time, researchers created a large catalog of mutant strains for these species, and the mutations were carefully studied, characterized, and mapped. Because of their well-characterized genetics, these species became **model organisms**, defined as organisms used for the study of basic biological processes. In later chapters, we will see how discoveries in model organisms are shedding light on many aspects of biology, including aging, cancer, and behavior.

1.7 遗传学研究依赖于模式生物的使用

在 1900 年孟德尔的研究工作被重新发现后，人们广泛使用多种生物开展科学研究并证实孟德尔所描述的遗传规律在动物和植物中均具有普遍性。遗传学家逐渐将注意力集中于少数生物物种上，包括果蝇和小鼠。这一趋势的形成主要有两个原因：首先，大多数生物体的遗传机制很显然是相同的；其次，这些生物具有尤其适于遗传研究的特征，它们易于生长，生命周期相对较短，产生的后代很多，并且它们的遗传分析相对简单明了。随着时间的推移，研究人员为这些物种创建了大型突变体名录，并对这些突变进行了深入研究、鉴定和基因定位。由于充分的遗传学研究，这些物种已成为**模式生物**，即可用于研究基本生物学过程的生物。在随后的章节中，我们将看到模式生物中的发现如何为生物学研究的诸多方面，包括衰老、癌症和行为机制的揭示披荆斩棘。

The Modern Set of Genetic Model Organisms

Gradually, geneticists added other species to their collection of model organisms: viruses (such as the T phages and lambda phage) and microorganisms (the bacterium *Escherichia coli* and the yeast *Saccharomyces cerevisiae*).

More recently, additional species have been developed as model organisms. Each species was chosen to allow study of some aspect of embryonic development. The nematode *Caenorhabditis elegans* was chosen as a model system to study the development and function of the nervous system because its nervous system contains only a few hundred cells and the developmental fate of these and all other cells in the body has been mapped out. *Arabidopsis thaliana*, a small plant with a short life cycle, has become a model organism for the study of many aspects of plant biology. The zebrafish, *Danio rerio*, is used to study vertebrate development: it is small, it reproduces rapidly, and its egg, embryo, and larvae are all transparent.

现代常见的遗传模式生物

遗传学家陆续将其他一些物种也加入模式生物的行列中：病毒（例如，T 噬菌体和 λ 噬菌体）与微生物（大肠杆菌和酿酒酵母）。

最近，其他一些物种也被开发为模式生物。这些物种可用于研究胚胎发育的某些方面。秀丽隐杆线虫作为模式生物可用于研究神经系统的发育和功能，因为其神经系统仅含几百个细胞，并且这些细胞及其体内其他细胞的发育过程均已被确定。拟南芥是一种小型植物，生命周期短，已成为植物生物学许多研究领域的模式生物。斑马鱼用于研究脊椎动物的发育，它个体小、易于繁殖，并且它的卵、胚胎和幼虫都是透明的。

Model Organisms and Human Diseases

The development of recombinant DNA technology and the results of genome sequencing have confirmed that all life has a common origin. Because of this, genes with similar functions in different organisms tend to be similar or identical in structure and nucleotide sequence. Much of what scientists learn by studying the genetics of model organisms can therefore be applied to humans as the basis for understanding and

模式生物与人类疾病

重组 DNA 技术的发展和基因组测序成果已经证实所有的生命形式都有一个共同的起源。正因如此，不同生物中具有相似功能的基因在结构和核苷酸序列上趋于相似或相同。因此，科学家通过研究模式生物遗传学所获得的大多数信息均可应用于人类，并成为了解和治疗人类疾病的基础。此外，在物种之间通过转移基因创建转基因生物也使科学家能够在细菌、真菌、植

treating human diseases. In addition, the ability to create transgenic organisms by transferring genes between species has enabled scientists to develop models of human diseases in organisms ranging from bacteria to fungi, plants, and animals (Table 1.1).

物和动物等各种不同生物中构建人类疾病模型（表1.1）。

TABLE 1.1 Model Organisms Used to Study Some Human Diseases
表1.1 用于研究人类疾病的模式生物

Organism	Human Diseases
E. coli	Colon cancer and other cancers
S. cerevisiae	Cancer, Werner syndrome
D. melanogaster	Disorders of the nervous system, cancer
C. elegans	Diabetes
D. rerio	Cardiovascular disease
M. musculus	Lesch-Nyhan syndrome, cystic fibrosis, fragile-X syndrome, and many other diseases

The idea of studying a human disease such as colon cancer by using *E. coli* may strike you as strange, but the basic steps of DNA repair (a process that is defective in some forms of colon cancer) are the same in both organisms, and a gene involved in DNA repair (*mutL* in *E. coli* and *MLH1* in humans) is found in both organisms. More importantly, *E. coli* has the advantage of being easier to grow (the cells divide every 20 minutes), and researchers can easily create and study new mutations in the bacterial *mutL* gene in order to figure out how it works. This knowledge may eventually lead to the development of drugs and other therapies to treat colon cancer in humans.

使用大肠杆菌来研究人类疾病如结肠癌的想法可能让你感到奇怪，但是DNA修复的基本过程（在某些类型的结肠癌中存在缺陷）在这两种生物中是相同的，并且在这两种生物中均发现了一种DNA修复相关基因，即大肠杆菌*mutL*基因和人类*MLH1*基因。更为重要的是，大肠杆菌具有易于生长的优势，大约每20分钟细胞就会分裂一次，因此研究人员可以轻松地构建和研究大肠杆菌*mutL*基因的各种新型突变，从而揭示该基因的工作机制。这些研究结果最终有助于人类结肠癌的治疗药物和其他疗法的研发。

The fruit fly, *Drosophila melanogaster*, is also being used to study a number of human diseases. Mutant genes have been identified in *D. melanogaster* that produce phenotypes with structural abnormalities of the nervous

果蝇也被用于研究许多人类疾病。研究人员在果蝇中已鉴定出若干突变基因，它们能够导致神经系统结构异常以及成年后发病的神经系统退行性病变表型，而来自基因组测序计划的结果表明在人类基因

system and adult-onset degeneration of the nervous system. The information from genome-sequencing projects indicates that almost all these genes have human counterparts. For example, genes involved in a complex human disease of the retina called retinitis pigmentosa are identical to *Drosophila* genes involved in retinal degeneration. Study of these mutations in *Drosophila* is helping to dissect this complex disease and identify the function of the genes involved.

组中几乎存在所有与之相对应的基因。例如，人类视网膜的一种复杂疾病——视网膜色素变性，其相关基因与果蝇视网膜变性的相关基因是相同的。对果蝇中这些突变的研究有助于我们剖析这种复杂疾病，并确定相关基因的功能。

Another approach to studying diseases of the human nervous system is to transfer mutant human disease genes into *Drosophila* using recombinant DNA technology. The transgenic flies are then used for studying the mutant human genes themselves, other genes that affect the expression of the human disease genes, and the effects of therapeutic drugs on the action of those genes—all studies that are difficult or impossible to perform in humans. This gene transfer approach is being used to study almost a dozen human neurodegenerative disorders, including Huntington disease, Machado – Joseph disease, myotonic dystrophy, and Alzheimer disease.

研究人类神经系统疾病的另一种方法是使用重组 DNA 技术将突变的人类疾病基因转入果蝇中，然后利用转基因果蝇研究人类突变基因本身、影响人类疾病基因表达的其他基因以及治疗药物对这些基因作用的影响——所有这些研究都是在人体中难以进行或不可能进行的。这种基因转移方法正被用于研究十多种人类神经系统变性疾病，包括亨廷顿病、马查多 - 约瑟夫病、强直性肌营养不良和阿尔茨海默病。

Throughout the following chapters, you will encounter these model organisms again and again. Remember each time you meet them that they not only have a rich history in basic genetics research but are also at the forefront in the study of human genetic disorders and infectious diseases.

在接下来的章节中，你将不断地遇到这些模式生物。每次与它们会面时，请记住它们不仅在基础遗传学研究中拥有丰富的历史，而且也处于人类遗传疾病和传染性疾病科学研究的最前沿。

ESSENTIAL POINT

The study of model organisms for

基本要点

通过模式生物探索人类健康和疾病发

understanding human health and disease is one of the many ways genetics and biotechnology are changing everyday life.

生机制是遗传学和生物技术正在改变日常生活的众多方式之一。

1.8 Genetics Has Had a Profound Impact on Society

1.8 遗传学对社会已经产生了深远影响

Mendel described his decade-long project on inheritance in pea plants in an 1865 paper presented at a meeting of the Natural History Society of Brünn in Moravia. Less than 100 years later, the 1962 Nobel Prize was awarded to James Watson, Francis Crick, and Maurice Wilkins for their work on the structure of DNA. This time span encompassed the years leading up to the acceptance of Mendel's work, the discovery that genes are on chromosomes, the experiments that proved DNA encodes genetic information, and the elucidation of the molecular basis for DNA replication. The rapid development of genetics from Mendel's monastery garden to the Human Genome Project and beyond is summarized in a timeline in Figure 1.4.

1865 年，在摩拉维亚召开的布隆自然历史学会的会议上，孟德尔发表了一篇论文，介绍他长达十年之久的豌豆遗传规律研究。其后不到 100 年，詹姆斯 · 沃森、弗朗西斯 · 克里克和莫里斯 · 威尔金斯因他们在 DNA 结构上的研究工作而获得了 1962 年的诺贝尔奖。这一时间跨度涵盖了孟德尔工作被科学界接受、基因位于染色体上的发现、证明 DNA 编码遗传信息的实验、DNA 复制的分子机制的阐明等事件。图 1.4 总结了遗传学从孟德尔的修道院花园到人类基因组计划的快速发展历程。

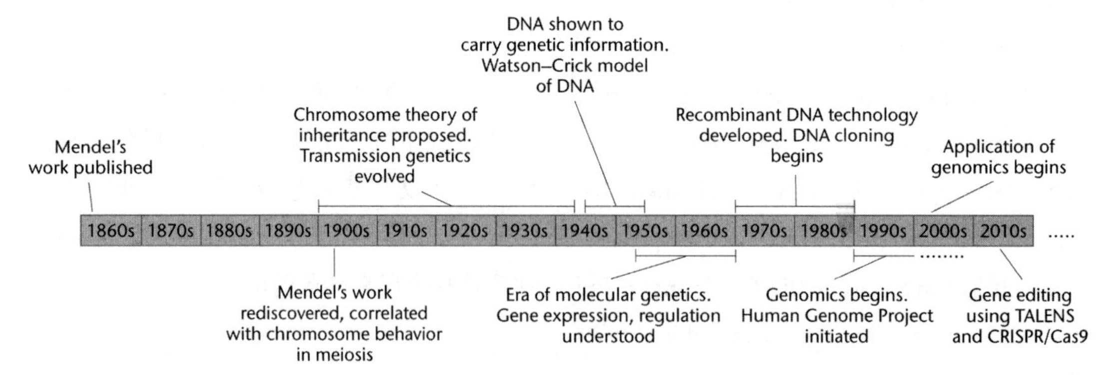

FIGURE 1.4 A timeline showing the development of genetics from Gregor Mendel's work on pea plants to the current era of genomics and its many applications in research, medicine, and society. Having a sense of the history of discovery in genetics should provide you with a useful framework as you proceed through this textbook.

图 1.4 遗传学发展时间轴：从孟德尔研究豌豆直至当今的基因组学时代，以及遗传学在科学研究、医学、社会中的应用。了解遗传学发展史能为你在阅读本书的过程中提供有用的结构框架。

The Nobel Prize and Genetics

No other scientific discipline has experienced the explosion of information and the level of excitement generated by the discoveries in genetics. This impact is especially apparent in the list of Nobel Prizes related to genetics, beginning with those awarded in the early and mid-twentieth century and continuing into the present. Nobel Prizes in Physiology or Medicine and Chemistry have been consistently awarded for work in genetics and related fields.The first such prize awarded was given to Thomas H. Morgan in 1933 for his research on the chromosome theory of inheritance.That award was followed by many others, including prizes for the discovery of genetic recombination, the relationship between genes and proteins, the structure of DNA, and the genetic code. This trend has continued throughout the twentieth and twenty-first centuries. The advent of genomic studies and the applications of such findings will most certainly lead the way for future awards.

诺贝尔奖与遗传学

没有任何一门其他学科经历过类似遗传学发现所带来的信息爆炸和科学兴奋。这种影响在与遗传相关的诺贝尔奖获奖清单中体现得尤为明显，这些获奖工作从 20 世纪初开始到 20 世纪中叶一直延续到现在。遗传学及其相关领域的研究屡获诺贝尔生理学或医学奖、诺贝尔化学奖。1933 年，托马斯 · H. 摩尔根因其在遗传的染色体学说方面的杰出工作而成为首位获得诺贝尔奖的遗传学家。随后，遗传学领域的许多研究工作陆续获得诺贝尔奖，包括遗传重组的发现、基因与蛋白质之间的关系、DNA 的结构以及遗传密码的发现。这一发展趋势在整个 20 世纪直至 21 世纪一直持续着。基因组研究的到来以及这些发现的应用无疑为将来的获奖开辟了道路。

Genetics, Ethics, and Society

Just as there has never been a more exciting time to study genetics, the impact of this discipline on society has never been more profound. Genetics and its applications in biotechnology are developing much faster than the social conventions, public policies, and laws required to regulate their use. As a society, we are grappling with a host of sensitive genetics-related issues, including concerns about prenatal testing,

遗传、伦理与社会

正如从未有过比研究遗传学更激动人心的时刻一样，遗传学对于社会的影响也从未如此深远。遗传学及其在生物技术中的应用发展远远快于相应的可用于规范其应用的社会公约、公共政策和相关法律的发展。作为一个社会，我们正在努力解决遗传学相关的许多敏感问题，包括对产前检查、遗传歧视、基因所有权、基因治疗实施和安全性，以及遗传隐私等诸多问题

genetic discrimination, ownership of genes, access to and safety of gene therapy, and genetic privacy. This emphasis on ethics reflects the growing concern and dilemmas that advances in genetics pose to our society and the future of our species. It is our hope that upon the completion of your study of genetics, you will become an informed, active participant in future debates that arise.

的担忧。对伦理学的重视反映了人们对于遗传学进展给人类社会和人类未来带来的影响而产生的越来越多的担忧和困境。我们希望在完成遗传学课程的学习后，你将成为知识渊博、积极参与未来辩论的一分子。

ESSENTIAL POINT

Genetic technology is having a profound effect on society, while raising many ethical dilemmas.

基本要点

遗传技术对社会已经产生了深远影响，同时也引发了许多伦理困境。

2 Mitosis and Meiosis

2　有丝分裂和减数分裂

CHAPTER CONCEPTS

■ Genetic continuity between generations of cells and between generations of sexually reproducing organisms is maintained through the processes of mitosis and meiosis, respectively.

■ Diploid eukaryotic cells contain their genetic information in pairs of homologous chromosomes, with one member of each pair being derived from the maternal parent and one from the paternal parent.

■ Mitosis provides a mechanism by which chromosomes, having been duplicated, are distributed into progeny cells during cell reproduction.

■ Mitosis converts a diploid cell into two diploid daughter cells.

■ Meiosis provides a mechanism by which one member of each homologous pair of chromosomes is distributed into each gamete or spore, thus reducing the diploid chromosome number to the haploid chromosome number.

■ Meiosis generates genetic variability by distributing various combinations of maternal and paternal members into gametes or spores.

■ During the stages of mitosis and meiosis, the genetic material is condensed into discrete structures called chromosomes.

本章速览

■ 细胞或进行有性繁殖的生物体在世代交替过程中分别通过有丝分裂和减数分裂保持遗传物质的连续性和稳定性。

■ 二倍体真核细胞将遗传信息储存于同源染色体对中，成对存在的同源染色体一条来自母本，另外一条来自父本。

■ 有丝分裂为细胞繁殖过程中如何将已复制的染色体分配至子代细胞提供了一种机制。

■ 通过有丝分裂，一个二倍体细胞分裂为两个二倍体子代细胞。

■ 减数分裂提供了一种分配机制，将每对同源染色体中的一条分配至子代配子或子代孢子中，从而使二倍体细胞的染色体数量减半至单倍染色体数量。

■ 减数分裂将母本或父本来源的染色体通过不同的自由组合方式随机分配到子代配子或子代孢子中，从而产生遗传多样性。

■ 在有丝分裂或减数分裂过程中，遗传物质被压缩成相互独立、不连续的染色体结构。

Every living thing contains a substance described as the genetic material.Except in certain viruses, this material is composed of the nucleic acid DNA. DNA has an underlying linear structure possessing segments called genes, the products of which direct the metabolic activities of cells. An organism's DNA, with its arrays of genes, is organized into structures called **chromosomes**, which serve as vehicles for transmitting genetic information. The manner in which chromosomes are transmitted from one generation of cells to the next and from organisms to their descendants must be exceedingly precise. In this chapter we consider exactly how genetic continuity is maintained between cells and organisms.

所有生物体内都含有遗传物质，除一些病毒外，遗传物质均由核酸 DNA 组成。DNA 分子具有线性结构，包含被称为基因的片段，这些基因的产物指导细胞内各项代谢活动顺利进行。基因沿着生物体的 DNA 线性排列，有序组织形成名为**染色体**的结构，从而担当遗传信息传递的载体。在细胞或生物体世代交替过程中，染色体的传递过程必须非常精确。在这一章中，我们将讨论在细胞和生物中是如何保证遗传物质稳定传递的。

Two major processes are involved in the genetic continuity of nucleated cells: **mitosis** and **meiosis**. Although the mechanisms of the two processes are similar in many ways, the outcomes are quite different. Mitosis leads to the production of two cells, each with the same number of chromosomes as the parent cell. In contrast, meiosis reduces the genetic content and the number of chromosomes by precisely half. This reduction is essential if sexual reproduction is to occur without doubling the amount of genetic material in each new generation. Strictly speaking, mitosis is that portion of the cell cycle during which the hereditary components are equally partitioned into daughter cells. Meiosis is part of a special type of cell division that leads to the production of sex cells: **gametes** or **spores**. This process is an essential step in the

有核细胞的遗传物质稳定传递主要包含两种过程：**有丝分裂**和**减数分裂**。尽管两种过程在许多方面有相似之处，但是它们的最终结果却截然不同。有丝分裂最终产生两个子细胞，并且每个子细胞所含染色体数目均与它们的亲本细胞相同。与此不同的是，减数分裂中遗传物质含量和染色体数目精确减半，这种遗传物质减半的过程对于有性生殖十分必要，可以保证通过有性生殖产生的后代遗传物质不发生加倍。严格地说，有丝分裂是细胞周期中遗传组分被等量分配至子代细胞的部分。减数分裂则是一种特殊细胞分裂类型中的一部分，能够产生性细胞（**配子**或**孢子**），并且是生物体世代交替中遗传信息传递的必要步骤。

transmission of genetic information from an organism to its offspring.

Normally, chromosomes are visible only during mitosis and meiosis. When cells are not undergoing division, the genetic material making up chromosomes unfolds and uncoils into a diffuse network within the nucleus, generally referred to as **chromatin**. Before describing mitosis and meiosis, we will briefly review the structure of cells, emphasizing components that are of particular significance to genetic function. We will also compare the structural differences between the nonnucleated cells of bacteria and the eukaryotic cells of higher organisms. We then devote the remainder of the chapter to the behavior of chromosomes during cell division.

通常，染色体只在有丝分裂和减数分裂过程中明显可见。当细胞不进行分裂时，组成染色体的遗传物质将解旋展开、松散地存在于细胞核内，通常被称为**染色质**。在描述有丝分裂和减数分裂之前，我们将简要回顾细胞结构，尤其是那些与遗传功能密切相关、意义重大的细胞组分。我们还将比较细菌的无核细胞与高等生物真核细胞的结构差异，然后继续在本章讨论细胞分裂过程中的染色体行为。

2.1 Cell Structure Is Closely Tied to Genetic Function

2.1 细胞结构与遗传功能紧密相关

Before 1940, our knowledge of cell structure was limited to what we could see with the light microscope. Around 1940, the transmission electron microscope was in its early stages of development, and by 1950, many details of cell ultrastructure had emerged. Under the electron microscope, cells were seen as highly varied, highly organized structures whose form and function are dependent on specific genetic expression by each cell type. A new world of whorled membranes, organelles, microtubules, granules, and filaments was revealed. These discoveries revolutionized thinking in the entire field of biology. Many cell components, such as the nucleolus, ribosome,

在1940年之前，人们对细胞结构的了解仅限于光学显微镜下的可见部分。1940年左右，透射电子显微镜刚刚起步。直至1950年，人们才可以观察到细胞超显微结构中的许多细节。在电子显微镜下，人们观察到细胞内的结构具有高度多样性和高度组织性，并且这些结构的形式和功能均取决于每种细胞类型特定的基因表达特点。一个囊括了漩涡状膜、细胞器、微管、颗粒和丝状体的崭新世界从而被揭开。这些发现也彻底改变了人们对于整个生物学领域的认识。许多细胞组分如核仁、核糖体和中心粒都直接或间接地参与遗传过程，其他组分如线粒体和叶绿体则拥有各自独特的遗传信息。在这里，我们将主要关注

and centriole, are involved directly or indirectly with genetic processes. Other components—the mitochondria and chloroplasts—contain their own unique genetic information. Here, we will focus primarily on those aspects of cell structure that relate to genetic study. The generalized animal cell shown in Figure 2.1 illustrates most of the structures we will discuss.

那些与遗传学研究息息相关的细胞结构。图 2.1 所示的动物细胞展示了我们将要讨论的大多数细胞结构。

All cells are surrounded by a **plasma membrane**, an outer covering that defines the cell boundary and delimits the cell from

所有细胞均被一层**质膜**所包围，这是一层明确细胞边界并将细胞与其直接接触的外部环境分隔的外层保护膜。该膜并不

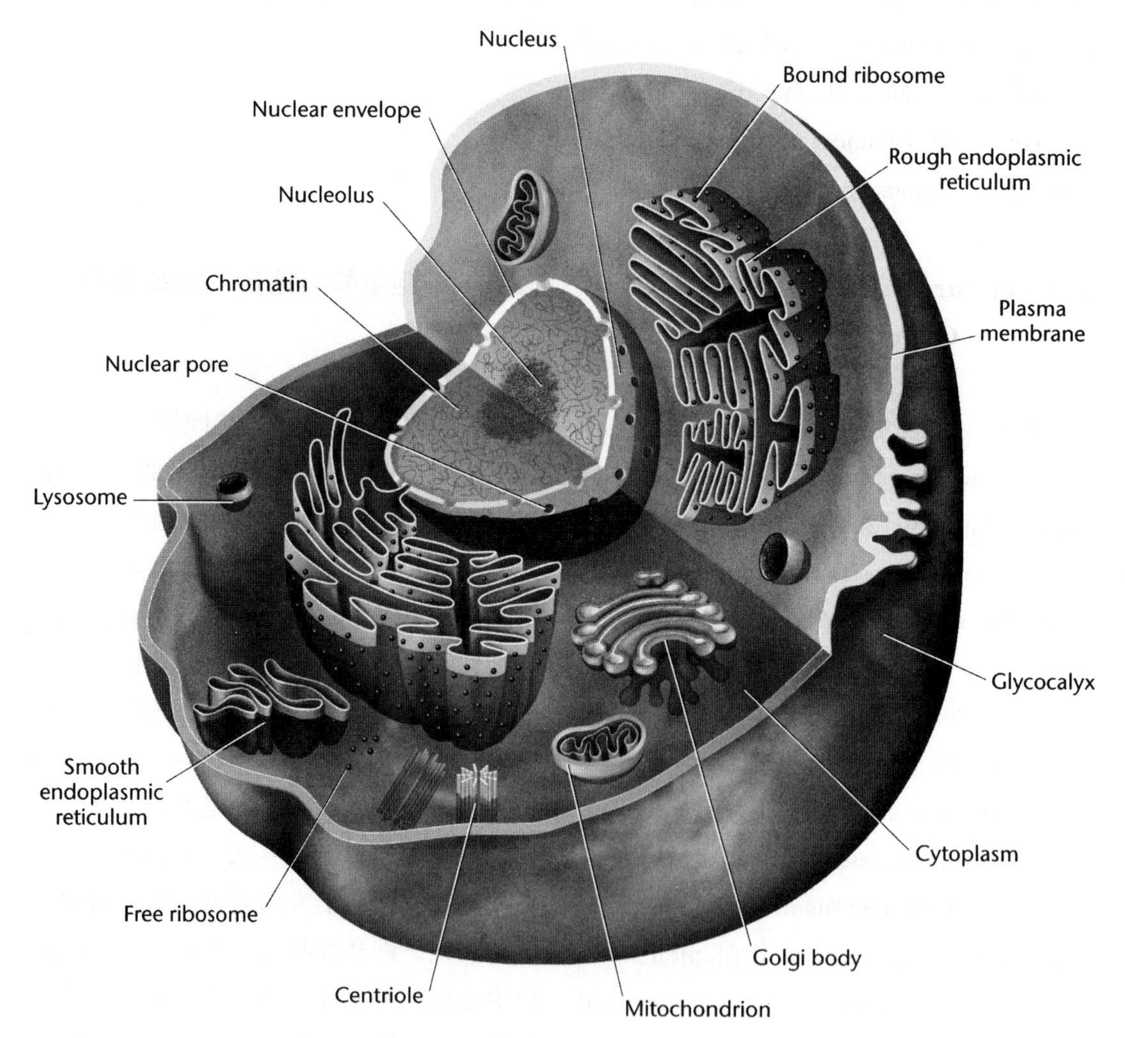

FIGURE 2.1 A generalized animal cell. The cellular components discussed in the text are emphasized here.

图 2.1 一个广义的动物细胞。书中讨论的细胞结构均在图中进行了标示。

its immediate external environment. This membrane is not passive but instead actively controls the movement of materials into and out of the cell. In addition to this membrane, plant cells have an outer covering called the **cell wall** whose major component is a polysaccharide called **cellulose**.

是被动存在的，而是能够主动控制物质进出细胞的运动。除了这层质膜之外，植物细胞还具有一层被称为**细胞壁**的外壳，其主要成分是一种多糖——**纤维素**。

Many, if not most, animal cells have a covering over the plasma membrane, referred to as the **glycocalyx**, or **cell coat**. Consisting of glycoproteins and polysaccharides, this covering has a chemical composition that differs from comparable structures in either plants or bacteria. The glycocalyx, among other functions, provides biochemical identity at the surface of cells, and the components of the coat that establish cellular identity are under genetic control. For example, various cell identity markers that you may have heard of—the AB, Rh, and MN antigens—are found on the surface of red blood cells, among other cell types. On the surface of other cells, histocompatibility antigens, which elicit an immune response during tissue and organ transplants, are present. Various *receptor molecules* are also found on the surfaces of cells. These molecules act as recognition sites that transfer specific chemical signals across the cell membrane into the cell.

即使不是大多数，至少也有很多动物细胞在质膜外还有一层包被，被称为**糖萼**或**细胞外被**。这层结构由糖蛋白和多糖组成，其化学组成在植物或细菌的相应结构中也存在差异。糖萼除具有其他功能外，还在细胞表面提供生化特性识别标识，并且这些提供细胞身份标识的组分均受遗传调控。例如，在不同类型的细胞表面存在多种多样的细胞识别标签，你可能听说过位于红细胞表面的各种抗原分子——AB 抗原、Rh 抗原和 MN 抗原，它们即为细胞识别标签。在其他类型细胞表面存在的组织相容性抗原，在组织和器官移植过程中会引发免疫反应。细胞表面还存在各种受体分子，这些分子作为信号识别位点可以将特定的化学信号跨膜传递至细胞内。

Living organisms are categorized into two major groups depending on whether or not their cells contain a nucleus. The presence of a nucleus and other membranous organelles is the defining characteristic of **eukaryotes**. The **nucleus** in eukaryotic cells is a membrane-bound structure that houses the genetic material, DNA, which

根据细胞内是否含细胞核可以将生物体分为两大类。细胞核和其他含膜细胞器的存在是**真核生物**最典型的特征。真核细胞的**细胞核**具有被膜所包被的结构，内部含遗传物质 DNA，并且该遗传物质与一系列酸性蛋白质和碱性蛋白质相结合形成细丝状结构。在细胞周期的非分裂阶段，这

is complexed with an array of acidic and basic proteins into thin fibers. During nondivisional phases of the cell cycle, the fibers are uncoiled and dispersed into chromatin (as mentioned above). During mitosis and meiosis, chromatin fibers coil and condense into chromosomes. Also present in the nucleus is the **nucleolus**, an amorphous component where ribosomal RNA (rRNA) is synthesized and where the initial stages of ribosomal assembly occur. The portions of DNA that encode rRNA are collectively referred to as the **nucleolus organizer region**, or the **NOR**.

些丝状物不发生卷曲并以染色质形式分散存在（如前所述）。在有丝分裂和减数分裂期，染色质丝卷曲、聚集形成染色体结构。**核仁**也存在于细胞核中，这是一种无固定形状的组分，核糖体 RNA（rRNA）在核仁中合成并进行核糖体的初始组装。编码 rRNA 的 DNA 部分被统称为**核仁组织区**（**NOR**）。

Prokaryotes, of which there are two major groups, lack a nuclear envelope and membranous organelles. For the purpose of our brief discussion here, we will consider the **eubacteria**, the other group being the more ancient bacteria referred to as **archaea**. In eubacteria, such as *Escherichia coli*, the genetic material is present as a long, circular DNA molecule that is compacted into an unenclosed region called the **nucleoid**. Part of the DNA may be attached to the cell membrane, but in general the nucleoid extends through a large part of the cell. Although the DNA is compacted, it does not undergo the extensive coiling characteristic of the stages of mitosis, during which the chromosomes of eukaryotes become visible. Nor is the DNA associated as extensively with proteins as is eukaryotic DNA. Prokaryotic cells do not have a distinct nucleolus but do contain genes that specify rRNA molecules.

原核生物主要分为两大类，它们缺乏细胞核膜和含膜细胞器。为了方便，这里进行简要讨论，我们将以**真细菌**为讨论对象，而原核生物的另一类则是更为古老的细菌，即**古细菌**。在真细菌如大肠杆菌中，遗传物质以长的、环形 DNA 分子形式存在，该分子相对集中地存在于细胞内被称为**拟核**的非封闭区域。DNA 分子的部分区域可能附着于细胞膜上，但总的说来，拟核分散存在于细胞内的大部分区域。尽管原核细胞的 DNA 也会压缩聚集，但它并没有类似有丝分裂阶段中形成显微镜下肉眼可见的真核生物染色体结构的高度卷曲特征。原核细胞的 DNA 也不像真核细胞的 DNA 那样广泛地与蛋白质相互作用。原核细胞没有明显的核仁，但已确定含有编码 rRNA 分子的基因。

The remainder of the eukaryotic cell within the plasma membrane, excluding the nucleus, is

除细胞核外，被细胞质膜包被的真核细胞内其他组分被称为**细胞质**，其中含有

referred to as **cytoplasm** and includes a variety of extranuclear cellular organelles. In the cytoplasm, a nonparticulate, colloidal material referred to as the **cytosol** surrounds and encompasses the cellular organelles. The cytoplasm also includes an extensive system of tubules and filaments, comprising the **cytoskeleton**, which provides a lattice of support structures within the cell. Consisting primarily of **microtubules**, which are made of the protein **tubulin**, and **microfilaments**, which derive from the protein **actin**, this structural framework maintains cell shape, facilitates cell mobility, and anchors the various organelles.

多种核外细胞器。在细胞质中，一种被称为**胞质溶胶**的非颗粒状胶体物质包裹环绕着细胞器。细胞质中还包含广泛存在的微管和微丝系统，以构成**细胞骨架**，并为细胞提供框架状支撑结构。该结构框架主要由**微管蛋白**组成的**微管**和由**肌动蛋白**衍生而来的**微丝**构成，可以维持细胞形状，促进细胞移动，并有助于锚定各种细胞器。

One organelle, the membranous **endoplasmic reticulum** (**ER**), compartmentalizes the cytoplasm, greatly increasing the surface area available for biochemical synthesis. The ER appears smooth in places where it serves as the site for synthesizing fatty acids and phospholipids; in other places, it appears rough because it is studded with ribosomes. **Ribosomes** serve as sites where genetic information contained in messenger RNA (mRNA) is translated into proteins.

膜状**内质网**（ER）作为一种细胞器将细胞质区域进行分隔，从而大大增加了可用于生化合成的表面积。作为脂肪酸和磷脂合成场所的内质网，表面光滑；而在其他区域，内质网表面由于布满核糖体而显得粗糙。**核糖体**是携带遗传信息的信使RNA（mRNA）被翻译成蛋白质的场所。

Three other cytoplasmic structures are very important in the eukaryotic cell's activities: mitochondria, chloroplasts, and centrioles. **Mitochondria** are found in most eukaryotes, including both animal and plant cells, and are the sites of the oxidative phases of cell respiration. These chemical reactions generate large amounts of the energy-rich molecule adenosine triphosphate (ATP). **Chloroplasts**, which are found in plants, algae, and some

另外三种在真核细胞活动中十分重要的胞质内结构分别是线粒体、叶绿体和中心粒。**线粒体**存在于大多数真核生物中，包括动物细胞和植物细胞，是细胞呼吸中氧化代谢所在的场所。这些化学反应产生大量的高能分子——三磷酸腺苷（ATP）。**叶绿体**存在于植物、藻类和一些原生动物细胞中，与地球上的主要能量捕获过程——光合作用相关。线粒体和叶绿体均含有与细胞核内存在形式不同的DNA，它们能够

protozoans, are associated with photosynthesis, the major energy-trapping process on Earth. Both mitochondria and chloroplasts contain DNA in a form distinct from that found in the nucleus. They are able to duplicate themselves and transcribe and translate their own genetic information.

自我复制，并且转录和翻译自身携带的遗传信息。

Animal cells and some plant cells also contain a pair of complex structures called **centrioles**. These cytoplasmic bodies, each located in a specialized region called the **centrosome**, are associated with the organization of spindle fibers that function in mitosis and meiosis. In some organisms, the centriole is derived from another structure, the basal body, which is associated with the formation of cilia and flagella (hair-like and whip-like structures for propelling cells or moving materials).

动物细胞和一些植物细胞还含有一对具有复杂结构的**中心粒**。这些细胞质小体中的每个均位于特定区域——**中心体**中，并与在有丝分裂和减数分裂中发挥作用的纺锤丝有关。在一些生物中，中心粒来源于另一种名为基体的结构，该结构与纤毛和鞭毛的形成有关。纤毛和鞭毛类似毛发状或鞭状结构，可以推进细胞或移动物质。

The organization of **spindle fibers** by the centrioles occurs during the early phases of mitosis and meiosis. These fibers play an important role in the movement of chromosomes as they separate during cell division. They are composed of arrays of microtubules consisting of polymers of the protein tubulin.

由中心体形成**纺锤丝**组织发生在有丝分裂和减数分裂的早期阶段。细胞分裂时，这些纤维丝在染色体彼此分离的运动中发挥着重要作用。它们由微管排列组成，而微管又由微管蛋白的聚合物构成。

ESSENTIAL POINT

Most components of cells are involved directly or indirectly with genetic processes.

基本要点

细胞中的绝大多数组分均直接或间接地参与遗传过程。

2.2 Chromosomes Exist in Homologous Pairs in Diploid Organisms

2.2 二倍体生物中染色体以同源染色体对的形式存在

As we discuss the processes of mitosis and meiosis, it is important that you understand the

在讨论有丝分裂和减数分裂的过程中，准确理解同源染色体的概念十分重要，并

concept of homologous chromosomes. Such an understanding will also be of critical importance in our future discussions of Mendelian genetics. Chromosomes are most easily visualized during mitosis. When they are examined carefully, distinctive lengths and shapes are apparent. Each chromosome contains a constricted region called the **centromere**, whose location establishes the general appearance of each chromosome. Figure 2.2 shows chromosomes with centromere placements at different distances along their length. Extending from either side of the centromere are the arms of the chromosome. Depending on the position of the centromere, different arm ratios are produced. As Figure 2.2 illustrates, chromosomes are classified as

且对于我们未来讨论孟德尔遗传学也非常重要。染色体在有丝分裂过程中十分易于观察。当对其进行仔细研究时，每条染色体的长度和形态均存在明显差异。每条染色体均含有一个浓缩聚集的区域，称为**着丝粒**。着丝粒的位置确定了每条染色体整体的形态特征。图 2.2 显示了着丝粒位置各不相同的染色体。着丝粒两侧直至染色体的末端形成了染色体的两条臂，根据着丝粒位置形成不同的臂长比。正如图 2.2 所示，染色体依据着丝粒位置可分为：**中着丝粒染色体**、**近中着丝粒染色体**、**近端着丝粒染色体**、**端着丝粒染色体**。按照惯例，图中位于着丝粒上方较短的臂被称为 **p 臂**（p 源自英文单词 petite“娇小的”首字母）；图中位于着丝粒下方较长的臂被称为 **q 臂**

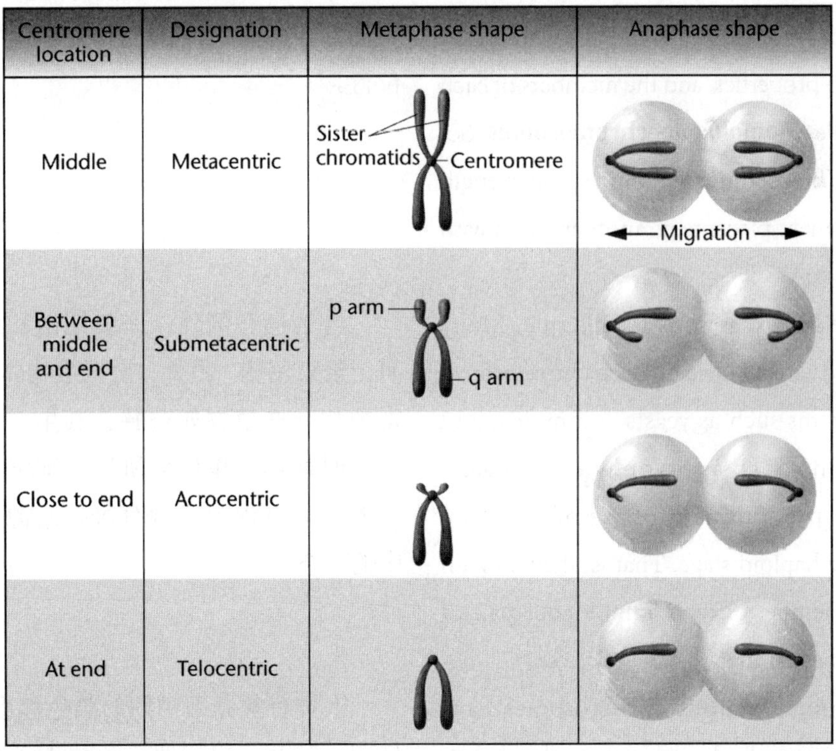

Centromere location	Designation	Metaphase shape	Anaphase shape
Middle	Metacentric	Sister chromatids; Centromere	Migration
Between middle and end	Submetacentric	p arm; q arm	
Close to end	Acrocentric		
At end	Telocentric		

FIGURE 2.2 Centromere locations and the chromosome designations that are based on them. Note that the shape of the chromosome during anaphase is determined by the position of the centromere during metaphase.

图 2.2 着丝粒位置及依据着丝粒位置进行的染色体命名。注意后期的染色体形态由中期的着丝粒位置决定。

metacentric, **submetacentric**, **acrocentric**, or **telocentric** on the basis of the centromere location. The shorter arm, by convention, is shown above the centromere and is called the **p arm** (p, for "petite"). The longer arm is shown below the centromere and is called the **q arm** (q because it is the next letter in the alphabet).

（因为英文字母顺序中 q 位于 p 之后）。

In the study of mitosis, several other observations are of particular relevance. First, all somatic cells derived from members of the same species contain an identical number of chromosomes. In most cases, this represents what is referred to as the **diploid number** (**2*n***). When the lengths and centromere placements of all such chromosomes are examined, a second general feature is apparent. With the exception of sex chromosomes, they exist in pairs with regard to these two properties, and the members of each pair are called **homologous chromosomes**. So, for each chromosome exhibiting a specific length and centromere placement, another exists with identical features.

在有丝分裂研究中，一些现象具有尤其特殊的关联性。首先，同一物种个体的所有体细胞所含染色体数目完全相同。大多数情况下，该数目代表了**二倍染色体数目**（**2*n***）。其次，当研究所有这些染色体的长度、着丝粒位置时，另一特征则显而易见，即除性染色体之外，其他所有染色体根据染色体特征进行划分，均成对存在，并且每对染色体被称为**同源染色体**。因此，同源染色体具有相同的特定长度和着丝粒位置。

There are exceptions to this rule. Many bacteria and viruses have but one chromosome, and organisms such as yeasts and molds, and certain plants such as bryophytes (mosses), spend the predominant phase of their life cycle in the haploid stage. That is, they contain only one member of each homologous pair of chromosomes during most of their lives.

当然也有例外，许多细菌和病毒仅含有一条染色体。另外，有些生物体如酵母、霉菌和一些植物如苔藓，其生命周期中的主要时期均是单倍体阶段，即它们在大部分的生命过程中仅含有每种同源染色体对中的一条。

The human mitotic chromosomes have been photographed, cut out of the print, and matched up, creating a display called a **karyotype**. Humans have a $2n$ number of 46

人类有丝分裂中期的染色体经过拍照、打印、剪裁，再按照染色体的长度和着丝粒位置进行比对排列并进行展现，即为**核型**。人类体细胞含有 46 条染色体（$2n =$

chromosomes, which on close examination exhibit a diversity of sizes and centromere placements.

46），分别具有不同的大小和着丝粒位置。

The **haploid number (*n*)** of chromosomes is equal to one-half the diploid number. Collectively, the genetic information contained in a haploid set of chromosomes constitutes the **genome** of the species. This, of course, includes copies of all genes as well as a large amount of noncoding DNA. The examples listed in Table 2.1 demonstrate the wide range of *n* values found in plants and animals.

染色体的**单倍体数目**（***n***）等于二倍体数目的一半。总的说来，单倍染色体组中所包含的遗传信息构成了物种的**基因组**，当然，这包含了所有基因以及大量非编码DNA的拷贝。表2.1中所举示例展示了植物和动物中存在的各种不同的 *n* 值。

TABLE 2.1 The Haploid Number of Chromosomes for a Variety of Organisms

表 2.1 不同生物物种的单倍体染色体数目

Common Name	Scientific Name	Haploid Number
Black bread mold	*Aspergillus nidulans*	8
Broad bean	*Vicia faba*	6
Chimpanzee	*Pan troglodytes*	24
Corn	*Zea mays*	10
Cotton	*Gossypium hirsutum*	26
Fruit fly	*Drosophila melanogaster*	4
Garden pea	*Pisum sativum*	7
House mouse	*Mus musculus*	20
Human	*Homo sapiens*	23
Pink bread mold	*Neurospora crassa*	7
Roundworm	*Caenorhabditis elegans*	6
Yeast	*Saccharomyces cerevisiae*	16
Zebrafish	*Danio rerio*	25

Homologous chromosomes have important genetic similarities. They contain identical gene sites along their lengths; each site is called a **locus** (pl. loci). Thus, they are identical in the traits that they influence and in their genetic potential. In sexually reproducing organisms, one member of each pair is derived from the maternal parent (through the ovum) and the other member is derived from the

同源染色体具有十分重要的遗传相似性。它们沿着各自的染色体含有完全相同的基因位点，每个基因位点被称为**基因座**。因此，同源染色体所影响的性状和它们潜在的遗传功能完全一样。在有性繁殖生物体中，每对同源染色体中的一条通过卵子来自母本，另一条则通过精子来自父本。因此，每个二倍体生物均含有同种基因的两个拷贝，这也正是**双亲遗传**的结果，即

paternal parent (through the sperm). Therefore, each diploid organism contains two copies of each gene as a consequence of **biparental inheritance**, inheritance from two parents. As we shall see during our discussion of transmission genetics (Chapters 3 and 4), the members of each pair of genes, while influencing the same characteristic or trait, need not be identical. In a population of members of the same species, many different alternative forms of the same gene, called **alleles**, can exist.

遗传来自两个亲本。正如我们在随后传递遗传学（第3、4章）讨论中将看到的那样，每对基因的各个成员，即使影响同一特征或性状，它们的作用也并不完全相同。在同一种群成员中，同种基因存在多种基因形式，这些基因形式被称为**等位基因**。

The concepts of haploid number, diploid number, and homologous chromosomes are important for understanding the process of meiosis. During the formation of gametes or spores, meiosis converts the diploid number of chromosomes to the haploid number. As a result, haploid gametes or spores contain precisely one member of each homologous pair of chromosomes—that is, one complete haploid set. Following fusion of two gametes at fertilization, the diploid number is reestablished; that is, the zygote contains two complete haploid sets of chromosomes. The constancy of genetic material is thus maintained from generation to generation.

单倍体数目、二倍体数目和同源染色体的概念对于理解减数分裂过程十分重要。在配子或孢子的形成过程中，减数分裂将二倍染色体数目变为单倍染色体数目。结果便是单倍体配子或孢子将精确地含有每对同源染色体中的一个成员，即一套完整的单倍染色体组。受精过程中，两个配子发生融合，二倍染色体数目又重新建立，即合子含两套完整的单倍染色体组。因此，遗传物质得以持续稳定地世代相传。

There is one important exception to the concept of homologous pairs of chromosomes. In many species, one pair, consisting of the **sex-determining chromosomes**, is often not homologous in size, centromere placement, arm ratio, or genetic content. For example, in humans, while females carry two homologous X chromosomes, males carry one Y chromosome in addition to one X chromosome. These X and

同源染色体对的概念中有一个重要的例外。在许多物种中，由**性别决定染色体**组成的一对染色体通常在染色体大小、着丝粒位置、臂长比或遗传组成上并不同源。例如，在人类中，女性个体携带两条同源X染色体，而男性个体则携带一条X染色体和一条Y染色体。X染色体和Y染色体并非严格同源。Y染色体相对X染色体较小，并且缺乏大多数X染色体上所包含的基因

Y chromosomes are not strictly homologous. The Y is considerably smaller and lacks most of the gene loci contained on the X. Nevertheless, they contain homologous regions and behave as homologs in meiosis so that gametes produced by males receive either one X or one Y chromosome.

座。尽管如此，它们也含有相互同源的区域，并且减数分裂中的行为也类似同源染色体，因此男性个体产生的配子要么含一条 X 染色体，要么含一条 Y 染色体。

ESSENTIAL POINT

In diploid organisms, chromosomes exist in homologous pairs, where each member is identical in size, centromere placement, and gene sites. One member of each pair is derived from the maternal parent, and one is derived from the paternal parent.

基本要点

在二倍体生物中，染色体以同源染色体对的形式存在，每对染色体的成员在染色体大小、着丝粒位置和基因座上均完全相同。每对同源染色体中的一个成员来自母本，另一个成员来自父本。

2.3 Mitosis Partitions Chromosomes into Dividing Cells

2.3 有丝分裂将染色体分配至分裂细胞中

The process of mitosis is critical to all eukaryotic organisms. In some single-celled organisms, such as protozoans and some fungi and algae, mitosis (as a part of cell division) provides the basis for asexual reproduction. Multicellular diploid organisms begin life as single-celled fertilized eggs called **zygotes**. The mitotic activity of the zygote and the subsequent daughter cells is the foundation for the development and growth of the organism. In adult organisms, mitotic activity is the basis for wound healing and other forms of cell replacement in certain tissues. For example, the epidermal cells of the skin and the intestinal lining of humans are continuously sloughed off and replaced. Cell division also results in

有丝分裂过程对于所有的真核生物都是至关重要的。在一些单细胞生物中，如原生动物、某些真菌和藻类，有丝分裂（作为细胞分裂过程的一部分）为无性生殖提供了基础。多细胞二倍体生物的生命起始于被称为**合子**的单细胞受精卵。合子和随后子代细胞的有丝分裂行为是整个生物体发育和生长的基础。成年生物体中，有丝分裂是伤口愈合和某些肌体组织其他形式细胞新陈代谢的基础。例如，人类皮肤和肠壁的表皮细胞不断发生脱落更新。细胞分裂还连续不断地产生网织红细胞，这些细胞最终脱落细胞核并补充脊椎动物中红细胞的供应。在异常情况下，体细胞的细胞分裂可能失去控制并引发肿瘤。

the continuous production of reticulocytes that eventually shed their nuclei and replenish the supply of red blood cells in vertebrates. In abnormal situations, somatic cells may lose control of cell division and form a tumor.

The genetic material is partitioned into daughter cells during nuclear division, or **karyokinesis**. This process is quite complex and requires great precision. The chromosomes must first be exactly replicated and then accurately partitioned. The end result is the production of two daughter nuclei, each with a chromosome composition identical to that of the parent cell.

遗传物质在细胞核分裂或**核分裂**过程中被分配至子细胞。该过程非常复杂，并且要求非常精确。染色体必须首先被准确复制，然后被精确分配。最终结果是产生两个子代细胞核，每个子代细胞核中所含的染色体组成与亲代细胞的染色体组成完全相同。

Karyokines is is followed by cytoplasmic division, or **cytokinesis**. This less complex process requires a mechanism that partitions the volume into two parts and then encloses each new cell in a distinct plasma membrane. As the cytoplasm is reconstituted, organelles replicate themselves, arise from existing membrane structures, or are synthesized *de novo* (anew) in each cell.

紧随细胞核分裂之后的是细胞质分裂或**胞质分裂**。这一过程相对简单，所需机制是将细胞内空间分成两部分，然后用不同的质膜分别进行包裹从而形成两个新细胞。在细胞质的重构过程中，细胞器进行自我复制，或者从现有的膜结构中产生，抑或在每个细胞内从头合成。

Following cell division, the initial size of each new daughter cell is approximately one-half the size of the parent cell. However, the nucleus of each new cell is not appreciably smaller than the nucleus of the original cell. Quantitative measurements of DNA confirm that there is an amount of genetic material in the daughter nuclei equivalent to that in the parent cell.

细胞分裂后，每个新生子细胞的初始体积约为亲代细胞大小的一半。然而，每个新生细胞的细胞核并不比原来亲本细胞的细胞核小。DNA 的定量测定证实，子代细胞核中所含遗传物质的量与亲代细胞相同。

Interphase and the Cell Cycle

Many cells undergo a continuous alternation between division and nondivision. The events that occur from the completion of

间期和细胞周期

许多细胞会在细胞分裂期和不分裂期之间经历连续的交替变化。从一次细胞分裂完成直到下一次细胞分裂完成，在此之

one division until the completion of the next division constitute the **cell cycle** (Figure 2.3). We will consider **interphase**, the initial stage of the cell cycle, as the interval between divisions. It was once thought that the biochemical activity during interphase was devoted solely to the cell's growth and its normal function. However, we now know that another biochemical step critical to the ensuing mitosis occurs during interphase: the replication of the DNA of each chromosome. This period, during which DNA is synthesized, occurs before the cell enters mitosis and is called the **S phase**. The initiation and completion of synthesis can be detected by monitoring the incorporation of radioactive precursors into DNA.

间所发生的所有事件构成了**细胞周期**（图2.3）。我们将**间期**即细胞周期的初始阶段视为两次细胞分裂之间的间隔。人们曾经认为，间期阶段的生化活动仅对细胞生长及其正常功能有贡献。然而，我们现在已知间期中的另一项生化步骤对于随后的有丝分裂至关重要：每条染色体 DNA 的复制。DNA 合成发生的这一时期，在细胞进入有丝分裂之前，被称为 **S 期**。DNA 合成的起始和完成可通过监测整合入 DNA 的放射性前体来检测。

Investigations of this nature demonstrate two periods during interphase when no DNA synthesis occurs, one before and one after the S phase. These are designated **G1** (**gap I**) and **G2** (**gap II**), respectively. During both of these intervals, as well as during S, intensive metabolic activity, cell growth, and cell differentiation are evident. By the end of G2, the volume of the cell has roughly doubled, DNA has been replicated,

对间期的特性研究表明，间期中存在两个不发生 DNA 合成的阶段，一个位于 S 期之前，另一个位于 S 期之后，分别被称为 **G1 期**和 **G2 期**。在这两段间隔期以及 S 期中，进行着旺盛的新陈代谢活动，细胞生长和细胞分化均十分明显。在 G2 期即将结束时，细胞体积大致增加一倍，DNA 已完成复制，并且开始有丝分裂（M）。有丝分裂后，持续分裂的细胞将循环往复这个过程（G1，S，

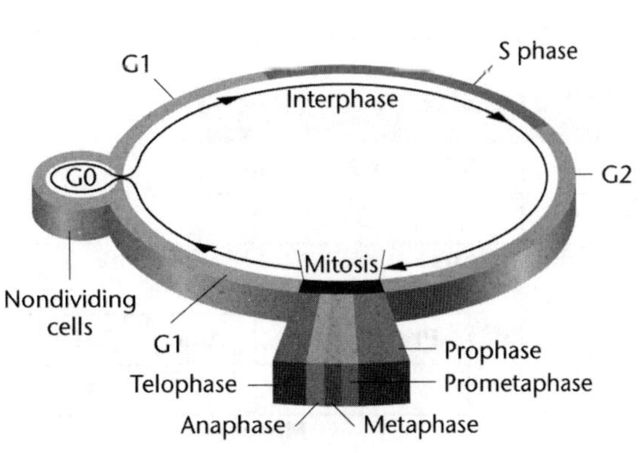

FIGURE 2.3 The stages comprising an arbitrary cell cycle. Following mitosis, cells enter the G1 stage of interphase, initiating a new cycle. Cells may become nondividing (G0) or continue through G1, where they become committed to begin DNA synthesis (S) and complete the cycle (G2 and mitosis). Following mitosis, two daughter cells are produced, and the cycle begins anew for both of them.

图 2.3 任意细胞周期所包含的阶段。在有丝分裂之后，细胞进入间期的 G1 期，从而开启新的细胞周期。细胞可以变为不分裂状态（G0）或者继续经过 G1 期开始 DNA 复制（S 期）并完成细胞周期（G2 期和有丝分裂）。有丝分裂结束后会产生两个子代细胞，并且每个子代细胞会重新开始新一轮的细胞周期。

and mitosis (M) is initiated. Following mitosis, continuously dividing cells then repeat this cycle (G1, S, G2, M) over and over, as shown in Figure 2.3.

G2，M），如图 2.3 所示。

Much is known about the cell cycle based on *in vitro* (literally, "in glass") studies. When grown in culture, many cell types in different organisms traverse the complete cycle in about 16 hours. The actual process of mitosis occupies only a small part of the overall cycle, often less than an hour. The lengths of the S and G2 phases of interphase are fairly consistent in different cell types. Most variation is seen in the length of time spent in the G1 stage. Figure 2.4 shows the relative length of these intervals as well as the length of the stages of mitosis in a human cell in culture.

人们对于细胞周期的了解大多基于体外研究（字面意思是“在试管中”）。当细胞在培养基中生长时，不同物种的许多细胞类型完成整个细胞周期大约需要 16 个小时。实际上，有丝分裂过程仅占整个细胞周期的一小部分，通常不足 1 小时。间期中的 S 期和 G2 期的时长在不同细胞类型中相当一致，绝大多数差异体现在 G1 期所花费的时间长度上。图 2.4 显示了人类细胞在体外培养时，这些间隔的相对长短以及有丝分裂各阶段的时长。

G1 is of great interest in the study of cell proliferation and its control. At a point during G1, all cells follow one of two paths. They either withdraw from the cycle, become quiescent, and enter the **G0 stage** (see Figure 2.3), or they become committed to proceed through G1, initiating DNA synthesis, and completing the cycle. Cells that enter G0 remain viable and metabolically active but are not proliferative. Cancer cells apparently avoid entering G0 or pass through it very quickly. Other cells enter G0

在细胞增殖及其调控研究中，G1 期引起了科学家的极大兴趣。在 G1 期内的某个时刻，所有细胞都将选择沿着以下两种不同路径中的一条继续前进：它们要么选择退出细胞周期，变为静止状态，然后进入 **G0 期**（见图 2.3）；要么选择继续沿着 G1 期前行，启动 DNA 合成并完成细胞周期。进入 G0 期的细胞依然保持细胞活力并代谢旺盛，但不进行细胞增殖。癌细胞则避免进入 G0 期或很快地越过 G0 期。有些细胞进入 G0 期后不再重返细胞周期，也有一些

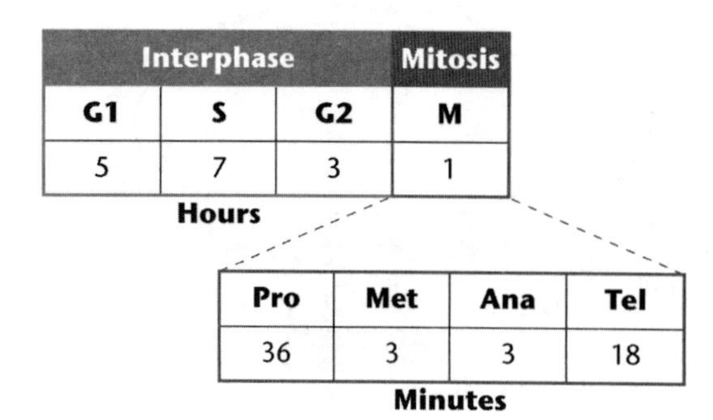

FIGURE 2.4 The time spent in each interval of one complete cell cycle of a human cell in culture. Times vary according to cell types and conditions.

图 2.4 体外培养的人体细胞一个完整的细胞周期中各阶段所需的时间。时间随细胞类型及培养条件的不同而有所变化。

and never reenter the cell cycle. Still other cells in G0 can be stimulated to return to G1 and thereby reenter the cell cycle.

位于 G0 期的细胞在受到信号刺激后重返 G1 期，从而再次进入细胞周期。

Cytologically, interphase is characterized by the absence of visible chromosomes. Instead, the nucleus is filled with chromatin fibers that are formed as the chromosomes uncoil and disperse after the previous mitosis [Figure 2.5(a)]. Once G1, S, and G2 are completed, mitosis is initiated. Mitosis is a dynamic period of vigorous and continual activity. For discussion purposes,

细胞学上，间期的特征在于不存在肉眼可见的染色体，或者细胞核内充满了染色质丝，它们由前一次有丝分裂完成后染色体发生解螺旋并分散而形成 [图 2.5(a)]。一旦 G1 期、S 期和 G2 期完成，有丝分裂随即开始。有丝分裂是一个持续的充满活力的动态时期，为了方便讨论，整个过程被分为几个独立阶段，每个阶段都包含一

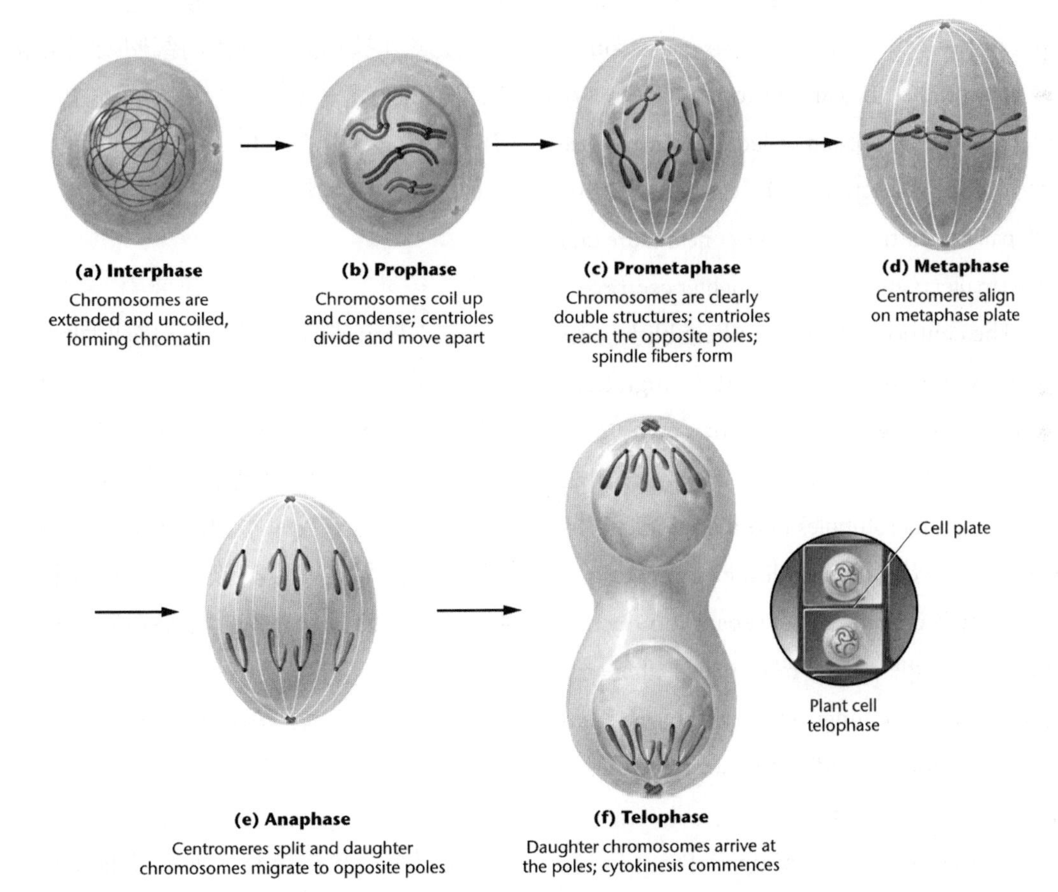

FIGURE 2.5 Drawings depicting mitosis in an animal cell with a diploid number of 4. The events occurring in each stage are described in the text. Of the two homologous pairs of chromosomes, one pair consists of longer, metacentric members and the other of shorter, submetacentric members. To the right of (f), a drawing of late telophase in a plant cell shows the formation of the cell plate and lack of centrioles.

图 2.5 二倍染色体数目为 4 的动物细胞有丝分裂示意图。每一阶段所发生的事件在文中均有描述。两对同源染色体对中，一对为较长的中部着丝粒染色体，另一对则为较短的近中着丝粒染色体。图 (f) 右侧展示了植物细胞有丝分裂末期即将结束时，细胞板的形成及中心粒的缺失。

the entire process is subdivided into discrete stages, and specific events are assigned to each one. These stages, in order of occurrence, are prophase, prometaphase, metaphase, anaphase, and telophase.

些特定事件。这些阶段按照发生的顺序依次为前期、前中期、中期、后期和末期。

Prophase

Often, over half of mitosis is spent in **prophase** [Figure 2.5(b)], a stage characterized by several significant occurrences. One of the early events in prophase of all animal cells is the migration of two pairs of centrioles to opposite ends of the cell. These structures are found just outside the nuclear envelope in an area of differentiated cytoplasm called the centrosome (introduced in Section 2.1). It is believed that each pair of centrioles consists of one mature unit and a smaller, newly formed daughter centriole.

The centrioles migrate and establish poles at opposite ends of the cell. After migration, the centrosomes, in which the centrioles are localized, are responsible for organizing cytoplasmic microtubules into the spindle fibers that run between these poles, creating an axis along which chromosomal separation occurs. Interestingly, the cells of most plants (there are a few exceptions), fungi, and certain algae seem to lack centrioles. Spindle fibers are nevertheless apparent during mitosis.

As the centrioles migrate, the nuclear envelope begins to break down and gradually disappears. In a similar fashion, the nucleolus disintegrates within the nucleus. While these events are taking place, the diffuse chromatin fibers have begun to condense, until distinct

前期

通常**前期**占据了有丝分裂整个过程一半以上的时间 [图 2.5(b)]，这一阶段包含几件特征性的重要事件。在所有动物细胞中，前期早期发生的事件之一即为两对中心粒分别向细胞的两端进行迁移。这些结构位于核膜之外被称为中心体的细胞质已分化区域（在 2.1 节中已介绍）中。每对中心粒均由一个成熟中心粒和一个较小的新形成的子代中心粒组成。

中心粒分别迁移至细胞相对的两端并建立两极后，中心粒所在的中心体将负责组织细胞质内的微管形成贯穿两极的纺锤丝，从而成为染色体可沿着进行分离的轴。有趣的是，大多数植物细胞（少数物种除外）、真菌和某些藻类似乎缺少中心粒。尽管如此，纺锤丝在有丝分裂期间依然清晰可见。

随着中心粒的迁移，细胞核的核膜开始破裂并逐渐消失，并且核仁也以类似的方式在核内瓦解。随着这些事件发生，弥散在细胞核内的染色质丝开始聚集浓缩，直到形成清晰的、肉眼可见的线状结构——染色体。在前期即将结束时，能明显地看

thread-like structures, the chromosomes, become visible. It becomes apparent near the end of prophase that each chromosome is actually a double structure split longitudinally except at a single point of constriction, the centromere. The two parts of each chromosome are called **sister chromatids** because the DNA contained in each of them is genetically identical, having formed from a single replicative event. Sister chromatids are held together by a multi-subunit protein complex called **cohesin**. This molecular complex is originally formed between them during the S phase of the cell cycle when the DNA of each chromosome is replicated. Thus, even though we cannot see chromatids in interphase because the chromatin is uncoiled and dispersed in the nucleus, the chromosomes are already double structures, which becomes apparent in late prophase. In humans, with a diploid number of 46, a cytological preparation of late prophase reveals 46 chromosomes randomly distributed in the area formerly occupied by the nucleus.

到每个染色体实际上具有除着丝粒这一单缢痕外纵向分裂的双重平行结构。每条染色体的两个部分被称为**姐妹染色单体**，因为它们所含的 DNA 在遗传组成上彼此等同，由单一复制事件而形成。姐妹染色单体通过一种被称为**黏连蛋白**的多亚基蛋白复合物捆绑在一起。这种分子复合物在姐妹染色单体之间的最初形成起始于每条染色体 DNA 进行复制的 S 期。因此，即使在间期染色单体由于未形成超螺旋结构而弥散在细胞核中无法可见时，每条染色体也已经是含两个 DNA 分子的双重结构，并在随后的前期稍晚阶段逐渐变得明显。人类二倍染色体数目为 46，利用前期稍晚阶段的人类细胞进行细胞制备即能发现 46 条染色体随机分布在先前细胞核所占据的区域中。

Prometaphase and Metaphase

The distinguishing event of the two ensuing stages is the migration of every chromosome, led by its centromeric region, to the equatorial plane. The equatorial plane, also referred to as the metaphase plate, is the midline region of the cell, a plane that lies perpendicular to the axis established by the spindle fibers. In some descriptions, the term **prometaphase** refers to the period of chromosome movement [Figure 2.5(c)], and the term **metaphase** is applied strictly to the chromosome configuration

前中期和中期

随后两个阶段中所发生的典型事件是每条染色体在其着丝粒区域的牵引下向赤道板迁移。赤道板也被称为中期板，位于整个细胞的中线区域，该平面垂直于纺锤丝所形成的轴线。在某些描述中会使用术语“**前中期**”描述染色体运动的时期 [图 2.5(c)]，而术语“**中期**”则严格适用于迁移后染色体所呈现的形态。

following migration.

Migration is made possible by the binding of spindle fibers to the chromosome's **kinetochore**, an assembly of multilayered plates of proteins associated with the centromere. This structure forms on opposite sides of each paired centromere, in intimate association with the two sister chromatids. Once properly attached to the spindle fibers, cohesin is degraded by an enzyme, appropriately named **separase**, and the sister chromatid arms disjoin, except at the centromere region. A unique protein family called **shugoshin** (from the Japanese meaning "guardian spirit") protects cohesin from being degraded by separase at the centromeric regions. The involvement of the cohesin and shugoshin complexes with a pair of sister chromatids during mitosis is depicted in Figure 2.6.

纺锤丝与染色体的**动粒**相结合从而使染色体迁移成为可能。动粒由着丝粒相关蛋白形成多层板状结构组装而成，该结构形成于每对着丝粒相对的两侧，与两条姐妹染色单体紧密结合。黏连蛋白一旦准确地附着在纺锤丝上，即会被**分离酶**所降解，然后除着丝粒区域外，姐妹染色单体的臂随之分开。独特的 **shugoshin**（日语，意为“守护神”）蛋白质家族能够保护黏连蛋白，使其不被位于着丝粒区域的分离酶降解。图 2.6 显示了有丝分裂过程中黏连蛋白和 shugoshin 形成的蛋白复合物与一对姐妹染色单体间的相互关系。

We know a great deal about the molecular interactions involved in kinetochore assembly along the centromere. This is of great interest because of the consequences when mutations alter the proteins that make up the kinetochore complex. Altered kinetochore function potentially leads to errors during chromosome migration, altering the diploid content of

我们对于动粒沿着着丝粒进行组装的过程所涉及的分子相互作用已经有了很多了解，并且人们对于动粒复合物的组成蛋白发生突变所引起的后果也抱有极大兴趣。动粒功能改变可能导致染色体在迁移过程中发生错误，从而改变子细胞内的二倍染色体含量。

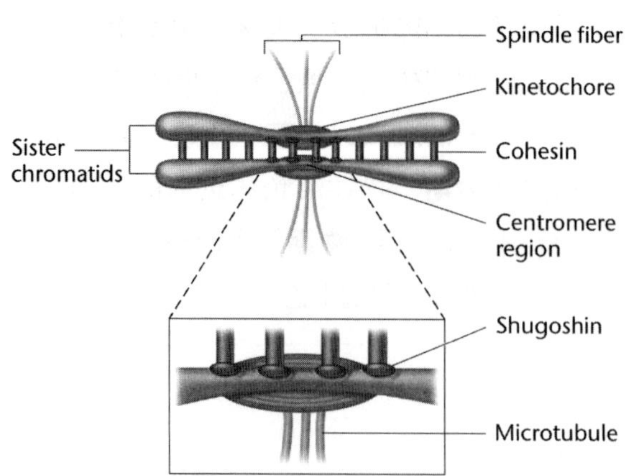

FIGURE 2.6 The depiction of the alignment, pairing, and disjunction of sister chromatids during mitosis, involving the molecular complexes cohesin and shugoshin and the enzyme separase.

图 2.6 有丝分裂过程中姐妹染色单体间对准、配对和联会示意图，包括黏连蛋白与 shugoshin 蛋白形成的复合物与分离酶。

daughter cells.

We also know a great deal about spindle fibers and the mechanism responsible for their attachment to the kinetochore. Spindle fibers consist of microtubules, which themselves consist of molecular subunits of the protein tubulin. Microtubules seem to originate and "grow" out of the two centrosome regions at opposite poles of the cell. They are dynamic structures that lengthen and shorten as a result of the addition or loss of polarized tubulin subunits. The microtubules most directly responsible for chromosome migration make contact with, and adhere to, kinetochores as they grow from the centrosome region. They are referred to as **kinetochore microtubules** and have one end near the centrosome region (at one of the poles of the cell) and the other end anchored to the kinetochore. The number of microtubules that bind to the kinetochore varies greatly between organisms. Yeast (*Saccharomyces*) has only a single microtubule bound to each plate-like structure of the kinetochore. Mitotic cells of mammals, at the other extreme, reveal 30 to 40 microtubules bound to each portion of the kinetochore.

我们对于纺锤丝及其与动粒相附着的机理也已经有了很多了解。纺锤丝由微管组成，微管本身又由微管蛋白的分子亚单位组成。微管似乎从位于细胞相对两极的两个中心体区域起源并由此“生长”延伸。微管是动态结构，随着极化的微管蛋白亚单位的增加或减少而发生延伸和缩短。微管从中心体区域生长延伸，与动粒接触并附着，直接负责染色体的迁移。它们被称为**动粒微管**，其一端靠近中心体区域（位于细胞一极），另一端则锚定在动粒上。在不同生物体中，锚定在动粒上的微管数量差异很大。酵母只有一条微管与动粒的板状结构相结合，而哺乳动物的有丝分裂细胞中则有 30—40 条微管与动粒的每个部位相结合。

At the completion of metaphase, each centromere is aligned at the metaphase plate with the chromosome arms extending outward in a random array. This configuration is shown in Figure 2.5(d).

在中期完成时，每个着丝粒均整齐地排列在赤道板上，染色体的臂以随机的方式向外展开。这种排布结构在图 2.5(d) 中进行了展示。

Anaphase

Events critical to chromosome distribution during mitosis occur during anaphase, the

后期

有丝分裂期间与染色体分布紧密相关的事件发生在**后期**，这也是有丝分裂中用

shortest stage of mitosis. During this phase, sister chromatids of each chromosome, held together only at their centromere regions, disjoin (separate) from one another—an event described as **disjunction**—and are pulled to opposite ends of the cell. For complete disjunction to occur: (1) shugoshin must be degraded, reversing its protective role; (2) the cohesin complex holding the centromere region of each sister chromosome is then cleaved by separase; and (3) sister chromatids of each chromosome are pulled toward the opposite poles of the cell (Figure 2.6). As these events proceed, each migrating chromatid is now referred to as a **daughter chromosome**.

时最短的阶段。在此时期，每条染色体中原本共用同一着丝粒区域的姐妹染色单体将彼此分开（该事件被称为**分离**），并被拉至细胞相对的两个末端。为了保证姐妹染色单体能够完全分离：（1）shugoshin 蛋白必须被降解，从而解除其保护作用；（2）连接姐妹染色单体着丝粒区域的黏连蛋白复合体被分离酶剪切；（3）每条染色体上的姐妹染色单体被拉向细胞相对的两极（图 2.6）。随着这些事件的进行，每条迁移的染色单体此时被称为**子代染色体**。

The location of the centromere determines the shape of the chromosome during separation, as you saw in Figure 2.2. The steps that occur during anaphase are critical in providing each subsequent daughter cell with an identical set of chromosomes. In human cells, there would now be 46 chromosomes at each pole, one from each original sister pair. Figure 2.5(e) shows anaphase prior to its completion.

着丝粒的位置决定了分离过程中染色体的形态，正如图 2.2 所示。后期发生的系列步骤对于随后产生的每个子细胞是否含有完全相同的染色体组成至关重要。在人类细胞中，细胞的每一极此时都含有 46 条染色体，每一条均来自最初的姐妹染色单体对。图 2.5(e) 展示了后期在完成前的情况。

Telophase

Telophase is the final stage of mitosis and is depicted in Figure 2.5(f). At its beginning, two complete sets of chromosomes are present, one set at each pole. The most significant event of this stage is cytokinesis, the division or partitioning of the cytoplasm. Cytokinesis is essential if two new cells are to be produced from one cell. The mechanism of cytokinesis differs greatly in plant and animal cells, but the end result is

末期

末期是有丝分裂的最后一个阶段并在图 2.5(f) 中进行了描述。末期开始时，细胞内存在两套完整的染色体组成，细胞的每一极各有一套。该阶段发生的最重要事件是胞质分裂，即细胞质的分裂或分配。在从一个细胞产生两个新细胞的过程中，胞质分裂是必不可少的。动物细胞和植物细胞胞质分裂的机理差异很大，但最终结果相同：即产生两个新细胞。植物细胞在赤

the same: Two new cells are produced. In plant cells, a cell plate is synthesized and laid down across the region of the metaphase plate. Animal cells, however, undergo a constriction of the cytoplasm, much as a loop of string might be tightened around the middle of a balloon.

道板区域会合成细胞板。然而，动物细胞的胞质分裂则类似将位于气球中部的线圈逐渐拉紧最终形成一个细胞质的缢痕。

It is not surprising that the process of cytokinesis varies in different organisms. Plant cells, which are more regularly shaped and structurally rigid, require a mechanism for depositing new cell wall material around the plasma membrane. The cell plate laid down during telophase becomes a structure called the **middle lamella**. Subsequently, the primary and secondary layers of the cell wall are deposited between the cell membrane and middle lamella in each of the resulting daughter cells. In animals, complete constriction of the cell membrane produces the cell furrow characteristic of newly divided cells.

胞质分裂过程在不同生物体中各有差异并不奇怪。植物细胞形态上更规则、结构上也更具有刚性，因此需要一种机制能够在质膜周围沉积新生细胞壁形成所需的材料。末期沉积而成的细胞板形成名为**胞间层**的结构。随后，在每个正在形成的子代细胞的细胞膜和胞间层之间不断沉积构成细胞壁的第一层和第二层。在动物细胞中，细胞膜完全收缩形成细胞沟，这是新生分裂细胞的特征。

Other events necessary for the transition from mitosis to interphase are initiated during late telophase. They generally constitute a reversal of events that occurred during prophase. In each new cell, the chromosomes begin to uncoil and become diffuse chromatin once again, while the nuclear envelope re-forms around them, the spindle fibers disappear, and the nucleolus gradually re-forms and becomes visible in the nucleus during early interphase. At the completion of telophase, the cell enters interphase.

在末期即将结束之时，从有丝分裂向间期过渡所必需的其他事件开始启动。它们通常是前期所发生事件的逆转事件。在每个新生细胞中，染色体开始解螺旋并再次变成弥散的染色质，同时核膜在其周围重新形成，纺锤丝消失，核仁逐渐重新形成并在间期早期的细胞核中变得可见。末期结束后，细胞将进入间期。

Cell-Cycle Regulation and Checkpoints

细胞周期调控与检查点

The cell cycle, culminating in mitosis,

以有丝分裂作为高潮的细胞周期在所

is fundamentally the same in all eukaryotic organisms. This similarity in many diverse organisms suggests that the cell cycle is governed by a genetically regulated program that has been conserved throughout evolution. Because disruption of this regulation may underlie the uncontrolled cell division characterizing malignancy, interest in how genes regulate the cell cycle is particularly strong.

有真核生物中基本相同。这种存在于不同生物中的相似性表明细胞周期由遗传调控过程所控制，并且在整个生物进化中是保守的。因为该调控机制的破坏可能引发以恶性肿瘤为特征的细胞分裂失控，所以人们对于基因如何参与调节细胞周期的兴趣尤其强烈。

A mammoth research effort over the past 20 years has paid high dividends, and we now have knowledge of many genes involved in the control of the cell cycle. This work was recognized by the awarding of the 2001 Nobel Prize in Physiology or Medicine to Lee Hartwell, Paul Nurse, and Tim Hunt. As with other studies of genetic control over essential biological processes, investigation has focused on the discovery of mutations that interrupt the cell cycle and on the effects of those mutations. As we shall return to this subject in much greater detail later, what follows is a very brief overview.

人们在过去 20 年的研究中付出的巨大努力已经获得了高回报，我们现在已经了解了许多有关细胞周期调控相关基因的知识。李 · 哈特韦尔、保罗 · 纳斯和蒂姆 · 亨特获得 2001 年诺贝尔生理学或医学奖正是对该项研究的认可。与其他基础生物学过程的基因调控研究一样，细胞周期调控机制的研究也集中于发现阻断细胞周期的突变以及这些突变的遗传效应。在稍后，我们将更详细地讨论该主题，这里仅做非常简短的概述。

Many mutations are now known that exert an effect at one or another stage of the cell cycle. First discovered in yeast, but now evident in all organisms, including humans, such mutations were originally designated as cell division cycle (cdc) mutations. The normal products of many of the mutated genes are enzymes called **kinases** that can add phosphates to other proteins. They serve as "master control" molecules functioning in conjunction with proteins called **cyclins**. Cyclins bind to these kinases (creating **cyclin dependent kinases**), activating them at appropriate times during the cell cycle. Activated

目前已知的许多突变在细胞周期的不同阶段发挥作用。这些突变最初被称为细胞分裂周期（cdc）突变，最早是在酵母中发现的，现在在包括人类在内的所有生物中均已得到证实。这些突变基因的正常产物大多是一类**激酶**，激酶可以将磷酸基团添加到其他蛋白质上。它们作为“主控”分子与**细胞周期蛋白**一起发挥功能。细胞周期蛋白与这些激酶相结合产生**周期蛋白依赖激酶**，并在细胞周期中的适当时刻对激酶进行激活。随后，被激活的激酶对其他参与细胞周期调控的靶蛋白进行磷酸化。有关细胞分裂周期突变的研究已发现细胞

kinases then phosphorylate other target proteins that regulate the progress of the cell cycle. The study of *cdc* mutations has established that the cell cycle contains at least three **cell-cycle checkpoints**, where the processes culminating in normal mitosis are monitored, or "checked," by these master control molecules before the next stage of the cycle is allowed to commence.

周期中至少存在三个**细胞周期检查点**。在这些检查点，正常有丝分裂的所有进程均会在细胞周期下一阶段进程起始之前受到这些主控分子的监控或“检查”。

The importance of cell-cycle control and these checkpoints can be demonstrated by considering what happens when this regulatory system is impaired. Let's assume, for example, that the DNA of a cell has incurred damage leading to one or more mutations impairing cell-cycle control. If allowed to proceed through the cell cycle, this genetically altered cell would divide uncontrollably—a key step in the development of a cancer cell. If, instead, the cell cycle is arrested at one of the checkpoints, the cell can repair the DNA damage or permanently stop the cell from dividing, thereby preventing its potential malignancy.

当细胞调控系统受损时，细胞周期控制及这些检查点的重要性随之得以显现。例如，假设细胞 DNA 受到破坏，导致一种或多种基因突变，从而引发细胞周期调控受损。在此情况下，如果细胞周期继续进行，则这种遗传信息发生改变的细胞的分裂将不受调节和控制——这也正是癌细胞发生的一个关键步骤。与此不同，如果细胞周期被阻滞在某一检查点，细胞能够修复 DNA 损伤或永久地阻止细胞分裂，那么就可以防止其潜在的恶化。

ESSENTIAL POINT

Mitosis is subdivided into discrete stages that initially depict the condensation of chromatin into the diploid number of chromosomes, each of which is initially a double structure, each composed of a pair of sister chromatids. During mitosis, sister chromatids are pulled apart and directed toward opposite poles, after which cytoplasmic division creates two new cells with identical genetic information.

基本要点

有丝分裂被分成相互独立的几个阶段，最初的染色质凝聚浓缩成二倍体数目的染色体，每条染色体最初具有由一对姐妹染色单体组成的双重结构。在有丝分裂期间，每对姐妹染色单体被拉开并向相反的细胞两极进行移动。在此之后，细胞质发生分裂，产生两个含完全相同遗传信息的新生子代细胞。

2.4 Meiosis Creates Haploid Gametes and Spores and Enhances Genetic Variation in Species

Whereas in diploid organisms, mitosis produces two daughter cells with full diploid complements, **meiosis** produces gametes or spores that are characterized by only one haploid set of chromosomes. During sexual reproduction, haploid gametes then combine at fertilization to reconstitute the diploid complement found in parental cells. Meiosis must be highly specific since haploid gametes or spores must contain precisely one member of each homologous pair of chromosomes. When successfully completed, meiosis provides the basis for maintaining genetic continuity from generation to generation.

Another major accomplishment of meiosis is to ensure that during sexual reproduction an enormous amount of genetic variation is produced among members of a species. Such variation occurs in two forms. First, meiosis produces haploid gametes with many unique combinations of maternally and paternally derived chromosomes. As we will see (Chapter 3), this process is the underlying basis of Mendel's principles of segregation and independent assortment. The second source of variation is created by the meiotic event referred to as **crossing over**, which results in genetic exchange between members of each homologous pair of chromosomes prior to one or the other finding its way into a haploid gamete or spore. This creates intact chromosomes that are mosaics

2.4 减数分裂产生单倍体配子和孢子并增加了物种的遗传变异

在二倍体生物中，有丝分裂产生两个具有完整二倍染色体组成的子代细胞，而**减数分裂**则产生以仅有一套单倍染色体组成为特征的配子或孢子。在有性生殖过程中，单倍体配子在受精时结合在一起，从而可以重建与亲代细胞同样的二倍染色体组成。减数分裂一定是具有高度特异性的，因为单倍体配子或孢子必须精确地包含每一同源染色体对中的一条。当减数分裂成功完成时，即为维持世代之间的遗传连续性奠定了基础。

减数分裂的另一个重要贡献是确保有性生殖过程中，物种个体间产生大量的遗传变异。这种变异的产生有两种形式。首先，减数分裂产生单倍体配子时，源自母本和父本的染色体可以形成许多独特的组合。正如我们将要看到的（第 3 章），该过程是孟德尔分离定律和自由组合定律的基础。遗传变异的第二个来源是被称为**交换**的减数分裂事件。该事件使得每个同源染色体对成员在分别进入不同的单倍体配子或孢子之前发生遗传交换，从而形成同时含有母本和父本同源染色体组分的镶嵌染色体。因此，有性生殖通过将遗传物质进行显著的重新组合来产生高度多样化的后代。

of the maternal and paternal homologs. Sexual reproduction therefore significantly reshuffles the genetic material, producing highly diverse offspring.

Meiosis: Prophase I

As in mitosis, the process in meiosis begins with a diploid cell duplicating its genetic material in the interphase stage preceding chromosome division. To achieve haploidy, two divisions are thus required. The meiotic achievements are largely dependent on the behavior of chromosomes during the initial stage of the first division, called prophase I. Recall that in mitosis the paternally and maternally derived members of each homologous pair of chromosomes behave autonomously during division. Each chromosome is duplicated, creating genetically identical **sister chromatids**, and subsequently, one chromatid of each pair is distributed to each new cell. The major difference in meiosis is that once the chromatin characterizing interphase has condensed into visible structures, the homologous chromosomes are not autonomous but are instead seen to be paired up, having undergone the process called **synapsis**. Figure 2.7 illustrates this process as well as the ensuing events of prophase I. Each synapsed pair of homologs is initially called a **bivalent**, and the number of bivalents is equal to the haploid number. In Figure 2.7, we have depicted two homologous pairs of chromosomes and thus two bivalents. As the homologs condense and shorten, each bivalent gives rise to a unit called a **tetrad**, consisting of two pairs of

减数分裂：前期 I

与有丝分裂一样，减数分裂过程也始于二倍体细胞，并且该细胞在染色体分离前的间期阶段已完成遗传物质的复制。为获得单倍体子代细胞，减数分裂需要两次细胞分裂过程。减数分裂的贡献主要基于第一次细胞分裂初期阶段（即前期 I）中的染色体行为。让我们回想一下，在有丝分裂中，每对同源染色体中的两条分别来自父本和母本，它们在细胞分裂过程中都各自具有行动自主性。每条染色体经 DNA 复制后产生两条遗传上完全相同的**姐妹染色单体**。随后，每对姐妹染色单体中的一条被分配到每个新生子代细胞中。减数分裂的主要区别在于一旦标志着间期的染色质开始聚集成显微镜下肉眼可见的结构，同源染色体将不再是自主的，而是开始进行配对，该过程被称为**联会**，图 2.7 显示了这一过程以及前期 I 的后续事件。最初形成的每一对发生联会配对的同源染色体被称为**二价体**，二价体数目等于单倍染色体数目。在图 2.7 中，我们绘制了两对同源染色体对，即两个二价体。随着同源染色体继续浓缩变短，每个二价体变为一个被称为**四联体**的单元，该单元由两对姐妹染色单体组成，每对姐妹染色单体共用一个着丝粒。需要提醒大家的是，其中一对姐妹染色单体来自母本，另一对则来自父本。四联体的存在是两条同源染色体均已进行复制的可视化证据。随着减数分裂前期 I 的进行，每个

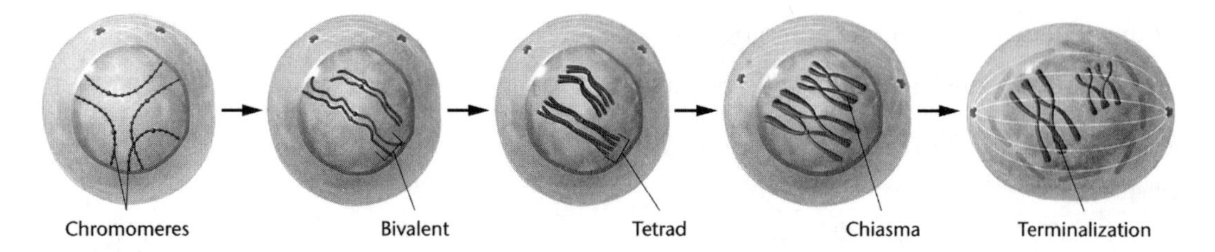

FIGURE 2.7 The events characterizing meiotic prophase I. In the first two frames, illustrating chromomeres and bivalents, each chromatid is actually a double structure, consisting of sister chromatids, which first becomes apparent in the ensuing tetrad stage.

图 2.7 发生在减数分裂前期 I 中的特征性事件。前两张图显示了染色体和二价体，每条染色体实际为含有两条姐妹染色单体的双重结构，该结构在随后的四联体阶段将首次变得清晰。

sister chromatids, each of which is joined at a common centromere. Remember that one pair of sister chromatids is maternally derived, and the other pair paternally derived. The presence of tetrads is visible evidence that both homologs have, in fact, duplicated. As prophase progresses within each tetrad, each pair of sister chromatids is seen to pull apart. However, one or more areas remain in contact where chromatids are intertwined. Each such area, called a **chiasma** (pl. chiasmata), is thought to represent a point where **nonsister chromatids** (one paternal and one maternal chromatid) have undergone genetic exchange through the process of crossing over. Since crossing over is thought to occur one or more times in each tetrad, mosaic chromosomes are routinely created during every meiotic event. During the final period of prophase I, the nucleolus and nuclear envelope break down, and the two centromeres of each tetrad attach to the recently formed spindle fibers.

四联体中的每对姐妹染色单体将发生分离。然而，染色质发生缠绕的区域中一处或多处仍然保持接触。这些依然保持接触的区域被称为**交叉**，代表着交换发生过程中**非姐妹染色单体**（一条来自父本，另一条来自母本）间遗传物质进行交换的位点。由于交换在每个四联体中发生一次或多次，因此每个减数分裂事件中总会产生融合有父本、母本遗传信息的镶嵌染色体。在前期 I 的最后阶段，核仁和核膜发生瓦解，每个四联体的两个着丝粒将与刚刚形成的纺锤丝相附着。

Metaphase I, Anaphase I, and Telophase I

The remainder of the meiotic process is depicted in Figure 2.8. After meiotic prophase

中期 I、后期 I 和末期 I

减数分裂过程的其余阶段如图 2.8 所示。在减数分裂前期 I 之后发生的步骤类似

I, steps similar to those of mitosis occur. In the first division, metaphase I, the chromosomes have maximally shortened and thickened. The terminal chiasmata of each tetrad are visible and appear to be the only factor holding the nonsister chromatids together. Each tetrad interacts with spindle fibers, facilitating movement to the metaphase plate. The alignment of each tetrad prior to the first anaphase is random. Half of each tetrad is pulled randomly to one or the other pole, and the other half then moves to the opposite pole.

于有丝分裂的相关步骤。在减数分裂第一次核分裂的中期 I，染色体最大限度地变短、变粗。每个四联体的末端交叉在显微镜下肉眼可见，并且这也是将非姐妹染色单体联系在一起的唯一方式。每个四联体与纺锤丝相互作用，促使四联体向赤道板移动，并且四联体的排列方式在后期 I 之前均是随机的。每个四联体的一半将被随机拉至细胞的一极，而另一半则被拉至相对的细胞另一极。

During the stages of meiosis I, a single centromere holds each pair of sister chromatids together. It does not divide. At anaphase I, one-half of each tetrad (the dyad) is pulled toward each pole of the dividing cell. This separation process is the physical basis of **disjunction**, the separation of chromosomes from one another. Occasionally, errors in meiosis occur and separation is not achieved. The term **nondisjunction** describes such an error. At the completion of a normal anaphase I, a series of dyads equal to the haploid number is present at each pole.

在减数分裂 I 期的所有阶段里，单个着丝粒将每对姐妹染色单体联系在一起，着丝粒不发生分裂。在后期 I，每个四联体的一半（即二分体）被拉向正在进行分裂的细胞的一极，这一过程是**分离**（即染色体彼此分开）的物理基础。有时，减数分裂会发生错误，无法实现染色体的正确分离。专业术语“**不分离**”则用于描述这样的错误。一次正常的后期 I 完成时，细胞的每一极都存在一系列二分体，并且二分体数目等于单倍染色体数目。

In many organisms, telophase I reveals a nuclear membrane forming around the dyads. Next, the nucleus enters into a short interphase period. If interphase occurs, the chromosomes do not replicate since they already consist of two chromatids. In other organisms, the cells go directly from anaphase I to meiosis II. In general, meiotic telophase is much shorter than the corresponding stage in mitosis.

在许多生物中，末期 I 时细胞的每一极在二分体周围会形成核膜。接下来，细胞核进入一个较短的间期阶段。如果有间期存在，此时的染色体不会发生复制，因为每条染色体均已由两条姐妹染色单体组成。而在另外一些生物中，细胞则直接从后期 I 进入减数分裂 II 期。通常情况下，减数分裂末期要比有丝分裂相应阶段的时间短得多。

The Second Meiotic Division

A second division, meiosis II, is essential if each gamete or spore is to receive only one chromatid from each original tetrad. The stages characterizing meiosis II are shown in the right half of Figure 2.8. During prophase II, each dyad is composed of one pair of sister chromatids attached by a common centromere. During metaphase II, the centromeres are positioned on the metaphase plate. When they divide, anaphase II is initiated, and the sister chromatids of each dyad are pulled to opposite poles. Because the number of dyads is equal to the haploid number,

减数分裂 II 期

减数分裂的第二次核分裂，即减数分裂 II 期，对于每个配子或孢子只获得原始四联体中的一条姐妹染色单体是十分必要的。图 2.8 的右半部分显示了减数分裂 II 期的各个阶段。在前期 II，每个二分体由共用一个着丝粒的一对姐妹染色单体组成。在中期 II，着丝粒位于赤道板上。当着丝粒发生分裂时，后期 II 开始了，每个二分体中的姐妹染色单体将被拉向细胞相对的两极。由于二分体数目等于单倍染色体数目，因此末期 II 时细胞的每一极均将含有每对同源染色体中的一条。这样，每条染色体就

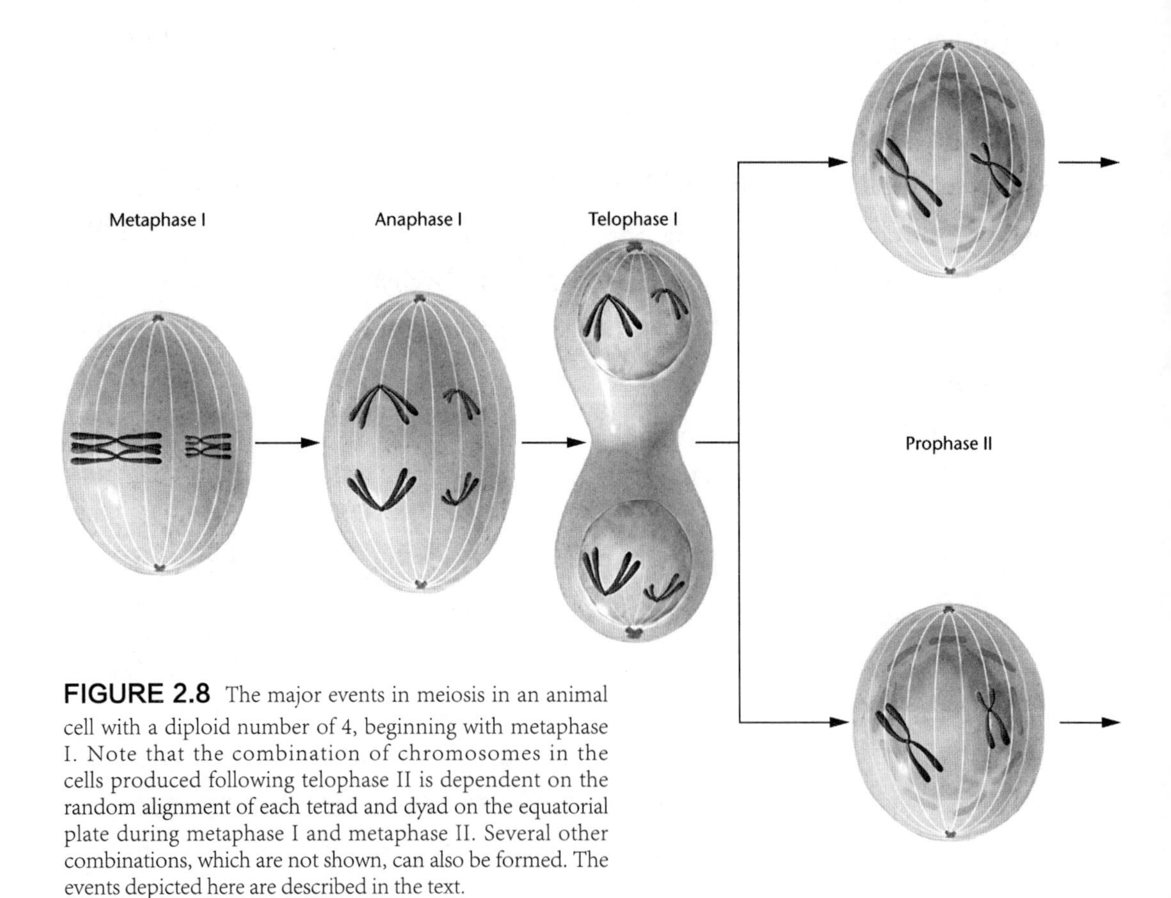

FIGURE 2.8 The major events in meiosis in an animal cell with a diploid number of 4, beginning with metaphase I. Note that the combination of chromosomes in the cells produced following telophase II is dependent on the random alignment of each tetrad and dyad on the equatorial plate during metaphase I and metaphase II. Several other combinations, which are not shown, can also be formed. The events depicted here are described in the text.

图 2.8 二倍染色体数目为 4 的动物细胞减数分裂过程中自中期 I 开始的主要事件。注意末期 II 后产生的细胞内染色体组合依赖于中期 I 和中期 II 中赤道板上每一个四联体和二分体的随机排列。其他的组合方式同样可以形成，图中未显示。这里展示的为书中所描述的各事件。

telophase II reveals one member of each pair of homologous chromosomes at each pole. Each chromosome is now a monad. Following cytokinesis in telophase II, four haploid gametes may result from a single meiotic event. At the conclusion of meiosis II, not only has the haploid state been achieved, but if crossing over has occurred, each monad is also a combination of maternal and paternal genetic information. As a result, the offspring produced by any gamete receives a mixture of genetic information originally present in his or her grandparents.

会成为一个单分体。在末期II的胞质分离后，通过单个减数分裂事件将产生四个单倍体配子。因此减数分裂II期结束时，不仅细胞会变为单倍体状态，而且如果有交换发生，则每个单分体也均会同时含有母本和父本的遗传信息。最终，任何配子参与产生的后代都将获得源自其祖父母遗传信息的组合。因此，减数分裂显著提高了后续每一世代的遗传变异水平。

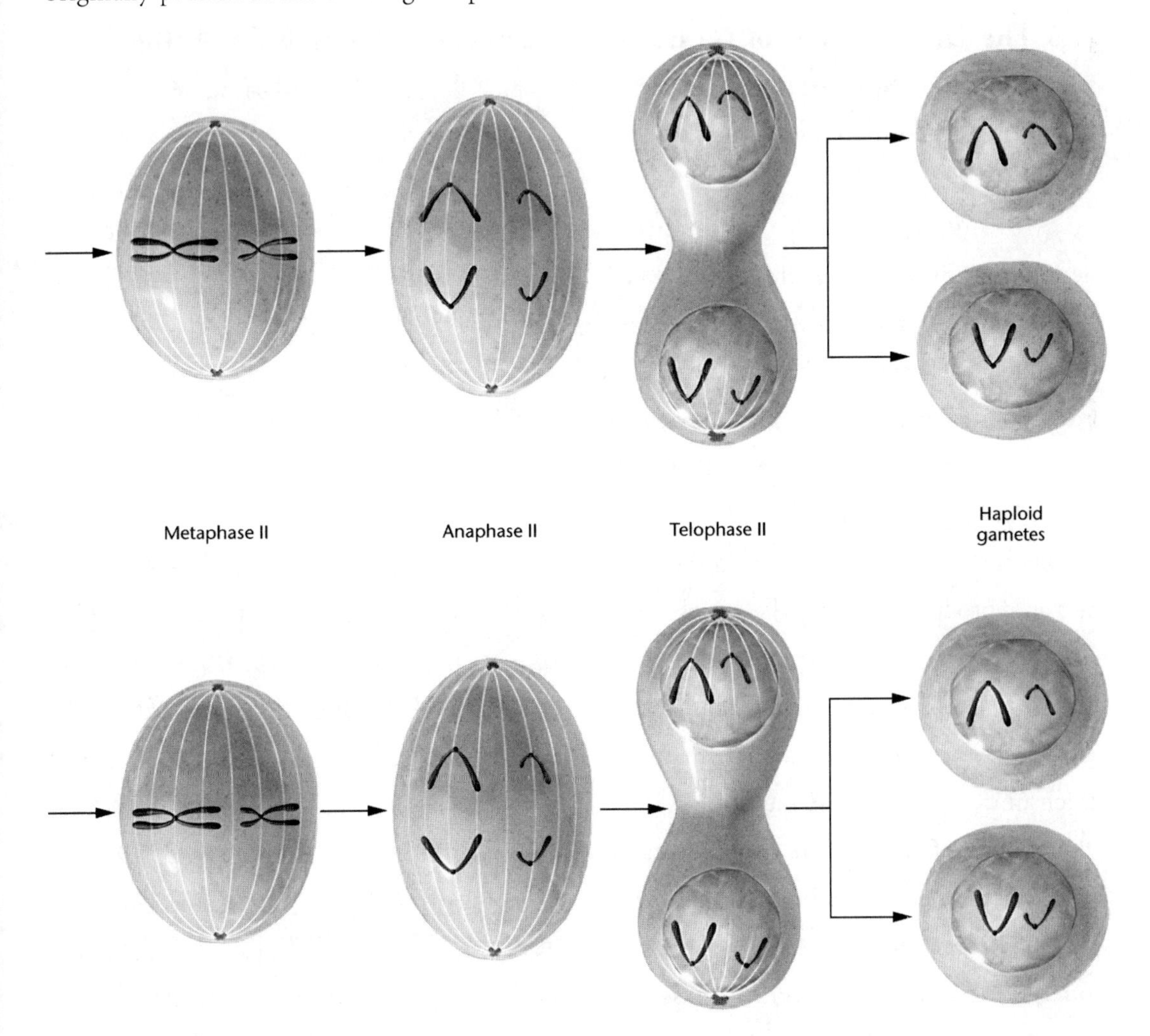

Meiosis thus significantly increases the level of genetic variation in each ensuing generation.

ESSENTIAL POINT

Meiosis converts a diploid cell into a haploid gamete or spore, making sexual reproduction possible. As a result of chromosome duplication and two subsequent meiotic divisions, each haploid cell receives one member of each homologous pair of chromosomes.

基本要点

减数分裂将二倍体细胞变为单倍体配子或孢子，从而使有性生殖成为可能。染色体复制和随后发生的两次核分裂过程使得每个子代单倍体细胞都含有每对同源染色体中的一条。

2.5 The Development of Gametes Varies in Spermatogenesis Compared to Oogenesis

2.5 精子发生与卵子发生中的配子发育差异

Although events that occur during the meiotic divisions are similar in all cells participating in gametogenesis in most animal species, there are certain differences between the production of a male gamete (spermatogenesis) and a female gamete (oogenesis). Figure 2.9 summarizes these processes.

尽管减数分裂期间发生的事件在大多数动物物种参与配子发生的所有细胞中都是相似的，但是雄性配子（精子发生）和雌性配子（卵子发生）的产生依然存在一定差异。图 2.9 对于这些过程进行了总结。

Spermatogenesis takes place in the testes, the male reproductive organs. The process begins with the enlargement of an undifferentiated diploid germ cell called a spermatogonium. This cell grows to become a primary spermatocyte, which undergoes the first meiotic division. The products of this division, called secondary spermatocytes, contain a haploid number of dyads. The secondary spermatocytes then undergo meiosis II, and each of these cells produces two haploid spermatids. Spermatids go through a series of developmental changes,

精子发生发生在雄性生殖器官——精巢中，该过程始于未分化的二倍体生殖细胞——精原细胞的体积变大。精原细胞发育成为初级精母细胞，并进入减数分裂 I 期，此次细胞分裂产生次级精母细胞，包含单倍体数目的二分体。次级精母细胞随后进入减数分裂 II 期，每个次级精母细胞会产生两个单倍体精细胞。精细胞还将经历一系列发育变化即精子发生，从而成为高度特异的、具有活力的精子。在精子发生过程中产生的所有精子均含有单倍染色体数目和等量的细胞质。

spermiogenesis, to become highly specialized, motile spermatozoa, or sperm. All sperm cells produced during spermatogenesis contain the haploid number of chromosomes and equal amounts of cytoplasm.

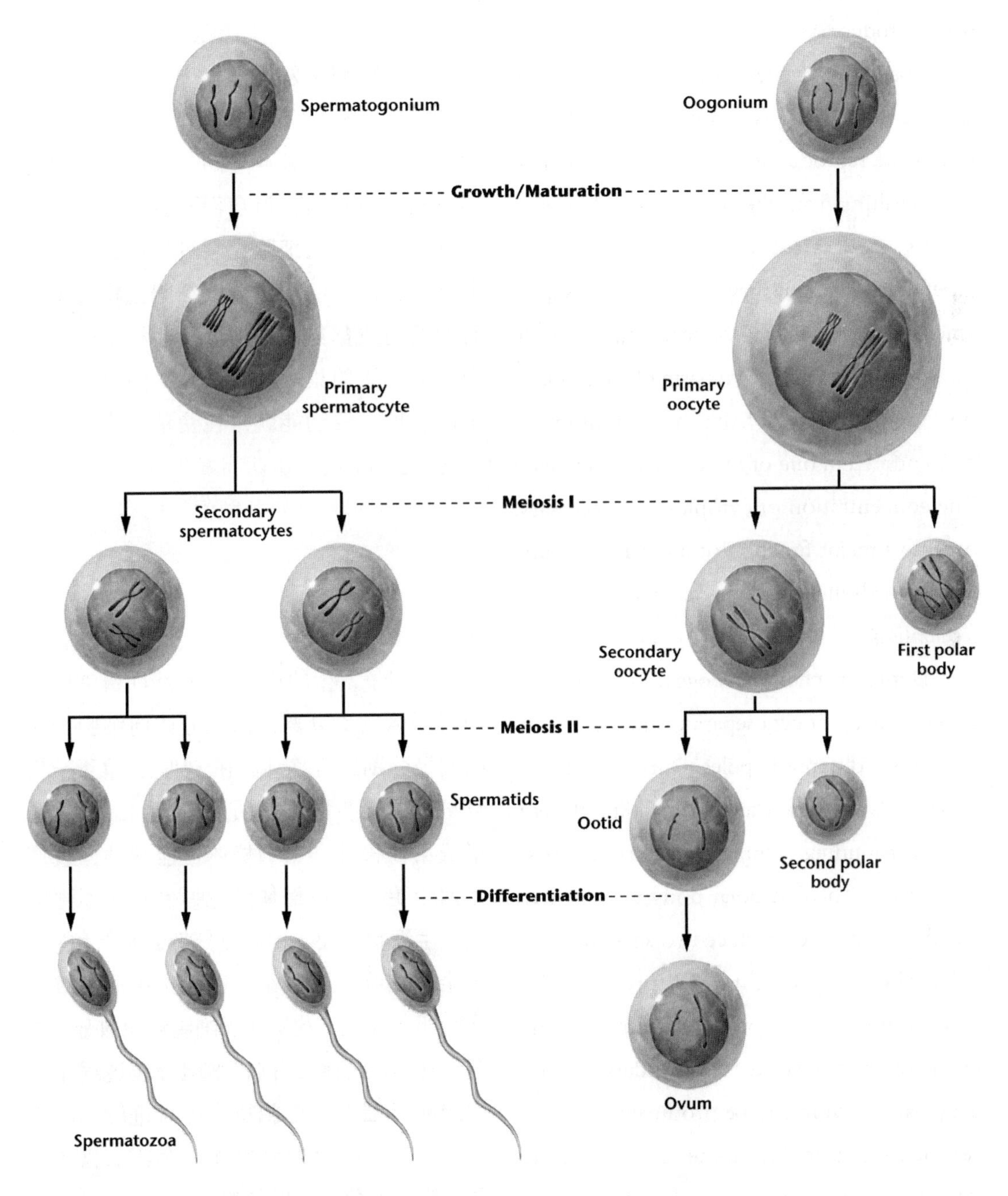

FIGURE 2.9 Spermatogenesis and oogenesis in animal cells.

图 2.9 动物细胞的精子发生和卵子发生。

Spermatogenesis may be continuous or may occur periodically in mature male animals; its onset is determined by the species' reproductive cycles. Animals that reproduce year-round produce sperm continuously, whereas those whose breeding period is confined to a particular season produce sperm only during that time.

在成年雄性动物中，精子发生可以是连续进行的，也可以是周期性进行的，这取决于物种的繁殖周期。全年繁殖的动物可以持续产生精子，然而那些繁殖期仅限于特定季节的动物则只在特定时期产生精子。

In animal oogenesis, the formation of ova (sing. ovum), or eggs, occurs in the ovaries, the female reproductive organs. The daughter cells resulting from the two meiotic divisions of this process receive equal amounts of genetic material, but they do not receive equal amounts of cytoplasm. Instead, during each division, almost all the cytoplasm of the primary oocyte, itself derived from the oogonium, is concentrated in one of the two daughter cells. The concentration of cytoplasm is necessary because a major function of the mature ovum is to nourish the developing embryo following fertilization.

在动物卵子发生过程中，卵的形成发生在雌性生殖器官——卵巢中。经减数分裂的两次细胞分裂产生的子代细胞会获得等量的遗传物质，但它们不会获得等量的细胞质。在每次细胞分裂过程中，由卵原细胞发育而来的初级卵母细胞会将几乎所有的细胞质都集中于两个子代细胞中的一个。细胞质的集中十分必要，因为成熟卵子的一项主要功能就是在受精后为胚胎发育提供营养。

During anaphase I in oogenesis, the tetrads of the primary oocyte separate, and the dyads move toward opposite poles. During telophase I, the dyads at one pole are pinched off with very little surrounding cytoplasm to form the first polar body. The first polar body may or may not divide again to produce two small haploid cells. The other daughter cell produced by this first meiotic division contains most of the cytoplasm and is called the secondary oocyte. The mature ovum will be produced from the secondary oocyte during the second meiotic division. During this division, the cytoplasm of the secondary oocyte again divides unequally,

在卵子发生的后期I，初级卵母细胞中的四联体发生分离，即二分体向细胞相对的两极分别进行移动。在末期I，位于一极且被周围极少的细胞质所包裹的二分体会形成第一极体。第一极体可能继续分裂产生两个更小的单倍体细胞，也可能不再继续发生细胞分裂。由减数分裂I期产生的另一个子细胞包含了绝大部分细胞质，成为次级卵母细胞。次级卵母细胞经减数分裂II期产生成熟的卵。在该细胞分裂过程中，次级卵母细胞的细胞质会再次进行不均等分裂，产生一个卵细胞和一个第二极体。卵细胞最终分化为成熟的卵子。

producing an ootid and a second polar body. The ootid then differentiates into the mature ovum.

Unlike the divisions of spermatogenesis, the two meiotic divisions of oogenesis may not be continuous. In some animal species, the second division may directly follow the first. In others, including humans, the first division of all oocytes begins in the embryonic ovary but arrests in prophase I. Many years later, meiosis resumes in each oocyte just prior to its ovulation. The second division is completed only after fertilization.

与精子发生的细胞分裂不同，卵子发生的两次减数分裂可以不连续。在某些动物物种中，减数分裂II期紧随减数分裂I期。而在包括人类在内的其他物种中，所有卵母细胞的减数分裂I期都开始于胚胎卵巢，但停滞于前期I。许多年后，减数分裂在每个卵母细胞排卵前继续进行，但减数分裂II期仅在受精后才完成。

ESSENTIAL POINT

There is a major difference between meiosis in males and in females. On the one hand, spermatogenesis partitions the cytoplasmic volume equally and produces four haploid sperm cells. Oogenesis, on the other hand, collects the bulk of cytoplasm in one egg cell and reduces the other haploid products to polar bodies. The extra cytoplasm in the egg contributes to zygote development following fertilization.

基本要点

雄性和雌性个体体内发生的减数分裂之间存在巨大差异。一方面，精子发生均等地分配细胞质，产生四个单倍体精子细胞。另一方面，卵子发生则将大部分细胞质集中于一个卵细胞中，并使其他单倍体子代细胞的个体变小成为极体。卵中超多的细胞质将在受精后为受精卵发育贡献力量。

2.6 Meiosis Is Critical to Sexual Reproduction in All Diploid Organisms

The process of meiosis is critical to the successful sexual reproduction of all diploid organisms. It is the mechanism by which the diploid amount of genetic information is reduced to the haploid amount. In animals, meiosis leads to the formation of gametes, whereas in plants haploid spores are produced, which in turn lead

2.6 减数分裂对于所有二倍体生物的有性生殖都至关重要

减数分裂过程对于所有二倍体生物进行成功的有性生殖都至关重要。这一机制将二倍体数量的遗传信息减少至单倍体数量。减数分裂在动物中促使单倍体配子形成，在植物中则产生单倍体孢子并继而形成单倍体配子。

to the formation of haploid gametes.

Each diploid organism stores its genetic information in the form of homologous pairs of chromosomes. Each pair consists of one member derived from the maternal parent and one from the paternal parent. Following meiosis, haploid cells potentially contain either the paternal or the maternal representative of every homologous pair of chromosomes. However, the process of crossing over, which occurs in the first meiotic prophase, further reshuffles the alleles between the maternal and paternal members of each homologous pair, which then segregate and assort independently into gametes. These events result in the great amount of genetic variation present in gametes.

每个二倍体生物均以同源染色体对的形式存储其遗传信息。同源染色体对中一条来自母本，另一条来自父本。减数分裂后，单倍体细胞将随机含有每对同源染色体中来自父本或来自母本的一条染色体。但是，减数分裂前期 I 中发生的交换过程将每对同源染色体中母本和父本的等位基因进一步进行了重组，然后再经过分离和自由组合分配至配子中。这些事件的发生最终使得配子中产生大量的遗传变异。

It is important to touch briefly on the significant role that meiosis plays in the life cycles of fungi and plants. In many fungi, the predominant stage of the life cycle consists of haploid vegetative cells. They arise through meiosis and proliferate by mitotic cell division. In multicellular plants, the life cycle alternates between the diploid sporophyte stage and the haploid gametophyte stage. While one or the other predominates in different plant groups during this “alternation of generations”, the processes of meiosis and fertilization constitute the “bridges” between the sporophyte and gametophyte stages. Therefore, meiosis is an essential component of the life cycle of plants.

简要地对减数分裂在真菌和植物生命周期中所起的重要作用进行讨论是十分重要的。在许多真菌中，生命周期的主要阶段以单倍体营养细胞的形式存在。它们由减数分裂产生，并通过有丝分裂方式进行增殖。在多细胞植物中，生命周期则在二倍孢子体阶段和单倍配子体阶段之间交替存在。尽管这种“世代交替”过程在不同植物群体中占主导地位的具体情形有所不同，但是减数分裂和受精构成了联系孢子体和配子体阶段的“桥梁”。因此，减数分裂是植物生命周期的重要组成部分。

ESSENTIAL POINT

Meiosis results in extensive genetic variation by virtue of the exchange during crossing over

基本要点

减数分裂通过母本和父本染色单体之间发生的遗传物质交换和随机分离至不同

between maternal and paternal chromatids and their random segregation into gametes. In addition, meiosis plays an important role in the life cycles of fungi and plants, serving as the bridge between alternating generations.

配子而产生广泛的遗传多样性。此外，减数分裂在真菌和植物的生命周期中作为世代交替间的桥梁而扮演着重要角色。

2.7 Electron Microscopy Has Revealed the Physical Structure of Mitotic and Meiotic Chromosomes

2.7 电子显微镜揭示了有丝分裂和减数分裂过程中的染色体结构

Thus far in this chapter, we have focused on mitotic and meiotic chromosomes, emphasizing their behavior during cell division and gamete formation. An interesting question is why chromosomes are invisible during interphase but visible during the various stages of mitosis and meiosis. Studies using electron microscopy clearly show why this is the case.

Recall that, during interphase, only dispersed chromatin fibers are present in the nucleus. Once mitosis begins, however, the fibers coil and fold, condensing into typical mitotic chromosomes. If the fibers comprising a mitotic chromosome are loosened, the areas of greatest spreading reveal individual fibers similar to those seen in interphase chromatin. Very few fiber ends seem to be present, and in some cases, none can be seen. Instead, individual fibers always seem to loop back into the interior. Such fibers are obviously twisted and coiled around one another, forming the regular pattern of folding in the mitotic chromosome. Starting in late telophase of mitosis and continuing during G1 of interphase, chromosomes unwind to form the long fibers characteristic of chromatin,

到目前为止，我们在本章中始终关注于有丝分裂和减数分裂过程中的染色体，尤其是它们在细胞分裂和配子形成过程中的行为。这里有一个有趣的问题，即在间期时，染色体在显微镜下为什么不可见，而在有丝分裂和减数分裂的各个阶段时，显微镜下的染色体是可见的。使用电子显微镜开展的系列研究清楚地回答了这个问题。

回想一下，在间期时，细胞核中仅分散存在着染色质丝。然而，一旦有丝分裂开始，染色质丝即螺旋折叠，聚集压缩成典型的有丝分裂染色体。如果形成有丝分裂染色体的染色质丝发生松散，则每条染色质丝最大限度分散的区域即类似于间期所看到的染色质情形。染色质丝末端几乎不可见，在某些情况下，人们无法看到染色质丝末端。反之，每条染色质丝似乎总是趋向于向内部盘旋。这些染色质丝明显地相互缠绕和盘旋，形成有丝分裂染色体常规的折叠模式。从有丝分裂末期即将结束时开始，一直持续到间期的 G1 期，染色体发生解螺旋，形成具有染色质特征的长纤维丝状。该染色质丝由 DNA 及相关蛋白质组成，尤其是组蛋白。正是这些物理过程使得 DNA 能在转录和复制的过程中极高

which consist of DNA and associated proteins, particularly proteins called histones. It is in this physical arrangement that DNA can most efficiently function during transcription and replication.

效地发挥作用。

Electron microscopic observations of metaphase chromosomes in varying degrees of coiling led Ernest DuPraw to postulate the **folded-fiber model**. During metaphase, each chromosome consists of two sister chromatids joined at the centromeric region. Each arm of the chromatid appears to be a single fiber wound much like a skein of yarn. The fiber is composed of tightly coiled double-stranded DNA and protein. An orderly coiling-twisting-condensing process appears to facilitate the transition of the interphase chromatin into the more condensed mitotic chromosomes. Geneticists believe that during the transition from interphase to prophase, a 5000-fold compaction occurs in the length of DNA within the chromatin fiber! This process must be extremely precise given the highly ordered and consistent appearance of mitotic chromosomes in all eukaryotes. Note particularly in the micrographs the clear distinction between the sister chromatids constituting each chromosome. They are joined only by the common centromere that they share prior to anaphase.

欧内斯特·杜普劳领导的研究小组在电子显微镜下观察不同卷曲程度的中期染色体并提出了**折叠纤维模型**。在中期，每条染色体由共用着丝粒区域的两条姐妹染色单体组成。染色单体的每条臂呈现为一条缠绕程度类似一束线的单独纤维。每条纤维由紧密缠绕的双链 DNA 分子和蛋白质组成。有序的卷曲 - 缠绕 - 压缩过程促进了间期染色质向结构更加紧密的有丝分裂染色体进行过渡。遗传学家认为，从间期到前期的过渡过程中，染色质纤维内的 DNA 长度发生了 5000 倍的压缩！鉴于观察到的所有真核生物中有丝分裂染色体所具有的高度有序、一致的外形，这一过程必须非常精确。尤其要注意在显微照片中，构成每条染色体的两条姐妹染色单体之间的明显区别。它们在后期之前仅由一个共同的着丝粒相连。

ESSENTIAL POINT

Mitotic chromosomes are produced as a result of the coiling and condensation of chromatin fibers characteristic of interphase and are thus visible only during cell division.

基本要点

有丝分裂染色体是由标志着间期的染色质丝发生卷曲和压缩而形成的，因此仅在细胞分裂过程中可见。

3 Mendelian Genetics

3 孟德尔遗传学

CHAPTER CONCEPTS

■ Inheritance is governed by information stored in discrete unit factors called genes.

■ Genes are transmitted from generation to generation on vehicles called chromosomes.

■ Chromosomes, which exist in pairs in diploid organisms, provide the basis of biparental inheritance.

■ During gamete formation, chromosomes are distributed according to postulates first described by Gregor Mendel, based on his nineteenth-century research with the garden pea.

■ Mendelian postulates prescribe that homologous chromosomes segregate from one another and assort independently with other segregating homologs during gamete formation.

■ Genetic ratios, expressed as probabilities, are subject to chance deviation and may be evaluated statistically.

■ The analysis of pedigrees allows predictions concerning the genetic nature of human traits.

本章速览

■ 遗传由储存在相互独立且不连续的单位因子即基因中的信息所决定。

■ 基因以染色体为载体在世代交替过程中进行传递。

■ 在二倍体生物中成对存在的染色体是实现双亲遗传的基础。

■ 在配子形成过程中，染色体按照格雷戈尔·孟德尔首次提出的假说进行分配，这些假说均基于其 19 世纪利用豌豆作为实验材料开展的科学研究。

■ 孟德尔定律指出在配子形成过程中，同源染色体相互分离，非同源染色体间进行自由组合。

■ 遗传比可用概率表示，受随机偏差影响，并可通过统计学方法进行评估。

■ 系谱分析可用于对人类性状的遗传特性进行预测。

Although inheritance of biological traits has been recognized for thousands of years, the first significant insights into how it takes place only occurred about 150 years ago. In 1866, Gregor Johann Mendel published the results of a series

尽管人类意识到生物性状具有遗传性已有数千年历史，但是直到大约 150 年前，生物性状如何进行遗传的机制才首次被揭示。1866 年，格雷戈尔·约翰·孟德尔发表了一系列实验结果，并为遗传学定律奠

of experiments that would lay the foundation for the formal discipline of genetics. Mendel's work went largely unnoticed until the turn of the twentieth century, but eventually, the concept of the gene as a distinct hereditary unit was established. Since then, the ways in which genes, as segments of chromosomes, are transmitted to offspring and control traits have been clarified. Research continued unabated throughout the twentieth century and into the present—indeed, studies in genetics, most recently at the molecular level, have remained at the forefront of biological research since the early 1900s.

定了基础。孟德尔的研究工作直到 20 世纪初才引起科学界的关注，基因最终作为独立遗传单位的概念得以确立。从那时起，基因作为染色体的一部分传递给后代并控制性状的方式得到了阐明。相关研究一直持续了整个 20 世纪直至今日——事实上，自 20 世纪初以来，遗传学研究，包括当今绝大多数分子水平的研究，一直处于生物学研究的前沿领域。

When Mendel began his studies of inheritance using *Pisum sativum*, the garden pea, chromosomes and the role and mechanism of meiosis were totally unknown. Nevertheless, he determined that discrete units of inheritance exist and predicted their behavior in the formation of gametes. Subsequent investigators, with access to cytological data, were able to relate their own observations of chromosome behavior during meiosis and Mendel's principles of inheritance. Once this correlation was recognized, Mendel's postulates were accepted as the basis for the study of what is known as **transmission genetics** — how genes are transmitted from parents to offspring. These principles were derived directly from Mendel's experimentation.

当孟德尔使用豌豆开始遗传研究时，人们对于染色体以及减数分裂的功能及其机制尚一无所知。然而，孟德尔认为遗传过程中存在相对独立的遗传单位，并预测了它们在配子形成过程中的行为。随后的研究人员利用已获得的细胞生物学数据将减数分裂研究过程中所观察到的染色体行为与孟德尔提出的遗传规律联系了起来。一旦确立二者之间所存在的相关性，人们就接纳了孟德尔假说为**传递遗传学**的基础。传递遗传学即研究基因如何从亲本遗传至后代的规律。孟德尔假说直接源自孟德尔的实验研究。

3.1 Mendel Used a Model Experimental Approach to Study Patterns of Inheritance

3.1 孟德尔使用模型实验方法研究遗传模式

Johann Mendel was born in 1822 to a

约翰 · 孟德尔 1822 年出生于位于中欧

peasant family in the Central European village of Heinzendorf. An excellent student in high school, he studied philosophy for several years afterward and in 1843, taking the name Gregor, was admitted to the Augustinian Monastery of St. Thomas in Brno, now part of the Czech Republic. In 1849, he was relieved of pastoral duties, and from 1851 to 1853, he attended the University of Vienna, where he studied physics and botany. He returned to Brno in 1854, where he taught physics and natural science for the next 16 years. Mendel received support from the monastery for his studies and research throughout his life.

的村庄海因策多夫的一个农民家庭。他高中阶段成绩优秀，并在此后多年内一直学习哲学。1843 年，他以格雷戈尔的名字考入现在隶属捷克共和国布尔诺的圣 · 托马斯奥古斯丁修道院。1849 年，他不再担任牧师的职务，从 1851 年到 1853 年进入维也纳大学学习物理学和植物学。1854 年，他重返布尔诺，并在接下来的 16 年里教授物理学和自然科学。孟德尔的学习和研究终其一生都得到了修道院的支持。

In 1856, Mendel performed his first set of hybridization experiments with the garden pea, launching the research phase of his career. His experiments continued until 1868, when he was elected abbot of the monastery. Although he retained his interest in genetics, his new responsibilities demanded most of his time. In 1884, Mendel died of a kidney disorder. The local newspaper paid him the following tribute:

1856 年，孟德尔用豌豆进行了一组杂交实验，从此开启了他职业生涯中的研究阶段。他的实验工作一直持续到 1868 年，同年他被推选为修道院院长。尽管他对遗传学依然保有兴趣，但是新的工作职责占据了他的大部分时间。1884 年，孟德尔由于肾脏疾病而去世。当地报纸刊登了以下文字以示对他的敬意:

> His death deprives the poor of a benefactor, and mankind at large of a man of the noblest character, one who was a warm friend, a promoter of the natural sciences, and an exemplary priest.

> 他的去世使得穷人失去了一位恩人，使得整个人类失去了一位品格高尚的人，他是一位热情的朋友、一位自然科学的倡导者、一位杰出的牧师。

Mendel first reported the results of some simple genetic crosses between certain strains of the garden pea in 1865. Although his findings went unappreciated until the turn of the century, well after his death, his work was not the first attempt to provide experimental evidence pertaining to inheritance. Mendel's success

孟德尔于 1865 年首次发表了他在不同豌豆株系之间进行的简单遗传杂交实验的结果。他的发现直到 20 世纪初他去世后才得到人们的认可，他的工作并不是人类为遗传学研究提供实验证据的首次尝试。孟德尔在其他人均失败的情况下能够取得成功至少可以部分归因于他杰出的实验设计

where others had failed can be attributed, at least in part, to his elegant experimental design and analysis.

Mendel showed remarkable insight into the methodology necessary for good experimental biology. First, he chose an organism that was easy to grow and to hybridize artificially. The pea plant is self-fertilizing in nature, but it is easy to cross-breed experimentally. It reproduces well and grows to maturity in a single season. Mendel followed seven visible features (we refer to them as characters, or characteristics), each represented by two contrasting forms, or **traits** (Figure 3.1). For the character stem height, for example, he experimented with the traits tall and

和分析方法。

孟德尔对出色的实验生物学所必需的方法学有着不同寻常的洞察力。首先，他选择了一种易于种植并且方便进行人工授精的植物。豌豆在天然情况下自花授粉，但是也便于进行人工杂交。它易于繁殖，并且在一个季节内就可以成熟。孟德尔跟踪了七类肉眼可见的外部特点（我们称之为特征），每类特征都有两种相对形式或**性状**（图 3.1）。例如，对于植株的高度这一特征，孟德尔针对高、矮两种性状进行了实验研究。同时他还选择了另外六对相对性状，涉及种子的形状和颜色、豆荚的

Character	Contrasting traits	F_1 results	F_2 results	F_2 ratio
Seed shape	round/wrinkled	all round	5474 round 1850 wrinkled	2.96:1
Seed color	yellow/green	all yellow	6022 yellow 2001 green	3.01:1
Pod shape	full/constricted	all full	882 full 299 constricted	2.95:1
Pod color	green/yellow	all green	428 green 152 yellow	2.82:1
Flower color	violet/white	all violet	705 violet 224 white	3.15:1
Flower position	axial/terminal	all axial	651 axial 207 terminal	3.14:1
Stem height	tall/dwarf	all tall	787 tall 277 dwarf	2.84:1

FIGURE 3.1 Seven pairs of contrasting traits and the results of Mendel's seven monohybrid crosses of the garden pea (*Pisum sativum*). In each case, pollen derived from plants exhibiting one trait was used to fertilize the ova of plants exhibiting the other trait. In the F_1 generation, one of the two traits was exhibited by all plants. The contrasting trait reappeared in approximately 1/4 of the F_2 plants.

图 3.1 孟德尔利用豌豆 (*Pisum sativum*) 进行的 7 个单因子杂交所涉及的 7 对相对性状及其实验结果。在每个杂交实验中，花粉来自表现其中一种性状的植株，并用于受精表现另一种性状的植株的卵细胞。在 F_1 代中，所有植株均仅表现两种性状中的一种，未能表现出的相对性状在 F_2 代中以大约 1/4 的比例重现。

dwarf. He selected six other contrasting pairs of traits involving seed shape and color, pod shape and color, and flower color and position. From local seed merchants, Mendel obtained true-breeding strains, those in which each trait appeared unchanged generation after generation in self-fertilizing plants.

形状和颜色、花的颜色和位置。孟德尔还从当地种子商人那里得到了纯种系植株，即所研究的性状在这些植株的自交后代中世代保持不变。

There were several other reasons for Mendel's success. In addition to his choice of a suitable organism, he restricted his examination to one or very few pairs of contrasting traits in each experiment. He also kept accurate quantitative records, a necessity in genetic experiments. From the analysis of his data, Mendel derived certain postulates that have become the principles of transmission genetics.

孟德尔能够取得成功还有许多其他原因。除了他选择了合适的生物物种外，他在每个实验中仅研究一对或少量几对相对性状。他还进行了准确的数据记录，这对于遗传实验必不可少。通过对实验数据进行分析，孟德尔提出了一些假设并最终成为传递遗传学的定律。

3.2 The Monohybrid Cross Reveals How One Trait Is Transmitted from Generation to Generation

3.2 单因子杂交揭示了单一性状世代传递的规律

Mendel's simplest crosses involved only one pair of contrasting traits. Each such experiment is called a **monohybrid cross**. A monohybrid cross is made by mating true-breeding individuals from two parent strains, each exhibiting one of the two contrasting forms of the character under study. Initially, we examine the first generation of offspring of such a cross, and then we consider the offspring of **selfing**, that is, of self-fertilization of individuals from this first generation. The original parents constitute the $\mathbf{P_1}$, or **parental generation**; their offspring are the $\mathbf{F_1}$, or first **filial generation**; the individuals resulting from the selfed F_1 generation

孟德尔实验中最简单的杂交实验只涉及一对相对性状，每一个类似的杂交实验被称为**单因子杂交**。单因子杂交是将两个亲本植株的纯种系个体进行杂交，每个个体仅表现出待研究的相对性状中的一种。先观察该单因子杂交的第一代后代，然后将第一代后代个体进行**自交**并观察其自交后代。在整个单因子杂交体系中，最初的亲本植株构成了 $\mathbf{P_1}$ **代**，或被称为**亲代**；它们的后代是 $\mathbf{F_1}$ **代**，或被称为**子一代**；由 F_1 代自交产生的后代是 $\mathbf{F_2}$ **代**，或被称为**子二代**；以此类推。

are the $\mathbf{F_2}$, or **second filial generation**; and so on.

The cross between true-breeding pea plants with tall stems and dwarf stems is representative of Mendel's monohybrid crosses. Tall and dwarf are contrasting traits of the character of stem height. Unless tall or dwarf plants are crossed together or with another strain, they will undergo self-fertilization and breed true, producing their respective traits generation after generation. However, when Mendel crossed tall plants with dwarf plants, the resulting F_1 generation consisted of only tall plants. When members of the F_1 generation were selfed, Mendel observed that 787 of 1064 F_2 plants were tall, while 277 of 1064 were dwarf. Note that in this cross (Figure 3.1), the dwarf trait disappeared in the F_1 generation, only to reappear in the F_2 generation.

使用高植株豌豆纯种系、矮植株豌豆纯种系进行的杂交实验是孟德尔单因子杂交的代表实验之一。植株高、矮是针对豌豆植株高度这一特征的一对相对性状。如果高植株、矮植株之间不进行杂交或者也不与其他品系进行人工杂交，那么这些豌豆植株将进行严格的自花授粉，最终得到纯种系，并且世代均只表现它们各自的特征性状。然而，当孟德尔将高植株豌豆纯种系与矮植株豌豆纯种系进行杂交时，得到的 F_1 代却均为高植株。当 F_1 代个体进行自交后，孟德尔发现 F_2 代个体共有 1064 株，其中 787 株为高植株，277 株为矮植株。请大家注意在这个杂交实验中（图 3.1），植株矮这一性状在 F_1 代中消失了，而在 F_2 代中又再次出现了。

Genetic data are usually expressed and analyzed as ratios. In this particular example, many identical P_1 crosses were made and many F_1 plants—all tall—were produced. As noted, of the 1064 F_2 offspring, 787 were tall and 277 were dwarf—a ratio of approximately 2.8:1.0, or about 3:1.

遗传数据通常以比例形式进行表示和分析。在这一特定例子中，孟德尔进行了许多完全相同的 P_1 代杂交实验，并由此得到许多 F_1 代植株，并且都是高植株。如前所述，在 1064 株 F_2 代个体中，787 株为高植株，277 株为矮植株，比例约为 2.8 ∶ 1.0，即约 3 ∶ 1。

Mendel made similar crosses between pea plants exhibiting each of the other pairs of contrasting traits; the results of these crosses are shown in Figure 3.1. In every case, the outcome was similar to the tall/dwarf cross just described. For the character of interest, all F_1 offspring expressed the same trait exhibited by one of the parents, but in the F_2 offspring, an approximate ratio of 3:1 was obtained. That is, three-fourths looked like the F_1 plants, while one-fourth

孟德尔针对豌豆其他成对存在的相对性状也进行了类似的杂交实验，这些杂交实验的结果如图 3.1 所示。在每组实验中，实验结果与刚刚描述的针对植株高 / 矮这对相对性状的杂交实验结果相似。对于所研究的性状，所有 F_1 代植株均仅表现出与其中一个亲本相同的性状，而在 F_2 代中两种性状则呈现出大约为 3 ∶ 1 的比例。也就是说，F_2 代个体中，3/4 的表型与 F_1 代类似，而 1/4 则表现出与之相对且在 F_1 代中消失

exhibited the contrasting trait, which had disappeared in the F_1 generation.

的那种表型。

We note one further aspect of Mendel's monohybrid crosses. In each cross, the F_1 and F_2 patterns of inheritance were similar regardless of which P_1 plant served as the source of pollen (sperm) and which served as the source of the ovum (egg). The crosses could be made either way— pollination of dwarf plants by tall plants, or vice versa. Crosses made in both these ways are called **reciprocal crosses**. Therefore, the results of Mendel's monohybrid crosses were not sex dependent.

我们还注意到孟德尔单因子杂交的另一特点。在每组杂交实验中，无论是 P_1 代植株的哪种性状个体提供花粉（精子）或提供卵子（卵），F_1 代和 F_2 代的遗传模式都是相似的。杂交可以通过两种不同方式进行：高植株纯种系个体向矮植株纯种系个体进行授粉，反之亦然。这两种杂交方式被称为**正反交**。因此，孟德尔单因子杂交的结果与亲本性别无关。

To explain these results, Mendel proposed the existence of particulate unit factors for each trait. He suggested that these factors serve as the basic units of heredity and are passed unchanged from generation to generation, determining various traits expressed by each individual plant. Using these general ideas, Mendel proceeded to hypothesize precisely how such factors could account for the results of the monohybrid crosses.

为了解释这些实验结果，孟德尔提出每种性状都是由独立存在的单位因子所决定的。他认为这些因子是遗传的基本单位，在世代传递过程中保持不变，并且决定了每个植物个体所表现的不同性状。运用这些基本观点，孟德尔进一步对这些单位因子决定单因子杂交结果的机制进行了精确地预测。

Mendel's First Three Postulates

孟德尔的前三条假设

Using the consistent pattern of results in the monohybrid crosses, Mendel derived the following three postulates, or principles, of inheritance.

基于单因子杂交结果的一致性，孟德尔提出以下三条遗传假设或原理。

1. UNIT FACTORS IN PAIRS

1. 单位因子成对存在

Genetic characters are controlled by unit factors existing in pairs in individual organisms.

遗传性状由生物体中成对存在的单位因子控制。

In the monohybrid cross involving tall and dwarf stems, a specific unit factor exists for each trait. Each diploid individual receives one

在针对植株高、矮这对相对性状的单因子杂交中，每种特定性状均由一种特定的单位因子所决定。每个二倍体个体从其每个亲

factor from each parent. Because the factors occur in pairs, three combinations are possible: two factors for tall stems, two factors for dwarf stems, or one of each factor. Every individual possesses one of these three combinations, which determines stem height.

本中分别获得一个单位因子。因为这些单位因子成对存在，所以有三种可能的组合方式：两个单位因子均决定植株高性状、两个单位因子均决定植株矮性状、以上两种单位因子各含一个。每个个体含有以上三种组合中的任意一种，并由此决定植株高度。

2. DOMINANCE/RECESSIVENESS

When two unlike unit factors responsible for a single character are present in a single individual, one unit factor is dominant to the other, which is said to be recessive.

In each monohybrid cross, the trait expressed in the F_1 generation is controlled by the dominant unit factor. The trait not expressed is controlled by the recessive unit factor. The terms dominant and recessive are also used to designate traits. In this case, tall stems are said to be dominant over recessive dwarf stems.

2. 显性与隐性

当同一个体中决定某一性状的两种不同形式单位因子同时存在时，其中一个单位因子对于另一个呈显性，也就是说另一个单位因子为隐性。

在每个单因子杂交中，F_1 代表现的性状由显性单位因子控制，F_1 代未能表现的性状则由隐性单位因子控制。显性和隐性这对术语也可用于描述性状，即植株高相对于植株矮是显性的。

3. SEGREGATION

During the formation of gametes, the paired unit factors separate, or segregate, randomly so that each gamete receives one or the other with equal likelihood.

If an individual contains a pair of like unit factors (e.g., both specific for tall), then all its gametes receive one of that same kind of unit factor (in this case, tall). If an individual contains unlike unit factors (e.g., one for tall and one for dwarf), then each gamete has a 50 percent probability of receiving either the tall or the dwarf unit factor.

3. 分离

在配子形成过程中，成对存在的单位因子相互分离，每个配子将随机得到每对单位因子中的一个，并且概率相等。

如果一个个体含有一对相同的单位因子（如均为决定植株高的因子），那么其所有配子均会获得相同种类的单位因子（在这个例子中即均为决定植株高的因子）。如果一个个体含有不同的单位因子（如一个决定植株高，一个决定植株矮），则其所产生的每个配子分别获得这两种不同单位因子的概率各为 50%。

These postulates provide a suitable explanation for the results of the monohybrid crosses. Let's use the tall/dwarf cross to illustrate. Mendel reasoned that P_1 tall plants contained

这些假说为单因子杂交结果提供了合理的解释。下面我们使用植株高 / 矮杂交实验进行说明。孟德尔认为 P_1 亲代中高植株含有成对存在的同种单位因子，P_1 亲代中矮植株

identical paired unit factors, as did the P_1 dwarf plants. The gametes of tall plants all receive one tall unit factor as a result of segregation. Similarly, the gametes of dwarf plants all receive one dwarf unit factor. Following fertilization, all F_1 plants receive one unit factor from each parent—a tall factor from one and a dwarf factor from the other—reestablishing the paired relationship, but because tall is dominant to dwarf, all F_1 plants are tall.

也是如此。高植株产生的配子根据分离假说均会获得一个决定植株高的单位因子。同样，所有矮植株产生的配子均会获得一个决定植株矮的单位因子。受精后，所有 F_1 代植株分别从每个亲本获得一个单位因子——从植株高的亲本获得一个决定植株高的单位因子，从植株矮的亲本获得另一个决定植株矮的单位因子，从而重新组成成对存在的单位因子，但是由于植株高相对于植株矮是显性，因此所有的 F_1 代植株都是高的。

When F_1 plants form gametes, the postulate of segregation demands that each gamete randomly receives either the tall *or* dwarf unit factor. Following random fertilization events during F_1 selfing, four F_2 combinations will result with equal frequency:

(1) tall/tall
(2) tall/dwarf
(3) dwarf/tall
(4) dwarf/dwarf

当 F_1 代植株产生配子时，依据分离假说每个配子将随机获得一个决定植株高的单位因子或一个决定植株矮的单位因子。在 F_1 代自交的随机受精过程中，四种 F_2 代单位因子组合将等比例产生：

（1）高 / 高
（2）高 / 矮
（3）矮 / 高
（4）矮 / 矮

Combinations (1) and (4) will clearly result in tall and dwarf plants, respectively. According to the postulate of dominance/recessiveness, combinations (2) and (3) will both yield tall plants. Therefore, the F_2 is predicted to consist of 3/4 tall and 1/4 dwarf, or a ratio of 3:1. This is approximately what Mendel observed in his cross between tall and dwarf plants. A similar pattern was observed in each of the other monohybrid crosses (Figure 3.1).

组合（1）和（4）显然分别将产生高植株和矮植株。根据显性 / 隐性假说，组合（2）和（3）都将产生高植株。因此，可以预计 F_2 代由 3/4 高植株和 1/4 矮植株组成，表型比例为 3 ∶ 1。这大约就是孟德尔在其针对植株高 / 矮所进行的单因子杂交中所观察到的。在其他的单因子杂交中也观察到了类似的遗传模式（图 3.1）。

ESSENTIAL POINT

Mendel's postulates help describe the basis for the inheritance of phenotypic traits. He hypothesized that unit factors exist in pairs

基本要点

孟德尔假说有助于描述性状表型遗传的理论基础。他认为单位因子成对存在，并且在决定性状表达时表现出显性 / 隐性关

and exhibit a dominant/recessive relationship in determining the expression of traits. He further postulated that unit factors segregate during gamete formation, such that each gamete receives one or the other factor, with equal probability.

系。他还进一步假设成对存在的单位因子在配子形成过程中发生分离，从而使每个配子以相同概率获得其中任何一个单位因子。

Modern Genetic Terminology

To analyze the monohybrid cross and Mendel's first three postulates, we must first introduce several new terms as well as a symbol convention for the unit factors. Traits such as tall or dwarf are physical expressions of the information contained in unit factors. The physical expression of a trait is the **phenotype** of the individual. Mendel's unit factors represent units of inheritance called **genes** by modern geneticists. For any given character, such as plant height, the phenotype is determined by alternative forms of a single gene, called **alleles**. For example, the unit factors representing tall and dwarf are alleles determining the height of the pea plant.

现代遗传术语

为了分析单因子杂交和孟德尔的前三条假设，我们必须首先介绍一些新的术语以及单位因子的符号约定。植株高或植株矮等性状是蕴含在单位因子中的遗传信息的生理体现。一种性状的生理体现被称为个体的**表型**。孟德尔的单位因子代表了现代遗传学中被称为**基因**的遗传单元。对于任何一种性状，如植株高度，其表型由基因的不同形式所决定，这些可相互替代的不同基因形式被称为**等位基因**。例如，决定植株高或矮的两种不同单位因子即为决定豌豆植株高度的等位基因。

Geneticists have several different systems for using symbols to represent genes. Later in the text (see Chapter 4), we will review a number of these conventions, but for now, we will adopt one to use consistently throughout this chapter. According to this convention, the first letter of the recessive trait symbolizes the character in question; in lowercase italic, it designates the allele for the recessive trait, and in uppercase italic, it designates the allele for the dominant trait. Thus for Mendel's pea plants, we use *d* for the dwarf allele and *D* for the tall allele. When alleles are written in pairs to represent the two

遗传学家采用多种符号系统来表示基因。在随后章节中（请参阅第4章），我们将介绍这些约定，但现在，我们将采用其中一种约定并在本章中持续使用。根据该约定，表示隐性性状的英文单词首字母被用于表示所研究的性状；该字母的小写斜体形式代表决定隐性性状的等位基因，而其大写斜体形式则代表决定显性性状的等位基因。因此在孟德尔豌豆杂交实验中，我们用 *d* 表示决定植株矮的等位基因，用 *D* 表示决定植株高的等位基因。当等位基因成对书写时则代表个体中存在的两个单位因子（*DD*、*Dd* 或 *dd*），该符号也被称

unit factors present in any individual (*DD*, *Dd*, or *dd*), the resulting symbol is called the **genotype**. The genotype designates the genetic makeup of an individual for the trait or traits it describes, whether the individual is haploid or diploid. By reading the genotype, we know the phenotype of the individual: *DD* and *Dd* are tall, and *dd* is dwarf. When both alleles are the same (*DD* or *dd*), the individual is **homozygous** for the trait, or a **homozygote**; when the alleles are different (*Dd*), we use the terms **heterozygous** and **heterozygote**. These symbols and terms are used in Figure 3.2 to describe the monohybrid cross.

为**基因型**。基因型用于描述所讨论的个体性状的遗传组成，无论该个体是单倍体还是二倍体。通过基因型，我们可以知道个体的表型：*DD* 和 *Dd* 是高植株，而 *dd* 是矮植株。当个体所含的两个等位基因相同（*DD* 或 *dd*）时，该个体的性状是**纯合的**，或被称为**纯合子**；当个体所含的两个等位基因不同（*Dd*）时，我们则使用术语**杂合的**或**杂合子**来进行描述。这些符号和术语在图 3.2 中用于描述单因子杂交。

Punnett Squares

The genotypes and phenotypes resulting from combining gametes during fertilization can be easily visualized by constructing a diagram called a **Punnett square**, named after the person who first devised this approach, Reginald C. Punnett. Figure 3.3 illustrates this method of analysis for our $F_1 \times F_1$ monohybrid cross. Each of the possible gametes is assigned a column or a row; the vertical columns represent those of the female parent, and the horizontal rows represent those of the male parent. After assigning the gametes to the rows and columns, we predict the new generation by entering the male and female gametic information into each box and thus producing every possible resulting genotype. By filling out the Punnett square, we are listing all possible random fertilization events. The genotypes and phenotypes of all potential offspring are ascertained by reading the

庞氏表（棋盘法）

庞氏表以最初设计这种方法的科学家雷金纳德 · C. 庞尼特的名字命名。构建庞氏表便于将受精过程中配子组合所形成的后代基因型和表型进行可视化。图 3.3 展示了用这种方法来分析 $F_1 \times F_1$ 单因子杂交的过程。每种可能的配子都被列于指定一列或一行；垂直列代表雌性亲本提供的配子，水平行则代表雄性亲本提供的配子。将配子与行和列进行一一对应后，我们将雄性和雌性的配子信息输入每个格中，由此产生每种可能的基因型，从而对新生一代进行预测。通过填写庞氏表，我们列出了所有可能的随机受精事件。通过分析表中组合可以确定后代所有潜在的基因型和表型。

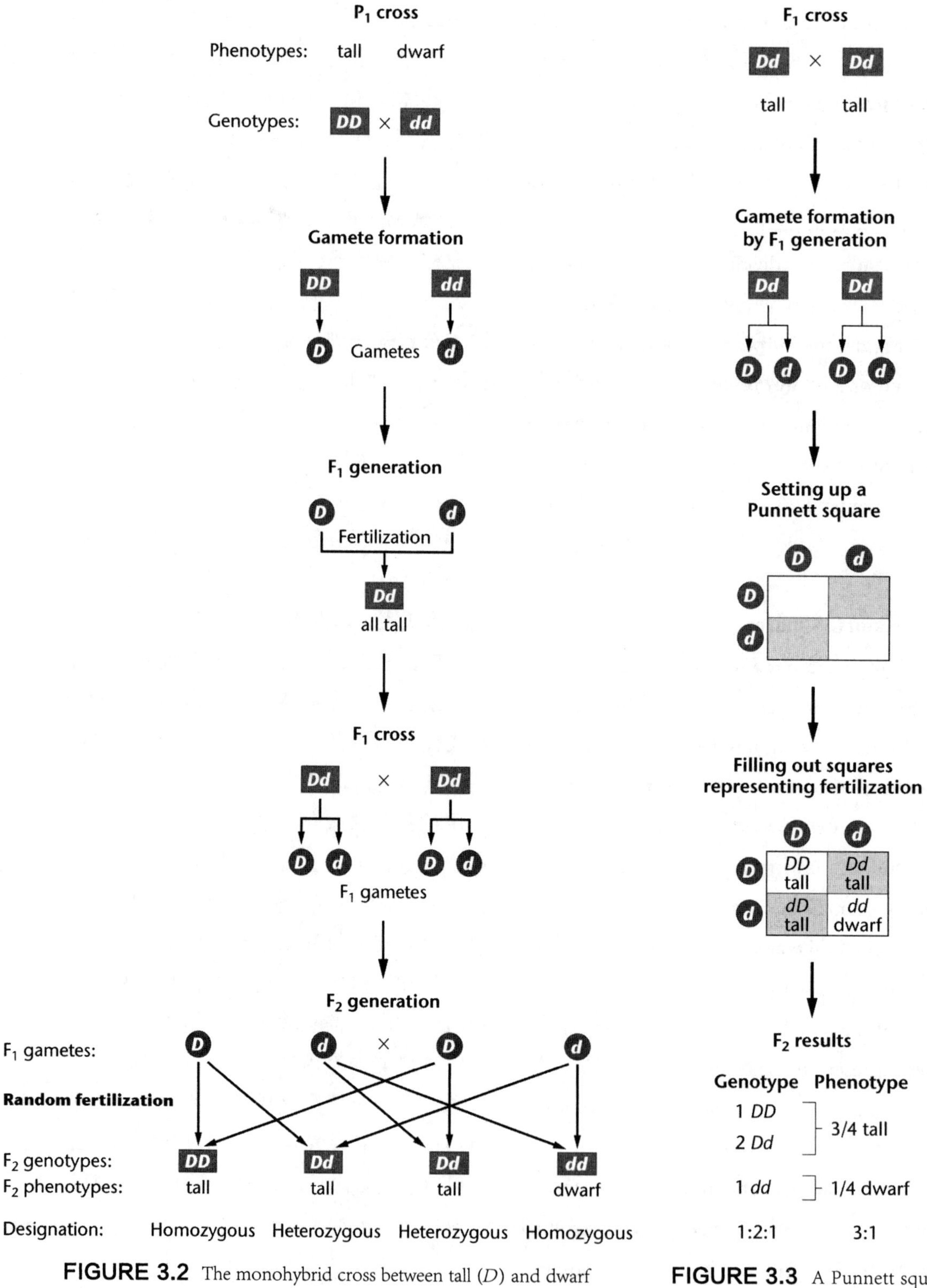

FIGURE 3.2 The monohybrid cross between tall (D) and dwarf (d) pea plants. Individuals are shown in rectangles, and gametes are shown in circles.

图 3.2 豌豆高植株 (D) 与矮植株 (d) 间的单因子杂交。个体基因型在正方形框中列出，配子基因型在圆形框中列出。

FIGURE 3.3 A Punnett square generating the F_2 ratio of the $F_1 \times F_1$ cross shown in Figure 3.2.

图 3.3 利用庞氏表计算图 3.2 所示 $F_1 \times F_1$ 杂交所得的 F_2 代比例。

combinations in the boxes.

The Punnett square method is particularly useful when you are first learning about genetics and how to solve genetics problems. Note the ease with which the 3:1 phenotypic ratio and the 1:2:1 genotypic ratio may be derived for the F_2 generation in Figure 3.3.

当你首次学习遗传学并学习如何解决遗传问题时，庞氏表方法尤其有用。大家可以注意到图 3.3 使用庞氏表即可十分轻松地得到 F_2 代的表型分离比 3 ： 1 和基因型分离比 1 ： 2 ： 1。

The Testcross: One Character

Tall plants produced in the F_2 generation are predicted to have either the *DD* or the *Dd* genotype. You might ask if there is a way to distinguish the genotype. Mendel devised a rather simple method that is still used today to discover the genotype of plants and animals: the **testcross**. The organism expressing the dominant phenotype but having an unknown genotype is crossed with a known homozygous recessive individual. For example, as shown in Figure 3.4(a), if a tall plant of genotype *DD* is testcrossed with a dwarf plant, which must have the *dd* genotype, all offspring will be tall phenotypically and *Dd* genotypically. However, as shown in Figure 3.4(b), if a tall plant is *Dd* and is crossed with a dwarf plant (*dd*), then one-half

测交：单一性状

F_2 代中的高植株个体，根据预测其基因型可能是 *DD* 或 *Dd*。你可能会问是否有办法来区分二者的基因型呢。孟德尔设计了一种相对简单并且至今仍在使用的方法来鉴定植物和动物的基因型：**测交**，即基因型未知的显性表型个体与已知隐性纯合个体进行杂交。如图 3.4 (a) 所示，如果基因型 *DD* 的高植株与基因型 *dd* 的矮植株进行测交，那么所有后代的表型均为高植株且基因型均为 *Dd*。然而，如图 3.4 (b) 所示，如果基因型 *Dd* 的高植株与矮植株（*dd*）进行测交，那么后代中一半将是高植株（*Dd*），另一半将是矮植株（*dd*）。因此，高 / 矮表型比例为 1 ： 1 则表明基因型未知的高植株是杂合的。测交结果也进一步证实了孟德尔的结论：性状由独立的单位因子控制。

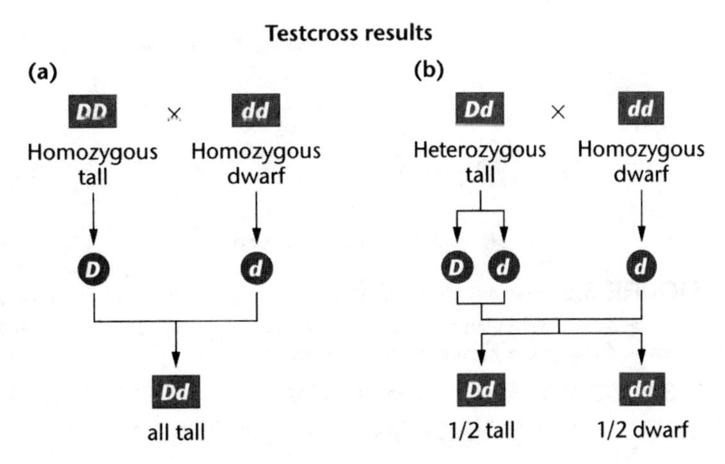

FIGURE 3.4 Testcross of a single character. In (a), the tall parent is homozygous, but in (b), the tall parent is heterozygous. The genotype of each tall P_1 plant can be determined by examining the offspring when each is crossed with the homozygous recessive dwarf plant.

图 3.4 单一性状的测交。(a) 中高植株亲本是纯合个体，但是 (b) 中高植株亲本是杂合个体。每个 P_1 代高植株个体的基因型均可通过与纯合隐性矮植株进行杂交来确定。

of the offspring will be tall (*Dd*) and the other half will be dwarf (*dd*). Therefore, a 1:1 tall/dwarf ratio demonstrates the heterozygous nature of the tall plant of unknown genotype. The results of the testcross reinforced Mendel's conclusion that separate unit factors control traits.

3.3 Mendel's Dihybrid Cross Generated a Unique F_2 Ratio

As a natural extension of the monohybrid cross, Mendel also designed experiments in which he examined two characters simultaneously. Such a cross, involving two pairs of contrasting traits, is a **dihybrid cross**, or a two-factor cross. For example, if pea plants having yellow seeds that are round were bred with those having green seeds that are wrinkled, the results shown in Figure 3.5 would

3.3 孟德尔双因子杂交产生独特的 F_2 代比例

作为单因子杂交的自然拓展，孟德尔还设计了同时研究两种特征的实验。这样的杂交实验同时研究两对相对性状，被称为**双因子杂交**。例如，如果将种子黄色圆滑的豌豆植株与种子绿色皱缩的豌豆植株进行杂交，其结果如图3.5所示：F_1 代所有个体的种子全部为黄色圆滑。因此显而易见，黄色相对于绿色是显性，而圆滑相对于皱缩是显性。当 F_1 代个体进行自交时，

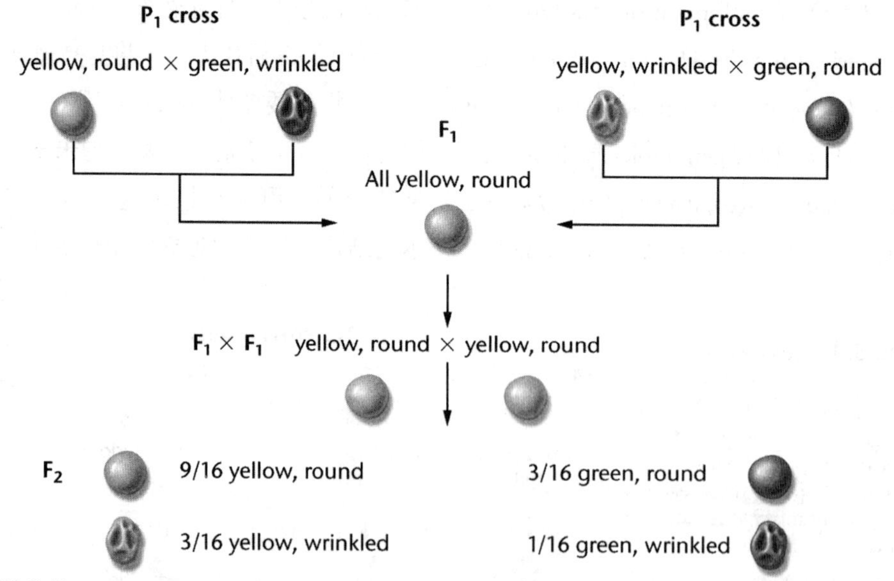

FIGURE 3.5 F_1 and F_2 results of Mendel's dihybrid crosses in which the plants on the top left with yellow, round seeds are crossed with plants having green, wrinkled seeds, and the plants on the top right with yellow, wrinkled seeds are crossed with plants having green, round seeds.

图3.5 孟德尔双因子杂交的 F_1 代和 F_2 代结果，左上亲本分别为种子黄色圆滑的植株与种子绿色皱缩的植株，右上亲本分别为种子黄色皱缩的植株与种子绿色圆滑的植株。

occur: the F_1 offspring would all be yellow and round. It is therefore apparent that yellow is dominant to green and that round is dominant to wrinkled. When the F_1 individuals are selfed, approximately 9/16 of the F_2 plants express the yellow and round traits, 3/16 express yellow and wrinkled, 3/16 express green and round, and 1/16 express green and wrinkled.

在所得到的 F_2 代中，大约 9/16 的个体种子表现为黄色圆滑、3/16 为黄色皱缩、3/16 为绿色圆滑、1/16 则为绿色皱缩。

A variation of this cross is also shown in Figure 3.5. Instead of crossing one P_1 parent with both dominant traits (yellow, round) to one with both recessive traits (green, wrinkled), plants with yellow, wrinkled seeds are crossed with those with green, round seeds. In spite of the change in the P_1 phenotypes, both the F_1 and F_2 results remain unchanged. Why this is so will become clear below.

图 3.5 还显示了该杂交实验的另一种设计。与亲本 P_1 代分别采用双显性性状亲本（黄色圆滑）和双隐性性状亲本（绿色皱缩）不同，图 3.5 中另一杂交实验的两个亲本分别为种子黄色皱缩的植株与种子绿色圆滑的植株。尽管 P_1 亲代表型发生了变化，但 F_1 代和 F_2 代均保持不变。我们将在下面对该原因给予清楚的解释。

Mendel's Fourth Postulate: Independent Assortment

孟德尔的第四条假设：自由组合

We can most easily understand the results of a dihybrid cross if we consider it theoretically as consisting of two monohybrid crosses conducted separately. Think of the two sets of traits as being inherited independently of each other; that is, the chance of any plant having yellow or green seeds is not at all influenced by the chance that this plant will have round or wrinkled seeds. Thus, because yellow is dominant to green, all F_1 plants in the first theoretical cross would have yellow seeds. In the second theoretical cross, all F_1 plants would have round seeds because round is dominant to wrinkled. When Mendel examined the F_1 plants of the dihybrid cross, all were yellow and round, as our theoretical crosses predict.

如果把双因子杂交在理论上看成由两个独立的单因子杂交组成，那么就能非常容易地理解双因子杂交的实验结果。如果两组性状彼此之间相互独立地进行遗传，那么每个后代植株种子颜色是黄色或绿色的概率完全不受该植株种子是圆滑或皱缩概率的影响。因此，由于黄色性状相对于绿色是显性，所以在第一个理论假设实验中，所有 F_1 代植株种子都是黄色的。又由于圆滑性状相对于皱缩是显性，所以在第二个理论假设实验中，所有 F_1 代植株种子都是圆滑的。当孟德尔观察双因子杂交的 F_1 代植株时，所有植株种子均为黄色圆滑，结果与我们理论假设的杂交实验预测的一样。

The predicted F_2 results of the first cross are 3/4 yellow and 1/4 green. Similarly, the second cross would yield 3/4 round and 1/4 wrinkled. Figure 3.5 shows that in the dihybrid cross, 12/16 F_2 plants are yellow, while 4/16 are green, exhibiting the expected 3:1 (3/4:1/4) ratio. Similarly, 12/16 of all F_2 plants have round seeds, while 4/16 have wrinkled seeds, again revealing the 3:1 ratio.

在第一个假设杂交实验的 F_2 代中，3/4 为黄色种子、1/4 为绿色种子。类似地，在第二个假设杂交实验的 F_2 代中，3/4 种子圆滑、1/4 种子皱缩。图 3.5 所示双因子杂交中，12/16 的 F_2 代植株种子为黄色，而 4/16 的 F_2 代植株种子为绿色，表现出预期的 3 ∶ 1（3/4 ∶ 1/4）比例。类似地，所有 F_2 代植株中 12/16 的种子圆滑，而 4/16 的种子皱缩，再次呈现出 3 ∶ 1 的比例。

These numbers demonstrate that the two pairs of contrasting traits are inherited independently, so we can predict the frequencies of all possible F_2 phenotypes by applying the **product law** of probabilities: the probability of two or more independent events occurring simultaneously is equal to the product of their individual probabilities. For example, the probability of an F_2 plant having yellow and round seeds is (3/4)(3/4), or 9/16, because 3/4 of all F_2 plants should be yellow and 3/4 of all F2 plants should be round.

这些数字表明这两对相对性状独立地进行遗传，因此我们可以利用概率中的**相乘定律**来预测 F_2 代所有可能表型的出现频率。相乘定律为：两个或多个独立事件同时发生的概率等于各事件单独发生概率的乘积。例如，F_2 代植株中种子同时具有黄色和圆滑特征的概率为（3/4）×（3/4）或 9/16，因为所有 F_2 代植株中 3/4 的种子应该是黄色性状，同时 3/4 的种子应该是圆滑性状。

In a like manner, the probabilities of the other three F_2 phenotypes can be calculated: yellow (3/4) and wrinkled (1/4) are predicted to be present together 3/16 of the time; green (1/4) and round (3/4) are predicted 3/16 of the time; and green (1/4) and wrinkled (1/4) are predicted 1/16 of the time. These calculations are shown in Figure 3.6.

使用类似方式可以计算出其他三种 F_2 代表型的概率：黄色（3/4）和皱缩（1/4）同时出现的概率预计是 3/16；绿色（1/4）和圆滑（3/4）同时出现的概率预计是 3/16；绿色（1/4）和皱缩（1/4）同时出现的概率预计是 1/16。这些计算如图 3.6 所示。

It is now apparent why the F_1 and F_2 results are identical whether the initial cross is yellow, round plants bred with green, wrinkled plants, or whether yellow, wrinkled plants are bred with green, round plants. In both crosses, the F_1 genotype of all offspring is identical. As a result,

现在可以很清楚地理解无论初始参与杂交的 P_1 亲代是黄色圆滑植株与绿色皱缩植株，还是黄色皱缩植株与绿色圆滑植株，它们的 F_1 代和 F_2 代结果均完全相同的原因。在这两组杂交实验中，所有 F_1 代个体的基因型都完全相同。因此，F_2 代在两组杂交

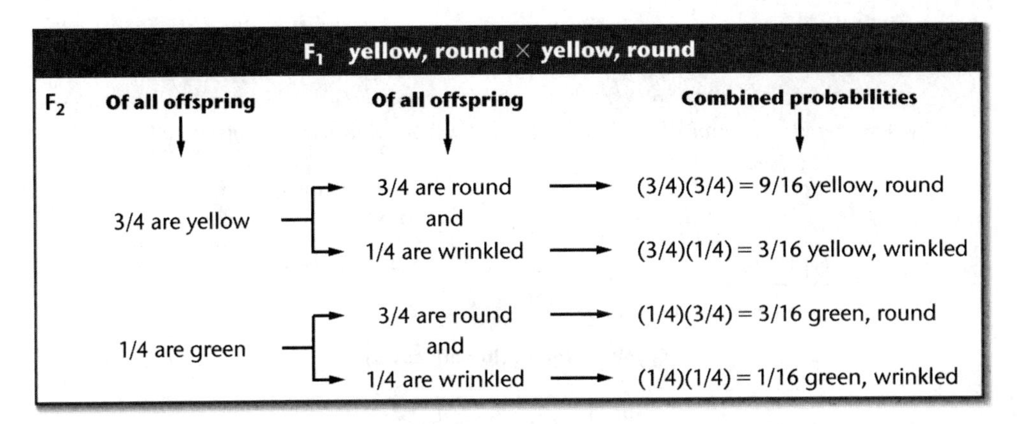

FIGURE 3.6 Computation of the combined probabilities of each F_2 phenotype for two independently inherited characters. The probability of each plant being yellow or green is independent of the probability of it bearing round or wrinkled seeds.

图 3.6 两种独立遗传性状的每种 F_2 代表型的概率计算。每个植株种子是黄色或绿色的概率与其种子是圆滑或皱缩的概率彼此独立、互不影响。

the F_2 generation is also identical in both crosses.

On the basis of similar results in numerous dihybrid crosses, Mendel proposed a fourth postulate:

实验中也完全相同。

基于大量双因子杂交的类似结果，孟德尔提出了第四条假设：

4. INDEPENDENT ASSORTMENT

During gamete formation, segregating pairs of unit factors assort independently of each other.

This postulate stipulates that segregation of any pair of unit factors occurs independently of all others. As a result of random segregation, each gamete receives one member of every pair of unit factors. For one pair, whichever unit factor is received does not influence the outcome of segregation of any other pair. Thus, according to the postulate of independent assortment, all possible combinations of gametes should be formed in equal frequency.

4. 自由组合

在配子形成过程中，成对存在的单位因子在进行分离时，决定不同性状的单位因子将发生自由组合。

该假设指出任何一对单位因子的相互分离都是独立的，并且不受其他单位因子的影响。由于分离是随机的，所以每个配子可以获得每对单位因子中的任何一个。对于一对单位因子，配子获得其中任何一个单位因子均不影响其他成对存在的单位因子的分离。因此，根据自由组合假说，所有可能的配子类型组合均应该等比例产生。

The Punnett square in Figure 3.7 shows how independent assortment works in the formation of the F_2 generation. Examine the formation of gametes by the F_1 plants; segregation prescribes

图 3.7 的庞氏表显示了在 F_2 代形成过程中自由组合是如何发生的。以 F_1 代植株的配子形成为例，分离使得每个配子将获得等位基因 G 或 g 中的任何一个，以及 W

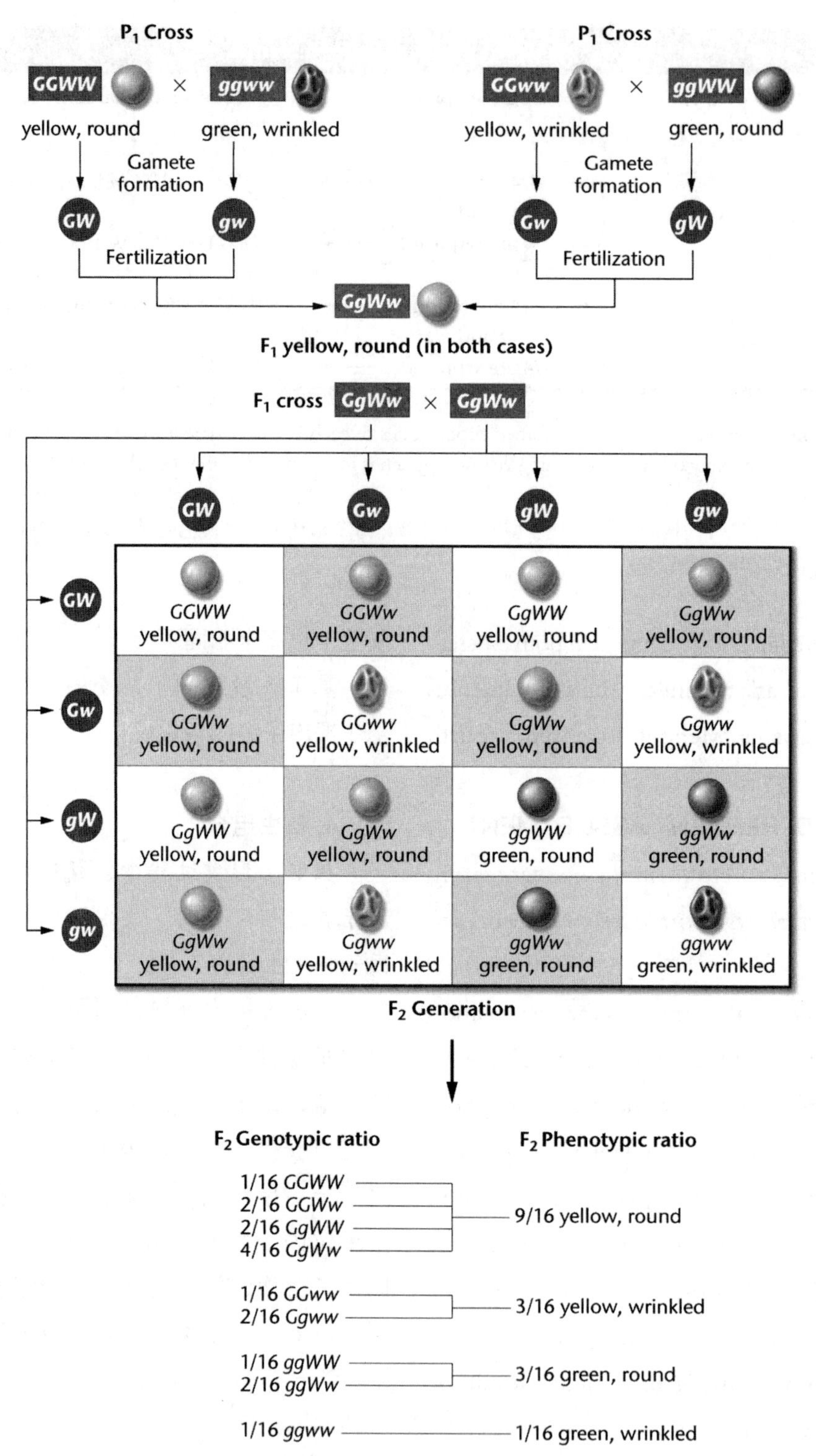

FIGURE 3.7 Analysis of the dihybrid crosses shown in Figure 3.5. The F_1 heterozygous plants are self-fertilized to produce an F_2 generation, which is computed using a Punnett square. Both the phenotypic and genotypic F_2 ratios are shown.

图 3.7 对图 3.5 所示双因子杂交的分析。F_1 代杂合个体自交产生 F_2 代，图中使用庞氏表进行计算。F_2 代表型比例和基因型比例均已列出。

that every gamete receives either a *G* or *g* allele and a *W* or *w* allele. Independent assortment stipulates that all four combinations (*GW*, *Gw*, *gW*, and *gw*) will be formed with equal probabilities.

或 *w* 中的任何一个。由自由组合假说可知，所有四种组合的配子类型（*GW*、*Gw*、*gW* 和 *gw*）将等比例形成。

In every $F_1 \times F_1$ fertilization event, each zygote has an equal probability of receiving one of the four combinations from each parent. If many offspring are produced, 9/16 have yellow, round seeds, 3/16 have yellow, wrinkled seeds, 3/16 have green, round seeds, and 1/16 have green, wrinkled seeds, yielding what is designated as **Mendel's 9:3:3:1 dihybrid ratio**. This is an ideal ratio based on probability events involving segregation, independent assortment, and random fertilization. Because of deviation due strictly to chance, particularly if small numbers of offspring are produced, actual results are highly unlikely to match the ideal ratio.

在每个 $F_1 \times F_1$ 受精事件中，每个合子都能从每个亲本获得四种配子类型中的一种，并且概率相同。如果产生大量后代，那么 9/16 的后代种子是黄色圆滑、3/16 的后代种子是黄色皱缩、3/16 的后代种子是绿色圆滑、1/16 的后代种子是绿色皱缩，从而产生**孟德尔 9 ∶ 3 ∶ 3 ∶ 1 双因子杂种率**。这是一个基于包含了分离、自由组合和随机受精三个概率事件的理想比例。由于偶然性导致的偏差，特别是后代数量少，实际结果极有可能与该理想比例相去甚远。

ESSENTIAL POINT

Mendel's postulate of independent assortment states that each pair of unit factors segregates independently of other such pairs. As a result, all possible combinations of gametes are formed with equal probability.

基本要点

孟德尔自由组合假说指出每对单位因子的相互分离都独立于其他对单位因子，互不影响。因此，所有可能的配子类型将等比例形成。

3.4 The Trihybrid Cross Demonstrates That Mendel's Principles Apply to Inheritance of Multiple Traits

3.4 三因子杂交表明孟德尔定律适用于多性状遗传

Thus far, we have considered inheritance of up to two pairs of contrasting traits. Mendel demonstrated that the processes of segregation and independent assortment also apply to three

到目前为止，我们已经考虑了两对相对性状的遗传。孟德尔还通过实验证明了分离过程和自由组合过程同样适用于三对相对性状的遗传，即所谓的**三因子杂交**。

pairs of contrasting traits, in what is called a **trihybrid cross**, or three-factor cross.

Although a trihybrid cross is somewhat more complex than a dihybrid cross, its results are easily calculated if the principles of segregation and independent assortment are followed. For example, consider the cross shown in Figure 3.8 where the allele pairs of theoretical contrasting traits are represented by the symbols *A, a, B, b, C,* and *c*. In the cross between *AABBCC* and *aabbcc* individuals, all F_1 individuals are heterozygous for all three gene pairs. Their genotype, *AaBbCc*, results in the phenotypic expression of the dominant *A*, *B*, and *C* traits. When F_1 individuals serve as parents, each produces eight different gametes in equal frequencies. At this point, we could construct a Punnett square with 64 separate boxes and read out the phenotypes—but such a method is cumbersome in a cross involving so many factors. Therefore, another method has been devised to calculate the predicted ratio.

尽管三因子杂交比双因子杂交更为复杂，但是如果同样遵循分离定律和自由组合定律，那么则可以轻松地计算出该杂交实验的结果。例如，我们来考虑图 3.8 所示的杂交实验，其中相对性状的等位基因分别用符号 *A*、*a*、*B*、*b*、*C* 和 *c* 表示。在个体 *AABBCC* 和 *aabbcc* 的杂交中，所有 F_1 代个体的该三对基因都是杂合的，并且基因型 *AaBbCc* 使得显性 *A*、*B* 和 *C* 性状得以表达。当 F_1 代个体作为亲本时，每个个体将等比例地产生八种不同的配子类型。此时，我们可以构建一个包含 64 个独立单元格的庞氏表来分析表型，但是该方法在涉及多因子杂交时比较麻烦。因此，另一种方法也被用来计算和预测比例。

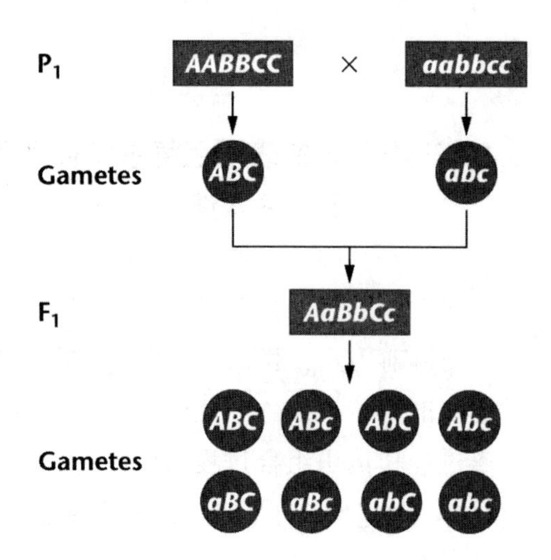

FIGURE 3.8 Formation of P_1 and F_1 gametes in a trihybrid cross.

图 3.8 三因子杂交中 P_1 代和 F_1 代的配子形成。

The Forked-Line Method, or Branch Diagram

It is much less difficult to consider each contrasting pair of traits separately and then to combine these results by using the **forked-line method**, first shown in Figure 3.6. This method, also called a **branch diagram**, relies on the simple application of the laws of probability established for the dihybrid cross. Each gene pair is assumed to behave independently during gamete formation.

When the monohybrid cross $AA \times aa$ is made, we know that:

1. All F_1 individuals have the genotype Aa and express the phenotype represented by the A allele, which is called the A phenotype in the discussion that follows.

2. The F_2 generation consists of individuals with either the A phenotype or the a phenotype in the ratio of 3:1.

The same generalizations can be made for the $BB \times bb$ and $CC \times cc$ crosses. Thus, in the F_2 generation, 3/4 of all organisms will express phenotype A, 3/4 will express B, and 3/4 will express C. Similarly, 1/4 of all organisms will express a, 1/4 will express b, and 1/4 will express c. The proportions of organisms that express each phenotypic combination can be predicted by assuming that fertilization, following the independent assortment of these three gene pairs during gamete formation, is a random process. We apply the product law of probabilities once again. Figure 3.9 uses the forked-line method to calculate the phenotypic proportions of the F_2 generation. They fall into the trihybrid ratio

分叉线法或分枝图

使用**分叉线法**需要先分别考虑每对相对性状，然后再整合这些结果进行分析计算，如图 3.6 所示。该方法也被称为**分枝图**，它依赖于概率定律在双因子杂交中的简单应用，并且假定每对基因在配子形成过程中的行为都是相互独立的。

当进行单因子杂交 $AA \times aa$ 时，我们知道：

1. 所有 F_1 代个体基因型均为 Aa，并表达由 A 等位基因所决定的表型，在以下讨论中我们将称之为 A 表型。

2. F_2 代由具有 A 表型或 a 表型的个体组成，并且比例为 3 ∶ 1。

$BB \times bb$ 和 $CC \times cc$ 杂交也可以进行同样的概括归纳。因此在 F_2 代中，3/4 的个体将表达 A 表型，3/4 的个体将表达 B 表型，并且 3/4 的个体将表达 C 表型。类似地，所有 F_2 个体的 1/4 将表达 a 表型，1/4 将表达 b 表型，且 1/4 将表达 c 表型。如果受精为随机过程，配子形成过程中三对基因遵循自由组合，那么即可预测表达每种表型组合的个体比例。我们再次应用概率相乘定律。图 3.9 使用分叉线法计算 F_2 代的表型比例。三因子杂交的 F_2 代表型比例为 27 ∶ 9 ∶ 9 ∶ 9 ∶ 3 ∶ 3 ∶ 3 ∶ 1。同样的方法也可用于分析涉及任意不同数量基因对的杂交实验，只要这些基因对在遗传过程中相互独立、互不影响。在随后章节中

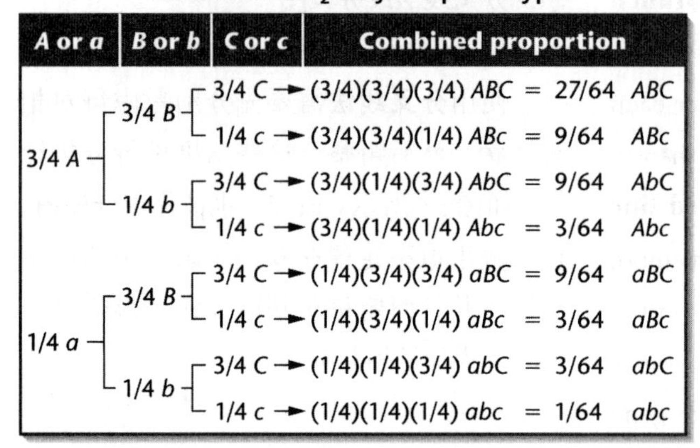

FIGURE 3.9 Generation of the F_2 trihybrid phenotypic ratio using the forked-line method. This method is based on the expected probability of occurrence of each phenotype.

图 3.9 使用分叉线法计算三因子杂交 F_2 代表型比例。该方法基于每种表型发生的预期概率。

of 27:9:9:9:3:3:3:1. The same method can be used to solve crosses involving any number of gene pairs, provided that all gene pairs assort independently from each other. We shall see later that gene pairs do not always assort with complete independence. However, it appeared to be true for all of Mendel's characters.

我们还将看到不同基因对间并不总是进行自由组合。然而，所有孟德尔性状都能自由组合。

ESSENTIAL POINT

The forked-line method is less complex than, but just as accurate as, the Punnett square in predicting the probabilities of phenotypes or genotypes from crosses involving two or more gene pairs.

基本要点

预测涉及两对或两对以上基因杂交实验的表型或基因型概率时，分叉线法与庞氏表同样精确，并且相对更为简便。

3.5 Mendel's Work Was Rediscovered in the Early Twentieth Century

3.5 孟德尔的工作在20世纪初被重新发现

Mendel published his work in 1866. While his findings were often cited and discussed, their significance went unappreciated for about 35 years. Then, in the latter part of the nineteenth century, a remarkable observation set the scene for the recognition of Mendel's work: Walther Flemming's discovery of chromosomes in the

孟德尔于1866年发表了他的研究成果。尽管他的发现常被引用和讨论，但其工作的重要性在随后的约35年中一直未能得到科学界的重视。19世纪下半叶的一项非凡发现为孟德尔的工作被认可奠定了基础：瓦尔特·弗莱明在蝾螈细胞的细胞核中发现了染色体。1879年，弗莱明描述了细胞

nuclei of salamander cells. In 1879, Flemming described the behavior of these thread-like structures during cell division. As a result of his findings and the work of many other cytologists, the presence of discrete units within the nucleus soon became an integral part of scientists' ideas about inheritance.

分裂过程中这些细线状结构的行为。根据他的发现和许多其他细胞生物学家的工作，细胞核中存在独立单元很快成为科学家关于遗传的相关思想中不可或缺的一部分。

In the early twentieth century, hybridization experiments similar to Mendel's were performed independently by three botanists, Hugo de Vries, Carl Correns, and Erich Tschermak. De Vries's work demonstrated the principle of segregation in several plant species. Apparently, he searched the existing literature and found that Mendel's work had anticipated his own conclusions! Correns and Tschermak also reached conclusions similar to those of Mendel.

在20世纪初，雨果·弗里斯、卡尔·科伦斯和埃里克·切尔马克三位植物学家分别独立地进行了与孟德尔杂交实验类似的研究工作。弗里斯在几种不同植物中均证明了分离定律。显然，他查询了已有文献，并发现孟德尔的工作已经准确预测了自己的实验结果！科伦斯和切尔马克同样也得到了与孟德尔类似的实验结论。

About the same time, two cytologists, Walter Sutton and Theodor Boveri, independently published papers linking their discoveries of the behavior of chromosomes during meiosis to the Mendelian principles of segregation and independent assortment. They pointed out that the separation of chromosomes during meiosis could serve as the cytological basis of these two postulates. Although they thought that Mendel's unit factors were probably chromosomes rather than genes on chromosomes, their findings reestablished the importance of Mendel's work and led to many ensuing genetic investigations. Sutton and Boveri are credited with initiating the **chromosome theory of inheritance**, the idea that the genetic material in living organisms is contained in chromosomes, which was developed during the

大约在同一时期，两位细胞生物学家沃尔特·萨顿和特奥多尔·博韦里分别发表论文将他们发现的减数分裂过程中染色体行为与孟德尔分离定律、自由组合定律相联系。他们指出减数分裂过程中染色体分离是这两条定律的细胞学基础。尽管他们认为孟德尔所说的单位因子可能是染色体，而不是染色体上的基因，但是他们的发现重新确立了孟德尔工作的重要性，并启发了随后的许多遗传研究。萨顿和博韦里被誉为创建了**遗传的染色体学说**，即生物体的遗传物质存在于染色体中，这一理论在此后20年间得到了进一步发展。正如我们在后续章节中将看到的那样，托马斯·H.摩尔根、艾尔弗雷德·H.斯特蒂文特、卡尔文·布里奇斯以及其他学者的工作毫无疑问地证实了萨顿和博韦里的假说是正确的。

next two decades. As we will see in subsequent chapters, work by Thomas H. Morgan, Alfred H. Sturtevant, Calvin Bridges, and others established beyond a reasonable doubt that Sutton's and Boveri's hypothesis was correct.

ESSENTIAL POINT

The discovery of chromosomes in the late 1800s, along with subsequent studies of their behavior during meiosis, led to the rebirth of Mendel's work, linking the behavior of his unit factors to that of chromosomes during meiosis.

基本要点

19 世纪后期染色体的发现以及随后对它们在减数分裂过程中行为的研究使得孟德尔的研究工作重见天日，并将孟德尔所描述的单位因子的行为与减数分裂过程中染色体的行为联系在了一起。

Unit Factors, Genes, and Homologous Chromosomes

Because the correlation between Sutton's and Boveri's observations and Mendelian postulates serves as the foundation for the modern description of transmission genetics, we will examine this correlation in some depth before moving on to other topics.

As we know, each species possesses a specific number of chromosomes in each somatic cell nucleus. For diploid organisms, this number is called the **diploid number** (**2*n***) and is characteristic of that species. During the formation of gametes (meiosis), the number is precisely halved (*n*), and when two gametes combine during fertilization, the diploid number is reestablished. During meiosis, however, the chromosome number is not reduced in a random manner. It was apparent to early cytologists that the diploid number of chromosomes is composed of homologous pairs identifiable by their morphological appearance and behavior.

单位因子、基因和同源染色体

由于萨顿和博韦里的发现与孟德尔假说之间的相关性构成了传递遗传学现代表述的基础，因此在继续探讨其他主题之前，我们将对此相关性展开进一步探讨。

众所周知，每个物种在其体细胞细胞核中都含有特定数目的染色体。对于二倍体生物来说，该数目被称为**二倍体数**（**2*n***），并且是该物种的特征。配子形成(减数分裂)过程中，染色体数目精确地减半（*n*）。当受精过程中两个配子相结合时，二倍体数目又将重建。然而，减数分裂过程中，染色体数目不会以随机的方式减少。早期细胞生物学家清楚地了解二倍染色体是由形态和行为均可识别的同源染色体对组成的。配子仅含每对同源染色体中的一条，因此配子的染色体组成非常明确，其所含染色体数目等于单倍体染色体数目。

The gametes contain one member of each pair—thus the chromosome complement of a gamete is quite specific, and the number of chromosomes in each gamete is equal to the haploid number.

With this basic information, we can correlate the behavior of unit factors and chromosomes and genes. Unit factors are really genes located on homologous pairs of chromosomes. Members of each pair of homologs separate, or segregate, during gamete formation.

利用这些基本信息，我们可以将单位因子、染色体、基因的行为相联系。单位因子实际上是位于同源染色体对上的基因。在配子形成过程中，每对同源染色体的成员会分开或分离。

To illustrate the principle of independent assortment, it is important to distinguish between members of any given homologous pair of chromosomes. One member of each pair is derived from the **maternal parent**, whereas the other comes from the **paternal parent**. Following independent segregation of each pair of homologs, each gamete receives one member from each pair of chromosomes. All possible combinations are formed with equal probability.

为了说明自由组合定律，首先需要区分任何指定同源染色体对中的两个成员。每对同源染色体对中，一个成员来自**母本**，而另一个成员则来自**父本**。每对同源染色体经独立分离后，每个配子将从每对染色体中获得一个成员，并且所有可能的组合等比例产生。

Observations of the phenotypic diversity of living organisms make it logical to assume that there are many more genes than chromosomes. Therefore, each homolog must carry genetic information for more than one trait. The currently accepted concept is that a chromosome is composed of a large number of linearly ordered, information-containing genes. Mendel's paired unit factors (which determine tall or dwarf stems, for example) actually constitute a pair of genes located on one pair of homologous chromosomes. The location on a given chromosome where any particular gene occurs is called its **locus** (pl. loci). The different

观察所见生物体表型的多元化，我们有理由认为生物体内所含的基因数目远比染色体数目多得多。因此，每条同源染色体必然携带不止一种性状的遗传信息。当前已被接受的观点是染色体由大量线性有序排列、富含遗传信息的基因组成。孟德尔所说的成对存在的单位因子（如决定植株高或矮的单位因子）实际上是位于同源染色体对上的一对基因。给定染色体上任何特定基因所在的位置被称为**基因座**。给定基因的不同等位基因形式（例如 *G* 和 *g*）所包含的遗传信息(绿色或黄色)略有不同，但都决定了同一特征（在这一例子中为种子颜色）。尽管目前我们所涉及的基因只

alleles of a given gene (for example, *G* and *g*) contain slightly different genetic information (green or yellow) that determines the same character (seed color in this case). Although we have examined only genes with two alternative alleles, most genes have more than two allelic forms. We conclude this section by reviewing the criteria necessary to classify two chromosomes as a homologous pair:

有两种可相互替代的等位基因，但是大多数基因都具有两种以上的等位基因形式。最后，我们通过回顾将两条染色体定义为同源染色体对所必需的标准来对本节进行总结：

1. During mitosis and meiosis, when chromosomes are visible in their characteristic shapes, both members of a homologous pair are the same size and exhibit identical centromere locations. The sex chromosomes (e.g., the X and the Y chromosomes in mammals) are an exception.

1. 在有丝分裂和减数分裂过程中，当染色体的特征形态清晰可见时，同源染色体对的两个成员染色体大小相同、着丝粒位置相同。性染色体（如哺乳动物的X染色体和Y染色体）则例外。

2. During early stages of meiosis, homologous chromosomes form pairs, or synapse.

2. 在减数分裂早期阶段，同源染色体进行配对，即发生联会。

3. Although it is not generally visible under the microscope, homologs contain the identical linear order of gene loci.

3. 尽管在显微镜下通常不可见，但同源染色体含有同样的线性排列的基因座顺序。

3.6 Independent Assortment Leads to Extensive Genetic Variation

3.6 自由组合产生广泛的遗传变异

One consequence of independent assortment is the production by an individual of genetically dissimilar gametes. Genetic variation results because the two members of any homologous pair of chromosomes are rarely, if ever, genetically identical. As the maternal and paternal members of all pairs are distributed to gametes through independent assortment, all possible chromosome combinations are produced, leading to extensive genetic diversity.

自由组合的结果之一就是个体产生大量遗传上不同的配子。由于任何同源染色体对的两个成员在遗传上都不可能完全相同，因此产生了遗传变异。当所有同源染色体对的母本来源成员和父本来源成员按照自由组合方式分配至配子中时，所有可能的染色体组合即会产生，从而形成广泛的遗传多样性。

We have seen that the number of possible gametes, each with different chromosome compositions, is 2^n, where n equals the haploid number. Thus, if a species has a haploid number of 4, then 2^4, or 16, different gamete combinations can be formed as a result of independent assortment. Although this number is not high, consider the human species, where $n = 23$. When 2^{23} is calculated, we find that in excess of 8×10^6, or over 8 million, different types of gametes are possible through independent assortment. Because fertilization represents an event involving only one of approximately 8×10^6 possible gametes from each of two parents, each offspring represents only one of $(8 \times 10^6)^2$ or one of only 64×10^{12} potential genetic combinations. Given that this probability is less than one in one trillion, it is no wonder that, except for identical twins, each member of the human species exhibits a distinctive set of traits—this number of combinations of chromosomes is far greater than the number of humans who have ever lived on Earth! Genetic variation resulting from independent assortment has been extremely important to the process of evolution in all sexually reproducing organisms.

我们已经明白可能形成的含有不同染色体组成的配子种类数为 2^n，其中 n 等于单倍染色体数目。因此，如果一个物种的单倍染色体数为 4，由于自由组合，则可以形成的不同配子种类数为 2^4 或 16。尽管该数值并不大，但是当考虑人类时，$n = 23$。通过计算 2^{23}，我们发现由于自由组合，单一个体将产生超过 8×10^6 或超过 800 万种不同组合类型的配子。由于受精仅代表涉及分别来自两个亲本约 8×10^6 种可能配子中的一个进行随机结合的过程，因此每个后代仅代表 $(8 \times 10^6)^2$ 或 64×10^{12} 种潜在遗传组合中的一个。鉴于这种可能性小于万亿分之一，因此毫无疑问，除了同卵双生的双胞胎之外，每个人类个体均将表现出各种独特的性状。这一染色体组合数目远远超过曾经在地球上生活过的人类总数！源于自由组合所产生的遗传变异对于所有有性繁殖生物的进化过程都具有极为重要的意义。

3.7 Laws of Probability Help to Explain Genetic Events

3.7 概率理论有助于解释遗传学事件

Recall that genetic ratios—for example, 3/4 tall:1/4 dwarf— are most properly thought of as probabilities. These values predict the outcome of each fertilization event, such that the probability of each zygote having the genetic

之前所介绍的遗传比“3/4 植株高：1/4 植株矮”被理解为概率最为合适。这些数值预测了每个受精事件的结果，即每个合子含有的遗传信息决定植株高的概率为 3/4，而决定植株矮的概率为 1/4。概率的数

potential for becoming tall is 3/4, whereas the potential for its being a dwarf is 1/4. Probabilities range from 0.0, where an event is certain not to occur, to 1.0, where an event is certain to occur. In this section, we consider the relation of probability to genetics. When two or more events with known probabilities occur independently but at the same time, we can calculate the probability of their possible outcomes occurring together. This is accomplished by applying the **product law**, which states that the probability of two or more independent events occurring simultaneously is equal to the product of their individual probabilities (see Section 3.3). Two or more events are independent of one another if the outcome of each one does not affect the outcome of any of the others under consideration.

值范围从 0.0（事件肯定不发生）到 1.0（事件肯定发生）。在本节中，我们将考察概率与遗传学之间的关系。当两个或多个已知概率的事件同时发生且相互独立时，我们可以计算它们同时发生的概率。该计算将应用**相乘定律**。相乘定律指两个或多个独立事件同时发生的概率等于各事件单独发生概率的乘积（请参见 3.3 节）。如果每个事件的发生结果不影响同时考察的其他事件的发生结果，那么这两个或多个事件即彼此独立。

To illustrate the product law, consider the possible results if you toss a penny (P) and a nickel (N) at the same time and examine all combinations of heads (H) and tails (T) that can occur. There are four possible outcomes:

$(P_H : N_H) = (1/2)(1/2) = 1/4$

$(P_T : N_H) = (1/2)(1/2) = 1/4$

$(P_H : N_T) = (1/2)(1/2) = 1/4$

$(P_T : N_T) = (1/2)(1/2) = 1/4$

The probability of obtaining a head or a tail in the toss of either coin is 1/2 and is unrelated to the outcome for the other coin. Thus, all four possible combinations are predicted to occur with equal probability.

为了说明相乘定律，我们可以举抛硬币的例子。如果我们同时抛两枚硬币，一枚一分硬币（P）和一枚一角硬币（N），并观察所有可能出现的正面（H）和背面（T）的组合，那么将有四种可能结果：

$(P_H : N_H) = (1/2)(1/2) = 1/4$

$(P_T : N_H) = (1/2)(1/2) = 1/4$

$(P_H : N_T) = (1/2)(1/2) = 1/4$

$(P_T : N_T) = (1/2)(1/2) = 1/4$

其中任何一枚硬币在抛掷中得到正面或反面的概率均为 1/2，并且与另一枚硬币的抛掷结果无关。因此，我们可以预测所有四种可能组合将等比例发生。

If we want to calculate the probability when the possible outcomes of two events are independent of one another but can be

如果两个独立事件可能发生的方式不止一种，那么计算它们发生的概率将应用**相加定律**。例如，如果我们抛掷两枚硬币，

accomplished in more than one way, we can apply the **sum law**. For example, what is the probability of tossing our penny and nickel and obtaining one head and one tail? In such a case, we do not care whether it is the penny or the nickel that comes up heads, provided that the other coin has the alternative outcome. As we saw above, there are two ways in which the desired outcome can be accomplished, each with a probability of 1/4. The sum law states that the probability of obtaining any single outcome, where that outcome can be achieved by two or more events, is equal to the sum of the individual probabilities of all such events. Thus, according to the sum law, the overall probability in our example is equal to

$$(1/4) + (1/4) = 1/2$$

One-half of all two-coin tosses are predicted to yield the desired outcome.

These simple probability laws will be useful throughout our discussions of transmission genetics and for solving genetics problems. In fact, we already applied the product law when we used the forked-line method to calculate the phenotypic results of Mendel's dihybrid and trihybrid crosses. When we wish to know the results of a cross, we need only calculate the probability of each possible outcome. The results of this calculation then allow us to predict the proportion of offspring expressing each phenotype or each genotype.

An important point to remember when you deal with probability is that predictions of possible outcomes are based on large sample sizes. If we predict that 9/16 of the offspring of

一枚一分硬币和一枚一角硬币，同时得到一个正面和一个背面的概率是多少？在这种情况下，我们不必考虑显示正面的是分币还是角币，只需另一枚硬币显示的是背面即可。正如我们之前所描述的，共有两种情形可以满足题目所需，每种情形的概率都是 1/4。相加定律指通过两种或多种途径均可实现的事件的概率等于每种途径实现概率的加和。因此，根据相加定律，该例中“同时获得一个正面和一个背面”的总概率等于

$$(1/4) + (1/4) = 1/2$$

即预计所有两枚硬币同时投掷的事件中有一半将产生题目所要求达到的结果。

这些简单的概率定律在我们讨论传递遗传学以及解决遗传学问题的整个过程中都将十分有用。实际上，当使用分叉线法计算孟德尔双因子杂交和三因子杂交的表型结果时，我们就已经在应用相乘定律。当想要了解某个杂交实验的结果时，我们只需计算每种可能结果的概率即可。该计算结果将帮助我们预测后代中表达每种表型或每种基因型所占的比例。

在进行概率计算时有一点十分重要并请牢记，那就是利用概率预测可能的结果时需要基于大量的样本数量。如果我们预测双因子杂交后代中有 9/16 同时表达两种

a dihybrid cross will express both dominant traits, it is very unlikely that, in a small sample, exactly 9 of every 16 will express this phenotype. Instead, our prediction is that, of a large number of offspring, approximately 9/16 will do so. The deviation from the predicted ratio in smaller sample sizes is attributed to chance, a subject we examine in our discussion of statistics in Section 3.8. As you shall see, the impact of deviation due strictly to chance diminishes as the sample size increases.

显性性状，那么在一个小样本体系中，不可能出现每 16 个个体中有 9 个呈现双显表型的情形。相反，我们的预测是在足够大量的后代中大约有 9/16 的个体将是这样。在较小的样本系统中出现的这种与预测比例的偏差归因于偶然性，这个话题我们将在 3.8 节统计学的相关内容中进行探讨。正如你将看到的，这种严格意义上由偶然性引起的偏差的影响将随着样本量的增加而减小。

ESSENTIAL POINT

Since genetic ratios are expressed as probabilities, deriving outcomes of genetic crosses requires an understanding of the laws of probability.

基本要点

由于遗传比可以以概率的方式表示，因此在遗传杂交实验的结果推导中需要了解概率定律。

3.8 Chi-Square Analysis Evaluates the Influence of Chance on Genetic Data

3.8 卡方分析评估偶然性对于遗传数据的影响

Mendel's 3:1 monohybrid and 9:3:3:1 dihybrid ratios are hypothetical predictions based on the following assumptions: (1) each allele is dominant or recessive, (2) segregation is unimpeded, (3) independent assortment occurs, and (4) fertilization is random. The final two assumptions are influenced by chance events and therefore are subject to random fluctuation. This concept of **chance deviation** is most easily illustrated by tossing a single coin numerous times and recording the number of heads and tails observed. In each toss, there is a probability of 1/2 that a head will occur and a probability of 1/2 that a tail will occur. Therefore, the expected

孟德尔的 3 ∶ 1 单因子杂种率和 9 ∶ 3 ∶ 3 ∶ 1 双因子杂种率均是基于以下假设进行的预测：（1）每个等位基因或者是显性的或者是隐性的；（2）分离顺利进行，不受阻碍；（3）发生自由组合；（4）受精过程随机发生。最后两个假设由于受偶然事件的影响，因此可能发生随机波动。多次抛掷单枚硬币并记录所观察到的正面和反面的数量，这一例子可以非常简单地说明**机会偏差**的概念。在每次抛掷中，呈现正面的概率为 1/2，呈现背面的概率也为 1/2。因此，多次投掷过程中呈现正面和背面的预期比例为 1/2 ∶ 1/2 或 1 ∶ 1。如果将硬币抛掷 1000 次，通常将观察到大约

ratio of many tosses is 1/2:1/2, or 1:1. If a coin is tossed 1000 times, usually about 500 heads and 500 tails will be observed. Any reasonable fluctuation from this hypothetical ratio (e.g., 486 heads and 514 tails) is attributed to chance.

500 次正面和 500 次背面。该假设比例的任何合理波动 (例如 486 次正面和 514 次背面) 均归因于偶然性。

As the total number of tosses is reduced, the impact of chance deviation increases. For example, if a coin is tossed only four times, you would not be too surprised if all four tosses resulted in only heads or only tails. For 1000 tosses, however, 1000 heads or 1000 tails would be most unexpected. In fact, you might believe that such a result would be impossible. Actually, all heads or all tails in 1000 tosses can be predicted to occur with a probability of $(1/2)^{1000}$. Since $(1/2)^{20}$ is less than one in a million times, an event occurring with a probability as small as $(1/2)^{1000}$ is virtually impossible. Two major points to keep in mind when predicting or analyzing genetic outcomes are:

随着投掷总次数的减少，机会偏差的影响将随之增加。例如，如果只抛掷 4 次硬币，那么所有 4 次全为正面或全为反面，你也不会感到惊讶。然而，对于 1000 次抛掷来说，1000 次正面或 1000 次背面将是十分令人意外的。实际上，你可能认为这样的结果绝不可能发生。事实上，可以预测 1000 次抛掷中全部呈现为正面或全为背面的发生概率仅为 $(1/2)^{1000}$。由于 $(1/2)^{20}$ 小于百万分之一，因此发生概率为 $(1/2)^{1000}$ 的事件实际上即为不可能发生事件。预测或分析遗传结果时请务必记住以下两个关键点：

1. The outcomes of independent assortment and fertilization, like coin tossing, are subject to random fluctuations from their predicted occurrences as a result of chance deviation.

1. 自由组合和受精的结果就如同抛掷硬币，由于机会偏差的存在，可能导致偏离预测事件而发生随机波动。

2. As the sample size increases, the average deviation from the expected results decreases. Therefore, a larger sample size diminishes the impact of chance deviation on the final outcome.

2. 随着样本量的增加，与预期结果的平均偏差将会减小。因此，较大的样本量将减少机会偏差对于最终结果的影响。

Chi-Square Calculations and the Null Hypothesis

卡方计算和零假设

In genetics, being able to evaluate observed deviation is a crucial skill. When we assume that data will fit a given ratio such as 1:1, 3:1, or 9:3:3:1, we establish what is called the **null**

在遗传学中，能够评估观测偏差是一项至关重要的技能。当我们假设数据符合某给定比例（例如 1 ∶ 1，3 ∶ 1 或者 9 ∶ 3 ∶ 3 ∶ 1）时，那么我们即建立了**零**

hypothesis (H_0). It is so named because the hypothesis assumes that there is no real difference between the measured values (or ratio) and the predicted values (or ratio). Any apparent difference can be attributed purely to chance. The validity of the null hypothesis for a given set of data is measured using statistical analysis. Depending on the results of this analysis, the null hypothesis may either (1) be rejected or (2) fail to be rejected. If it is rejected, the observed deviation from the expected result is judged not to be attributable to chance alone. In this case, the null hypothesis and the underlying assumptions leading to it must be reexamined. If the null hypothesis fails to be rejected, any observed deviations are attributed to chance.

假设（H_0）。如此命名是因为该假设认为测量值（或测量比例）与预测值（或预测比例）之间不存在实际差异。任何明显的差异都可以纯粹归因于偶然性。对于给定数据的有效性，零假设可以通过统计分析来进行衡量。根据此分析结果，零假设可能 (1) 被推翻或 (2) 不被推翻。如果被推翻，则认为观察到的与预期结果的偏差并非仅归因于偶然性。在这种情况下，零假设及其所基于的假设均必须被重新考虑。如果零假设不能被推翻，那么任何观察到的偏差均归因于偶然性。

One of the simplest statistical tests for assessing the goodness of fit of the null hypothesis is **chi-square** (**χ^2**) **analysis**. This test takes into account the observed deviation in each component of a ratio (from what was expected) as well as the sample size and reduces them to a single numerical value. The value for χ^2 is then used to estimate how frequently the observed deviation can be expected to occur strictly as a result of chance. The formula used in chi-square analysis is

$$\chi^2 = \Sigma \frac{(o-e)^2}{e}$$

where o is the observed value for a given category, e is the expected value for that category, and Σ (the Greek letter sigma) represents the sum of the calculated values for each category in the ratio. Because $(o-e)$ is the deviation (d) in each case, the equation reduces to

评估零假设适合度的一种最简单的统计学检验方法是**卡方**（χ^2）**分析**。该方法不仅考虑预期比例中每个组分的观测与预期结果间的偏差，而且同时考虑样本量的大小，并将这些因素简化为单个数值。χ^2 数值用于评价仅由于偶然性造成的观测偏差的频率。卡方分析中使用的公式为

$$\chi^2 = \Sigma \frac{(o-e)^2}{e}$$

其中，o 是指定类别的观测值，e 是该类别的预期值，Σ（希腊字母西格玛）表示比例中每个类别的计算值之和。因为 $(o-e)$ 是每种类别的偏差（d），所以该公式可以简化为

$$\chi^2 = \Sigma \frac{d^2}{e}$$

Table 3.1(a) shows the steps in the χ^2 calculation for the F_2 results of a hypothetical monohybrid cross. To analyze the data obtained from this cross, work from left to right across the table, verifying the calculations as appropriate. Note that regardless of whether the deviation d is positive or negative, d^2 always becomes positive after the number is squared. In Table 3.1(b) F_2 results of a hypothetical dihybrid cross are analyzed. Make sure that you understand how each number was calculated in this example.

$$\chi^2 = \Sigma \frac{d^2}{e}$$

表 3.1（a）列出了假设的单因子杂交中 F_2 代 χ^2 值计算的步骤。为了分析该杂交实验所得数据，请从表格左侧至右侧进行操作，并对计算结果进行验证。请注意，无论偏差 d 是正数还是负数，对 d 求平方后，d^2 必定变为正数。在表 3.1（b）中分析了假设的双因子杂交 F_2 代结果。请确保你能够理解该示例中每个数字是如何计算获得的。

TABLE 3.1 Chi-Square Analysis

表 3.1 卡方分析

(a) Monohybrid Cross Expected Ratio	**Observed (*o*)**	**Expected (*e*)**	**Deviation (*o*–*e*=*d*)**	**Deviation²**	***d*²/*e***
3/4	740	3/4(1000)=750	740 – 750=–10	$(-10)^2$=100	100/750 = 0.13
1/4	260	1/4(1000)=250	260 – 250=+10	$(+10)^2$=100	100/250 = 0.40
	Total=1000				χ^2=0.53
					p=0.48
(b) Dihybrid Cross Expected Ratio	**Observed (*o*)**	**Expected (*e*)**	**Deviation (*o*–*e*=*d*)**	**Deviation²**	***d*²/*e***
9/16	587	567	+20	400	0.71
3/16	197	189	+8	64	0.34
3/16	168	189	–21	441	2.33
1/16	56	63	–7	49	0.78
	Total=1008				χ^2=4.16
					p=0.26

The final step in chi-square analysis is to interpret the χ^2 value. To do so, you must initially determine a value called the **degrees of freedom** (***df***), which is equal to $n - 1$, where n is the number of different categories into which the data are divided, in other words, the number of possible outcomes. For the 3:1 ratio, n=2,

卡方分析的最后一步是诠释 χ^2 值。为此，你必须首先确定**自由度**（***df***）的数值，该数值等于 $n-1$，其中 n 是数据被划分为不同类别的总类别数，换句话说，是可能结果的总类别数。对于比例 3 ∶ 1，n=2，因此 df=1；对于比例 9 ∶ 3 ∶ 3 ∶ 1，n=4，因此 df=3。类别数目越大，由于偶然性的

so df=1. For the 9:3:3:1 ratio, n=4 and df=3. Degrees of freedom must be taken into account because the greater the number of categories, the more deviation is expected as a result of chance.

存在，偏差将越大，所以必须考虑自由度。

Once you have determined the degrees of freedom, you can interpret the χ^2 value in terms of a corresponding **probability value (*p*)**. Since this calculation is complex, we usually take the p value from a standard table or graph. Figure 3.10 shows a wide range of χ^2 values and the corresponding p values for various degrees of freedom in both a graph and a table. Let's use the graph to explain how to determine the p value. The caption for Figure 3.10(b) explains how to

一旦确定自由度后，就可以根据相应的**概率值**（***p***）来诠释 χ^2 值。由于该计算复杂，因此我们通常从标准表或标准图中获取 p 值。图 3.10 显示了在标准图和标准表的一定范围内，χ^2 值在不同自由度下所对应的 p 值。我们利用标准图来说明如何确定 p 值。图 3.10（b）的图注解释了如何使用该表。

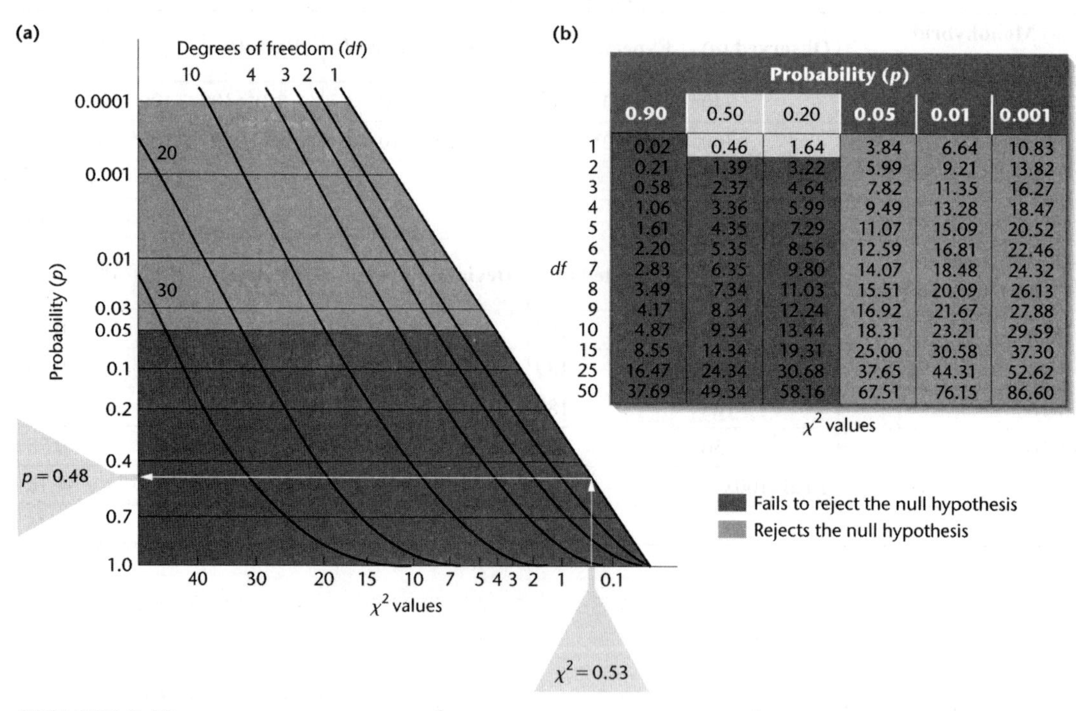

	Probability (p)					
df	0.90	0.50	0.20	0.05	0.01	0.001
1	0.02	0.46	1.64	3.84	6.64	10.83
2	0.21	1.39	3.22	5.99	9.21	13.82
3	0.58	2.37	4.64	7.82	11.35	16.27
4	1.06	3.36	5.99	9.49	13.28	18.47
5	1.61	4.35	7.29	11.07	15.09	20.52
6	2.20	5.35	8.56	12.59	16.81	22.46
7	2.83	6.35	9.80	14.07	18.48	24.32
8	3.49	7.34	11.03	15.51	20.09	26.13
9	4.17	8.34	12.24	16.92	21.67	27.88
10	4.87	9.34	13.44	18.31	23.21	29.59
15	8.55	14.34	19.31	25.00	30.58	37.30
25	16.47	24.34	30.68	37.65	44.31	52.62
50	37.69	49.34	58.16	67.51	76.15	86.60

χ^2 values

FIGURE 3.10 (a) Graph for converting χ^2 values to p values. (b) Table of χ^2 values for selected values of df and p. χ^2 values that lead to a p value of 0.05 or greater (darker areas) justify failure to reject the null hypothesis. Values leading to a p value of less than 0.05 (lighter areas) justify rejecting the null hypothesis. For example, the table in part (b) shows that for χ^2=0.53 with 1 degree of freedom, the corresponding p value is between 0.20 and 0.50. The graph in (a) gives a more precise p value of 0.48 by interpolation. Thus, we fail to reject the null hypothesis.

图 3.10 （a）χ^2 值与 p 值转换图。（b）特定 df 值和 p 值所对应的 χ^2 值表。由 χ^2 值得出的 p 值如果是 0.05 或者大于 0.05（深色区域）则说明无法推翻零假设。得出的 p 值如果小于 0.05（浅色区域）则说明应该推翻零假设。例如，图中（b）表格显示 χ^2 = 0.53，自由度为 1 时，相应的 p 值介于 0.20 和 0.50 之间。（a）图则利用插值给出更为精确的 p 值，为 0.48。因此，我们无法推翻零假设。

use the table.

To determine p using the graph, execute the following steps:

1. Locate the χ^2 value on the abscissa (the horizontal axis, or x-axis).

2. Draw a vertical line from this point up to the line on the graph representing the appropriate *df*.

3. From there, extend a horizontal line to the left until it intersects the ordinate (the vertical axis, or y-axis).

4. Estimate, by interpolation, the corresponding p value.

利用标准图确定 p 值，请按照以下步骤进行：

1. 在横坐标（水平轴或 x 轴）上找到 χ^2 值。

2. 从该点绘制一条垂直于横坐标的直线，直至与图中相应 *df* 的曲线相交为止。

3. 从交点开始，向左延伸绘制一条水平线，直至与纵坐标（垂直轴或 y 轴）相交。

4. 通过插值估计相应的 p 值。

We used these steps for the monohybrid cross in Table 3.1(a) to estimate the p value of 0.48, as shown in Figure 3.10(a). Now try this method to see if you can determine the p value for the dihybrid cross [Table 3.1(b)]. Since the χ^2 value is 4.16 and $df = 3$, an approximate p value is 0.26. Checking this result in the table confirms that p values for both the monohybrid and dihybrid crosses are between 0.20 and 0.50.

我们利用这些步骤对表 3.1（a）中的单因子杂交预测 p 值，如图 3.10（a）所示所估 p 值为 0.48。现在请你尝试用该方法来确定表 3.1（b）中双因子杂交的 p 值。由于 χ^2 值为 4.16，并且 $df = 3$，因此近似的 p 值为 0.26。由标准表确认这里的单因子杂交和双因子杂交的 p 值都在 0.20 和 0.50 之间。

Interpreting Probability Values

So far, we have been concerned with calculating χ^2 values and determining the corresponding p values. These steps bring us to the most important aspect of chi-square analysis: understanding the meaning of the p value. It is simplest to think of the p value as a percentage. Let's use the example of the dihybrid cross in Table 3.1(b) where $p = 0.26$, which can be thought of as 26 percent. In our example, the p value indicates that if we repeat the same

诠释概率值

至此，我们一直关注 χ^2 值的计算以及确定相应的 p 值。这些步骤将我们带到了卡方分析中最为重要的部分：理解 p 值的含义。最简单的理解是将 p 值视为百分比。让我们以表 3.1（b）中的双因子杂交为例，其中 $p = 0.26$，可以理解为 26%。在我们的示例中，p 值表示如果我们多次重复同一实验，预计其中 26% 的重复实验将呈现出与初始实验相同或比之更大的机会偏差。与此相反，由于偶然性的存在，74% 的重复实

experiment many times, 26 percent of the trials would be expected to exhibit chance deviation as great as or greater than that seen in the initial trial. Conversely, 74 percent of the repeats would show less deviation than initially observed as a result of chance. Thus, the *p* value reveals that a null hypothesis (concerning the 9:3:3:1 ratio, in this case) is never proved or disproved absolutely. Instead, a relative standard is set that we use to either reject or fail to reject the null hypothesis. This standard is most often a *p* value of 0.05. When applied to chi-square analysis, a *p* value less than 0.05 means that the observed deviation in the set of results will be obtained by chance alone less than 5 percent of the time. Such a *p* value indicates that the difference between the observed and predicted results is substantial and requires us to reject the null hypothesis.

验呈现的偏差要比初始实验中所观察到的少。因此，*p* 值表明所设的零假设（该示例中为9∶3∶3∶1比例）不能被绝对否定。相对标准的设置有助于我们用来权衡拒绝或接受零假设，该标准通常设 *p* 值为0.05。当应用于卡方分析时，*p* 值小于0.05意味着仅有不到5%的事件会由于偶然性的存在而出现所得结果中观察到的偏差。这样的 *p* 值意味着观测值与预测值之间的差异巨大，因此我们应该否定原零假设。

On the other hand, *p* values of 0.05 or greater (0.05 to 1.0) indicate that the observed deviation will be obtained by chance alone 5 percent or more of the time. This conclusion allows us not to reject the null hypothesis (when we are using $p = 0.05$ as our standard). Thus, with its *p* value of 0.26, the null hypothesis that independent assortment accounts for the results fails to be rejected. Therefore, the observed deviation can be reasonably attributed to chance.

另一方面，当我们使用 $p = 0.05$ 作为标准时，如果 *p* 值大于或等于0.05（0.05—1.0）时，则表明由于偶然性的存在5%或更多的事件将出现类似的观测偏差，该结论使我们不能推翻该零假设。所以 *p* 值为0.26时，自由组合的零假设无法被否定。因此，观测偏差可以合理地归因于偶然性。

A final note is relevant here concerning the case where the null hypothesis is rejected, that is, where $p \leqslant 0.05$. Suppose we had tested a dataset to assess a possible 9:3:3:1 ratio, as in Table 3.1(b), but we rejected the null hypothesis based on our calculation. What are alternative interpretations of the data? Researchers will

最后在这里考虑与零假设被推翻的相关情况，即 $p \leqslant 0.05$。假设我们检验一组数据以评估可能的9∶3∶3∶1比例，如表3.1（b）所示，但是基于计算我们推翻了该零假设，那么该如何解读这组数据呢？研究人员将会重新评估零假设所基于的假定。在双因子杂交中，我们假定两对基因均进

reassess the assumptions that underlie the null hypothesis. In our dyhibrid cross, we assumed that segregation operates faithfully for both gene pairs. We also assumed that fertilization is random and that the viability of all gametes is equal regardless of genotype— that is, all gametes are equally likely to participate in fertilization. Finally, we assumed that, following fertilization, all preadult stages and adult offspring are equally viable, regardless of their genotype. If any of these assumptions is incorrect, then the original hypothesis is not necessarily invalid.

行准确无误的分离过程。我们同时还假定受精随机，并且不管基因型如何，所有配子的生存力是相等的，也就是说，所有配子参与受精过程的概率相等。最后，我们假定受精后，无论个体基因型如何，所有后代个体在成年前阶段及成年后都具有同等的生存能力。如果这些假定中任何一个不成立，那么原假设就不一定无效。

An example will clarify this point. Suppose our null hypothesis is that a dihybrid cross between fruit flies will result in 3/16 mutant wingless flies. However, perhaps fewer of the mutant embryos are able to survive their preadult development or young adulthood compared to flies whose genotype gives rise to wings. As a result, when the data are gathered, there will be fewer than 3/16 wingless flies. Rejection of the null hypothesis is not in itself cause for us to reject the validity of the postulates of segregation and independent assortment, because other factors we are unaware of may also be affecting the outcome.

一个例子可以阐明这一点。假设依据我们的零假设，果蝇双因子杂交将产生 3/16 突变体无翅果蝇。但是，与有翅基因型果蝇胚胎相比，也许突变体胚胎顺利存活至成年前或成年早期的数量更少。因此，收集数据时，无翅果蝇将少于 3/16。拒绝接受零假设究其本质并非让我们否定分离和自由组合假说的有效性，因为可能同时存在我们尚未可知的其他因素影响实验结果。

ESSENTIAL POINT

Chi-square analysis allows us to assess the null hypothesis, which states that there is no real difference between the expected and observed values. As such, it tests the probability of whether observed variations can be attributed to chance deviation.

基本要点

卡方分析使我们能够评估零假设，即由该假设得到的预期值和观察值之间没有实际差异。就此而言，它对于观测偏差是否能够归因于机会偏差进行了评估。

3.9 Pedigrees Reveal Patterns of Inheritance of Human Traits

We now explore how to determine the mode of inheritance of phenotypes in humans, where experimental matings are not made and where relatively few offspring are available for study. The traditional way to study inheritance has been to construct a family tree, indicating the presence or absence of the trait in question for each member of each generation. Such a family tree is called a **pedigree**. By analyzing a pedigree, we may be able to predict how the trait under study is inherited—for example, is it due to a dominant or recessive allele? When many pedigrees for the same trait are studied, we can often ascertain the mode of inheritance.

Pedigree Conventions

Figure 3.11 illustrates some of the conventions geneticists follow in constructing pedigrees. Circles represent females and squares designate males. If the sex of an individual is unknown, a diamond is used. Parents are generally connected to each other by a single horizontal line, and vertical lines lead to their offspring. If the parents are related—that is, **consanguineous**—such as first cousins, they are connected by a double line. Offspring are called **sibs** (short for **siblings**) and are connected by a horizontal **sibship line**. Sibs are placed in birth order from left to right and are labeled with Arabic numerals. Parents also receive an Arabic number designation. Each generation is indicated by a Roman numeral. When a pedigree

3.9 系谱图揭示了人类性状的遗传模式

现在，我们开始探索如何确定人类表型的遗传模式，在此过程中无法进行交配实验，并且可用于研究的后代数量相对较少。研究人类遗传的传统方法是构建家谱树，以展示每一世代中每一成员是否表现出所研究的性状。这样的家谱树也被称为**系谱**。通过分析系谱，我们可以预测所研究性状的遗传方式，例如，它是由显性等位基因决定还是由隐性等位基因决定？当同时研究涉及同一性状的多个系谱时，我们通常还可以确定该性状的遗传模式。

系谱惯例

图3.11展示了遗传学家在构建系谱时所遵循的一些惯例。圆形代表女性，正方形代表男性。如果个体性别未知，则使用菱形表示。父母之间通常通过一条单一水平线段相连，并使用垂直线段与他们的后代相连。如果父亲与母亲有亲缘关系，例如表兄妹关系，即为**近亲**，则使用两条平行线段将他们相连。后代个体之间被称为**同胞**，通过水平的**同胞线**相连。兄弟姐妹按照出生顺序从左到右进行排列，并用阿拉伯数字标记顺序。父母也使用阿拉伯数字进行顺序标记。每一世代使用罗马数字标记顺序。当一个系谱仅研究单一性状时，如果携带该性状，则表示相应个体的圆形、正方形和菱形被阴影覆盖，否则表示为非阴影。在某些系谱中，那些未表现出某隐

traces only a single trait, the circles, squares, and diamonds are shaded if the phenotype being considered is expressed and unshaded if not. In some pedigrees, those individuals that fail to express a recessive trait but are known with certainty to be heterozygous carriers have a shaded dot within their unshaded circle or square. If an individual is deceased and the phenotype is unknown, a diagonal line is placed over the circle or square.

性性状但可以确定其为杂合子的携带者，在相应个体的圆形或正方形内使用阴影点标识。如果个体已经去世并且表型未知，则在相应个体的圆形或正方形上画对角线表示。

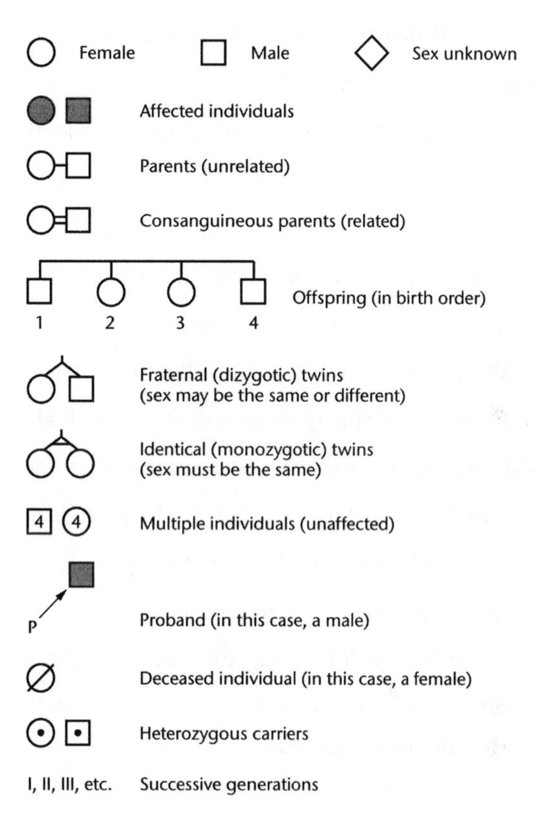

FIGURE 3.11 Conventions commonly encountered in human pedigrees.

图 3.11 人类系谱中常用的惯例符号。

Twins are indicated by diagonal lines stemming from a vertical line connected to the sibship line. For identical, or **monozygotic**, twins, the diagonal lines are linked by a horizontal line. Fraternal, or **dizygotic**, twins lack this connecting line. A number within one of the symbols represents that number of sibs of the same sex and of the same or unknown phenotypes. The individual whose phenotype first brought attention to the family is called the **proband** and is indicated by an arrow connected to the designation p. This term applies to either a male or a female.

双胞胎由源自一条同胞垂直线段的两条共顶点斜线表示。对于**同卵双生**双胞胎，这两条共顶点斜线之间用一条水平线段相连；**双卵双生**双胞胎则没有此连接线。圆形、正方形或菱形符号内用数字表示性别相同、表型相同或表型未知的同胞数。该家族中首个携带关注性状的个体被称为**先证者**，在系谱中用与字母 p 相连的箭头进行标识，该术语同时适用于男性或女性。

Pedigree Analysis

In Figure 3.12, two pedigrees are shown. The first is a representative pedigree for a trait that demonstrates autosomal recessive inheritance, such as **albinism**, where synthesis of the pigment melanin in obstructed. The male parent of the first generation (I-1) is affected. Characteristic of a situation in which a parent has a rare recessive trait, the trait "disappears" in the offspring of the next generation. Assuming recessiveness, we might predict that the unaffected female parent (I-2) is a homozygous normal individual because none of the offspring show the disorder. Had she been heterozygous, one-half of the offspring would be expected to exhibit albinism, but none do. However, such a small sample (three offspring) prevents our knowing for certain.

系谱分析

图 3.12 展示了两张系谱。第一张是常染色体隐性遗传性状的代表性系谱，例如由于黑色素合成受阻所导致的**白化病**。第一世代的男性亲本（I-1）是患病个体。亲本携带罕见隐性性状的遗传特点是该性状在其下一世代后代中可能“消失”。假设该性状为隐性，我们可以预测未患病的女性亲本（I-2）是纯合正常个体，因为其后代无一表现出这种疾病。假如她是杂合子，那么将有一半的后代表现白化病，但系谱中显示无患病后代。然而，如此小的样本量（仅有 3 个后代）也使我们很难确定具体情况。

Further evidence supports the prediction of a recessive trait. If albinism were inherited

进一步证据可用于支持隐性性状的相关预测。如果白化病以显性性状进行遗传，

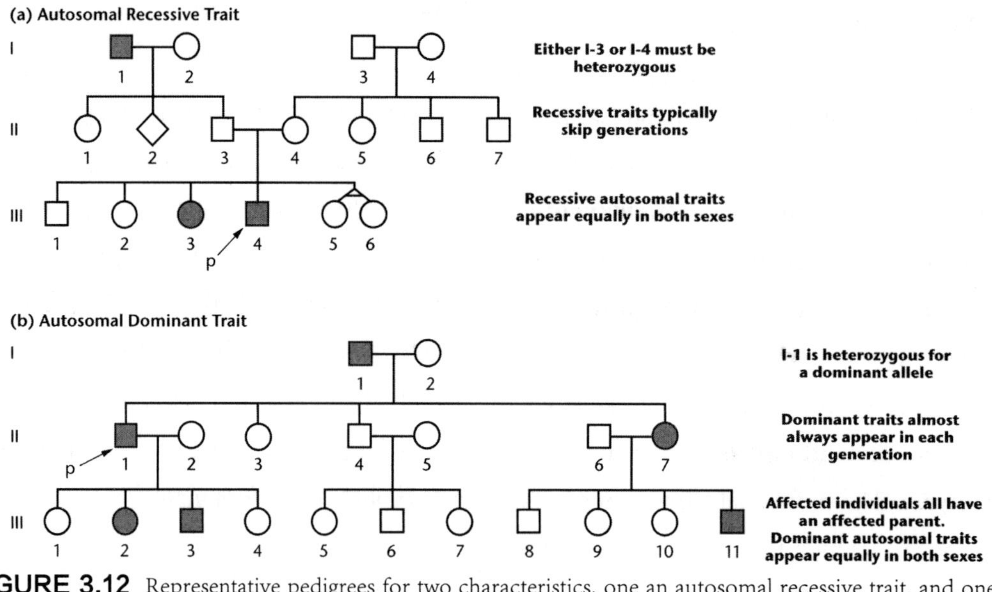

FIGURE 3.12 Representative pedigrees for two characteristics, one an autosomal recessive trait, and one an autosomal dominant trait, both followed through three generations.

图 3.12 两张记录三个世代的系谱，分别涉及两种性状（一个是常染色体隐性性状，另一个是常染色体显性性状）。

as a dominant trait, individual II-3 would have to express the disorder in order to pass it to his offspring (III-3 and III-4), but he does not. Inspection of the offspring constituting the third generation (row III) provides still further support for the hypothesis that albinism is a recessive trait. If it is, parents II-3 and II-4 are both heterozygous, and approximately one-fourth of their offspring should be affected. Two of the six offspring do show albinism. This deviation from the expected ratio is not unexpected in crosses with few offspring. Once we are confident that albinism is inherited as an autosomal recessive trait, we could portray the II-3 and II-4 individuals with a shaded dot within their larger square and circle. Finally, we can note that, characteristic of pedigrees for autosomal traits, both males and females are affected with equal probability. Later in the text (see Chapter 4), we will examine a pedigree representing a gene located on the sex-determining X chromosome. We will see certain patterns characteristic of the transmission of X-linked traits, such as that these traits are more prevalent in male offspring and are never passed from affected fathers to their sons.

那么个体 II-3 必定表现这种疾病才能将其传给后代（III-3 和 III-4），但系谱显示其并未患病。研究第三世代（第 III 行）个体也为“白化病是一种隐性性状”这一假设提供了进一步支持。如果假设正确，那么亲本 II-3 和 II-4 都是杂合子，其后代中约有 1/4 的个体将患病。系谱显示其 6 个后代中确有 2 个患白化病。这种与预期比例存在偏离的情况出现在后代很少的杂交实验中并不意外。一旦我们确定白化病是一种常染色体隐性遗传疾病，那么我们就可以在表示 II-3 和 II-4 个体的正方形和圆形中用阴影点进行标识。最后，我们还会注意到，常染色体性状系谱的另一特征是男性和女性具有同等的患病概率。在随后章节（请参阅第 4 章）中，我们将研究位于参与性别决定的 X 染色体上的基因系谱。我们将看到 X 连锁性状遗传模式的特征，例如这些性状在男性后代中的出现概率更为普遍，并且绝不可能从患病父亲传给他们的儿子。

The second pedigree illustrates the pattern of inheritance for a trait such as **Huntington disease**, which is caused by an autosomal dominant allele. The key to identifying a pedigree that reflects a dominant trait is that all affected offspring will have a parent that also expresses the trait. It is also possible, by chance, that none of the offspring will inherit the dominant allele. If so, the trait will cease to exist in future generations. Like recessive traits, provided that the gene is autosomal, both males

第二张系谱展示了由常染色体显性等位基因引起的性状遗传模式，如**亨廷顿病**。确定显性性状遗传系谱的关键是所有患病后代必定都有一个同样患病的亲本。在偶然情况下也可能出现后代无显性患病个体的情形。如果是这样，那么该性状在此后世代中将不再出现。与隐性性状一样，只要该基因是常染色体上的基因，那么男性和女性都会具有同等的患病概率。

and females are equally affected.

When a given autosomal dominant disease is rare within the population, and most are, then it is highly unlikely that affected individuals will inherit a copy of the mutant gene from both parents. Therefore, in most cases, affected individuals are heterozygous for the dominant allele. As a result, approximately one-half of the offspring inherit it. This is borne out in the second pedigree in Figure 3.12. Furthermore, when a mutation is dominant, and a single copy is sufficient to produce a mutant phenotype, homozygotes are likely to be even more severely affected, perhaps even failing to survive. An illustration of this is the dominant gene for **familial hypercholesterolemia**. Heterozygotes display a defect in their receptors for low-density lipoproteins, the so-called LDLs (known popularly as "bad cholesterol"). As a result, too little cholesterol is taken up by cells from the blood, and elevated plasma levels of LDLs result. Without intervention, such heterozygous individuals usually have heart attacks during the fourth decade of their life, or before. While heterozygotes have LDL levels about double that of a normal individual, rare homozygotes have been detected. They lack LDL receptors altogether, and their LDL levels are nearly ten times above the normal range. They are likely to have a heart attack very early in life, even before age 5, and almost inevitably before they reach the age of 20.

当指定的常染色体显性遗传疾病在人类群体中十分罕见时，大多数情况下患病个体不可能同时从父母双方各获得一个突变体基因拷贝。因此，大多数情况下，患病个体的显性致病基因是杂合的。所以，其后代中大约一半将继承该显性致病基因，正如图 3.12 中第二个系谱所示。此外，当突变是显性并且单个拷贝足以产生突变体表型时，纯合个体可能患病情形更严重，甚至无法存活。**家族性高胆固醇血症**就是这种情况。杂合个体的低密度脂蛋白（LDLs，俗称“坏胆固醇”）受体存在缺陷，因此细胞从血液中吸收的胆固醇太少，导致血浆中低密度脂蛋白含量升高。在不进行干预的情况下，这些杂合个体通常在 40 多岁甚至更早时罹患心脏病，其体内低密度脂蛋白水平约为正常个体的两倍。目前纯合个体很少被检测发现，他们完全缺失低密度脂蛋白受体，其低密度脂蛋白水平比正常范围高出近十倍。他们很可能在生命早期甚至 5 岁之前即发作心脏病，并且几乎不可避免地在 20 岁之前发作心脏病。

Pedigree analysis of many traits has historically been an extremely valuable research technique in human genetic studies. However, the approach does not usually provide the certainty of the conclusions obtained through experimental crosses

在人类遗传学研究历史中，许多性状的系谱分析一直是极具价值的研究手段。然而，该方法通常不能像通过杂交实验获得大量后代数据那样提供确定的结论。尽管如此，当同时分析针对同一性状或疾病

yielding large numbers of offspring. Nevertheless, when many independent pedigrees of the same trait or disorder are analyzed, consistent conclusions can often be drawn. Table 3.2 lists numerous human traits and classifies them according to their recessive or dominant expression.

的多个独立系谱时，通常能够得出一致的结论。表 3.2 列出了众多人类性状，并根据它们的显隐性表达方式进行了分类。

TABLE 3.2 Representative Recessive and Dominant Human Traits

表 3.2 具有代表性的人类隐性性状和人类显性性状

Recessive Traits	Dominant Traits
Albinism	Achondroplasia
Alkaptonuria	Brachydactyly
Color blindness	Ehler-Danlos syndrome
Cystic fibrosis	Hypotrichosis
Duchenne muscular dystrophy	Huntington disease
Galactosemia	Hypercholesterolemia
Hemophilia	Marfan syndrome
Lesch-Nyhan syndrome	Myotonic dystrophy
Phenylketonuria	Neurofibromatosis
Sickle-cell anemia	Phenylthiocarbamide tasting
Tay-Sachs disease	Porphyria (some forms)

3.10 Tay-Sachs Disease: The Molecular Basis of a Recessive Disorder in Humans

3.10 泰 - 萨克斯病：人类隐性遗传疾病的分子基础

We conclude this chapter by examining a case where the molecular basis of normal and mutant genes and their resultant phenotypes have now been revealed. This discussion expands your understanding of how genes control phenotypes.

在本章的最后，我们将探讨一个正常基因、突变体基因及其表型发生的分子基础已被揭示的例子。这将有助于你更为深入地了解基因是如何控制表型的。

Of particular interest are cases where a single mutant gene causes multiple effects associated with a severe disorder in humans. Let's consider the modern explanation of the gene that causes **Tay-Sachs disease (TSD)**, a devastating recessive disorder involving unalterable destruction of the central nervous system. Infants with TSD are unaffected at birth and appear to develop normally until they are about 6

单个突变体基因引发人类严重疾病相关的多种基因效应的情况特别令科研人员感兴趣。下面我们将讨论引发**泰 - 萨克斯病**（TSD）的致病基因的现代机制。该疾病是一种具有毁灭性的隐性遗传疾病，会造成中枢神经系统产生不可修复的破坏。患有 TSD 的婴儿在出生时并无患病症状，并且直至出生后 6 个月左右，个体发育均表现正常。随后，其智力和身体发育出现渐

months old. Then, a progressive loss of mental and physical abilities occurs. Afflicted infants eventually become blind, deaf, intellectually disabled, and paralyzed, often within only a year or two, seldom living beyond age 5. Typical of rare autosomal recessive disorders, two unaffected heterozygous parents, who most often have no family history of the disorder, have a probability of one in four of having a Tay-Sachs child.

进性丧失。患病婴儿通常在一年或两年内最终变得眼盲、耳聋、智力障碍和身体瘫痪，寿命很少能超过5岁。这是一种典型的罕见常染色体隐性遗传疾病。如果父母双方均为未患病的杂合个体，他们大多也没有该病的家族史，那么他们的后代约有1/4的概率罹患泰-萨克斯病。

We know that proteins are the end products of the expression of most all genes. The protein product involved in TSD has been identified, and we now have a clear understanding of the underlying molecular basis of the disorder. TSD results from the loss of activity of a single enzyme, **hexosaminidase A** (**Hex-A**). Hex-A, normally found in lysosomes within cells, is needed to break down the ganglioside GM2, a lipid component of nerve cell membranes. Without functional Hex-A, gangliosides accumulate within neurons in the brain and cause deterioration of the nervous system. Heterozygous carriers of TSD with one normal copy of the gene produce only about 50 percent of the normal amount of Hex-A, but they show no symptoms of the disorder. The observation that the activity of only one gene (one wild-type allele) is sufficient for the normal development and function of the nervous system explains and illustrates the molecular basis of recessive mutations. Only when both genes are disrupted by mutation is the mutant phenotype evident. The responsible gene is located on chromosome 15 and codes for the alpha subunit of the Hex-A enzyme. More than 50 different mutations within the gene have been identified that lead to TSD phenotypes.

我们知道蛋白质是绝大多数基因表达的最终产物。与TSD疾病相关的蛋白质产物已被成功鉴定，并且我们现在对该疾病的发病分子机制也有了清楚的了解。TSD是由单一**氨基己糖苷酶A**（**Hex-A**）的活性丧失所致。Hex-A通常存在于细胞内的溶酶体中，用于降解神经细胞膜的脂质组分——神经节苷脂GM2。如果Hex-A酶功能异常，神经节苷脂将在大脑神经元内积累并导致神经系统恶化。TSD致病基因的杂合携带者含有一个拷贝的正常基因，仅产生正常个体所含Hex-A酶量的50%左右，但是他们不会呈现患病症状。仅一个野生型等位基因拷贝的活性足以满足神经系统的正常发育和功能所需，这一研究结果解释并展示了隐性突变的分子基础。只有当两个基因拷贝均因突变被破坏时，突变体表型才明显呈现。该基因位于人类第15号染色体上，并编码Hex-A酶的α亚基。目前已鉴定出50种以上的该基因不同突变形式，并且均能导致TSD患病表型。

4 Modification of Mendelian Ratios

4 孟德尔比率的扩展

CHAPTER CONCEPTS

■ While alleles are transmitted from parent to offspring according to Mendelian principles, they sometimes fail to display the clear-cut dominant/recessive relationship observed by Mendel.

■ In many cases, in contrast to Mendelian genetics, two or more genes are known to influence the phenotype of a single characteristic.

■ Still another exception to Mendelian inheritance is the presence of genes on sex chromosomes, whereby one of the sexes contains only a single member of that chromosome.

■ Phenotypes are often the combined result of both genetics and the environment within which genes are expressed.

■ The result of the various exceptions to Mendelian principles is the occurrence of phenotypic ratios that differ from those resulting from standard monohybrid, dihybrid, and trihybrid crosses.

■ Extranuclear inheritance, resulting from the expression of genes present in the DNA found in mitochondria and chloroplasts, modifies Mendelian inheritance patterns. Such genes are most often transmitted through the female gamete.

本章速览

■ 在等位基因按照孟德尔定律从亲代向子代进行传递的过程中，有时并不出现类似孟德尔所观察到的清晰的显性/隐性关系。

■ 在很多情况下，与孟德尔遗传学不同，单个性状的表型同时受两种或多种基因的共同影响。

■ 孟德尔遗传的另一种例外则是基因位于性染色体上，其中一种性别仅含一条该种染色体。

■ 表型通常是遗传和基因表达环境共同影响的结果。

■ 各种孟德尔定律例外的结果是产生与标准单因子杂交、双因子杂交、三因子杂交不同的子代表型比例。

■ 源自线粒体或叶绿体 DNA 上基因表达的核外遗传使得孟德尔遗传模式发生变化。这些基因通常通过雌性配子传递给子代。

In Chapter 3, we discussed the fundamental principles of transmission genetics. We saw that genes are present on homologous chromosomes and that these chromosomes segregate from each other and assort independently with other segregating chromosomes during gamete formation.These two postulates are the basic principles of gene transmission from parent to offspring. However, when gene expression does not adhere to a simple dominant/recessive mode or when more than one pair of genes influences the expression of a single character, the classic 3:1 and 9:3:3:1 ratios are usually modified. In this and the next several chapters, we consider more complex modes of inheritance. In spite of the greater complexity of these situations, the fundamental principles set down by Mendel still hold.

在第 3 章中，我们讨论了传递遗传学的基本原理。我们发现基因存在于同源染色体上，这些同源染色体在配子形成过程中彼此分离并且与其他相互分离的同源染色体进行自由组合。这两条假说是基因从亲代传给子代过程中所遵循的两个基本原理。然而，当基因表达不是简单的显性 / 隐性模式或单一性状的表达受到一对以上的基因共同影响时，经典的 3 ∶ 1 和 9 ∶ 3 ∶ 3 ∶ 1 的表型比例通常会发生变化。在本章和接下来的几章中，我们将考虑更为复杂的遗传模式。尽管这些情况更为复杂，但是孟德尔发现的基本定律仍然成立。

In this chapter, we restrict our initial discussion to the inheritance of traits controlled by only one set of genes. In diploid organisms, which have homologous pairs of chromosomes, two copies of each gene influence such traits. The copies need not be identical because alternative forms of genes (alleles) occur within populations. How alleles influence phenotypes is our primary focus. We will then consider gene interaction, a situation in which a single phenotype is affected by more than one set of genes. Numerous examples will be presented to illustrate a variety of heritable patterns observed in such situations.

在本章中，我们首先讨论仅由一组基因控制的性状遗传。二倍体生物含有成对存在的同源染色体，每种基因的两个拷贝影响着这些性状。这两个拷贝并不一定完全相同，因为种群内每种基因都存在多种基因形式（等位基因）。我们首先关注等位基因是如何影响表型的。随后，我们将考虑基因的相互作用。在这种情况下，每种表型受一组以上基因的共同影响。我们还将提供大量示例展示这些情形下不同的遗传模式。

Thus far, we have restricted our discussion to chromosomes other than the X and Y pair. By examining cases where genes are present on

到目前为止，我们的讨论仅限于 X、Y 染色体以外的其他染色体。通过研究 X 染色体上基因的遗传，即 X 连锁，我们还将

the X chromosome, illustrating X-linkage, we will see yet another modification of Mendelian ratios. Our discussion of modified ratios also includes the consideration of sex-limited and sex-influenced inheritance, cases where the sex of the individual, but not necessarily genes on the X chromosome, influences the phenotype.We will also consider how a given phenotype often varies depending on the overall environment in which a gene, a cell, or an organism finds itself. This discussion points out that phenotypic expression depends on more than just the genotype of an organism. Finally, we conclude with a discussion of extranuclear inheritance, cases where DNA within organelles influences an organism's phenotype.

看到孟德尔比率的另一类变化。我们对比率变化的讨论还包括限性遗传和从性遗传。在这些情形中，个体性别——且并不一定是位于X染色体上的基因——将影响表型。我们还将探讨某特定表型如何随着基因、细胞或个体所处整体环境的变化而变化。这一讨论指出，表型的表达并不仅仅取决于生物体的基因型。最后，我们将对核外遗传进行探讨，即细胞器内DNA如何影响生物体表型。

4.1 Alleles Alter Phenotypes in Different Ways

4.1 等位基因通过不同途径影响表型

After Mendel's work was rediscovered in the early 1900s, researchers focused on the many ways in which genes influence an individual's phenotype. Each type of inheritance was more thoroughly investigated when observations of genetic data did not conform precisely to the expected Mendelian ratios, and hypotheses that modified and extended the Mendelian principles were proposed and tested with specifically designed crosses. The explanations were in accord with the principle that a phenotype is under the control of one or more genes located at specific loci on one or more pairs of homologous chromosomes.

20世纪初，孟德尔的工作重新被发现后，研究人员开始关注基因影响个体表型的多种方式。当所得的遗传数据与预期孟德尔比率不能精确地保持一致时，人们即对该遗传模式进行了更为深入的研究，提出对孟德尔定律进行修正和扩展的新假说，并进一步专门设计杂交实验对假说进行验证。这些解释均符合“表型由位于同源染色体特定位点上的一对或多对基因共同控制”这一原理。

To understand the various modes of

为了了解不同的遗传模式，我们必须

inheritance, we must first examine the potential function of alleles. Alleles are alternative forms of the same gene. The allele that occurs most frequently in a population, the one that we arbitrarily designate as normal, is called the **wild-type allele**. This is often, but not always, dominant. Wild-type alleles are responsible for the corresponding wild-type phenotype and are the standards against which all other mutations occurring at a particular locus are compared.

首先理解等位基因的潜在功能。等位基因是同种基因的不同形式。种群中出现频率最高的等位基因，通常被认为是功能正常的基因形式，我们称之为**野生型等位基因**。野生型等位基因通常是显性的，但也并非总是如此。野生型等位基因决定相应的野生型表型，并且作为标准用于对比特定基因座上存在的所有其他突变。

A mutant allele contains modified genetic information and often specifies an altered gene product. For example, in human populations, there are many known alleles of the gene that encodes the β chain of human hemoglobin. All such alleles store information necessary for the synthesis of the β-chain polypeptide, but each allele specifies a slightly different form of the same molecule. Once the allele's product has been manufactured, the function of the product may or may not be altered.

突变体等位基因包含变化了的遗传信息，通常产生变化了的基因产物。例如，在人类群体中，有许多已知的编码人血红蛋白 β 链的不同等位基因，所有这些等位基因所含信息均为 β 链多肽合成所必需，但是每种等位基因均决定了同种分子略有差异的不同形式。一旦该等位基因的基因产物被合成，那么该基因产物的功能就可能发生改变，也可能不发生改变。

The process of mutation is the source of alleles. For a new allele to be recognized when observing an organism, it must cause a change in the phenotype. A new phenotype results from a change in functional activity of the cellular product specified by that gene. Often, the mutation causes the diminution or the loss of the specific wild-type function.For example, if a gene is responsible for the synthesis of a specific enzyme, a mutation in that gene may ultimately change the conformation of this enzyme and reduce or eliminate its affinity for the substrate. Such a case is designated as a **loss-of-function mutation**. If the loss is complete, the mutation

突变是产生等位基因的源泉。当观察生物体并识别出新的等位基因时，新的等位基因必须产生表型的变化。新表型的产生源于受该基因控制的细胞产物的功能活性发生了改变。通常，突变会导致特定野生型功能的减少或丧失。例如，如果一个基因负责一种特定酶的合成，则该基因的突变最终可能改变该酶的构象并降低或消除其对底物的亲和力。这种情况被称为**功能失去突变**。如果功能完全丧失，则该突变将产生所谓的**无效等位基因**。

has resulted in what is called a **null allele**.

Conversely, other mutations may enhance the function of the wild-type product. Most often when this occurs, it is the result of increasing the quantity of the gene product. In such cases, the mutation may be affecting the regulation of transcription of the gene under consideration. Such cases are designated **gain-of-function mutations**, which generally result in dominant alleles since one copy in a diploid organism is sufficient to alter the normal phenotype. Examples of gain-of-function mutations include the genetic conversion of proto-oncogenes, which regulate the cell cycle, to oncogenes, where regulation is overridden by excess gene product. The result is the creation of a cancerous cell.

相反，有些突变可以增强野生型产物的功能。这一情况的发生大多是由于基因产物产量的增加。在这种情况下，突变可能影响相关基因的转录调控。此类情形被称为**功能获得突变**，并通常产生显性等位基因，因为二倍体生物中一个拷贝的显性等位基因即足以改变正常表型。功能获得突变的例子包括调节细胞周期的原癌基因向致癌基因的遗传转变。在这种情况下，基因产物过量表达使得基因调控失效，最终导致癌细胞的产生。

Having introduced the concept of gain- or loss-of-function mutations, it is important to note the possibility that a mutation will create an allele where no change in function can be detected. In this case, the mutation would not be immediately apparent since no phenotypic variation would be evident. However, such a mutation could be detected if the DNA sequence of the gene was examined directly. These are sometimes referred to as **neutral mutations** because the gene product presents no change to either the phenotype or the evolutionary fitness of the organism.

介绍了功能获得突变、功能失去突变的概念后，需要重点提醒的是突变也可能产生功能变化无法检测的等位基因。在这种情况下，由于没有明显的表型变异，因此突变不会立即明显地显现出来。但是，如果直接研究该基因的 DNA 序列，则可以检测到这些突变。有时这些突变被称为**中性突变**，因为其基因产物不仅对生物体表型而且对生物的进化适应性均未产生影响。

Finally, we note here that while a phenotypic trait may be affected by a single mutation in one gene, traits are often influenced by more than one gene. For example, enzymatic reactions are most often part of complex

最后，我们在这里需要提醒的是，虽然表型性状可能由单个基因的单一突变所致，但是性状通常受到一种以上基因的共同影响。例如，酶促反应通常是合成最终产物如氨基酸这一复杂代谢途径中的一部

metabolic pathways leading to the synthesis of an end product, such as an amino acid. Mutations in any of the various reactions have a common effect—the failure to synthesize the end product. Therefore, phenotypic traits related to the end product are often influenced by more than one gene.

分。不同反应中的任何突变均会产生一种共同效应，即导致终产物合成的失败。因此，与终产物相关的表型性状通常受一种以上基因的影响。

4.2 Geneticists Use a Variety of Symbols for Alleles

4.2 遗传学家使用多种符号表示等位基因

In Chapter 3, we learned a standard convention that is used to symbolize alleles for very simple Mendelian traits. The initial letter of the name of a recessive trait, lowercased and italicized, denotes the recessive allele, and the same letter in uppercase refers to the dominant allele. Thus, in the case of *tall* and *dwarf*, where *dwarf* is recessive, *D* and *d* represent the alleles responsible for these respective traits. Mendel used upper- and lowercase letters such as these to symbolize his unit factors.

Another useful system was developed in genetic studies of the fruit fly *Drosophila melanogaster* to discriminate between wild-type and mutant traits. This system uses the initial letter, or a combination of two or three letters, of the name of the mutant trait. If the trait is recessive, lowercase is used; if it is dominant, uppercase is used. The contrasting wild-type trait is denoted by the same letter, but with a superscript + . For example, *ebony* is a recessive body color mutation in *Drosophila*. The normal wild-type body color is gray. Using this system, we denote *ebony* by the symbol *e*, and we

在第3章，我们学习了一种用于表示决定简单孟德尔性状等位基因的标准惯例。选择隐性性状名称的首字母，小写、斜体表示隐性等位基因，同一字母的大写、斜体则表示显性等位基因。因此，在植株高和矮这对相对性状中，矮是隐性性状，所以选择“*D*”和“*d*”分别表示决定这两种性状的等位基因。孟德尔也正是使用了类似的大写和小写字母来表示性状的单位因子。

人们在黑腹果蝇的遗传研究中同样开发了一套适用的命名系统用以区分野生型性状和突变体性状。该命名系统使用突变体性状名称的首字母、前两个字母或前三个字母的组合。如果该性状是隐性性状，则使用小写字母形式；如果该性状是显性性状，则使用大写字母形式。相对应的野生型性状则使用相同形式的字母表示，同时在右上角标记“+”。例如，黑檀体是果蝇的一种隐性体色突变，而果蝇正常的野生型体色是灰色。使用该命名系统时，我们用符号“*e*”表示黑檀体，用“e^+”表示灰体。决定该性状的基因座可能被野生型

denote gray by e^+. The responsible locus may be occupied by either the wild-type allele (e^+) or the mutant allele (e). A diploid fly may thus exhibit one of three possible genotypes:

e^+/e^+	gray homozygote (wild type)
e^+/e	gray heterozygote (wild type)
e/e	ebony homozygote (mutant)

等位基因（e^+）占据，也可能被突变体等位基因（e）占据。因此，二倍体果蝇有三种可能的基因型：

e^+/e^+	灰体纯合子（野生型）
e^+/e	灰体杂合子（野生型）
e/e	黑檀体纯合子（突变体）

The slash between the letters indicates that the two allele designations represent the same locus on two homologous chromosomes. If we instead consider a dominant wing mutation such as *Wrinkled* (Wr) wing in *Drosophila*, the three possible designations are Wr^+/Wr^+, Wr^+/Wr, and Wr/Wr. The latter two genotypes express the wrinkled-wing phenotype.

字母之间的斜线表明分别位于两条同源染色体上相同基因座上的两个等位基因名称。我们再来考虑一种显性果蝇翅膀突变，例如果蝇的皱翅（Wr），有三种可能的基因型组合：Wr^+/Wr^+、Wr^+/Wr和Wr/Wr，后两种基因型决定的表型均为皱翅。

One advantage of this system is that further abbreviation can be used when convenient: the wild-type allele may simply be denoted by the + symbol. With *ebony* as an example, the designations of the three possible genotypes become

$+/+$	gray homozygote (wild type)
$+/e$	gray heterozygote (wild type)
e/e	ebony homozygote (mutant)

该命名系统的一个优点在于还可以进一步缩写，从而更方便：野生型等位基因可以简单地用符号“+”表示。以黑檀体为例，三种可能的基因型可以进一步简化为：

$+/+$	灰体纯合子（野生型）
$+/e$	灰体杂合子（野生型）
e/e	黑檀体纯合子（突变体）

Another variation is utilized when no dominance exists between alleles. We simply use uppercase italic letters and superscripts to denote alternative alleles (e.g., R^1 and R^2, L^M and L^N, I^A and I^B). Their use will become apparent later in this chapter.

当等位基因间不存在显隐性关系时，可以使用另一种命名办法。我们仅使用大写斜体字母和上标来表示各种等位基因形式（例如，R^1和R^2，L^M和L^N，I^A和I^B）。在本章后面的举例中将呈现它们的使用。

Many diverse systems of genetic nomenclature are used to identify genes in various organisms. Usually, the symbol selected

在不同生物体中人们会使用不同的遗传命名系统来表示基因。通常，选择的基因符号要么反映基因功能，要么反映由突

reflects the function of the gene or even a disorder caused by a mutant gene. For example, the yeast *cdk* is the abbreviation for the cyclin dependent kinase gene, whose product is involved in cell-cycle regulation. In bacteria, leu^- refers to a mutation that interrupts the biosynthesis of the amino acid leucine, where the wild-type gene is designated leu^+. The symbol *dnaA* represents a bacterial gene involved in DNA replication (and DnaA is the protein made by that gene). In humans, capital letters are used to name genes: *BRCA*1 represents the first gene associated with susceptibility to breast cancer. Although these different systems may seem complex, they are useful ways to symbolize genes.

变体基因引起的疾病。例如，酵母 *cdk* 基因符号是周期蛋白依赖激酶基因的英文全称首字母缩写，其基因产物参与细胞周期调控。在细菌中，leu^- 是阻断亮氨酸生物合成的突变，而野生型基因表示为 leu^+。符号 *dnaA* 代表参与 DNA 复制的一种细菌基因（而 DnaA 则是该基因产生的蛋白质）。在人类中，大写字母被用于基因命名：*BRCA*1 表示第一个被发现的与乳腺癌易感性相关的基因。尽管这些不同的命名系统看似复杂，但均为表示基因的有效方法。

4.3 Neither Allele Is Dominant in Incomplete, or Partial, Dominance

4.3 在不完全显性中没有等位基因是显性的

A cross between parents with contrasting traits may generate offspring with an intermediate phenotype. For example, if plants such as four-o'clocks or snapdragons with red flowers are crossed with white-flowered plants, the offspring have pink flowers. Some red pigment is produced in the F_1 intermediate pink-colored flowers. Therefore, neither red nor white flower color is dominant. This situation is known as **incomplete**, or **partial**, **dominance**.

分别具有一对相对性状的亲本进行杂交时，可能产生介于二者之间的中间过渡表型后代。例如，开红花的金鱼草与开白花的金鱼草进行杂交，其后代开粉色的花朵，即 F_1 代合成的红色色素使得花朵呈粉色。因此，红花或白花都不是显性性状，这种情形被称为**不完全显性**。

If this phenotype is under the control of a single gene and two alleles where neither is dominant, the results of the F_1 (pink) $\times F_1$ (pink) cross can be predicted. The resulting F_2 generation shown in Figure 4.1 confirms the

如果这一表型是由单基因控制的，并且两种等位基因均不是显性的，则可以预测 F_1（粉色）× F_1（粉色）杂交的结果。F_2 代结果（图 4.1）证实了该假设，即性状由一对等位基因控制。F_2 代基因型比例

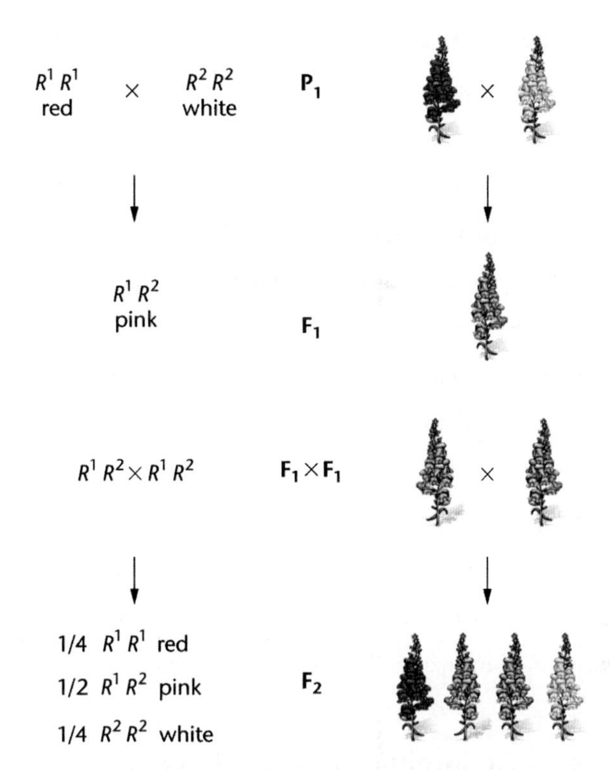

FIGURE 4.1 Incomplete dominance shown in the flower color of snapdragons.

图 4.1 金鱼草花朵颜色呈现的不完全显性。

hypothesis that only one pair of alleles determines these phenotypes. The genotypic ratio (1:2:1) of the F_2 generation is identical to that of Mendel's monohybrid cross. However, because neither allele is dominant, the phenotypic ratio is identical to the genotypic ratio. Note that because neither allele is recessive, we have chosen not to use upper and lowercase letters as symbols. Instead, we denoted the red and white alleles as R^1 and R^2, respectively. We could have used W^1 and W^2 or still other designations such as C^W and C^R, where C indicates "color" and the W and R superscripts indicate white and red.

（1 ∶ 2 ∶ 1）与孟德尔的单因子杂交相同。但是，由于两种等位基因都不是显性的，因此 F_2 代表型比例与基因型比例相同。请注意，由于这两种等位基因也都不是隐性的，因此我们不使用大写或小写字母作为基因符号，而是将红色等位基因和白色等位基因分别表示为 R^1 和 R^2，当然我们也可以使用 W^1 和 W^2 或其他符号，例如 C^W 和 C^R，其中 C 表示“颜色”，上标 W 和 R 分别表示白色和红色。

Clear-cut cases of incomplete dominance, which result in intermediate expression of the overt phenotype, are relatively rare. However, even when complete dominance seems apparent, careful examination of the gene product, rather than the phenotype, often reveals an

十分清晰的不完全显性，即产生中等程度表达的明显表型事例相对比较罕见。然而，即使有些显性性状看起来十分明显，但是如果仔细研究其基因产物，而不仅仅只观察其表型，那么通常会发现基因表达水平呈中等水平。一个例子是人类生化代

intermediate level of gene expression. An example is the human biochemical disorder **Tay-Sachs disease**, in which homozygous recessive individuals are severely affected with a fatal lipid storage disorder (see Chapter 3). There is almost no activity of the enzyme **hexosaminidase A** in those with the disease. Heterozygotes, with only a single copy of the mutant gene, are phenotypically normal but express only about 50 percent of the enzyme activity found in homozygous normal individuals. Fortunately, this level of enzyme activity is adequate to achieve normal biochemical function—a situation not uncommon in enzyme disorders.

谢疾病**泰-萨克斯病**，其中纯合隐性个体患有致命性脂肪储存紊乱（请参阅第3章），患者体内正常脂肪代谢所需的**氨基己糖苷酶A**几乎没有活性。仅含一个突变体基因拷贝的杂合个体虽然表型正常，但其仅表达纯合正常个体酶活性的约50%。幸运的是，50%的酶活性水平足以实现正常的生化功能——这种情况在酶代谢紊乱疾病中并不罕见。

4.4 In Codominance, the Influence of Both Alleles in a Heterozygote Is Clearly Evident

4.4 在共显性中，杂合子两种等位基因的影响十分明显

If two alleles of a single gene are responsible for producing two distinct, detectable gene products, a situation different from incomplete dominance or dominance/recessiveness arises. In this case, the joint expression of both alleles in a heterozygote is called **codominance**. The **MN blood group** in humans illustrates this phenomenon and is characterized by an antigen called a glycoprotein, found on the surface of red blood cells. In the human population, two forms of this glycoprotein exist, designated M and N; an individual may exhibit either one or both of them.

如果一种基因的两种等位基因形式分别决定合成两种不同的、可检测的基因产物，那么则会出现不同于不完全显性或显/隐性的情况。在这种情况下，杂合子中两种等位基因同时表达称为**共显性**。人类的**MN血型系统**展示了这种情形，该血型系统的表型由红细胞表面的糖蛋白抗原决定。在人类群体中，这种糖蛋白有两种存在形式，分别为M和N，个体可以表达其中一种或两种。

The MN system is under the control of an autosomal locus found on chromosome 4 and two alleles designated L^M and L^N. Humans are

MN血型系统由人类4号染色体上的一个常染色体基因座控制，其两种等位基因分别是L^M和L^N。人类是二倍体生物，因此

diploid, so three combinations are possible, each resulting in a distinct blood type:

可能存在三种基因组合，每种组合产生不同的血型：

Genotype	Phenotype
$L^M L^M$	M
$L^M L^N$	MN
$L^N L^N$	N

As predicted, a mating between two heterozygous MN parents may produce children of all three blood types, as follows:

所以可以预测，如果父母均为杂合的MN血型，那么他们的孩子可能出现三种血型，如下所示：

$$L^M L^N \times L^M L^N$$
$$\downarrow$$
$$1/4\ L^M L^M$$
$$1/2\ L^M L^N$$
$$1/4\ L^N L^N$$

Once again the genotypic ratio, 1:2:1, is upheld.

再次产生了1 ∶ 2 ∶ 1的基因型比例。

Codominant inheritance is characterized by distinct expression of the gene products of both alleles. This characteristic distinguishes it from incomplete dominance, where heterozygotes express an intermediate, blended phenotype. We shall see another example of codominance when we examine the ABO blood-type system in the following section.

共显性遗传的特点在于两种等位基因的基因产物均明显表达。这一特征将其与不完全显性相区分，在不完全显性中杂合子表达介乎两种纯合子表型之间的混合表型。在下一节ABO血型系统的讨论中，我们将看到另一个共显性的例子。

4.5 Multiple Alleles of a Gene May Exist in a Population

4.5 生物种群中可能存在复等位基因

The information stored in any gene is extensive, and mutations can modify this information in many ways. Each change produces a different allele. Therefore, for any specific gene, the number of alleles within members of a population need not be restricted to two. When three or more alleles of the same

任何基因均存储着丰富的信息，并且突变可以通过多种方式修饰这些遗传信息。每种变化均产生不同的等位基因。因此，对于任何特定基因，生物种群中等位基因的数目并不仅限于两种。当同一基因的三种或更多种等位基因被发现时，**复等位基因**出现，并呈现独特的遗传模式。需要重

gene are found, **multiple alleles** are present that create a unique mode of inheritance. It is important to realize that multiple alleles can be studied only in populations. An individual diploid organism has, at most, two homologous gene loci that may be occupied by different alleles of the same gene. However, among many members of a species, numerous alternative forms of the same gene can exist.

点指出的是，复等位基因只能在生物种群中进行研究。单个二倍体生物最多仅含两个可被同种基因的不同等位基因形式占据的同源基因座。然而，在同一物种的各成员中，同种基因的各种不同等位基因形式将共同存在。

The ABO Blood Group

The simplest case of multiple alleles is that in which three alternative alleles of one gene exist. This situation is illustrated by the **ABO blood group** in humans, discovered by Karl Landsteiner in the early 1900s. The ABO system, like the MN blood group, is characterized by the presence of antigens on the surface of red blood cells. The A and B antigens are distinct from MN antigens and are under the control of a different gene, located on chromosome 9. As in the MN system, one combination of alleles in the ABO system exhibits a codominant mode of inheritance.

When individuals are tested using antisera that contain antibodies against the A or B antigen, four phenotypes are revealed. Each individual has either the A antigen (A phenotype), the B antigen (B phenotype), the A and B antigens (AB phenotype), or neither antigen (O phenotype). In 1924, it was hypothesized that these phenotypes were inherited as the result of three alleles of a single gene. This hypothesis was based on studies of the blood types of many different families.

ABO 血型系统

最简单的复等位基因例子是一个基因存在三种不同等位基因形式。卡尔·兰德施泰纳在 20 世纪初发现的人类 **ABO 血型系统**就是这种情况。与 MN 血型系统类似，ABO 血型系统也是由红细胞表面存在的抗原决定的。A 和 B 抗原明显不同于 M、N 抗原，由一个位于 9 号染色体上的不同基因控制。与 MN 血型系统一样，ABO 血型系统中的一种等位基因组合表现出共显性遗传模式。

当个体测试所用抗血清能够用于 A 抗原或B抗原的检测时，个体将呈现四种表型。每一个体可能含 A 抗原（A 型血）、B 抗原（B 型血）、A 和 B 抗原（AB 型血）或不含任何抗原（O 型血）。1924 年，人们推测这些表型是单基因的三种等位基因遗传的结果。这一推测基于对许多不同家庭的血型进行研究的结果。

Although different designations can be used, we use the symbols I^A, I^B, and i to distinguish these three alleles; the i designation stands for *isoagglutinogen*, another term for antigen. If we assume that the I^A and I^B alleles are responsible for the production of their respective A and B antigens and that i is an allele that does not produce any detectable A or B antigens, we can list the various genotypic possibilities and assign the appropriate phenotype to each:

尽管可以使用不同的基因表示方式，但是我们使用符号 I^A、I^B 和 i 分别表示这三种不同的等位基因，i 源于抗原的另一个英文专业术语"*isoagglutinogen*"的首字母。如果我们假设 I^A、I^B 两种等位基因分别决定 A 抗原、B 抗原的合成，那么 i 则不产生任何可检测的 A 抗原或 B 抗原。因此我们可以列出各种可能的基因型组合及其对应的表型：

Genotype	Antigen	Phenotype
$I^A I^A$	A	A
$I^A i$	A	
$I^B I^B$	B	B
$I^B i$	B	
$I^A I^B$	A, B	AB
$i\,i$	Neither	O

In these assignments the I^A and I^B alleles are dominant to the i allele but are codominant to each other. Our knowledge of human blood types has several practical applications, the most important of which are compatible blood transfusions and organ transplantations.

在这些基因型与表型的关系中，I^A 和 I^B 两种等位基因对于 i 等位基因为显性，且二者彼此之间为共显性。我们对人类血型的了解有许多实际应用价值，其中最为重要的就是兼容输血和器官移植。

The Bombay Phenotype

The biochemical basis of the ABO blood-type system has been carefully worked out. The A and B antigens are actually carbohydrate groups (sugars) that are bound to lipid molecules (fatty acids) protruding from the membrane of the red blood cell. The specificity of the A and B antigens is based on the terminal sugar of the carbohydrate group. Both the A and B antigens are derived from a precursor molecule called the **H substance**, to which one or two terminal

孟买型

ABO 血型系统的生化机理已被研究得十分清楚。A 抗原和 B 抗原实际上是与红细胞膜上突起的脂肪酸分子相结合的碳水化合物基团，即糖分子基团。A 抗原和 B 抗原的特异性基于碳水化合物基团的末端糖基。A 抗原和 B 抗原均由前体分子 **H 物质**衍生而来，即在 H 抗原中添加了一个或两个末端糖基。

sugars are added.

In extremely rare instances, first recognized in a woman in Bombay in 1952, the H substance is incompletely formed. As a result, it is an inadequate substrate for the enzyme that normally adds the terminal sugar. This condition results in the expression of blood type O and is called the **Bombay phenotype**. Research has revealed that this condition is due to a rare recessive mutation at a locus separate from that controlling the A and B antigens. The gene is now designated *FUT1* (encoding an enzyme, fucosyl transferase), and individuals that are homozygous for the mutation cannot synthesize the complete H substance. Thus, even though they may have the I^A and/or I^B alleles, neither the A nor B antigen can be added to the cell surface. This information explains why the woman in Bombay expressed blood type O, even though one of her parents was type AB (thus she should not have been type O), and why she was able to pass the I^B allele to her children (Figure 4.2).

在极罕见的情况下，H 物质在人体内发生不完全合成，这种情形首次在 1952 年孟买的一名女性中被发现。因此在这种情形下，这样的前体分子不足以被生物酶识别，也无法正常继续添加末端糖基，从而表现为 O 型血，并称之为**孟买型**。研究表明，这种情况源自与决定 A、B 抗原合成基因不同的另一基因座上的罕见隐性突变。该基因现在被命名为 *FUT1*，编码岩藻糖基转移酶，该基因突变的纯合个体无法合成完整的 H 物质，因此，即使这些个体体内含有 I^A 和 / 或 I^B 等位基因，A 抗原或 B 抗原也不能在细胞表面正常合成。这些信息解释了为何这位孟买女性的一位亲本血型是 AB，而其本人却表现为 O 型血，虽然理论上她不可能为 O 型血；同时也解释了为何她能够将 I^B 基因传递给她的孩子（图 4.2）。

The *white* Locus in *Drosophila*

Many other phenotypes in plants and animals are known to be controlled by multiple allelic inheritance. In *Drosophila*, many alleles are known at practically every locus. The recessive mutation that causes white eyes,

果蝇的白眼基因座

在动植物中，人们已知许多其他表型同样受复等位基因的遗传控制。在果蝇中，一些已知特定基因座上均存在许多等位基因。托马斯 · H. 摩尔根和卡尔文 · 布里奇斯在 1912 年发现的白眼隐性突变就是该基

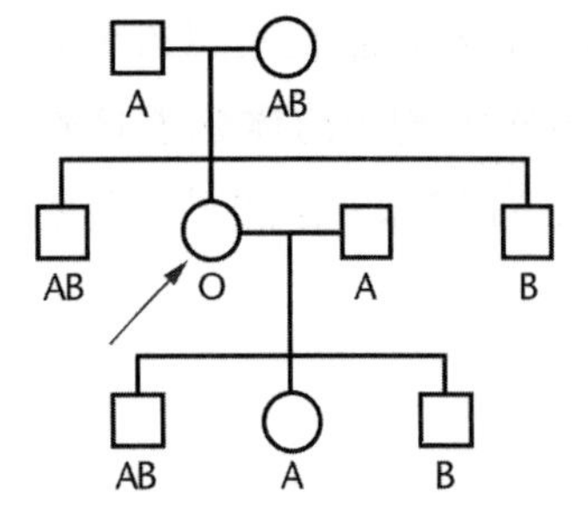

FIGURE 4.2 A partial pedigree of a woman with the Bombay phenotype. Functionally, her ABO blood group behaves as type O. Genetically, she is type B.

图 4.2 孟买表型女性的部分家族系谱图。该女性在功能上呈现 O 型血；但在基因组成上，她的血型为 B 型。

discovered by Thomas H. Morgan and Calvin Bridges in 1912, is one of over 100 alleles that can occupy this locus. In this allelic series, eye colors range from complete absence of pigment in the *white* allele to deep ruby in the *white-satsuma* allele, to orange in the *white-apricot* allele, to a buff color in the *white-buff* allele. These alleles are designated w, w^{sat}, w^{a}, and w^{bf}, respectively. In each case, the total amount of pigment in these mutant eyes is reduced to less than 20 percent of that found in the brick-red, wild-type eye.

因座上100多种等位基因中的一种。在该系列等位基因中，眼睛颜色从完全缺乏色素的 *white* 等位基因到深红宝石的 *white-satsuma* 等位基因，到橙色的 *white-apricot* 等位基因，再到浅黄色的 *white-buff* 等位基因。这些等位基因分别表示为 w、w^{sat}、w^{a} 和 w^{bf}。在每种情况下，这些突变体眼睛所含色素总量减少至不足野生型砖红色眼睛所含色素的20%。

4.6 Lethal Alleles Represent Essential Genes

4.6 致死等位基因体现必需基因

Many gene products are essential to an organism's survival. Mutations resulting in the synthesis of a gene product that is nonfunctional can often be tolerated in the heterozygous state; that is, one wild-type allele may be sufficient to produce enough of the essential product to allow survival. However, such a mutation behaves as a **recessive lethal allele**, and homozygous recessive individuals will not survive. The time of death will depend on when the product is essential. In mammals, for example, this might occur during development, early childhood, or even adulthood.

In some cases, the allele responsible for a lethal effect when homozygous may also result in a distinctive mutant phenotype when present heterozygously. It is behaving as a recessive lethal allele but is dominant with respect to the phenotype. For example, a mutation that causes a yellow coat in mice was discovered in the early

许多基因产物对于生物体的生存至关重要。突变导致合成的基因产物丧失生物学功能在基因突变杂合状态下依然可以接受；也就是说，一个野生型等位基因可能可以产生足够的必需基因产物来维持个体生存。然而，这种突变作为**隐性致死等位基因**，隐性纯合个体将无法存活，个体死亡时间则取决于该基因产物何时发挥其重要的生物学功能。例如，在哺乳动物中，致死可能发生在胚胎发育期、年幼期或者成年以后。

在某些情况下，具有纯合致死效应的等位基因在杂合状态时可能产生独特的突变体表型，即在作为隐性致死等位基因的同时，它在表型方面呈显性遗传。例如，21世纪初发现的引起小鼠产生黄色皮毛的突变。该黄色皮毛与正常野生型灰色皮毛表型不同。个体间各种不同的杂交组合得到

part of this century. The yellow coat varies from the normal agouti (wild-type) coat phenotype. Crosses between the various combinations of the two strains yield unusual results:

了不同寻常的结果：

Crosses

(A) agouti	×	agouti	→	all agouti
(B) yellow	×	yellow	→	2/3 yellow: 1/3 agouti
(C) agouti	×	yellow	→	1/2 yellow: 1/2 agouti

These results are explained on the basis of a single pair of alleles. With regard to coat color, the mutant *yellow* allele (A^Y) is dominant to the wild-type *agouti* allele (A), so heterozygous mice will have yellow coats. However, the *yellow* allele is also a homozygous recessive lethal. When present in two copies, the mice die before birth. Thus, there are no homozygous yellow mice.

In other cases, a mutation may behave as a **dominant lethal allele**. In such cases, the presence of just one copy of the allele results in the death of the individual. In humans, the disorder called **Huntington disease** is due to such an allele (H), where the onset of the disease and eventual lethality in heterozygotes (Hh) is delayed, usually well into adulthood. Affected individuals then undergo gradual nervous and motor degeneration until they die. This lethal disorder is particularly tragic because it has such a late onset, typically at about age 40, often after the affected individual has produced a family, where each child has a 50 percent probability of inheriting the lethal allele, transmitting the allele to his or her offspring, and eventually developing the disorder. Note that Huntington disease is

人们基于一对等位基因对这些结果进行了分析解释。关于皮毛颜色，突变体黄色等位基因（A^Y）相对于野生型灰色等位基因（A）为显性，因此杂合小鼠是黄色皮毛。然而，黄色等位基因也具有纯合隐性致死效应。当两个黄色突变体基因拷贝同时存在时，小鼠在出生前死亡。因此，子代中没有纯合黄色小鼠。

在另一些情况下，突变可能是**显性致死等位基因**。在这种情况下，仅存在一个拷贝的突变体等位基因即导致个体死亡。人类中，**亨廷顿病**即是源于这样的等位基因（H），在杂合个体（Hh）中，该疾病通常延迟至成年期发作并最终致死。患病个体逐渐出现神经退化和运动退化，直至死亡。这种致死性疾病是特别悲惨的，因为它发病晚，一般在 40 岁左右，患病个体通常已建立家庭，并且其子女均有 50%的可能性遗传获得该致死等位基因，最终也会患病。亨廷顿病是第 12 章的讨论主题，届时将以亨廷顿病为例介绍人类神经遗传学研究的重大进展。

the subject of the Chapter 12, where the disorder serves as the example of major advances in the study of neurogenetics in humans.

While Huntington disease is an exception, most all dominant lethal alleles are rarely observed. For these alleles to exist in a population, the affected individuals must reproduce before the lethal allele is expressed. If all affected individuals die before reaching reproductive age, the mutant gene will not be passed to future generations, and the mutation will disappear from the population unless it arises again as a result of a new mutation.

亨廷顿病是一个例外，绝大多数显性致死等位基因迄今极少被发现。因为这些等位基因存在于种群中，患病个体必须在致死基因表达之前繁衍后代。如果所有的患病个体均在育龄前死亡，那么该突变体基因将不会传递给后代，并且该突变也将从种群中消失，除非由于新的突变而再次出现。

ESSENTIAL POINT

Since Mendel's work was rediscovered, transmission genetics has been expanded to include many alternative modes of inheritance, including the study of incomplete dominance, codominance, multiple alleles, and lethal alleles.

基本要点

自孟德尔的工作重见天日后，传递遗传学已被扩展到包含多种遗传模式，包括不完全显性、共显性、复等位基因和致死等位基因。

4.7 Combinations of Two Gene Pairs with Two Modes of Inheritance Modify the 9:3:3:1 Ratio

4.7 两对基因通过两种遗传模式进行组合扩展了9：3：3：1比例

Each example discussed so far modifies Mendel's 3:1 F_2 monohybrid ratio. Therefore, combining any two of these modes of inheritance in a dihybrid cross will likewise modify the classical 9:3:3:1 ratio. Having established the foundation for the modes of inheritance of incomplete dominance, codominance, multiple alleles, and lethal alleles, we can now deal with the situation of two modes of inheritance occurring simultaneously.

到目前为止，我们讨论的所有例子均是对孟德尔3：1的F_2代单因子杂种率进行的扩展。因此，在双因子杂交中将这些遗传模式中的任何两种进行组合同样可以扩展经典的9：3：3：1表型比例。在建立不完全显性、共显性、复等位基因和致死等位基因的遗传模式理论基础之后，我们现在可以解决两种遗传模式同时发生的情况。孟德尔自由组合定律依然适用于这些情形，只要决定每种性状的基因不在同

Mendel's principle of independent assortment applies to these situations, provided that the genes controlling each character are not linked on the same chromosome—in other words, that they do not demonstrate what is called genetic linkage.

一条染色体上发生连锁，换句话说，它们不表现出所谓的遗传连锁现象。

Consider, for example, a mating that occurs between two humans who are both heterozygous for the autosomal recessive gene that causes albinism and who are both of blood type AB. What is the probability of a particular phenotypic combination occurring in each of their children? Albinism is inherited in the simple Mendelian fashion, and the blood types are determined by the series of three multiple alleles, I^A, I^B, and i. The solution to this problem is diagrammed in Figure 4.3,

例如，我们考虑两个人类个体，他们的血型均为AB型，并且白化病常染色体隐性基因均是杂合的。那么他们的子女中出现特定表型组合的可能性是多少呢？白化病遗传遵循简单的孟德尔规律，血型则由三种复等位基因I^A、I^B、和i决定。解决这个问题可以使用分叉线法，如图4.3所示。该双因子杂交不可能得到孟德尔经典的四种表型9∶3∶3∶1比例，而是将产生六种表型，且比例为3∶6∶3∶1∶2∶1，其中每项即为每种表型的预期概率。这也

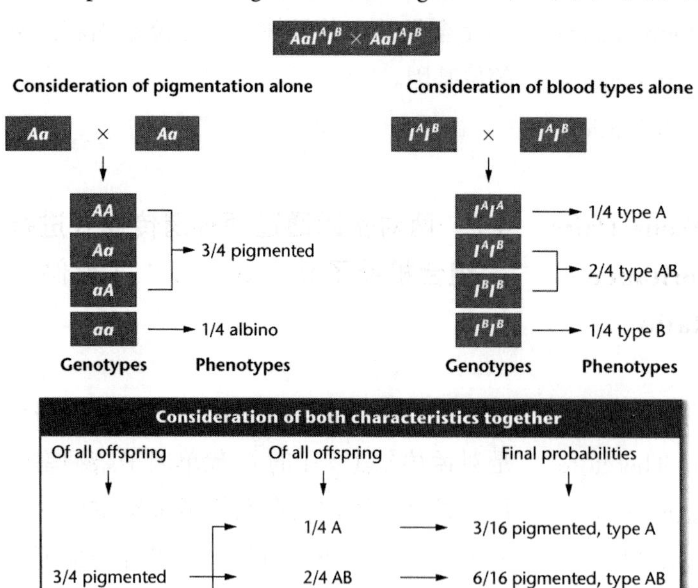

FIGURE 4.3 Calculation of the mating probabilities involving the ABO blood type and albinism in humans, using the forked-line method.

图4.3 使用分叉线法计算同时涉及ABO血型和白化病两种性状的人类遗传概率。

using the forked-line method. This dihybrid cross does not yield the classical four phenotypes in a 9:3:3:1 ratio. Instead, six phenotypes occur in a 3:6:3:1:2:1 ratio, establishing the expected probability for each phenotype. This is just one of the many variants of modified ratios that are possible when different modes of inheritance are combined.

仅仅是不同遗传模式相组合产生的诸多比例变化中的一种。

4.8 Phenotypes Are Often Affected by More Than One Gene

4.8 表型通常由一种以上的基因共同决定

Soon after Mendel's work was rediscovered, experimentation revealed that individual characteristics displaying discrete phenotypes are often under the control of more than one gene. This was a significant discovery because it revealed that genetic influence on the phenotype is often much more complex than Mendel had envisioned. Instead of single genes controlling the development of individual parts of the plant or animal body, it soon became clear that phenotypic characters can be influenced by the interactions of many different genes and their products.

孟德尔工作被重新发现后不久，实验发现，呈现不连续表型的个体性状通常受多种基因共同控制。这是意义十分重大的发现，因为它揭示了遗传对于表型的影响通常远比孟德尔预料的要复杂得多。人们很快清楚地发现，表型可以受许多不同基因及其产物相互作用的影响，而不仅仅由单基因控制植物或动物体各部分的发育。

The term **gene interaction** is often used to describe the idea that several genes influence a particular characteristic. This does not mean, however, that two or more genes, or their products, necessarily interact directly with one another to influence a particular phenotype. Rather, the cellular function of numerous gene products contributes to the development of a common phenotype. For example, the development of an organ such as the compound

专业术语**“基因互作”**通常用于描述由多种基因共同决定某一特定性状的现象。然而，这并不意味着两个或多个基因，或者它们的基因产物需要彼此间直接相互作用才能影响某一特定表型。的确，诸多基因产物的细胞功能共同促成了一种表型的发育。例如，昆虫复眼等器官的发育极其复杂，并形成具有多种表型特征的结构，如特定的大小、形状、质地和颜色。眼睛的形成由发育过程中一系列复杂、连续的

eye of an insect is exceedingly complex and leads to a structure with multiple phenotypic manifestations—such as specific size, shape, texture, and color. The formation of the eye results from a complex cascade of events during its development. This process exemplifies the developmental concept of **epigenesis**, whereby each step of development increases the complexity of this sensory organ and is under the control and influence of one or more genes.

事件促成。这一过程体现了**后成说**的发育思想，发育的每一步骤不断增加这一感觉器官的复杂性，并且受一种或多种基因的共同控制和影响。

An enlightening example of epigenesis and multiple gene interaction involves the formation of the inner ear in mammals. The inner ear consists of distinctive anatomical features to capture, funnel, and transmit external sound waves and to convert them into nerve impulses. During the formation of the ear, a cascade of intricate developmental events occur, influenced by many genes. Mutations that interrupt many of the steps of ear development lead to a common phenotype: **hereditary deafness**. In a sense, these many genes "interact" to produce a common phenotype. In such situations, the mutant phenotype is described as a **heterogeneous trait**, reflecting the many genes involved. In humans, while a few common alleles are responsible for the vast majority of cases of hereditary deafness, over 50 genes are involved in development of the ability to discern sound.

后成说和多基因互作的一个具有启发性的例子是哺乳动物内耳的形成。内耳具有独特的解剖特征，从而可以捕获、集中和传输外来的声波并将其转化为神经冲动。在耳朵的形成过程中，受多基因影响的连续、复杂发育事件依次发生。阻断耳朵发育诸多进程的突变都将导致共同的表型即**遗传性耳聋**。从某种意义上说，这些基因“相互作用”产生了共同的表型。在这种情况下，突变体表型被称为**异构型性状**，即该性状由许多基因共同参与决定。在人类中，绝大多数常见的遗传性耳聋病例由一些常见的等位基因决定，同时人们已发现50余种基因参与声音辨别能力的发育。

Epistasis

Some of the best examples of gene interaction are those that reveal the phenomenon of epistasis where the **expression** of one gene or gene pair masks or modifies the expression

上位效应

基因互作的最佳例子中有一些揭示了**上位效应**现象，即一个基因或一对基因的表达可以掩盖或修饰另一个基因或另一对基因的表达。有时决定同一表型性状表达

of another gene or gene pair. Sometimes the genes involved control the expression of the same general phenotypic characteristic in an antagonistic manner, as when masking occurs. In other cases, however, the genes involved exert their influence on one another in a complementary, or cooperative, fashion.

的基因之间存在相互拮抗的关系，此时掩盖效应即发生。而在另一些情况下，基因间也以互补或合作的方式相互影响。

For example, the homozygous presence of a recessive allele prevents or overrides the expression of other alleles at a second locus (or several other loci). In another example, a single dominant allele at the first locus influences the expression of the alleles at a second gene locus. In a third example, two gene pairs are said to complement one another such that at least one dominant allele at each locus is required to express a particular phenotype.

例如，纯合隐性等位基因的存在阻止或掩盖了另一个基因座（或多个其他基因座）上的其他等位基因的表达。又如，一个基因座上的单个显性等位基因影响了位于另一个基因座上的等位基因的表达。再如，两对基因的基因效应彼此互补，因此每个基因座上均至少存在一个显性等位基因才能表达特定表型。

The Bombay phenotype discussed earlier is an example of the homozygous recessive condition at one locus masking the expression of a second locus. There, we established that the homozygous presence of the mutant form of the *FUT1* gene masks the expression of the I^A and I^B alleles. Only individuals containing at least one wild-type *FUT1* allele can form the A or B antigen. As a result, individuals whose genotypes include the I^A or I^B allele and who lack a wild-type allele are of the type O phenotype, regardless of their potential to make either antigen. An example of the outcome of matings between individuals heterozygous at both loci is illustrated in Figure 4.4. If many such individuals have children, the phenotypic ratio of 3 A: 6 AB: 3 B: 4 O is expected in their offspring.

之前讨论的孟买型即为一个基因座上隐性纯合基因可以掩盖另一个基因座上的基因表达的例子。在这个例子中，*FUT1* 突变体基因纯合时会掩盖 I^A 和 I^B 等位基因的表达。个体中至少含有一个野生型 *FUT1* 等位基因拷贝才能形成 A 抗原或 B 抗原。因此，含有 I^A 和 I^B 等位基因、同时缺乏野生型 *FUT1* 等位基因的个体表现为 O 型血，无论其是否具有产生两种抗原的潜力。图 4.4 展示了两个基因座均为杂合的个体婚配后子女的血型情况。如果许多类似的个体均有孩子，那么这些后代的表型比则为 3A ∶ 6AB ∶ 3B ∶ 4O。

It is important to note the following points

当研究这一杂交并预测其表型比时，

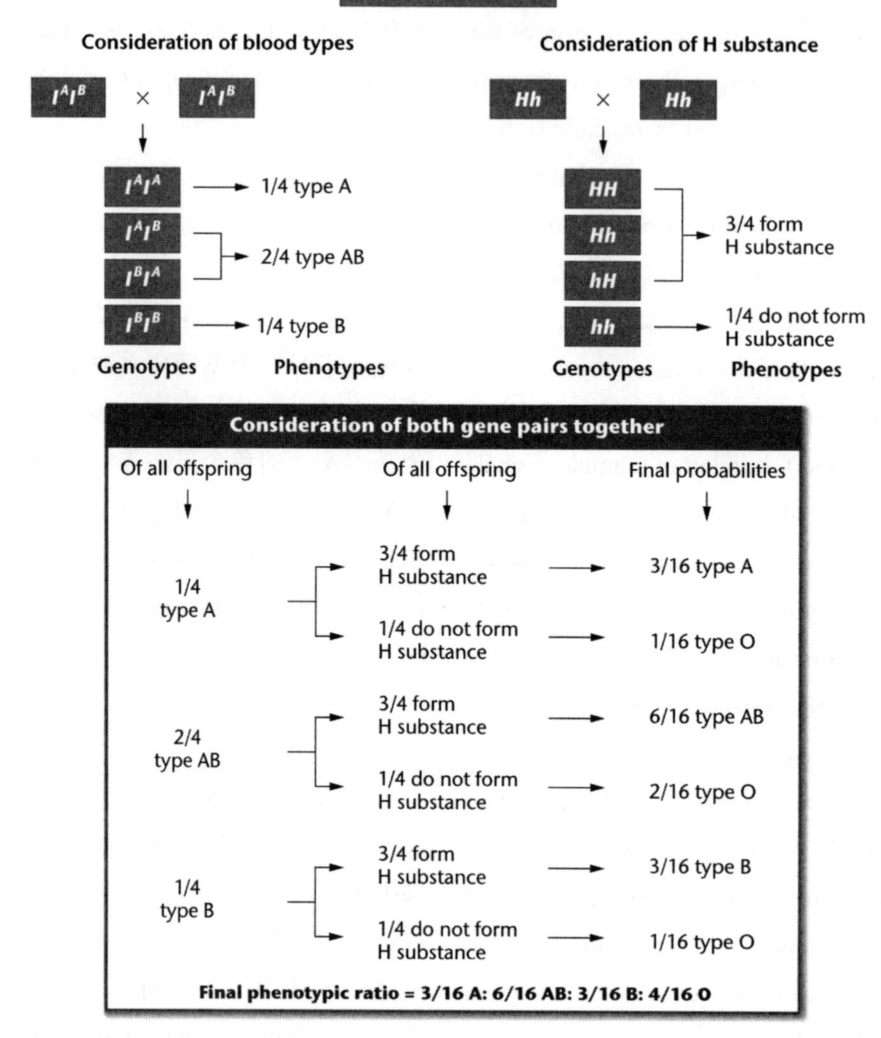

FIGURE 4.4 The outcome of a mating between individuals who are heterozygous at two genes determining their ABO blood type. Final phenotypes are calculated by considering both genes separately and then combining the results using the forked-line method.

图 4.4 决定人类 ABO 血型的两对基因均为杂合的个体婚配后子女血型的结果。本图显示了使用分叉线法计算两对基因分离和组合后得到的最终表型。

when examining this cross and the predicted phenotypic ratio:

1. A key distinction exists in this cross compared to the modified dihybrid cross shown in Figure 4.3: only one characteristic—blood type—is being followed. In the modified dihybrid cross of Figure 4.3, blood type and skin pigmentation are followed as separate

注意以下几点十分重要：

1. 与图 4.3 中所示的双因子杂交扩展相比，这一杂交存在一个关键区别：仅研究单一性状——血型。而图 4.3 的双因子杂交则同时研究相对独立的两种性状——血型和肤色。

phenotypic characteristics.

2. Even though only a single character was followed, the phenotypic ratio is expressed in sixteenths. If we knew nothing about the H substance and the genes controlling it, we could still be confident that a second gene pair, other than that controlling the A and B antigens, is involved in the phenotypic expression. When studying a single character, a ratio that is expressed in 16 parts (e.g.,3:6:3:4) suggests that two gene pairs are "interacting" during the expression of the phenotype under consideration.

2. 即使仅研究单一性状，表型比也依然可以分为16等份。如果我们对H物质及其决定基因一无所知，我们依然可以确信，除了决定A抗原和B抗原的基因之外，该表型的表达还涉及另一对基因。当研究单一性状时，如果表型比能以16等份呈现（例如3 ∶ 6 ∶ 3 ∶ 4），那么预示该表型表达中存在两对基因的“相互作用”。

The study of gene interaction reveals inheritance patterns that modify the classical Mendelian dihybrid F_2 ratio (9:3:3:1) in other ways as well. In these examples, epistasis combines one or more of the four phenotypic categories in various ways. The generation of these four groups is reviewed in Figure 4.5, along with several modified ratios.

基因互作研究揭示了经典的孟德尔双因子杂交F_2代表型比例（9 ∶ 3 ∶ 3 ∶ 1）可以扩展的其他遗传模式。在这些例子中，上位效应将四类表型类别中的一种或多种以不同方式进行组合。图4.5总结了这四类表型的形成以及扩展后的多种F_2代表型比。

As we discuss these and other examples, we will make several assumptions and adopt certain conventions:

在讨论这些及其他实例时，我们需进行以下几个假设并遵守相关约定：

1. In each case, distinct phenotypic classes are produced, each clearly discernible from all others. Such traits illustrate discontinuous variation, where phenotypic categories are discrete and qualitatively different from one another.

1. 在每种情况下，形成的表型类别相互之间差异明显，并且不同表型类别之间均可清晰地进行区别。这些性状展示了不连续变异，即表型类别间彼此不连续，有十分明显的不同。

2. The genes considered in each cross are not linked and therefore assort independently of one another during gamete formation. To allow you to easily compare the results of different crosses, we designated alleles as *A*, *a* and *B*, *b* in

2. 每个杂交实验中所涉及的基因之间不连锁，因此在配子形成过程中彼此之间能够独立地进行自由组合。为了能够方便地比较不同杂交实验的结果，我们在每个杂交实验中都将等位基因分别表示为*A*、*a*

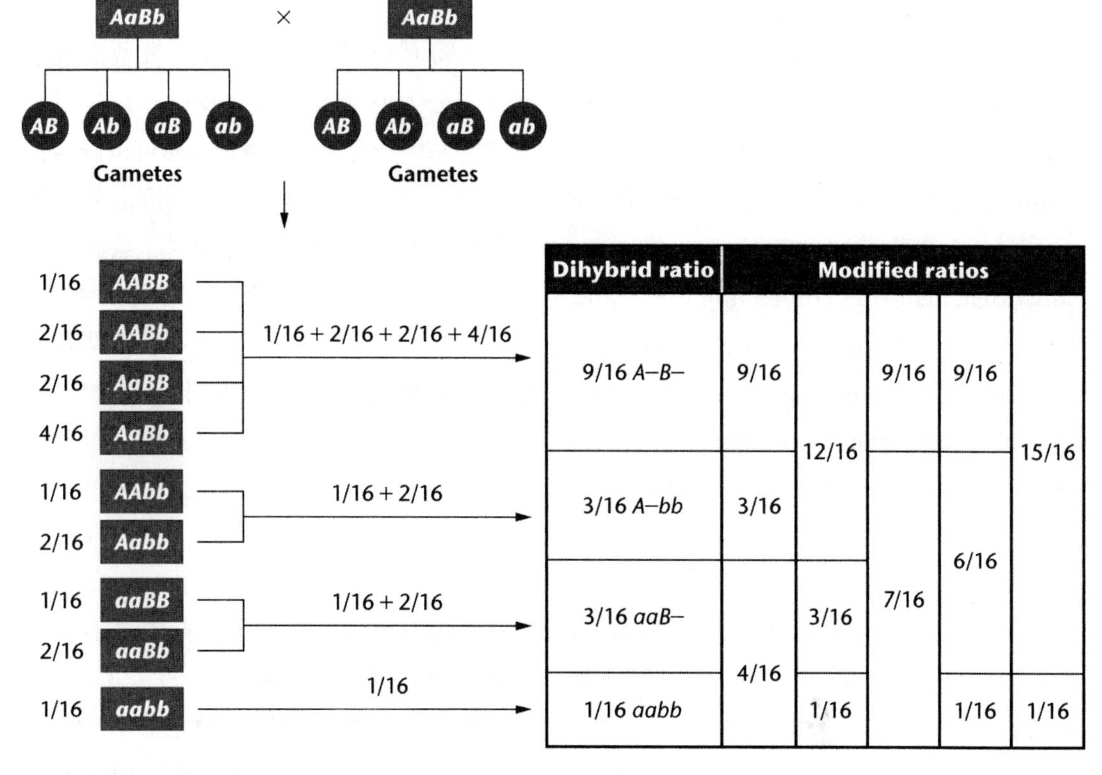

FIGURE 4.5 Generation of the various modified dihybrid ratios from the nine unique genotypes produced in a cross between individuals who are heterozygous at two genes.

图 4.5 两对基因均为杂合的个体进行杂交产生的 9 种不同基因型扩展形成的多种双因子杂交比例。

each case.

3. When we assume that complete dominance exists between the alleles of any gene pair, such that *AA* and *Aa* or *BB* and *Bb* are equivalent in their genetic effects, we use the designations *A*– or *B*– for both combinations, where the dash (–) indicates that either allele may be present, without consequence to the phenotype.

4. All P_1 crosses involve homozygous individuals (e.g., *AABB*×*aabb*, *AAbb*×*aaBB*, or *aaBB*×*AAbb*). Therefore, each F_1 generation consists of only heterozygotes of genotype AaBb.

5. In each example, the F_2 generation produced from these heterozygous parents is

和 *B*、*b*。

3. 我们假设每组基因对中的等位基因间均为完全的显隐性关系，那么 *AA*、*Aa* 的遗传效应相同，*BB*、*Bb* 的遗传效应相同，因此我们使用 *A*– 或 *B*– 分别表示这两组基因型，其中符号"–"表示可能存在的任何类型的等位基因，且对表型没有影响。

4. 所有参与 P_1 杂交的亲本均为纯合个体（例如，*AABB*×*aabb*、*AAbb*×*aaBB* 或 *aaBB*×*AAbb*）。因此，每个 F_1 代个体均为基因型为 *AaBb* 的杂合子。

5. 在每个实例中，由 F_1 代杂合子作为亲本产生的 F_2 代是我们分析的主要焦点。

our main focus of analysis. When two genes are involved (as in Figure 4.5), the F_2 genotypes fall into four categories: 9/16 *A–B–*, 3/16 *A–bb*, 3/16 *aaB–*, and 1/16 *aabb*. Because of dominance, all genotypes in each category have an equivalent effect on the phenotype.

Case 1 is the inheritance of coat color in mice (Figure 4.6). Normal wild-type coat color is agouti, a grayish pattern formed by alternating bands of pigment on each hair. Agouti is dominant to black (non-agouti) hair, which is caused by a recessive mutation, *a*. Thus, *A–* results in agouti, while aa yields black coat color. When it is homozygous, a recessive mutation, *b*, at a separate locus, eliminates pigmentation altogether, yielding albino mice (*bb*), regardless of the genotype at the other locus. The presence of at least one *B* allele allows pigmentation to occur in much the same way that the *H* allele in humans allows the expression of the ABO blood

当涉及两种基因时（如图 4.5 所示），F_2 代基因型将形成四类：9/16 *A–B–*、3/16 *A–bb*、3/16 *aaB–* 和 1/16 *aabb*。由于显性优势，因此每个类别中的所有基因型对于表型形成都具有同等效应。

例 1 是小鼠毛色的遗传（图 4.6）。正常野生型小鼠毛色是灰色，每根毛发都由交替存在的色素区带而形成灰色图案。灰色毛发相对于黑色（非灰色）毛发为显性，由隐性突变基因 *a* 决定。因此，基因型 *A–* 产生灰色毛发，*aa* 则产生黑色毛发。当另一个基因座上的隐性突变基因 *b* 纯合时，则会阻止色素沉积，从而产生白化小鼠（*bb*），且不论其他基因座上基因型的类型如何。至少存在一个显性等位基因 *B* 才能确保色素沉积发生，该机制与人类 *H* 等位基因确保 ABO 血型的表达相同。在灰色小鼠（*AABB*）和白化小鼠（*aabb*）的杂交中，F_1 代个体均为 *AaBb*，灰色毛发。两个 F_1 代

Case	Organism	Character	F_2 Phenotypes				Modified ratio
			9/16	3/16	3/16	1/16	
1	Mouse	Coat color	agouti	albino	black	albino	9:3:4
2	Squash	Color	white		yellow	green	12:3:1
3	Pea	Flower color	purple	white			9:7
4	Squash	Fruit shape	disc	sphere		long	9:6:1
5	Chicken	Color	white		colored	white	13:3
6	Mouse	Color	white-spotted	white	colored	white-spotted	10:3:3
7	Shepherd's purse	Seed capsule	triangular			ovoid	15:1
8	Flour beetle	Color	6/16 sooty and 3/16 red	black	jet	black	6:3:3:4

FIGURE 4.6 The basis of modified dihybrid F_2 phenotypic ratios, resulting from crosses between doubly heterozygous F_1 individuals. The four groupings of the F_2 genotypes shown in Figure 4.5 and across the top of this figure are combined in various ways to produce these ratios.

图 4.6 两对基因均为杂合的 F_1 代个体进行杂交后得到的扩展双因子杂交 F_2 代表型比例的基础。图 4.5 与本图顶部所显示的四组 F_2 代基因型通过不同的相互作用途径产生了这些比例。

types. In a cross between agouti (*AABB*) and albino (*aabb*), members of the F_1 are all *AaBb* and have agouti coat color. In the F_2 progeny of a cross between two F_1 heterozygotes, the following genotypes and phenotypes are observed:

杂合个体之间进行杂交产生的 F_2 代中，将观察到以下的基因型和表型：

F_1: *AaBb* × *AaBb*

↓

F_2 Ratio	Genotype	Phenotype	Final Phenotypic Ratio
9/16	*A–B–*	agouti	
3/16	*A–bb*	albino	9/16 agouti 3/16 black 4/16 albino
3/16	*aaB–*	black	
1/16	*aabb*	albino	

We can envision gene interaction yielding the observed 9:3:4 F_2 ratio as a two-step process:

我们可以假设基因互作产生所观察到的 9 ∶ 3 ∶ 4 的 F_2 代表型比通过两步生物合成途径：

	Gene *B*		Gene A	
Precursor Molecule (colorless)	↓ → *B–*	Black Pigment	↓ → *A–*	Agouti Pattern

In the presence of a *B* allele, black pigment can be made from a colorless substance. In the presence of an *A* allele, the black pigment is deposited during the development of hair in a pattern that produces the agouti phenotype. If the *aa* genotype occurs, all of the hair remains black. If the *bb* genotype occurs, no black pigment is produced, regardless of the presence of the *A* or *a* alleles, and the mouse is albino. Therefore, the *bb* genotype masks or suppresses the expression of the *A* gene. As a result, this is referred to as recessive epistasis.

当显性等位基因 *B* 存在时，黑色色素将以无色前体物质作为底物进行合成。在显性等位基因 *A* 存在时，黑色色素则在毛发发育过程中以产生灰色表型的模式进行沉积。如果基因型是 *aa*，则所有毛发均为黑色；如果基因型是 *bb*，则无论存在等位基因 *A* 还是 *a*，体内均不产生黑色色素，小鼠白化。因此，基因型 *bb* 掩盖或抑制了显性等位基因 *A* 的表达，这种情形被称为隐性上位效应。

A second type of epistasis, called dominant epistasis, occurs when a dominant allele at one genetic locus masks the expression of the alleles at a second locus. For instance, Case 2 of Figure

第二种上位效应是显性上位效应，即一个基因座上的显性等位基因掩盖了另一个基因座上的基因表达。例如，图 4.6 中的例 2 展示的南瓜果实颜色遗传。在该示例中，

4.6 deals with the inheritance of fruit color in summer squash. Here, the dominant allele *A* results in white fruit color regardless of the genotype at a second locus, *B*. In the absence of the dominant *A* allele (the *aa* genotype), *BB* or *Bb* results in yellow color, while *bb* results in green color. Therefore, if two white-colored double heterozygotes (*AaBb*) are crossed, this type of epistasis generates an interesting phenotypic ratio:

无论第二个基因座上*B*的基因型如何，显性等位基因*A*均导致白色的果皮颜色。在没有显性等位基因*A*（即为*aa*基因型）的情况下，*BB*或*Bb*则产生黄色果皮，而*bb*产生绿色果皮。因此，如果两个白色双杂合个体（*AaBb*）进行杂交，这种类型的上位效应将产生有趣的表型比：

F_1: *AaBb* × *AaBb*

↓

F_2 Ratio	Genotype	Phenotype	Final Phenotypic Ratio
9/16	*A–B–*	white	
3/16	*A–bb*	white	12/16 white
3/16	*aaB–*	yellow	3/16 yellow 1/16 green
1/16	*aabb*	green	

Of the offspring, 9/16 are *A–B–* and are thus white. The 3/16 bearing the genotypes *A–bb* are also white. Finally, 3/16 are yellow (*aaB–*) while 1/16 are green (*aabb*); and we obtain the modified ratio of 12:3:1.

在后代中，9/16基因型为*A–B–*，因此是白色果皮；3/16基因型为*A–bb*，也是白色果皮；3/16的个体是黄色果皮（*aaB–*）；1/16的个体是绿色果皮（*aabb*）；即扩展后的F_2代表型比例为12 ∶ 3 ∶ 1。

Our third type of gene interaction (Case 3 of Figure 4.6) was first discovered by William Bateson and Reginald Punnett (of Punnett square fame). It is demonstrated in a cross between two true-breeding strains of white-flowered sweet peas. Unexpectedly, the results of this cross yield all purple F_1 plants, and the F_2 plants occur in a ratio of 9/16 purple to 7/16 white. The proposed explanation suggests that the presence of at least one dominant allele of each of two gene pairs is essential for flowers to be purple. Thus, this cross represents a case of complementary gene interaction. All other genotype combinations

第三种基因互作（图4.6中的例3）由威廉·贝特森和雷金纳德·庞尼特（庞氏表发明者）首次发现。该杂交实验发生在两个纯种系白花甜豌豆品种之间。出人意料的是，该杂交产生的F_1代均为紫花；而F_2代植株中9/16的后代个体为紫花，7/16则为白花。人们推测在两种基因中，每种至少含一个显性等位基因才能形成紫花。因此，该杂交实验代表了互补基因互作。所有其他基因型组合均产生白花，因为任一纯合隐性等位基因均会掩盖另一个基因座上显性基因的表达。该杂交过程显示如下：

yield white flowers because the homozygous condition of either recessive allele masks the expression of the dominant allele at the other locus. The cross is shown as follows:

P_1: *AAbb* × *aaBB*
white white
↓
F_1: All *AaBb* purple
↓

F_2 Ratio	Genotype	Phenotype	Final Phenotypic Ratio
9/16	*A–B–*	purple	
3/16	*A–bb*	white	9/16 purple
3/16	*aaB–*	white	7/16 white
1/16	*aabb*	white	

We can now see how two gene pairs might yield such results:

现在我们可以看到两对基因可能产生这样的结果：

	Gene *A*		Gene *B*	
Precursor	↓	Intermediate	↓	Final
Substance	→	Product	→	Product
(colorless)	*A–*	(colorless)	*B–*	(purple)

At least one dominant allele from each pair of genes is necessary to ensure both biochemical conversions to the final product, yielding purple flowers. In our cross, this will occur in 9/16 of the F_2 offspring. All other plants (7/16) have flowers that remain white.

每对基因中至少需要一个显性等位基因才能确保合成最终产物的两个生化过程顺利进行，从而产生紫花。在该杂交中，F_2 代中将有 9/16 的后代个体产生紫花，而所有其他后代个体（7/16）则产生白花。

The preceding examples illustrate how the products of two genes “interact” to influence the development of a common phenotype. In other instances, more than two genes and their products are involved in controlling phenotypic expression.

前面的例子展示了两种基因的基因产物如何通过“相互作用”而影响同一表型的发育。在其他情况下，控制表型表达还可能涉及两种以上的基因及其产物。

Novel Phenotypes

Other cases of gene interaction yield novel, or new, phenotypes in the F_2 generation, in

产生新表型

基因互作除了产生不同的双因子杂交 F_2 代表型比例外，在其他一些情况下，在

addition to producing modified dihybrid ratios. Case 4 in Figure 4.6 depicts the inheritance of fruit shape in the summer squash *Cucurbita pepo*. When plants with disc-shaped fruit (*AABB*) are crossed to plants with long fruit (*aabb*), the F_1 generation all have disc fruit. However, in the F_2 progeny, fruit with a novel shape—sphere—appear, along with fruit exhibiting the parental phenotypes.

F_2代中还将产生新表型。图 4.6 中例 4 描述了美洲南瓜果实形状的遗传。当圆盘形果实植株（*AABB*）与长形果实植株（*aabb*）进行杂交时，F_1代均为圆盘形果实。但是，在F_2代后代中，除了具有亲本表型的后代个体外，还出现了一种新型形状——球形。

The F_2 generation, with a modified 9:6:1 ratio, is generated as follows:

F_2代产生了扩展的9 ∶ 6 ∶ 1表型比例，其原因如下：

F_1: *AaBb* × *AaBb*
disc disc
↓

F_2 Ratio	Genotype	Phenotype	Final Phenotypic Ratio
9/16	*A–B–*	disc	9/16 disc
3/16	*A–bb*	sphere	6/16 sphere
3/16	*aaB–*	sphere	1/16 long
1/16	*aabb*	long	

In this example of gene interaction, both gene pairs influence fruit shape equally. A dominant allele at either locus ensures a sphere-shaped fruit. In the absence of dominant alleles, the fruit is long. However, if both dominant alleles (*A* and *B*) are present, the fruit displays a flattened, disc shape.

在这个基因互作例子中，两对基因对于果实形状的影响效应是等同的。任何一个基因座上的显性等位基因均能决定球形果实的形成。在没有显性等位基因存在的情况下，则生成长形果实。但是，如果两种显性等位基因（*A* 和 *B*）同时存在，则生成圆盘形果实。

Other Modified Dihybrid Ratios

The remaining cases (5—8) in Figure 4.6 show additional modifications of the dihybrid ratio and provide still other examples of gene interactions. However, all eight cases have two things in common. First, we have not violated the principles of segregation and independent assortment to explain the inheritance pattern of

其他类型的双因子杂交 F_2 代表型比

图 4.6 中的其他几例（例 5—8）显示了双因子杂交F_2代表型比的另外一些扩展情形，并提供了其他一些基因互作的示例。但是，这八个例子具有两个共同特点。其一，在分析每一例的遗传模式时，我们均遵循孟德尔分离定律和自由组合定律。因此，在这些示例中所增加的遗传模式的复杂性

each case. Therefore, the added complexity of inheritance in these examples does not detract from the validity of Mendel's conclusions. Second, the F_2 phenotypic ratio in each example has been expressed in sixteenths. When similar observations are made in crosses where the inheritance pattern is unknown, it suggests to geneticists that two gene pairs are controlling the observed phenotypes. You should make the same inference in your analysis of genetics problems.

并不违背孟德尔定律。其二，每个例子中的 F_2 代表型比均为 16 等份。当在遗传模式未知的杂交实验中观察到类似的实验现象时，即向遗传学工作者暗示了所研究的表型同时受两对基因共同控制。在遗传学问题的分析中，你也应该能做出相同的推断。

ESSENTIAL POINT

Mendel's classic F_2 ratio is often modified in instances when gene interaction controls phenotypic variation. Such instances can be identified when the final ratio is divided into eighths or sixteenths.

基本要点

当基因互作控制表型变异时，孟德尔经典的 F_2 代表型比例通常发生变化。这种基因互作的情形可以通过观察最终的表型比是否可以被分为8等份或16等份而识别。

4.9 Complementation Analysis Can Determine if Two Mutations Causing a Similar Phenotype Are Alleles of the Same Gene

4.9 互补分析可以判断引发相似表型的突变是不是同一基因的不同等位基因

An interesting situation arises when two mutations, both of which produce a similar phenotype, are isolated independently. Suppose that two investigators independently isolate and establish a true-breeding strain of wingless *Drosophila* and demonstrate that each mutant phenotype is due to a recessive mutation. We might assume that both strains contain mutations in the same gene. However, since we know that many genes are involved in the formation of wings, mutations in any one of them might

当两个产生相似表型的突变分别独立地获得分离时，会产生一种有趣的现象。假设不同研究人员分别分离并建立了一株无翅果蝇的纯种系，并证明了这两个纯种系的突变体表型均由隐性突变所致。我们可以假设这两个纯种系的突变位于同一基因座上。然而，由于我们已知许多基因都参与果蝇翅膀的发育，因此其中任何一个基因发生突变都可能在发育过程中抑制翅膀的形成。**互补分析**这一实验方法可以帮助我们确定这两个突变是否发生在同种基

inhibit wing formation during development. The experimental approach called **complementation analysis** allows us to determine whether two such mutations are in the same gene—that is, whether they are alleles of the same gene or whether they represent mutations in separate genes.

因中，换句话说，它们是不是位于同一基因座上的等位基因，或者它们是否代表了不同基因座上的突变。

To repeat, our analysis seeks to answer this simple question: Are two mutations that yield similar phenotypes present in the same gene or in two different genes? To find the answer, we cross the two mutant strains and analyze the F_1 generation. Two alternative outcomes and interpretations of this cross are shown in Figure 4.7. We discuss both cases, using the designations m^a for one of the mutations and m^b for the other one. Now we will determine experimentally

或者说，我们的分析将试图回答以下这个简单的问题：产生相似表型的两个突变存在于同一基因中还是位于不同基因中？为了回答这一问题，我们将这两个突变体纯种系进行杂交并分析其 F_1 代个体。图 4.7 展示了得到的两种不同杂交结果并对结果进行了分析。在讨论中，我们将其中一种突变基因表示为 m^a，另一种突变基因表示为 m^b，那么现在即可通过实验确定 m^a 和 m^b 是否为同种基因的等位基因。

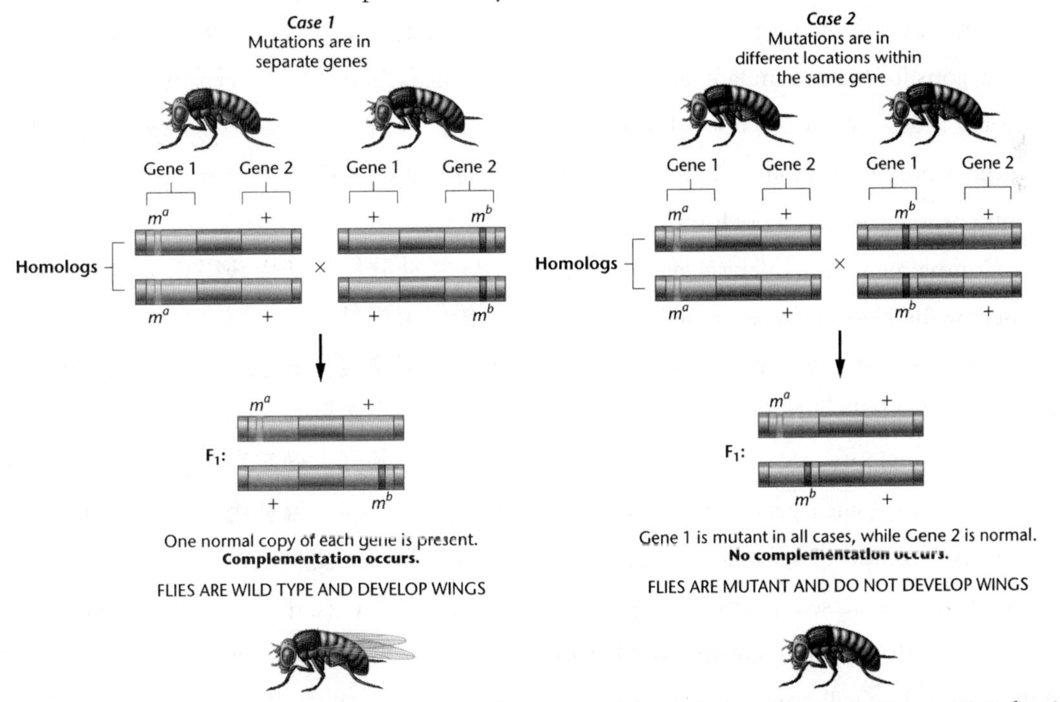

FIGURE 4.7 Complementation analysis of alternative outcomes of two wingless mutations in *Drosophila* (m^a and m^b). In Case 1, the mutations are not alleles of the same gene, whereas in Case 2, the mutations are alleles of the same gene.

图 4.7 对两种黑腹果蝇无翅突变体（m^a 和 m^b）的不同杂交结果进行的互补分析。在情况 1 中，突变分别位于不同基因内；在情况 2 中，突变是同一基因的不同等位基因。

whether or not m^a and m^b are alleles of the same gene.

Case 1. All offspring develop normal wings.

Interpretation: The two recessive mutations are in separate genes and are not alleles of one another. Following the cross, all F_1 flies are heterozygous for both genes. Complementation is said to occur. Since each mutation is in a separate gene and each F_1 fly is heterozygous at both loci, the normal products of both genes are produced (by the one normal copy of each gene), and wings develop.

Case 2. All offspring fail to develop wings.

Interpretation: The two mutations affect the same gene and are alleles of one another. Complementation does not occur. Since the two mutations affect the same gene, the F_1 flies are homozygous for the two mutant alleles (the m^a allele and the m^b allele). No normal product of the gene is produced, and in the absence of this essential product, wings do not form.

Complementation analysis, as originally devised by the Nobel Prize-winning *Drosophila* geneticist Edward B. Lewis, may be used to screen any number of individual mutations that result in the same phenotype. Such an analysis may reveal that only a single gene is involved or that two or more genes are involved. All mutations determined to be present in any single gene are said to fall into the same **complementation group**, and they will complement mutations in all other groups. When large numbers of mutations affecting the same trait are available and studied using complementation analysis, it

情况 1：所有后代均能进行正常的翅膀发育。

解释：这两种隐性突变分别位于不同基因中，彼此之间不是等位基因。杂交后，所有 F_1 代果蝇中这两种基因均是杂合的，因此发生了互补。由于每个突变均位于不同基因中，每个 F_1 代果蝇的这两个基因座均是杂合的，因此这两种基因的正常产物可通过每种基因的一个正常拷贝进行合成，翅膀发育正常。

情况 2：所有后代翅膀均无法正常发育。

解释：这两种突变影响的是同种基因，并且彼此之间互为等位基因，不发生互补。由于这两个突变影响的基因相同，因此 F_1 代果蝇的突变体等位基因是纯合的，分别为等位基因 m^a 和 m^b。没有正常的基因产物产生，并且在缺失该必需产物的情况下，翅膀无法正常发育。

互补分析最初由获得诺贝尔奖的果蝇遗传学家爱德华 · B. 刘易斯设计，用于筛选产生相同表型的个体突变。该分析可以揭示性状决定仅涉及单基因还是涉及两种或更多种基因。确定存在于单种基因的所有突变属于同一**互补组**，它们将与所有其他组的基因突变相互补。当影响同一性状的大量突变存在并可利用互补分析进行研究时，即可预测参与该性状决定的基因总数。

is possible to predict the total number of genes involved in the determination of that trait.

ESSENTIAL POINT

Complementation analysis determines whether independently isolated mutations producing similar phenotypes are alleles of one another or whether they represent separate genes.

基本要点

互补分析可用于确定独立获得的、产生相似表型的突变是同种基因的不同等位基因形式还是分别代表不同基因。

4.10 Expression of a Single Gene May Have Multiple Effects

4.10 单基因表达可以产生多种基因效应

While the previous sections have focused on the effects of two or more genes on a single characteristic, the converse situation, where expression of a single gene has multiple phenotypic effects, is also quite common. This phenomenon, which often becomes apparent when phenotypes are examined carefully, is referred to as **pleiotropy**. We will review two such cases involving human genetic disorders to illustrate this point.

尽管在前面的讨论中主要关注两个或多个基因对单一性状的影响，但与此相反的情况，即单一基因表达具有多种表型效应也十分普遍。这种现象被称为**基因多效性**，通常在仔细研究表型后会变得清晰。我们将回顾两个人类遗传疾病相关的示例以说明这一现象。

Marfan syndrome is a human malady resulting from an autosomal dominant mutation in the gene encoding the connective tissue protein fibrillin. Because this protein is widespread in many tissues in the body, one would expect multiple effects of such a defect. In fact, fibrillin is important to the structural integrity of the lens of the eye, to the lining of vessels such as the aorta, and to bones, among other tissues. As a result, the phenotype associated with Marfan syndrome includes lens dislocation, increased risk of aortic aneurysm, and lengthened long bones in limbs.

马方综合征是位于常染色体上编码结缔组织原纤蛋白的基因发生显性突变引起的人类疾病。由于该蛋白质广泛分布于人体的许多组织中，因此可以预测这种缺陷将产生多种效应。事实上，原纤蛋白对于眼睛晶状体的结构完整性、血管如主动脉内膜、骨骼以及其他组织均十分重要。因此，马方综合征相关症状包括晶状体脱位、主动脉瘤患病风险增加以及四肢长骨变长。该综合征在历史上即引发人们的关注，原因是很多猜测认为亚伯拉罕·林肯可能患有马方综合征。

This disorder is of historical interest in that speculation abounds that Abraham Lincoln had Marfan syndrome.

Our second example involves another human autosomal dominant disorder, **porphyria variegata**. Individuals with this disorder cannot adequately metabolize the porphyrin component of hemoglobin when this respiratory pigment is broken down as red blood cells are replaced. The accumulation of excess porphyrins is immediately evident in the urine, which takes on a deep red color. The severe features of the disorder are due to the toxicity of the buildup of porphyrins in the body, particularly in the brain. Complete phenotypic characterization includes abdominal pain, muscular weakness, fever, a racing pulse, insomnia, headaches, vision problems (that can lead to blindness), delirium, and ultimately convulsions. As you can see, deciding which phenotypic trait best characterizes the disorder is impossible.

第二个例子涉及另一种人类常染色体显性遗传疾病——**混合型卟啉病**。患有该疾病的个体无法充分代谢血红蛋白的卟啉组分，因为这种呼吸色素会随着红细胞的更新而分解。体内过量的卟啉会发生富集并立即在尿液中有所显现，使得尿液呈暗红色。该疾病严重的症状是由体内尤其是大脑中卟啉积聚的毒性所致。完整的表型特征包括腹痛、肌无力、高烧、心跳加速、失眠、头痛、视力问题（可导致失明）、精神错乱甚至惊厥。正如你所见，无法确定其中哪种表型性状最能代表该疾病。

Like Marfan syndrome, porphyria variegata is also of historical significance. George III, king of England during the American Revolution, is believed to have suffered from episodes involving all of the above symptoms. He ultimately became blind and senile prior to his death. We could cite many other examples to illustrate pleiotropy, but suffice it to say that if one looks carefully, most mutations display more than a single manifestation when expressed.

像马方综合征一样，混合型卟啉病同样具有历史意义。美国独立战争期间，英格兰国王乔治三世即遭受上述所有病症的折磨。在去世前，他最终失明并变得衰老。我们可以举出许多其他例子来说明基因多效性，但可以说如果仔细观察，绝大多数突变基因在表达时都呈现出不止一种表现性状。

ESSENTIAL POINT

Pleiotropy refers to multiple phenotypic

基本要点

基因多效性指由单一突变引发多种表

effects caused by a single mutation.

型效应的现象。

4.11 X-Linkage Describes Genes on the X Chromosome

4.11 X 连锁描述位于 X 染色体上的基因

In many animal and some plant species, one of the sexes contains a pair of unlike chromosomes that are involved in sex determination. In many cases, these are designated as the X and Y. For example, in both *Drosophila* and humans, males contain an X and a Y chromosome, whereas females contain two X chromosomes. While the Y chromosome must contain a region of pairing homology with the X chromosome if the two are to synapse and segregate during meiosis, much of the remainder of the Y chromosome in humans and other species is considered to be relatively inert genetically. Thus, it lacks most genes that are present on the X chromosome. As a result, genes present on the X chromosome exhibit unique patterns of inheritance in comparison with autosomal genes. The term **X-linkage** is used to describe these situations.

在许多动物和某些植物物种中，其中一种性别含有一对与性别决定相关的不同染色体。在许多情况下，它们被表示为 X 和 Y。例如，在果蝇和人类中，雄性个体含有一条 X 染色体和一条 Y 染色体，而雌性个体含有两条 X 染色体。Y 染色体上必须含有一段与 X 染色体同源的配对区域，以便二者在减数分裂过程中进行联会和分离。人类和其他物种 Y 染色体的大部分区域在遗传上是相对无效的，因此 Y 染色体缺乏与 X 染色体相对应的大多数基因。所以，与常染色体上的基因相比，X 染色体上的基因表现出独特的遗传模式。术语“**X 连锁**”即用于描述这种情形。

In the following discussion, we will focus on inheritance patterns resulting from genes present on the X but absent from the Y chromosome. This situation results in a modification of Mendelian ratios, the central theme of this chapter.

在接下来的讨论中，我们将关注 X 染色体上存在而 Y 染色体上缺失的基因的遗传模式。这种情况同样导致孟德尔比率的扩展，这也是本章讨论的主题。

X-Linkage in *Drosophila*

One of the first cases of X-linkage was documented by Thomas H. Morgan around 1920 during his studies of the *white* mutation in

果蝇的 X 连锁

1920 年左右，托马斯 · H. 摩尔根在研究果蝇白眼突变时记录了最早的 X 连锁现象。正常野生型红眼相对于白眼是显性性

the eyes of *Drosophila*. The normal wild type red eye color is dominant to white. We will use this case to illustrate X-linkage.

状。我们将使用这个例子来说明 X 连锁。

Morgan's work established that the inheritance pattern of the white-eye trait is clearly related to the sex of the parent carrying the mutant allele. Unlike the outcome of the typical monohybrid cross, reciprocal crosses between white and red-eyed flies did not yield identical results. In contrast, in all of Mendel's monohybrid crosses, F_1 and F_2 data were similar regardless of which P_1 parent exhibited the recessive mutant trait. Morgan's analysis led to the conclusion that the *white* locus is present on the X chromosome rather than on one of the autosomes. As such, both the gene and the trait are said to be X-linked.

摩尔根的工作证实，白眼性状的遗传模式显然与携带突变等位基因的亲本性别有关。与经典的单因子杂交结果不同，白眼果蝇和红眼果蝇的正反交实验结果不一样。与此不同的是，在所有孟德尔单因子杂交中，无论哪个 P_1 代亲本表现隐性突变性状，F_1 代和 F_2 代的性状统计数据均相似。摩尔根分析指出白眼基因座位于 X 染色体上，而不是任何一条常染色体上。在这样的情形中，所涉及的基因和性状均被称为是 X 连锁的。

Results of reciprocal crosses between white-eyed and red-eyed flies are shown in **Figure 4.8**.

图 4.8 显示了白眼果蝇和红眼果蝇之间的正反交杂交结果。F_1 代和 F_2 代的表型比

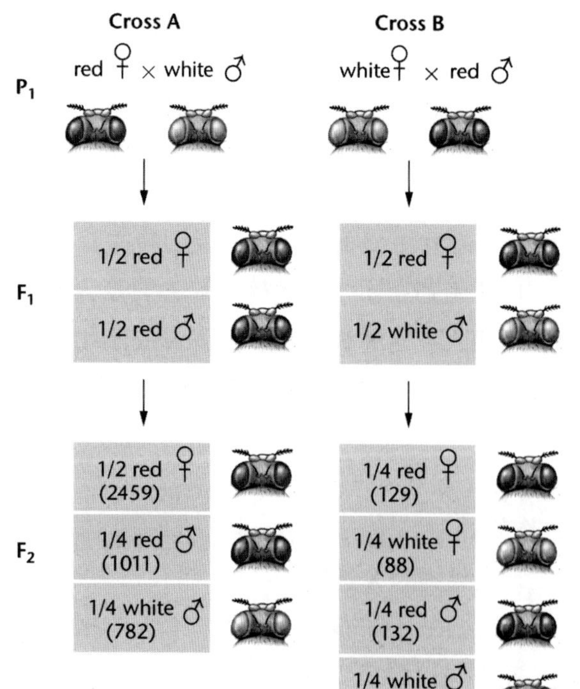

FIGURE 4.8 The F_1 and F_2 results of T. H. Morgan's reciprocal crosses involving the X-linked *white* mutation in *Drosophila melanogaster*. The actual F_2 data are shown in parentheses.

图 4.8 T. H. 摩尔根针对黑腹果蝇的 X 连锁白眼突变进行的正反交实验中 F_1 代和 F_2 代的实验结果，实际所得 F_2 代实验数据列于括弧内。

The obvious differences in phenotypic ratios in both the F_1 and F_2 generations are dependent on whether or not the P_1 white-eyed parent was male or female.

例存在明显差异并取决于 P_1 代白眼果蝇是雄性个体还是雌性个体。

Morgan was able to correlate these observations with the difference found in the sex-chromosome composition between male and female *Drosophila*. He hypothesized that the recessive allele for white eyes is found on the X chromosome, but its corresponding locus is absent from the Y chromosome. Females thus have two available gene sites, one on each X chromosome, whereas males have only one available gene site on their single X chromosome.

摩尔根将这些观察结果与雌雄果蝇的性染色体组成差异联系起来。他假设决定白眼的隐性等位基因位于 X 染色体上，并且相应的基因座在 Y 染色体上缺失。因此，雌蝇有两个相应的基因座，每条 X 染色体上分别含有一个，而雄蝇仅在其拥有的单条 X 染色体上含有一个相应的基因座。

Morgan's interpretation of X-linked inheritance, shown in Figure 4.9, provides a suitable theoretical explanation for his results. Since the Y chromosome lacks homology with most genes on the X chromosome, whatever alleles are present on the X chromosome of

摩尔根对 X 连锁遗传的解释如图 4.9 所示，从而为其实验结果提供了合适的理论解释。由于 Y 染色体与 X 染色体上大多数基因缺乏同源性，因此雄性 X 染色体上存在的等位基因将直接表达其所决定的表型。对于 X 连锁基因，雄性个体既不是纯合的

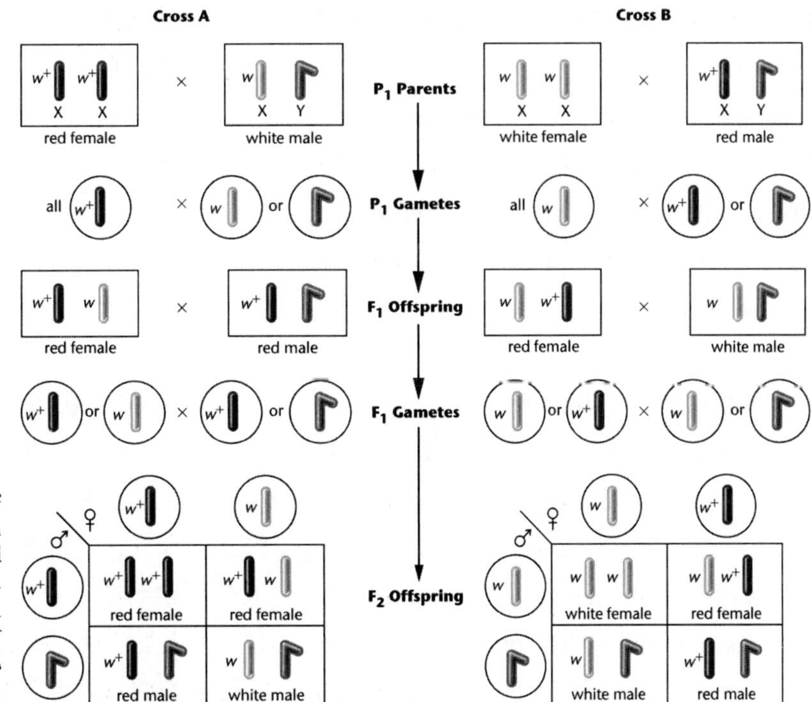

FIGURE 4.9 The chromosomal explanation of the results of the X-linked crosses shown in Figure 4.8.

图 4.9 对图 4.8 所示 X 连锁杂交实验结果的染色体遗传解释。

the males will be expressed directly in their phenotype. Males cannot be homozygous or heterozygous for X-linked genes, and this condition is referred to as being **hemizygous**.

也不是杂合的，这种情形被称为**半合子**。

One result of X-linkage is the **crisscross pattern of inheritance**, whereby phenotypic traits controlled by recessive X-linked genes are passed from homozygous mothers to all sons. This pattern occurs because females exhibiting a recessive trait carry the mutant allele on both X chromosomes. Because male offspring receive one of their mother's two X chromosomes and are hemizygous for all alleles present on that X, all sons will express the same recessive X-linked traits as their mother.

X 连锁的结果之一是**绞花式遗传模式**，即隐性 X 连锁基因控制的表型性状由纯合母亲传递给所有儿子。之所以出现这种模式，是因为具有隐性性状的雌性个体在两条 X 染色体上均携带突变型等位基因。因为雄性后代会获得来自母亲的两条 X 染色体中的一条，并且该 X 染色体上所有等位基因都是半合子，所以所有儿子将表达与其母亲相同的隐性 X 连锁性状。

Morgan's work has taken on great historical significance. By 1910, the correlation between Mendel's work and the behavior of chromosomes during meiosis had provided the basis for the **chromosome theory of inheritance**, first introduced in Chapter 3. Work involving the X chromosome around 1920 is considered to be the first solid experimental evidence in support of this theory. In the ensuing two decades, these findings inspired further research, which provided indisputable evidence in support of this theory.

摩尔根的工作具有重大的历史意义。到 1910 年，孟德尔的工作与减数分裂过程中染色体行为之间的相关性为第 3 章中首次介绍的**遗传染色体学说**提供了理论基础。1920 年左右所进行的 X 染色体相关研究则被认为是支持该学说的首个可靠的实验证据。在此后的 20 年中，这些发现激励着进一步的深入研究，并为这一理论提供了更多无可争议的证据。

X-Linkage in Humans

In humans, many genes and the respective traits they control are recognized as being linked to the X chromosome. These X-linked traits can be easily identified in a pedigree because of the crisscross pattern of inheritance. A pedigree for one form of human color blindness is shown in

人类 X 连锁现象

在人类中，许多基因及其控制的性状是与 X 染色体相关联的。由于绞花式遗传模式的存在，这些 X 连锁性状可以很容易地在家族系谱中被识别鉴定。图 4.10 所示是一种人类色盲性状的遗传系谱。世代 I 中的母亲会将该性状传递给她的所有儿子，

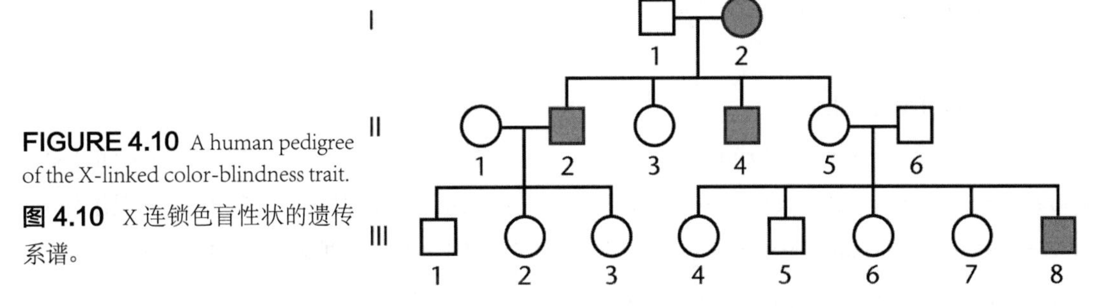

FIGURE 4.10 A human pedigree of the X-linked color-blindness trait.

图 4.10 X 连锁色盲性状的遗传系谱。

Figure 4.10.The mother in generation I passes the trait to all her sonsbut to none of her daughters. If the offspring in generation IImarry normal individuals, the color-blind sons will produceall normal male and female offspring (III-1, 2, and 3); thenormal-visioned daughters will produce normal-visionedfemale offspring (III-4, 6, and 7), as well as color-blind (III-8) and normal-visioned (III-5) male offspring.

而不会传递给她的女儿。如果世代 II 个体与视觉正常个体婚配，那么该世代中色盲男性个体的所有孩子均视觉正常，包括他的儿子和女儿（III-1，III-2 和 III-3）；视觉正常的女性个体的所有女儿视觉正常(III-4，III-6 和 III-7)，但儿子则有的是色盲(III-8)，有的视觉正常（III-5）。

The way in which X-linked genes are transmitted causes unusual circumstances associated with recessive X-linked disorders, in comparison to recessive autosomal disorders. For example, if an X-linked disorder debilitates or is lethal to the affected individual prior to reproductive maturation, the disorder occurs exclusively in males. This is so because the only sources of the lethal allele in the population are in heterozygous females who are "carriers" and do not express the disorder. They pass the allele to one-half of their sons, who develop the disorder because they are hemizygous but rarely, if ever, reproduce. Heterozygous females also pass the allele to one-half of their daughters, who become carriers but do not develop the disorder. An example of such an X-linked disorder is Duchenne muscular dystrophy. The disease has an onset prior to age 6 and is often lethal around

与常染色体隐性疾病相比，X 连锁基因的遗传方式会导致 X 连锁隐性疾病相关的异常情况。例如，如果 X 连锁疾病在性成熟前使患者个体衰弱或死亡，那么这种疾病仅会发生在男性中。之所以如此，是因为人群中该致死等位基因的唯一来源是杂合女性个体，她们是“携带者”并且不患病。她们会将该等位基因遗传给一半的儿子，她们的儿子由于是半合子而患病，并且几乎不生育。同时，杂合女性也会将该等位基因遗传给一半的女儿，后者会成为携带者且不患病。类似的 X 连锁疾病的一个例子是迪谢内肌营养不良。该病通常在 6 岁之前发作，20 岁左右致死，并且通常仅发生于男性中。

age 20. It normally occurs only in males.

ESSENTIAL POINT

Genes located on the X chromosome result in a characteristic mode of genetic transmission referred to as X-linkage, displaying so-called crisscross inheritance, whereby affected mothers pass X-linked traits to all of their sons.

基本要点

位于 X 染色体上的基因产生的遗传传递特征模式被称为 X 连锁，并表现出绞花式遗传模式，即患病母亲将 X 连锁性状传递给所有儿子。

4.12 In Sex-Limited and Sex-Influenced Inheritance, an Individual's Gender Influences the Phenotype

4.12 在限性遗传和从性遗传中，个体的性别会影响表型

In contrast to X-linked inheritance, patterns of gene expression may be affected by the gender of an individual even when the genes are not on the X chromosome. In some cases, the expression of a specific phenotype is absolutely limited to one gender; in others, the gender of an individual influences the expression of a phenotype but the phenotype is expressed in both genders. This distinction differentiates sex-limited inheritance from sex-influenced inheritance. In both types of inheritance, autosomal genes are responsible for the existence of contrasting phenotypes, but the expression of these genes is dependent on the hormone constitution of the individual. Thus, a heterozygous genotype may exhibit one phenotype in males and the contrasting one in females. In domestic fowl, for example, tail and neck plumage is often distinctly different in males and females, demonstrating **sex-limited inheritance**. Cock feathering is longer, more curved, and pointed, whereas hen feathering is shorter and less curved. Inheritance of these

与 X 连锁遗传不同，基因表达模式也可能受到个体性别的影响，即使该基因并不位于 X 染色体上。在一些情况下，特定表型的表达仅限于一种性别；而在另一些情况下，表型在两种性别中均会表达，但个体性别会对表型产生影响。这种明显的差异可以将限性遗传与从性遗传相区别。在这两种遗传模式中，常染色体基因决定了不同的相对性状，但这些基因的表达则依赖于个体的激素组成。因此，杂合基因型可能在男性中表现出一种表型，而在女性中表现出相对表型。例如，在家禽中，雄性和雌性尾羽和颈羽通常明显不同，从而表现出**限性遗传**。公鸡羽毛更长、更弯曲、更尖，而母鸡羽毛则更短、较少弯曲。这些羽毛表型的遗传由一对常染色体等位基因控制，并且该基因的表达受个体性激素的影响。

feather phenotypes is controlled by a single pair of autosomal alleles whose expression is modified by the individual's sex hormones.

As shown in the following chart, hen feathering is due to a dominant allele, *H*, but regardless of the homozygous presence of the recessive *h* allele, all females remain hen feathered. Only in males does the *hh* genotype result in cock feathering.

正如下表所示，母鸡羽毛由显性等位基因 *H* 决定，而且无论隐性等位基因 *h* 纯合与否，所有雌性个体均表现出母鸡羽毛特征。只有在雄性个体中，*hh* 基因型才会产生公鸡羽毛特征。

Genotype	**Phenotype**	
	Females	**Males**
HH	Hen-feathered	Hen-feathered
Hh	Hen-feathered	Hen-feathered
hh	Hen-feathered	Cock-feathered

In certain breeds of fowl, the hen-feathering or cock-feathering allele has become fixed in the population. In the Leghorn breed, all individuals are of the *hh* genotype; as a result, males always differ from females in their plumage. Sebright bantams are all *HH*, resulting in no sexual distinction in feathering phenotypes.

在某些禽类中，雌性羽型或雄性羽型等位基因在种群中已被固定下来。如来航鸡，所有个体均为 *hh* 基因型，结果便是雄性与雌性的羽型总是不同的。塞布赖特矮脚鸡均为 *HH* 基因型，因此在羽毛表型上没有性别差异。

Another example of sex-limited inheritance involves the autosomal genes responsible for milk yield in dairy cattle. Regardless of the overall genotype that influences the quantity of milk production, those genes are obviously expressed only in females.

限性遗传的另一个例子涉及决定奶牛产奶量的常染色体基因。无论影响奶产量的基因型组成如何，这些基因显然仅在雌性个体中表达。

Cases of **sex-influenced inheritance** include pattern baldness in humans, horn formation in certain breeds of sheep (e.g., Dorset Horn sheep), and certain coat-color patterns in cattle. In such cases, autosomal genes are again responsible for the contrasting phenotypes displayed by both males and females, but the

从性遗传的例子包括人类的秃发、某些绵羊品种（例如多塞特角羊）的角以及牛的某些皮毛花色。在这些例子中，常染色体基因再次决定雄性个体和雌性个体的相对性状，但是这些基因的表达取决于个体的激素组成。因此，杂合基因型在一种性别中表现出一种表型，而在另一种性别

expression of these genes is dependent on the hormonal constitution of the individual. Thus, the heterozygous genotype exhibits one phenotype in one sex and the contrasting one in the other. For example, **pattern baldness** in humans, where the hair is very thin on the top of the head, is inherited in this way:

中则表现出相对表型。例如，人类的**男性型秃发**，即头顶的头发非常少，就是通过这种方式遗传的：

Genotype	Phenotype	
	Females	Males
BB	Bald	Bald
Bb	Not bald	Bald
bb	Not bald	Not bald

Females can display pattern baldness, but this phenotype is much more prevalent in males. When females do inherit the *BB* genotype, the phenotype is less pronounced than in males and is expressed later in life.

女性可以表现出男性型秃发，但是这种表型在男性个体中更为普遍。而且即使女性确实继承了*BB*基因型，但是表型也不如男性个体明显，并且是在生命后期才进行表达的。

ESSENTIAL POINT

Sex-limited and sex-influenced inheritance occurs when the sex of the organism affects the phenotype controlled by a gene located on an autosome.

基本要点

当常染色体上基因决定的表型受到生物体性别影响时，则会出现限性遗传和从性遗传。

4.13 Genetic Background and the Environment Affect Phenotypic Expression

4.13 遗传背景与环境会影响表型表达

We now focus on phenotypic expression. In previous discussions, we assumed that the genotype of an organism is always directly expressed in its phenotype. For example, pea plants homozygous for the recessive *d* allele (*dd*) will always be dwarf. We discussed gene expression as though the genes operate in a

现在，我们关注表型表达。在前面的讨论中，我们假定生物的基因型总是直接表达为表型。例如，含纯合隐性等位基因*dd*的豌豆总是矮的。当我们讨论基因表达时，似乎基因在一个封闭系统中运行，其功能性产物的存在与否直接决定了个体的不同表型。实际情况其实要复杂得多。大

closed system in which the presence or absence of functional products directly determines the collective phenotype of an individual. The situation is actually much more complex. Most gene products function within the internal milieu of the cell, and cells interact with one another in various ways. Furthermore, the organism exists under diverse environmental influences. Thus, gene expression and the resultant phenotype are often modified through the interaction between an individual's particular genotype and the external environment. Here, we deal with several important variables that are known to modify gene expression.

多数基因产物在细胞内部环境中起作用，并且细胞会以各种方式彼此间相互作用。此外，生物体是在各种环境影响下存在的。因此，基因表达和产生的表型通常受到个体特定基因型与外部环境相互作用的影响。在这里，我们讨论几种已知的可影响基因表达的主要因素。

Penetrance and Expressivity

Some mutant genotypes are always expressed as a distinct phenotype, whereas others produce a proportion of individuals whose phenotypes cannot be distinguished from normal (wild type). The degree of expression of a particular trait can be studied quantitatively by determining the penetrance and expressivity of the genotype under investigation. The percentage of individuals who show at least some degree of expression of a mutant genotype defines the **penetrance** of the mutation. For example, the phenotypic expression of many mutant alleles in *Drosophila* can overlap with wild type. If 15 percent of mutant flies show the wild-type appearance, the mutant gene is said to have a penetrance of 85 percent.

外显率和基因表达度

一些突变基因型总是呈现出独特表型，而有些突变型则产生一定比例表型与正常（野生型）个体无法区分的个体。可以通过确定所研究基因型的外显率和基因表达度定量研究某特定性状表达的程度。突变基因型某种程度得以表达的个体百分比定义为突变的**外显率**。例如，果蝇中许多突变等位基因的表型表达与野生表型相重合，如果15%的突变型果蝇呈现出野生型表型，则该突变基因的外显率为85%。

By contrast, **expressivity** reflects the range of expression of the mutant genotype. Flies homozygous for the recessive mutant eyeless

相比之下，**基因表达度**则反映了突变基因型表达的程度。无眼基因隐性纯合的突变型果蝇表型从眼睛正常到复眼尺寸减

gene yield phenotypes that range from the presence of normal eyes to a partial reduction in size to the complete absence of one or both eyes. Although the average reduction of eye size is one-fourth to one-half, expressivity ranges from complete loss of both eyes to completely normal eyes.

小，再到单眼或双眼完全缺失。尽管复眼的大小平均减小 1/4 至 1/2，但是基因表达度的变化范围是从双眼完全丧失到眼睛完全正常。

Examples such as the expression of the *eyeless* gene provide the basis for experiments to determine the causes of phenotypic variation. If, on one hand, a laboratory environment is held constant and extensive phenotypic variation is still observed, other genes may be influencing or modifying the *eyeless* phenotype. On the other hand, if the genetic background is not the cause of the phenotypic variation, environmental factors such as temperature, humidity, and nutrition may be involved. In the case of the *eyeless* phenotype, experiments have shown that both genetic background and environmental factors influence its expression.

诸如无眼基因表达的例子为确定表型变异原因的实验提供了基础。一方面，在保持实验室环境恒定的条件下，如果仍能观察到广泛的表型变异，那么无眼表型可能受到其他基因的影响或修饰。另一方面，如果遗传背景不是表型变异的原因，那么环境因子，例如温度、湿度和营养可能影响表型。在无眼表型这一例子中，实验表明遗传背景和环境因素均影响该基因的表达。

Genetic Background: Position Effects

Although it is difficult to assess the specific effect of the **genetic background** and the expression of a gene responsible for determining a potential phenotype, one effect of genetic background has been well characterized, the **position effect**. In such instances, the physical location of a gene in relation to other genetic material may influence its expression. For example, if a region of a chromosome is relocated or rearranged (called a translocation or an inversion event), normal expression of genes in that chromosomal region may be modified.

遗传背景：位置效应

尽管很难准确评估**遗传背景**的特定效应和基因表达对于潜在表型的决定作用，但是一种遗传因素的效应，即**位置效应**，已经得到很好的研究。在这种情形下，基因在染色体上相对于其他遗传物质所处的位置可能影响基因的表达。例如，如果染色体某个区域发生重新定位或重新排列（即易位或倒位事件），那么该染色体区域内基因的正常表达可能被修饰。当基因被重新定位至染色体某些特定区域如**异染色质**区附近时更是如此，该区域的染色质高度浓缩并且基因表达不活跃。位置效应的一

This is particularly true if the gene is relocated to or near certain areas of the chromosome that are prematurely condensed and genetically inert, referred to as **heterochromatin**. An example of a position effect involves female *Drosophila* heterozygous for the X-linked recessive eye color mutant *white* (w). The w^+/w genotype normally results in a wild-type brick-red eye color. However, if the region of the X chromosome containing the wild-type w^+ allele is translocated so that it is close to a heterochromatic region, expression of the w^+ allele is modified. Instead of having a red color, the eyes are variegated, or mottled with red and white patches. Apparently, heterochromatic regions inhibit the expression of adjacent genes.

个例子是 X 连锁隐性白眼（w）突变基因的雌性杂合果蝇。w^+/w 基因型通常产生野生型砖红色眼睛颜色。但是，如果含野生型 w^+ 等位基因的 X 染色体区域发生易位，导致该基因邻近异染色质区域，那么 w^+ 等位基因的表达则会受到影响。果蝇眼睛颜色不再是砖红色，而呈现红白相间的斑驳颜色。显然，异染色质区域抑制了邻近基因的表达。

Temperature Effects—An Introduction to Conditional Mutations

Chemical activity depends on the kinetic energy of the reacting substances, which in turn depends on the surrounding temperature. We can thus expect temperature to influence phenotypes. An example is seen in the evening primrose, which produces red flowers when grown at 23℃ and white flowers when grown at 18℃. An even more striking example is seen in Siamese cats and Himalayan rabbits, which exhibit dark fur in certain regions where their body temperature is slightly cooler, particularly the nose, ears, and paws. In these cases, it appears that the enzyme normally responsible for pigment production is functional only at the lower temperatures present in the extremities, but it loses its catalytic function at the slightly higher temperatures

温度效应——简要介绍条件突变

化学活性取决于反应物质的动能，而动能又取决于环境温度，因此，可以预测温度将影响表型。在月见草中即可看到类似的例子，月见草在 23℃生长时开红花，而在 18℃生长时开白花。暹罗猫和喜马拉雅兔的例子则更加令人惊奇，它们体温略低的某些身体区域皮毛颜色较深，尤其是鼻子、耳朵和爪子。在这些例子中，负责色素合成的酶通常情况下仅在体温较低的肢体末端有生物学功能；而在体温稍高的身体其余部位则丧失催化功能。

found throughout the rest of the body.

Mutations whose expression is affected by temperature, called **temperature-sensitive mutations**, are examples of **conditional mutations**, whereby phenotypic expression is determined by environmental conditions. Examples of temperature-sensitive mutations are known in viruses and a variety of organisms, including bacteria, fungi, and *Drosophila*. In extreme cases, an organism carrying a mutant allele may express a mutant phenotype when grown at one temperature but express the wild-type phenotype when reared at another temperature. This type of temperature effect is useful in studying mutations that interrupt essential processes during development and are thus normally detrimental or lethal. For example, if bacterial viruses are cultured under permissive conditions of 25℃ , the mutant gene product is functional, infection proceeds normally, and new viruses are produced and can be studied. However, if bacterial viruses carrying temperature-sensitive mutations infect bacteria cultured at 42 ℃ —the restrictive condition—infection progresses up to the point where the essential gene product is required (e.g., for viral assembly) and then arrests. Temperature-sensitive mutations are easily induced and isolated in viruses, and have added immensely to the study of viral genetics.

表达受温度影响的突变被称为**温度敏感突变**，这是**条件突变**的一种。在条件突变中，表型表达由环境条件决定。在病毒和包括细菌、真菌、果蝇在内的多种生物中均存在温度敏感突变的例子。在极端情况下，携带突变型等位基因的生物在一个温度条件下生长时表达突变表型，而在另一个温度条件下生长时则表达野生型表型。这种温度效应对于研究在发育中中断基本发育步骤从而导致有害或致命生理效应的突变十分有用。例如，如果噬菌体在 25℃许可条件下进行培养，突变型基因产物有生理学功能，病毒感染可正常进行，也可以产生新生子代病毒并进行研究。但是，如果携带温度敏感突变的噬菌体感染 42℃限制性条件下培养的细菌，由于该条件是噬菌体感染宿主菌所需基本基因产物（如病毒包装）具有生物学功能的临界条件，因此宿主感染将停止。这些温度敏感突变易于在病毒中被诱导和分离，从而极大地推动了病毒遗传学的研究。

Onset of Genetic Expression

Not all genetic traits become apparent at the same time during an organism's life span. In most cases, the age at which a mutant gene

基因表达的开启

在生物体整个生命进程中，并非所有的遗传性状均在同一时间表达呈现。在大多数情况下，突变基因表现明显表型的时

exerts a noticeable phenotype depends on events during the normal sequence of growth and development. In humans, the prenatal, infant, preadult, and adult phases require different genetic information. As a result, many severe inherited disorders are not manifested until after birth. For example, as we saw in Chapter 3, **Tay-Sachs disease**, inherited as an autosomal recessive, is a lethal lipid metabolism disease involving an abnormal enzyme, hexosaminidase A. Newborns appear to be phenotypically normal for the first few months. Then developmental disability, paralysis, and blindness ensue, and most affected children die around the age of 3.

期取决于个体生长发育中正常顺序发展的过程。在人类中，产前期、婴儿期、成年前和成年期需表达的遗传信息不同。因此，许多严重的可遗传疾病直到出生后才显现出来。正如我们在第 3 章中所看到的常染色体隐性遗传疾病——**泰 - 萨克斯病**。它是一种致命的脂质代谢疾病，涉及氨基己糖苷酶 A 异常。新生儿出生后前几个月表型正常，然而之后会陆续出现发育障碍、瘫痪和失明，大多数患儿在 3 岁左右夭折。

Lesch-Nyhan syndrome, inherited as an X-linked recessive disease, is characterized by abnormal nucleic acid metabolism (biochemical salvage of nitrogenous purine bases), leading to the accumulation of uric acid in blood and tissues, intellectual disability, palsy, and self-mutilation of the lips and fingers. The disorder is due to a mutation in the gene encoding hypoxanthine-guanine phosphoribosyltransferase (HGPRT). Newborns are normal for six to eight months prior to the onset of the first symptoms.

莱施 - 奈恩综合征是一种 X 连锁隐性遗传病，其特征是核酸代谢（含氮嘌呤碱基的生化补救）异常，导致血液和组织中尿酸蓄积、智力发育出现障碍、瘫痪、唇部和手指残缺。该疾病是由于编码次黄嘌呤 - 鸟嘌呤磷酸核糖基转移酶（HGPRT）的基因发生突变所致。新生儿在出生后 6—8 个月表现正常，直至首次出现相关症状。

Still another example involves **Duchenne muscular dystrophy (DMD)**, an X-linked recessive disorder associated with progressive muscular wasting. It is not usually diagnosed until the child is 3 to 5 years old. Even with modern medical intervention, the disease is often fatal in the early 20s.

另一个例子是**迪谢内肌营养不良（DMD）**，这是一种进行性肌肉萎缩相关的 X 连锁隐性疾病，通常直至孩子 3—5 岁才被确诊。即使采用现代医学干预，患者寿命通常也只到 20 岁出头。

Perhaps the most age-variable of all inherited human disorders is **Huntington**

在所有人类遗传疾病中，患者年龄变化最大的可能是**亨廷顿病**。作为一种常染

disease. Inherited as an autosomal dominant, Huntington disease affects the frontal lobes of the cerebral cortex, where progressive cell death occurs over a period of more than a decade. Brain deterioration is accompanied by spastic uncontrolled movements, intellectual and emotional deterioration, and ultimately death. Onset of this disease is variable, but most frequently occurs around age 45.

色体显性遗传疾病，亨廷顿病影响大脑皮质额叶区，该区域会在长达10年的时间内发生进行性细胞死亡。脑部衰退会伴随产生痉挛性不自主运动、智力和情感障碍，直至最终死亡。该疾病的发作时间不一，最为常见的是发生于45岁左右。

These examples support the concept that the critical expression of genes varies throughout the life cycle of all organisms, including humans. Gene products may play more essential roles at certain life stages, and it is likely that the internal physiological environment of an organism changes with age.

这些例子均支持了这样的观点，即包括人类在内的所有生物的整个生命周期中基因的关键表达时期各有不同。基因产物在某些生命阶段可能扮演着更为重要的作用，并且有机体内部的生理环境也可能随着年龄而变化。

Genetic Anticipation

Interest in studying the genetic onset of phenotypic expression has intensified with the discovery of heritable disorders that exhibit a progressively earlier age of onset and an increased severity of the disorder in each successive generation. This phenomenon is called **genetic anticipation**.

遗传早现

随着一些遗传疾病在后续世代中发病年龄更早、发病程度更严重的现象被发现，人们对于研究表型表达的基因开启越来越感兴趣。而这些遗传病所呈现的遗传现象被称为**遗传早现**。

Myotonic dystrophy (DM), the most common type of adult muscular dystrophy, clearly illustrates genetic anticipation. Individuals with this autosomal dominant disorder exhibit extreme variation in the severity of symptoms. Mildly affected individuals develop cataracts as adults but have little or no muscular weakness. Severely affected individuals demonstrate more extensive myopathy and may be intellectually disabled. In its most extreme form, the disease is

强直性肌营养不良（DM）是最为常见的一种成年肌营养不良，它清楚地展示了遗传早现。患有这种常染色体显性遗传疾病的个体，症状的严重程度表现出极大差异：轻度患者成年后出现白内障，但几乎没有肌肉无力现象；严重患者表现出更多的肌肉疾病，并且可能出现智力障碍；在最严重的情况下，患者在出生后即死亡。1989年，C. J. 豪厄尔与同事证实了后续世代与病情严重程度增加和疾病发作提前之

fatal just after birth. In 1989, C. J. Howeler and colleagues confirmed the correlation of increased severity and earlier onset with successive generations. They studied 61 parent-child pairs, and in 60 cases, age of onset was earlier and more severe in the child than in his or her affected parent.

间的相关性。他们研究了 61 对亲子关系，其中 60 例孩子的发病年龄比其患病父母更早，病情也更严重。

In 1992, an explanation was put forward for the molecular cause of the mutation responsible for DM, as well as the basis of genetic anticipation. A particular region of the DM gene—a short trinucleotide DNA sequence—is repeated a variable number of times and is unstable. Normal individuals average about five copies of this region; minimally affected individuals have about 50 copies; and severely affected individuals have over 1000 copies. The most remarkable observation was that in successive generations, the size of the repeated segment increases. Although it is not yet clear how this expansion in size affects onset and phenotypic expression, the correlation is extremely strong. Several other inherited human disorders, including the fragile-X syndrome, Kennedy disease, and Huntington disease, also reveal an association between the size of specific regions of the responsible gene and disease severity.

1992 年，人们对引发强直性肌营养不良突变的分子机制及遗传早现的理论基础进行了阐释。强直性肌营养不良基因的一个含短三核苷酸 DNA 序列的特定区域存在多次且数目不稳定的序列重复现象。正常个体平均约含该区域的 5 个拷贝；轻度患者大约含 50 个拷贝；严重患者则含 1000 个以上的拷贝。最重要的发现是在后续子代中，序列重复片段的大小会增加。尽管尚不清楚这种片段大小增加如何影响疾病发作和表型表达，但二者之间具有非常强的相关性。其他几种人类遗传疾病，包括脆性 X 染色体综合征、肯尼迪病和亨廷顿病，同样显示出相关基因特定区域大小与疾病严重程度之间的关联性。

ESSENTIAL POINT

Phenotypic expression is not always the direct reflection of the genotype. A percentage of organisms may not express the expected phenotype at all, the basis of the penetrance of a mutant gene. In addition, the phenotype can be

基本要点

表型表达并不总是基因型的直接反映。一定比例的生物可能根本不表达预期表型，这是突变基因外显率的理论基础。此外，表型受到遗传背景、温度和营养的影响。基因表达的开启在生物体的一生中可能有

modified by genetic background, temperature, and nutrition. The onset of expression of a gene may vary during the lifetime of an organism, and in future generations, the onset may occur earlier and the symptoms increased in severity (genetic anticipation).

所不同，并且在此后世代中，基因开启可能发生得更早，症状也更严重(遗传早现)。

4.14 Extranuclear Inheritance Modifies Mendelian Patterns

4.14 核外遗传丰富了孟德尔遗传模式

Throughout the history of genetics, occasional reports have challenged the basic tenet of Mendelian transmission genetics—that the phenotype is determined solely by nuclear genes located on the chromosomes of both parents. In this final section of the chapter, we consider several examples of inheritance patterns that vary from those predicted by the traditional biparental inheritance of nuclear genes, phenomena that are designated as **extranuclear inheritance**. In the following cases, we will focus on two broad categories. In the first, an organism's phenotype is affected by the expression of genes contained in the DNA of mitochondria or chloroplasts rather than the nucleus, generally referred to as organelle heredity. In the second category, referred to as a maternal effect, an organism's phenotype is determined by genetic information expressed in the gamete of the mother—such that, following fertilization, the developing zygote's phenotype is influenced not by the individual's genotype, but by gene products directed by the genotype of the mother.

在整个遗传学发展史中，偶尔会出现对孟德尔传递遗传学基本原理——表型仅由位于父母双亲染色体上的核基因决定——提出挑战的报道。在本章的最后，我们将讨论与核基因的传统双亲遗传模式不同的一些示例，这些现象被称为**核外遗传**。在以下例子中，我们将重点关注两类情况：第一类，生物体的表型受线粒体或叶绿体 DNA 所含基因表达的影响，而不受细胞核内基因的影响，通常被称为细胞器遗传；第二类，称为母体效应，即生物体的表型由母亲配子中所表达的遗传信息决定，受精后，发育合子的表型不受自身基因型的影响，而受母亲基因型决定的基因产物的影响。

Initially, such observations met with skepticism. However, with increasing knowledge

最初，这些观察结果受到质疑。然而，随着人们对分子遗传学知识的积累以及线

of molecular genetics and the discovery of DNA in mitochondria and chloroplasts, the phenomenon of extranuclear inheritance came to be recognized as an important aspect of genetics.

粒体和叶绿体 DNA 的发现，核外遗传现象已被认为是遗传学的重要组成部分。

Organelle Heredity: DNA in Chloroplasts and Mitochondria

We begin by examining examples of inheritance patterns related to chloroplast and mitochondrial function. Before DNA was discovered in these organelles, the exact mechanism of transmission of the traits was not clear, except that their inheritance appeared to be linked to something in the cytoplasm rather than to genes in the nucleus. Furthermore, transmission was most often from the maternal parent through the ooplasm, causing the results of reciprocal crosses to vary. Such an extranuclear pattern of inheritance is now appropriately called **organelle heredity**.

细胞器遗传：叶绿体和线粒体中含有 DNA

我们首先从讨论叶绿体和线粒体功能相关的遗传模式的例子开始。在从这些细胞器中发现 DNA 之前，这些性状确切的遗传机理尚不清楚，人们只是觉得它们的遗传似乎与细胞质中存在的某些物质有关，而与细胞核基因无关。此外，性状遗传在绝大多数情况下源自母本卵母细胞的细胞质，从而导致正反交的实验结果不同。这一核外遗传模式现在被称为**细胞器遗传**。

Analysis of the inheritance patterns resulting from mutant alleles in chloroplasts and mitochondria has been difficult for two reasons. First, the function of these organelles is dependent on gene products from both nuclear and organelle DNA, making the discovery of the genetic origin of mutations affecting organelle function difficult. Second, many mitochondria and chloroplasts are contributed to each progeny. Thus, if only one or a few of the organelles contain a mutant gene in a cell among a population of mostly normal mitochondria, the corresponding mutant phenotype may not be revealed. This condition, referred to as

叶绿体和线粒体中突变等位基因的遗传模式分析难度大主要有两个原因：其一，这些细胞器的功能实现同时依赖于细胞核 DNA 和细胞器 DNA 的基因产物，从而使得影响细胞器功能的基因突变的真正遗传来源难以被发现。其二，每个子代均含有许多线粒体和叶绿体，因此，如果细胞中所含的绝大多数线粒体均正常，而仅有群体中一个或几个细胞器含有突变基因，那么相应的突变表型可能并不显现，这种情形被称为**异质性**。正是由于不含突变的细胞器为野生型功能实现提供了基础，从而形成正常细胞，因此细胞器的遗传分析要比孟德尔性状复杂得多。

heteroplasmy, may lead to normal cells since the organelles lacking the mutation provide the basis of wild-type function. Analysis is therefore much more complex than for Mendelian characters.

Chloroplasts: Variegation in Four-o'clock Plants

叶绿体：紫茉莉的花斑遗传

In 1908, Carl Correns (one of the rediscoverers of Mendel's work) provided the earliest example of inheritance linked to chloroplast transmission. Correns discovered a variant of the four-o'clock plant, *Mirabilis jalapa*, that had branches with either white, green, or variegated white-and-green leaves. The white areas in variegated leaves and in the completely white leaves lack chlorophyll that provides the green color to normal leaves. Chlorophyll is the light-absorbing pigment made within chloroplasts.

1908年，卡尔·科伦斯（重新发现孟德尔工作的科学家之一）提供了最早的与叶绿体遗传相关的实例。科伦斯发现了紫茉莉的一种变种，其枝条上的叶片是白色的、绿色的或绿白相间的。绿白相间叶片的白色区域和全白叶片的白色区域均缺少正常叶片呈现绿色所需的叶绿素。叶绿素为叶绿体内合成的光合色素。

Correns was curious about how inheritance of this phenotypic trait occurred. Inheritance in all possible combinations of crosses is strictly determined by the phenotype of the ovule source. For example, if the seeds (representing the progeny) were derived from ovules on branches with green leaves, all progeny plants bore only green leaves, regardless of the phenotype of the source of pollen. Correns concluded that inheritance was transmitted through the cytoplasm of the maternal parent because the pollen, which contributes little or no cytoplasm to the zygote, had no apparent influence on the progeny phenotypes.

科伦斯对于这种表型性状的遗传机制感到好奇。在所有可能的杂交组合中，性状遗传均严格由胚珠来源的亲本表型所决定。例如，如果种子（即后代）源自生长绿色叶片枝条上的胚珠，那么所有后代植株均只生长绿色叶片，而与花粉来源的亲本表型无关。因此科伦斯得出结论，遗传来自母本细胞质，因为花粉对于合子细胞质的贡献很少或者几乎没有贡献，从而对子代表型没有明显影响。

Since leaf coloration is related to the

由于叶片颜色与叶绿体有关，因此或

chloroplast, genetic information contained either in that organelle or somehow present in the cytoplasm and influencing the chloroplast must be responsible for the inheritance pattern. It now seems certain that the genetic "defect" that eliminates the green chlorophyll in the white patches on leaves is a mutation in the DNA housed in the chloroplast.

者是细胞器中所含的遗传信息，或者是以某种形式存在于细胞质中影响叶绿体功能的遗传信息必然决定了该种遗传模式。现在似乎可以肯定，导致叶片上白色区域中绿色叶绿素减少的遗传“缺陷”源自叶绿体所含 DNA 的一处突变。

Mitochondrial Mutations: *poky* in *Neurospora* and *petite* in *Saccharomyces*

Mutations affecting mitochondrial function have been discovered and studied, revealing that they too contain a distinctive genetic system. As with chloroplasts, mitochondrial mutations are transmitted through the cytoplasm. In our current discussion, we will emphasize the link between mitochondrial mutations and the resultant extranuclear inheritance patterns.

线粒体突变：链孢霉缓慢生长突变（*poky*）和酵母缓慢生长突变（*petite*）

影响线粒体功能的突变已陆续被发现和研究，并且研究表明它们同样含有独特的遗传系统。与叶绿体类似，线粒体突变也通过细胞质进行遗传。在当前讨论中，我们将强调线粒体突变与产生的核外遗传模式之间的联系。

In 1952, Mary B. Mitchell and Hershel K. Mitchell studied the bread mold *Neurospora crassa*. They discovered as low-growing mutant strain and named it *poky*. Slow growth is associated with impaired mitochondrial function, specifically in relation to certain cytochromes essential for electron transport. Results of genetic crosses between wild-type and *poky* strains suggest that *poky* is an extranuclear trait inherited through the cytoplasm. If one mating type is *poky* and the other is wild type, all progeny colonies are *poky*. The reciprocal cross, where *poky* is transmitted by the other mating type, produces normal wild-type colonies.

1952 年，玛丽 · B. 米切尔和赫谢尔 · K. 米切尔研究面包霉时发现了生长缓慢的突变菌株，并将其命名为“*poky*”。生长缓慢与线粒体功能受损有关，尤其与电子传递中所必需的某些细胞色素有关。野生型和 *poky* 菌株的遗传杂交结果表明 *poky* 是通过细胞质遗传的一种核外遗传性状。如果 *poky* 交配型菌株与野生型菌株进行杂交，其所有后代菌落表型均为 *poky*。在反交实验中，*poky* 菌株与其他交配型菌株进行杂交，则所有后代菌落表型均为野生型。

Another extensive study of mitochondrial mutations has been performed with the yeast

另一例对线粒体突变开展的深入研究工作是利用酿酒酵母进行的。1956 年，鲍

Saccharomyces cerevisiae. The first such mutation, described by Boris Ephrussi and his coworkers in 1956, was named *petite* because of the small size of the yeast colonies. Many independent *petite* mutations have since been discovered and studied, and all have a common characteristic—a deficiency in cellular respiration involving abnormal electron transport. The majority of them demonstrate cytoplasmic transmission, indicating mutations in the DNA of the mitochondria. This organism is a facultative anaerobe and can grow by fermenting glucose through glycolysis; thus, it may survive the loss of mitochondrial function by generating energy anaerobically.

里斯·埃弗吕西和他的同事首次描述了此类突变，由于酵母菌落小，因此他们将其命名为“*petite*”。此后许多类似的 *petite* 突变陆续被发现和研究，这些突变均具有一个共同特征——涉及异常电子传递的细胞呼吸缺陷。它们中的绝大多数表现出细胞质遗传，表明线粒体 DNA 存在突变。酿酒酵母属于兼性厌氧菌，可以通过糖酵解发酵葡萄糖进行生长，因此酿酒酵母在线粒体功能丧失的情况下可以通过厌氧产生能量而生存。

The complex genetics of *petite* mutations has revealed that a small proportion are the result of nuclear DNA changes. They exhibit Mendelian inheritance and illustrate that mitochondria function depends on both nuclear and organellar gene products.

petite 突变的复杂遗传学研究已揭示了小部分源自核 DNA 改变的结果。它们遵循孟德尔遗传，表明线粒体功能同时取决于核基因产物和细胞器基因产物。

Mitochondrial Mutations: Human Genetic Disorders

线粒体突变：人类遗传疾病

Our knowledge of the genetics of mitochondria has now greatly expanded. The DNA found in human mitochondria has been completely sequenced and contains 16,569 base pairs. Mitochondrial gene products have been identified and include the following:

我们现在对于线粒体遗传学的了解已大大扩展。人类线粒体基因组 DNA 已完成测序，包含 16 569 个碱基对。线粒体基因产物已被鉴定并包括：

13 proteins, required for aerobic cellular respiration

细胞有氧呼吸所需的 13 种蛋白质

22 transfer RNAs (tRNAs), required for translation

翻译所需的 22 种转移 RNA（tRNA）

2 ribosomal RNAs (rRNAs), required for

2 种翻译所需的核糖体 RNA（rRNA）

translation

Because a cell's energy supply is largely dependent on aerobic cellular respiration, disruption of any mitochondrial gene by mutation may potentially have a severe impact on that organism, such as we saw in our previous discussion of the *petite* mutation in yeast. In fact, mtDNA is particularly vulnerable to mutations for two possible reasons. First, the ability to repair mtDNA damage does not appear to be equivalent to that of nuclear DNA. Second, the concentration of highly mutagenic free radicals generated by cell respiration that accumulate in such a confined space very likely raises the mutation rate in mtDNA.

由于细胞能量供应主要依赖于细胞有氧呼吸，因此突变导致的任何线粒体基因破坏均可能对该生物造成严重影响，例如我们之前所讨论的酿酒酵母 *petite* 突变。事实上，mtDNA 尤其脆弱，易发生突变。其可能的原因有两个：首先，mtDNA 的损伤修复能力似乎不及核 DNA；其次，由细胞呼吸产生的高浓度、易致突变的自由基积聚在如此狭小的线粒体空间内，很可能提高了 mtDNA 的突变概率。

Fortunately, a zygote receives a large number of organelles through the egg, so if only one organelle or a few of them contain a mutation (an illustration of heteroplasmy), the impact is greatly diluted by the many mitochondria that lack the mutation and function normally. If a deleterious mutation arises or is present in the initial population of organelles, adults will have cells with a variable mixture of both normal and abnormal organelles. From a genetic standpoint, this condition of heteroplasmy makes analysis quite difficult.

幸运的是，合子从卵子获得了大量的细胞器，因此，如果其中只有一个或几个细胞器含突变（即异质性），那么该突变效应可被许多不含突变的线粒体稀释，从而大大削弱突变的影响，使得细胞功能正常。如果有害突变出现或存在于细胞器的初始群体中，则成年个体所含细胞将是同时包含正常细胞器和异常细胞器的混合物。从遗传学角度来看，这种异质性的存在使得遗传分析变得相当困难。

Many disorders in humans are known to be due to mutations in mitochondrial genes. For example, **myoclonic epilepsy and ragged-red fiber disease (MERRF)** demonstrates a pattern of inheritance consistent with maternal transmission. Only the offspring of affected mothers inherit this disorder, while the offspring of affected fathers are normal. Individuals with

人类许多疾病现已知是由线粒体基因突变造成的。例如，**肌阵挛性癫痫伴破碎红纤维综合征（MERRF）**表现出与母体遗传一致的遗传模式，即只有患病母亲的后代会遗传该疾病，而患病父亲的后代则均正常。患有这种罕见疾病的个体表现为共济失调（缺乏肌肉协调能力）、耳聋、痴呆和癫痫发作。该疾病因存在特征性“参

this rare disorder express ataxia (lack of muscular coordination), deafness, dementia, and epileptic seizures. The disease is named for the presence of "ragged-red" skeletal-muscle fibers that exhibit blotchy red patches resulting from the proliferation of aberrant mitochondria. Brain function, which has a high energy demand, is also affected in this disorder, leading to the neurological symptoms described above.

差不齐的破碎红色”骨骼肌纤维而得名。骨骼肌纤维由于异常线粒体增殖而出现斑点状红色斑块。高能量需求的脑功能同样受到该疾病的影响，导致上述神经系统症状的发生。

Analysis of mtDNA from patients with MERRF has revealed a mutation in one of the 22 mitochondrial genes encoding a transfer RNA. Specifically, the gene encoding $tRNA^{Lys}$ (the tRNA that delivers lysine during translation) contains an altered DNA sequence. This genetic alteration interferes with the capacity for translation within the organelle, which in turn leads to the various manifestations of the disorder.

MERRF患者mtDNA分析表明，22种编码tRNA的线粒体基因中有一处突变。具体说来，编码$tRNA^{Lys}$（翻译过程中携带赖氨酸的tRNA）的基因含一处DNA序列突变。这种基因突变会干扰细胞器内的蛋白质翻译能力，进而导致该疾病的各种异常症状。

The cells of affected individuals exhibit the condition called heteroplasmy, containing a mixture of normal and abnormal mitochondria. Different patients contain different proportions of the two, and even different cells from the same patient exhibit various levels of abnormal mitochondria.Were it not for heteroplasmy, the mutation would very likely be lethal, testifying to the essential nature of mitochondrial function and its reliance on the genes encoded by mtDNA within the organelle.

患者细胞是同时含有正常线粒体和异常线粒体的混合组成，表现出异质性。不同患者细胞内这二者的组成比例不同，甚至同一患者不同细胞内异常线粒体的含量也不同。如果不是异质性，这种突变很可能是致命的，这证实了线粒体功能的基本特性及其对细胞器内mtDNA编码基因的依赖。

Mitochondria, Human Health, and Aging

线粒体、人类健康与衰老

The study of hereditary mitochondrial-based disorders provides insights into the critical

基于线粒体可遗传疾病的研究为揭示正常发育过程中该细胞器的至关重要性提

importance of this organelle during normal development. In fact, mitochondrial dysfunction seems to be implicated in a large number of major human disease conditions, including anemia, blindness, Type 2 (late-onset) diabetes, autism, atherosclerosis, infertility, neurodegenerative diseases such as Parkinson, Alzheimer, and Huntington disease, schizophrenia and bipolar disorders, and a variety of cancers. It is becoming evident, for example, that mutations in mtDNA are present in such human malignancies as skin, colorectal, liver, breast, pancreatic, lung, prostate, and bladder cancers.

供了更深的见解。实际上，线粒体功能障碍似乎与许多主要的人类疾病均有关，包括贫血、失明、2 型（迟发性）糖尿病、孤独症、动脉粥样硬化、不孕症、神经系统变性疾病（如帕金森病、阿尔茨海默病和亨廷顿病）、精神分裂症、双相情感障碍以及各种癌症，并且越来越明显，mtDNA 突变存在于人类癌症中，例如皮肤癌、结直肠癌、肝癌、乳腺癌、胰腺癌、肺癌、前列腺癌和膀胱癌。

Over 400 mtDNA mutations associated with more than 150 distinct mtDNA-based genetic syndromes have been identified. Genetic tests for detecting mutations in the mtDNA genome that may serve as early-stage disease markers have been developed. However, it is still unclear whether mtDNA mutations are causative effects contributing to development of malignant tumors or whether they are the consequences of tumor formation. Nonetheless, there is an interesting link between mtDNA mutations and cancer, including data suggesting that many chemical carcinogens have significant mutation effects on mtDNA.

150 种以上基于 mtDNA 的不同遗传综合征相关的 400 多种 mtDNA 突变已被鉴定，并且检测可作为早期疾病标志物的 mtDNA 基因组突变的遗传检测技术也已得到开发。但是，人们目前尚不清楚 mtDNA 突变到底是癌症发展的原因还是形成的后果。尽管如此，mtDNA 突变与癌症之间还是存在着有趣的关联，包括数据表明许多化学致癌物对 mtDNA 有显著的突变效应。

The study of hereditary mitochondrial-based disorders has also suggested a link between the progressive decline of mitochondrial function and the aging process. It has been hypothesized that the accumulation of sporadic mutations in mtDNA leads to an increased prevalence of defective mitochon-dria (and the concomitant decrease in the supply of ATP) in

基于线粒体可遗传疾病的研究还表明，线粒体功能的逐步下降与衰老进程也存在联系。据推测，在一生中，mtDNA 中零星突变的积累会导致细胞中缺陷线粒体的普遍增加，同时伴随着 ATP 供给减少。这种情况在衰老过程中扮演着十分重要的作用。

cells over a lifetime. This condition in turn plays a significant role in aging.

Many studies have now documented that aging tissues contain mitochondria with increased levels of DNA damage. The major question is whether such changes are simply biomarkers of the aging process or whether they lead to a decline in physiological function, which in turn, contributes significantly to aging. In support of the latter hypothesis, one study links age-related muscle fiber atrophy in rats to deletions in mtDNA and electron transport abnormalities. Such deletions appear to be present in the mitochondria of atrophied muscle fibers, but are absent from fibers in regions of normal tissue. It is important to note that mutations in the nuclear genome also impact mitochondrial function and human disease and aging. For example, another study involving genetically altered mice is most revealing. Such mice have a nuclear gene altered that diminishes proofreading during the replication of mtDNA. These mice display reduced fertility and accumulate mutations over time at a much higher rate than is normal. These mice also show many characteristics of premature aging, as observed by loss and graying of hair, reduction in bone density and muscle mass, decline in fertility, anemia, and reduced life span.

许多已发表的研究表明衰老组织中线粒体 DNA 的损伤水平会升高。对此存在的主要疑问在于这些变化仅仅是衰老过程的生物标志物还是由于这些变化导致生理功能下降，进而显著促进了衰老。一项研究大鼠衰老相关的肌纤维萎缩症与 mtDNA 缺失和电子传递异常间联系的工作支持了后一种猜想。这些缺失似乎存在于发生萎缩的肌纤维线粒体中，而在正常组织区域的纤维中不存在。需要引起注意的是，核基因组突变同样影响线粒体功能以及人类疾病和衰老。例如，另一项涉及转基因小鼠的研究最具启发性。这些小鼠的一个核基因发生改变，使 mtDNA 复制过程中的校对能力消失。这些小鼠的生育力降低，并且随着时间推移积累突变的速度比正常小鼠高很多。这些小鼠也表现出许多过早衰老的特征，如毛发脱落和变白、骨密度和肌肉量下降、生育力下降、贫血和寿命缩短。

These, and other studies, continue to speak to the importance of normal mitochondrial function. As cells undergo genetic damage, which appears to be a natural phenomenon, their function declines, which may be an underlying factor in aging as well as in the progression of

这些以及其他研究陆续证明了正常线粒体功能的重要性。随着细胞不断遭受似乎是自然现象的遗传损伤，其功能不断下降，这可能正是衰老以及年龄相关疾病发展的潜在因素。

age-related disorders.

Maternal Effect

In **maternal effect**, also referred to as maternal influence, an offspring's phenotype for a particular trait is under the control of the mother's nuclear gene products present in the egg. This is in contrast to biparental inheritance, where both parents transmit information on genes in the nucleus that determines the offspring's phenotype. In cases of maternal effect, the nuclear genes of the female gamete are transcribed, and the genetic products (either proteins or yet untranslated mRNAs) accumulate in the egg ooplasm. After fertilization, these products are distributed among newly formed cells and influence the patterns or traits established during early development. The following example will illustrate such an influence of the maternal genome on particular traits.

母性效应

在**母性效应**，即母性影响中，后代特定性状的表型由受精卵中母本的核基因产物所决定。与双亲遗传不同，双亲遗传中父母双方共同传递决定子代表型的细胞核基因信息；在母性效应中，雌性配子中的核基因被转录，其基因产物（蛋白质或尚未翻译的mRNA）在卵细胞质中进行累积。受精后，这些产物分布于新形成的细胞中，并在胚胎早期发育过程中影响发育模式或性状。下面的例子将展示母体基因组信息对于特定性状的影响。

Embryonic Development in *Drosophila*

A recently documented example that illustrates maternal effect involves various genes that control embryonic development in *Drosophila melanogaster*. The genetic control of embryonic development in *Drosophila* is a fascinating story. The protein products of the maternal-effect genes function to activate other genes, which may in turn activate still other genes. This cascade of gene activity leads to a normal embryo whose subsequent development yields a normal adult fly. The extensive work by Edward B. Lewis, Christiane Nüsslein-Volhard, and Eric Wieschaus (who shared the 1995

果蝇的胚胎发育

最近报道的一个展示母性效应的例子是参与控制果蝇胚胎发育的各种基因。果蝇胚胎发育的遗传控制是一个令人着迷的故事。母性效应基因的蛋白质产物具有激活其他基因的功能，而这些基因又可以继续激活更多的其他基因。基因活性的这种级联激活过程引导正常胚胎随后的发育，最终形成正常的成年果蝇。爱德华·B.刘易斯、克里斯蒂亚娜·尼斯莱因-福尔哈德和埃里克·威绍斯开展了深入广泛的工作，阐明了这些基因及其他基因的功能（他们正是由于这一发现分享了1995年的诺贝尔生理学或医学奖）。表现母性效应的基因，

Nobel Prize for Physiology or Medicine for their findings) has clarified how these and other genes function. Genes that illustrate maternal effect have products that are synthesized by the developing egg and stored in the oocyte prior to fertilization. Following fertilization, these products specify molecular gradients that determine spatial organization as development proceeds.

其基因产物由发育的卵合成并在受精前储存于卵母细胞中。受精后，这些基因产物具体的分子梯度在胚胎发育进程中决定着胚胎的空间组织。

For example, the gene *bicoid* (*bcd*) plays an important role in specifying the development of the anterior portion of the fly. Embryos derived from mothers who are homozygous for this mutation (bcd^-/bcd^-) fail to develop anterior areas that normally give rise to the head and thorax of the adult fly. Embryos whose mothers contain at least one wild-type allele (bcd^+) develop normally, even if the genotype of the embryo is homozygous for the mutation. Consistent with the concept of maternal effect, the genotype of the female parent, not the genotype of the embryo, determines the phenotype of the offspring. Nusslein-Volhard and Wieschaus, using large-scale mutant screens, discovered many other maternal-effect genes critical to early development in *Drosophila*.

例如，基因 *bicoid*（*bcd*）在果蝇前部发育中起着重要作用。母亲为该突变纯合（bcd^-/bcd^-）的胚胎无法正常发育形成成年果蝇头部和胸部的前部区域。如果母亲至少含有一个野生型等位基因（bcd^+），那么胚胎将正常发育，即使该胚胎的基因型为纯合突变基因。与母性效应的概念一致，雌性亲本的基因型而不是胚胎自身的基因型决定了后代的表型。尼斯莱因 - 福尔哈德和威绍斯通过大规模的突变体筛选，发现了对果蝇早期发育至关重要的许多其他母性效应基因。

ESSENTIAL POINT

When patterns of inheritance vary from that expected due to biparental transmission of nuclear genes, phenotypes are often found to be under the control of DNA present in mitochondria or chloroplasts, or are influenced during development by the expression of the maternal genotype in the egg.

基本要点

当遗传模式与预期的核基因双亲遗传模式不同时，通常会发现表型受线粒体 DNA 或叶绿体 DNA 决定，或者在发育过程中受卵细胞中母本基因型表达的影响。

5 Sex Determination and Sex Chromosomes

5　性别决定与性染色体

CHAPTER CONCEPTS

■ A variety of mechanisms have evolved that result in sexual differentiation, leading to sexual dimorphism and greatly enhancing the production of genetic variation within species.

■ Often, specific genes, usually on a single chromosome, cause maleness or femaleness during development.

■ In humans, the presence of extra X or Y chromosomes beyond the diploid number may be tolerated but often leads to syndromes demonstrating distinctive phenotypes.

■ While segregation of sex-determining chromosomes should theoretically lead to a one-to-one sex ratio of males to females, in humans the actual ratio favors males at conception.

■ In mammals, females inherit two X chromosomes compared to one in males, but the extra genetic information in females is compensated for by random inactivation of one of the X chromosomes early in development.

■ In some reptilian species, temperature during incubation of eggs determines the sex of offspring.

本章速览

■ 多种多样的性别分化机制使物种形成两性异形，并大大促进了物种遗传变异的产生。

■ 通常，个体发育过程中引发雄性发育或雌性发育的特定基因位于单一染色体上。

■ 在人类中，如果二倍染色体数目之外同时存在额外的 X 染色体或 Y 染色体，那么这样的个体可以存活，但通常会引发呈现独特表型的综合征。

■ 虽然性别决定染色体的分离在理论上会产生 1 ∶ 1 的性别比，但是事实上人类在受精过程中男性后代的比例会更高。

■ 在哺乳动物中，雌性个体遗传获得两条 X 染色体，雄性个体只获得一条 X 染色体，但是雌性个体在胚胎发育早期会通过随机失活其中一条 X 染色体来使额外的遗传信息实现剂量补偿。

■ 在一些爬行类动物中，受精卵孵化时的温度决定了后代性别。

In the biological world, a wide range of reproductive modes and life cycles are observed.

在生物界中，我们可以观察到各种各样的繁殖模式和生命周期。有些生物完全

Some organisms are entirely asexual, displaying no evidence of sexual reproduction. Other organisms alternate between short periods of sexual reproduction and prolonged periods of asexual reproduction. In most diploid eukaryotes, however, sexual reproduction is the only natural mechanism for producing new members of the species. The perpetuation of all sexually reproducing organisms depends ultimately on an efficient union of gametes during fertilization. In turn, successful fertilization depends on some form of **sexual differentiation** in the reproductive organisms. Even though it is not overtly evident, this differentiation occurs in organisms as low on the evolutionary scale as bacteria and single-celled eukaryotic algae.In more complex forms of life, the differentiation of the sexes is more evident as phenotypic dimorphism of males and females. The ancient symbol for iron and for Mars, depicting a shield and spear (♂), and the ancient symbol for copper and for Venus, depicting a mirror (♀), have also come to symbolize maleness and femaleness, respectively.

是无性生殖的，丝毫没有有性生殖的迹象。有些生物具有短暂的有性生殖和长期的无性生殖相互交替的生活周期。然而，在大多数二倍体真核生物中，有性生殖依然是物种产生新成员的唯一自然机制。所有有性生殖生物的生生不息最终都依赖于受精过程中配子的有效结合。反过来，成功的受精依赖于具有繁殖能力的生物体的**性别分化**形式。这种分化在进化维度处于低等生物的细菌和单细胞真核藻类中也同样存在，尽管并不十分明显。在更为复杂的生命形式中，性别分化则更为明显地表现为雄性和雌性表型二态现象，古代标志铁和火星的符号“盾和矛”（♂）、标志铜和金星的符号“镜子”（♀）现在分别用于代表雄性和雌性。

Dissimilar, or **heteromorphic, chromosomes**, such as the XY pair in mammals, characterize one sex or the other in a wide range of species, resulting in their label as **sex chromosomes**. Nevertheless, it is genes, rather than chromosomes, that ultimately serve as the underlying basis of **sex determination**. As we will see, some of these genes are present on sex chromosomes, but others are autosomal. Extensive investigation has revealed a wide variation in sex-chromosome systems—even

哺乳动物体内的XY染色体对形态不同，也称为**异形染色体**，它们在不同物种中分别表征不同性别，因此也被称为**性染色体**。然而，**性别决定**的基础归根结底是基因而不是染色体。正如将要看到的那样，这些基因中一些位于性染色体上，而另一些则位于常染色体上。广泛深入的研究已经揭示了多种多样的性染色体系统，即使进化关系密切的物种，性染色体系统也存在差异，这些均表明控制性别决定的机制在生命进化历程中已经经历了多次快速进

in closely related organisms—suggesting that mechanisms controlling sex determination have undergone rapid evolution many times in the history of life.

化。

In this chapter, we delve more deeply into what is known about the genetic basis for the determination of sexual differences, with a particular emphasis on two organisms: our own species, representative of mammals; and *Drosophila*, on which pioneering sex-determining studies were performed.

在本章中，我们将更加深入地探究决定性别差异的遗传基础，尤其重点关注两种生物：哺乳动物代表——人类自身，以及在性别决定研究中取得开创性成果的物种——果蝇。

5.1 X and Y Chromosomes Were First Linked to Sex Determination Early in the Twentieth Century

5.1 20世纪初，X和Y染色体首次确定与性别决定相关

How sex is determined has long intrigued geneticists. In 1891, Hermann Henking identified a nuclear structure in the sperm of certain insects, which he labeled the X-body. Several years later, Clarence McClung showed that some of the sperm in grasshoppers contain an unusual genetic structure, called a *heterochromosome*, but the remainder of the sperm lack this structure. He mistakenly associated the presence of the heterochromosome with the production of male progeny. In 1906, Edmund B. Wilson clarified Henking and McClung's findings when he demonstrated that female somatic cells in the butterfly *Protenor* contain 14 chromosomes, including two X chromosomes. During oogenesis, an even reduction occurs, producing gametes with seven chromosomes, including one X chromosome. Male somatic cells, on the other hand, contain only 13 chromosomes, including

遗传学家对于性别决定机制的研究兴趣由来已久。1891年，赫尔曼·亨金在一些昆虫精细胞中鉴定出核结构，并将其标记为X小体。几年后，克拉伦斯·麦克朗指出蝗虫的一些精细胞中含有不同的遗传结构，并称之为异染色体，而其他精细胞中缺少类似结构。他错误地将异染色体的存在与雄性后代的产生联系在一起。1906年，埃德蒙·B. 威尔逊阐明了亨金和麦克朗的发现，他证明凤蝶属蝴蝶的雌性体细胞中含14条染色体，包括2条X染色体。在卵子发生过程中，染色体数目发生均等的减半，产生含7条染色体的配子，并包含1条X染色体。而雄性体细胞仅含13条染色体，包含1条X染色体。在精子发生过程中，产生的配子或者仅含6条染色体，不含X染色体；或者含7条染色体，其中1条是X染色体。含X染色体的精子参与受精将产生雌性后代，缺少X染色体的精子

one X chromosome. During spermatogenesis, gametes are produced containing either six chromosomes, without an X, or seven chromosomes, one of which is an X. Fertilization by X-bearing sperm results in female offspring, and fertilization by X-deficient sperm results in male offspring [Figure 5.1(a)].

参与受精则产生雄性后代［图 5.1（a）］。

The presence or absence of the X chromosome in male gametes provides an efficient mechanism for sex determination in this species and also produces a 1:1 sex ratio in the resulting offspring.

雄性配子中 X 染色体的存在与否为该物种的性别决定提供了有效机制，并且也在后代中产生了 1 ∶ 1 的性别比。

Wilson also experimented with the milkweed bug *Lygaeus turcicus*, in which both sexes have 14 chromosomes. Twelve of these are autosomes (A). In addition, the females have two X chromosomes, while the males have only a single X and a smaller heterochromosome labeled the **Y chromosome**. Females in this species produce only gametes of the (6A + X) constitution, but males produce two types of gametes in equal proportions, (6A + X) and (6A + Y). Therefore, following random fertilization, equal numbers of male and female progeny

威尔逊同时还对以马利筋属植物为食的蝽进行了研究。蝽的两种性别均含 14 条染色体，其中 12 条是常染色体（A）。此外，雌性个体含 2 条 X 染色体，而雄性个体仅含 1 条 X 染色体和 1 条形态较小、被标记为 **Y 染色体**的异染色体。该物种的雌性个体只产生组成为 (6A+X) 的配子，而雄性个体等量产生组成分别为 (6A+X) 和 (6A+Y) 的两种配子。因此，在随机受精后，将产生数量相等、染色体组成不同的雄性后代和雌性后代［图 5.1（b）］。

(a) XX Female (12A + 2X)　XO Male (12A + X)

Gamete formation　Gamete formation

6A　6A + X

6A + X　Male (12A + X)　Female (12A + 2X)

1:1 sex ratio

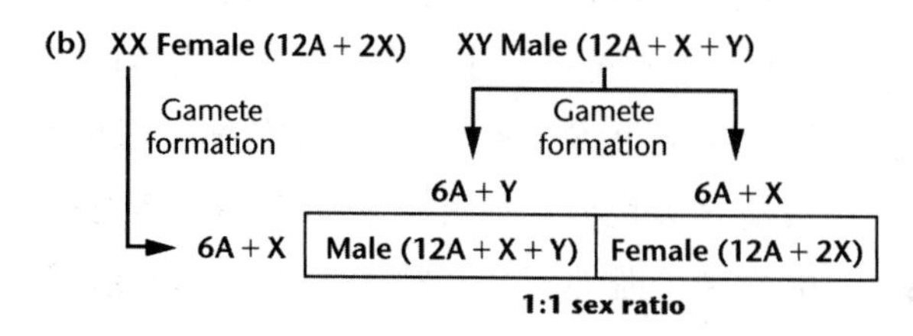

FIGURE 5.1 (a) Sex determination where the heterogametic sex (the male in this example) is XO and produces gametes with or without the X chromosome; (b) sex determination, where the heterogametic sex (again, the male in this example) is XY and produces gametes with either an X or a Y chromosome. In both cases, the chromosome composition of the offspring determines its sex.

图 5.1　(a) 异配性别为 XO（本例中雄性是异配性别）的性别决定，产生的配子要么含 X 染色体，要么不含 X 染色体；(b) 异配性别为 XY（本例中依然是雄性是异配性别）的性别决定，产生的配子要么含 X 染色体，要么含 Y 染色体。在这两种情况中，后代染色体的组成决定其性别。

will be produced with distinct chromosome complements [Figure 5.1(b)].

In *Protenor* and *Lygaeus* insects, males produce unlike gametes. As a result, they are described as the **heterogametic sex**, and in effect, their gametes ultimately determine the sex of the progeny in those species. In such cases, the female, which has like sex chromosomes, is the **homogametic sex**, producing uniform gametes with regard to chromosome numbers and types.

在凤蝶属和蝽属昆虫中，雄性个体产生组成不同的配子，因此，它们被称为**异配性别**，并且实际上，它们的配子最终决定了这些物种的后代性别。在这种情况下，雌性个体由于含相同的性染色体而被称为**同配性别**，并产生染色体数目和种类均相同的均一配子。

The male is not always the heterogametic sex. In some organisms, the female produces unlike gametes, exhibiting either the *Protenor* XX/XO or *Lygaeus* XX/XY mode of sex determination. Examples include certain moths and butterflies, some fish, reptiles, amphibians, at least one species of plants (*Fragaria orientalis*), and most birds. To immediately distinguish situations in which the female is the heterogametic sex, some geneticists use the notation **ZZ/ZW**, where ZZ is the homogametic male and ZW is the heterogametic female, instead of the XX/XY notation. For example, chickens are so denoted.

雄性个体并非总是异配性别。在某些生物中，雌性个体产生不同的配子类型，表现出类似凤蝶属 XX/XO 或者蝽属 XX/XY 的性别决定模式。这样的例子包括某些蛾类和蝴蝶类，一些鱼类、爬行动物、两栖动物，至少一种植物（东方草莓）和大多数鸟类。为了快速区分雌性个体是异配性别的情形，一些遗传学家不再使用 XX/XY 表示性染色体组成，而是使用符号 **ZZ/ZW** 来表示，其中 ZZ 代表同配性别雄性，而 ZW 则代表异配性别雌性。例如，鸡即是如此。

ESSENTIAL POINT

Specific sex chromosomes contain genetic information that controls sex determination and sexual differentiation.

基本要点

特定性染色体上包含控制性别决定和性别分化的遗传信息。

5.2 The Y Chromosome Determines Maleness in Humans

5.2 Y 染色体决定人类雄性发育

The first attempt to understand sex determination in our own species occurred

早在大约 100 年前，人们就试图开始了解人类自身的性别决定方式，并且对分裂

almost 100 years ago and involved the visual examination of chromosomes in dividing cells. Efforts were made to accurately determine the diploid chromosome number of humans, but because of the relatively large number of chromosomes, this proved to be quite difficult. Then, in 1956, Joe Hin Tjio and Albert Levan discovered an effective way to prepare chromosomes for accurate viewing. This technique led to a strikingly clear demonstration of metaphase stages showing that 46 was indeed the human diploid number. Later that same year, C. E. Ford and John L. Hamerton, also working with testicular tissue, confirmed this finding.

细胞中的染色体进行了可视化研究。人们为精确确定人类的二倍染色体数目付出了巨大努力，但是由于人类染色体数量相对较多，事实证明这项工作难度很大。随后，在 1956 年，蒋有兴和阿尔贝特 · 莱万发现一种可以使染色体便于准确观察的有效途径。这项技术可以十分清晰地展示细胞分裂中期阶段的染色体形态，并且表明人类二倍染色体数目确为 46 条。同年晚些时候，C.E. 福特和约翰 · L. 哈默顿通过研究睾丸组织同样证实了这一发现。

Of the normal 23 pairs of human chromosomes, one pair was shown to vary in configuration in males and females. These two chromosomes were designated the X and Y sex chromosomes. The human female has two X chromosomes, and the human male has one X and one Y chromosome.

在人类正常的 23 对染色体中，有一对染色体在男性和女性中形态有所不同，这两条染色体分别被称为 X 性染色体和 Y 性染色体。人类女性含两条 X 染色体，而人类男性则含一条 X 染色体和一条 Y 染色体。

We might believe that this observation is sufficient to conclude that the Y chromosome determines maleness. However, several other interpretations are possible. The Y could play no role in sex determination; the presence of two X chromosomes could cause femaleness; or maleness could result from the lack of a second X chromosome. The evidence that clarified which explanation was correct came from study of the effects of human sex-chromosome variations, described in the following section. As such investigations revealed, the Y chromosome does indeed determine maleness in humans.

我们可能认为这一观察足以得出 Y 染色体决定雄性发育这一结论，然而多种其他解释也是可能的。比如，Y 染色体在性别决定中不发挥作用；两条 X 染色体的存在决定雌性发育；或者第二条 X 染色体的缺失导致雄性发育。为了阐明这些解释中哪种正确，接下来介绍的人类性染色体异常所致效应的相关研究为此提供了证据。正如研究结果所表明的，Y 染色体的确决定了人类的雄性发育。

Klinefelter and Turner Syndromes

Around 1940, scientists identified two human abnormalities characterized by aberrant sexual development, **Klinefelter syndrome (47,XXY)** and **Turner syndrome (45,X)**. Individuals with Klinefelter syndrome are generally tall and often have long arms and legs. They usually have genitalia and internal ducts that are male, but their testes are reduced in size. Although 50 percent of affected individuals do produce sperm, a low sperm count renders most individuals sterile. At the same time, feminine sexual development is not entirely suppressed. Slight enlargement of the breasts (gynecomastia) is common, and the hips are often rounded. Individuals with Klinefelter syndrome most often show no cognitive reduction, and many individuals are unaware of having the disorder until they are treated for infertility.

克兰费尔特综合征和特纳综合征

1940 年左右，科学家发现两例性别发育异常的人类个体，即**克兰费尔特综合征（47,XXY）**和**特纳综合征（45,X）**。克兰费尔特综合征患者通常身材高大、四肢较长，他们常具有男性生殖器和输精管，但其睾丸缩小。尽管 50％的患者个体确实可以产生精子，但精子数量少，致使大多数个体不育。与此同时，女性性别发育并未被完全抑制，其乳房轻微变大形成男性女型乳房是常见现象，并且其臀部通常也表现为类似于女性的圆臀。患有克兰费尔特综合征的个体绝大多数不表现认知能力下降，许多人甚至直到接受不育治疗时才意识到自己患有该种疾病。

In Turner syndrome, the affected individual has female external genitalia and internal ducts, but the ovaries are rudimentary. Other characteristic abnormalities include short stature (usually under 5 feet), skin flaps on the back of the neck, and underdeveloped breasts. A broad, shieldlike chest is sometimes noted. Intelligence is usually normal.

在特纳综合征中，患病个体具有女性外生殖器和输卵管，但卵巢基本退化。其他异常特征包括身材矮小(通常 5 英尺以下，1 英尺为 30.48 厘米)，有颈蹼，乳房不发育，有时其胸廓呈盾状，智力水平通常正常。

In 1959, the karyotypes of individuals with these syndromes were determined to be abnormal with respect to the sex chromosomes. Individuals with Klinefelter syndrome have more than one X chromosome. Most often they have an XXY complement in addition to 44 autosomes, which is why people with this karyotype are designated 47,XXY. Individuals

1959 年，这些综合征患者的染色体核型得到确定并显示其性染色体组成异常。克兰费尔特综合征患者所含 X 染色体不止一条，他们中的绝大多数个体除含有 44 条常染色体外，还含 XXY 染色体，因此这些个体的核型用 47,XXY 表示。特纳综合征个体通常仅含 45 条染色体，包括单一的一条 X 染色体，因此，这些个体的核型用 45,X

with Turner syndrome most often have only 45 chromosomes, including just a single X chromosome; thus, they are designated 45,X. Note the convention used in designating these chromosome compo-sitions. The number states the total number of chromosomes present, and the information after the comma indicates the deviation from the normal diploid content. Both conditions result from **nondisjunction**, the failure of the sex chromosomes to segregate properly during meiosis (nondisjunction is described in Chapter 6 and illustrated in Figure 6.1).

表示。请注意染色体组成的书面表达惯例：数字表示细胞中所含染色体总数目，逗号后的信息则表示与正常二倍体组成相比有差异的信息。以上两种情况都是由**不分离**所致，即性染色体在减数分裂过程中不能正确地进行分离（不分离如图 6.1 所示，并在第 6 章中进行了介绍）。

These Klinefelter and Turner karyotypes and their corresponding sexual phenotypes led scientists to conclude that the Y chromosome determines maleness in humans. In its absence, the person's sex is female, even if only a single X chromosome is present. The presence of the Y chromosome in the individual with Klinefelter syndrome is sufficient to determine maleness, even though male development is not complete. Similarly, in the absence of a Y chromosome, as in the case of individuals with Turner syndrome, no masculinization occurs. Note that we cannot conclude anything regarding sex determination under circumstances where a Y chromosome is present without an X because Y-containing human embryos lacking an X chromosome (designated 45,Y) do not survive.

克兰费尔特和特纳核型及其相应的性别表型使得科学家确定 Y 染色体决定人类的雄性发育。当 Y 染色体缺失时，个体性别是女性，即使只存在一条 X 染色体。克兰费尔特综合征个体中 Y 染色体的存在足够决定其雄性发育，即使雄性发育并不完全。类似地，当 Y 染色体缺失时，如特纳综合征，则不发生男性化。这里需要说明的是，我们无法获得含一条 Y 染色体而不含 X 染色体时的性别决定情况，因为缺失 X 染色体而仅含 Y 染色体的胚胎（45,Y）无法存活。

Klinefelter syndrome occurs in about 1 of every 660 male births. The karyotypes **48,XXXY**, **48,XXYY**, **49,XXXXY**, and **49,XXXYY** are similar phenotypically to 47,XXY, but manifestations are often more

克兰费尔特综合征的发病率是每 660 名男性出生婴儿中有 1 名患病。核型为 **48,XXXY**，**48,XXYY**，**49,XXXXY** 和 **49,XXXYY** 的个体在表型上与 47,XXY 相似，但含 X 染色体数量越多的个体异常表现通

severe in individuals with a greater number of X chromosomes.

常越严重。

Turner syndrome can also result from karyotypes other than 45,X, including individuals called **mosaics**, whose somatic cells display two different genetic cell lines, each exhibiting a different karyotype. Such cell lines result from a mitotic error during early development, the most common chromosome combinations being **45,X/46,XY** and **45,X/46,XX**. Thus, an embryo that began life with a normal karyotype can give rise to an individual whose cells show a mixture of karyotypes and who exhibits varying aspects of this syndrome.

特纳综合征也可能由除 45,X 以外的其他核型引起，包括**镶嵌体**，即体细胞由两种不同遗传组成的细胞系构成，每种细胞系含有不同的核型。这些细胞系是由发育早期有丝分裂异常所致，最常见的染色体组合是 **45,X/46,XY** 和 **45,X/46,XX**。因此，起始核型正常的胚胎可能产生不同核型细胞混合的个体，该个体也将不同程度地表现出该综合征的相关特征。

Turner syndrome is observed in about 1 in 2000 female births, a frequency much lower than that for Klinefelter syndrome. One explanation for this difference is the observation that a substantial majority of 45,X fetuses die *in utero* and are aborted spontaneously. Thus, a similar frequency of the two syndromes may occur at conception.

特纳综合征的发病率是每 2000 名女性出生婴儿中大约有 1 名患病，其发生频率远低于克兰费尔特综合征。该差异的一种解释是人们观察到大多数核型为 45,X 的胎儿在子宫内便会死亡并发生自然流产。因此，这两种综合征在怀孕时的发生频率可能是相似的。

47,XXX Syndrome

The abnormal presence of three X chromosomes along with a normal set of autosomes (**47,XXX**) results in female differentiation. The highly variable syndrome that accompanies this genotype, often called **triplo-X**, occurs in about 1 of 1000 female births. Frequently, 47,XXX women are perfectly normal and may remain unaware of their abnormality in chromosome number unless a karyotype is done. In other cases, underdeveloped secondary sex characteristics, sterility, delayed development

47,XXX 综合征

正常常染色体组成与三条 X 染色体同时存在的异常现象（**47,XXX**）会导致雌性分化。这种基因型伴随的高度可变综合征通常被称为 **X- 三体**，发病率大约是每 1000 名女性出生个体中有 1 名患病。47,XXX 的女性表型完全正常，通常情况下除非进行核型鉴定，否则她们始终不知道自己的染色体数目异常。在其他一些病例中，可能出现第二性征发育不全、不育、语言和运动能力发育滞后以及智力发育迟缓。在极少数病例中，也报道有 **48,XXXX**（X- 四体）

of language and motor skills, and intellectual disability may occur. In rare instances, **48,XXXX** (tetra-X) and **49,XXXXX** (penta-X) karyotypes have been reported. The syndromes associated with these karyotypes are similar to but more pronounced than the 47,XXX syndrome. Thus, in many cases, the presence of additional X chromosomes appears to disrupt the delicate balance of genetic information essential to normal female development.

和 **49,XXXXX**（X- 五体）的核型组成。与这些核型相关的综合征与 47,XXX 综合征表型相似但外在表现更为明显。因此，在多数情况下，额外的 X 染色体存在似乎打破了正常雌性发育过程中必需遗传信息的微妙平衡。

47,XYY Condition

Another human condition involving the sex chromosomes is **47,XYY**. Studies of this condition, in which the only deviation from diploidy is the presence of an additional Y chromosome in an otherwise normal male karyotype, were initiated in 1965 by Patricia Jacobs. She discovered that 9 of 315 males in a Scottish maximum security prison had the 47,XYY karyotype. These males were significantly above average in height and had been incarcerated as a result of dangerous, violent, or criminal propensities. Of the nine males studied, seven were of subnormal intelligence, and all suffered personality disorders. Several other studies produced similar findings.

47,XYY 情况

涉及性染色体的另一种人类情形是 **47,XYY**。该疾病与二倍体的唯一差别在于比正常男性核型组成额外多出一条 Y 染色体。帕特里夏 · 雅各布斯于 1965 年开始研究这种疾病。她发现在苏格兰最高防御监狱中，315 名男性中有 9 名具有 47,XYY 核型。这些男性的身高明显高于平均水平，并且由于具有危险性、暴力或犯罪倾向而被监禁。在所研究的 9 名男性中，7 名智力发育低下，并且所有个体均患有人格障碍。其他一些研究也得出了类似发现。

The possible correlation between this chromosome composition and criminal behavior piqued considerable interest, and extensive investigation of the phenotype and frequency of the 47,XYY condition in both criminal and noncriminal populations ensued. Above-average height (usually over 6 feet) and subnormal intelligence were substantiated,

这种染色体组成与犯罪行为之间可能存在的相关性引起了人们的极大兴趣，随后人们对犯罪人群和非犯罪人群中 47,XYY 疾病的表型和频率开展了广泛研究。这些个体具有高于平均水平的身高（通常超过 6 英尺）和智力低下的特征被证实，并且相对于未被监禁人群，在刑事监禁所和精神病院中具有该种核型的男性个体比例确

and the frequency of males displaying this karyotype was indeed revealed to be higher in penal and mental institutions compared with unincarcerated populations (one study showed 29 XYY males when 28,366 were examined [0.10%]). A particularly relevant question involves the characteristics displayed by the XYY males who are not incarcerated. The only nearly constant association is that such individuals are over 6 feet tall.

实更高（一项研究表明在 28 366 人中有 29 位 XYY 男性，所占比例约为 0.10%）。一个高度与之相关的问题还涉及未被监禁的 XYY 男性所表现的特征，唯一保持不变的关联是这些个体的身高都超过了 6 英尺。

A study to further address this issue was initiated in 1974 to identify 47,XYY individuals at birth and to follow their behavioral patterns during preadult and adult development. While the study was considered unethical and soon abandoned, it has became clear that there are many XYY males present in the population who do not exhibit antisocial behavior and who lead normal lives. Therefore, we must conclude that there is a high, but not constant, correlation between the extra Y chromosome and the predisposition of these males to exhibit behavioral problems.

1974 年，人们开始了更进一步的研究，在出生时对 47,XYY 个体进行鉴定，并在成年前和成年发育过程中对其行为方式进行追踪研究。这项研究由于被认为违反伦理道德而很快被终止，同时诸多事实表明，很多存在于人群中的 XYY 男性并无反社会行为，过着正常的生活。因此，我们必须指出额外存在的 Y 染色体与男性个体是否表现行为问题存在高度相关性，但并非始终存在必然联系。

Sexual Differentiation in Humans

Once researchers had established that, in humans, it is the Y chromosome that houses genetic information necessary for maleness, they attempted to pinpoint a specific gene or genes capable of providing the "signal" responsible for sex determination. Before we delve into this topic, it is useful to consider how sexual differentiation occurs in order to better comprehend how humans develop into sexually dimorphic males and females. During early

人类性别分化

研究人员一旦确定人类 Y 染色体上确实含有雄性发育所必需的遗传信息，便会尝试定位能够提供性别决定所需“信号”的特定基因或基因群。在我们深入探讨该问题之前，有必要先了解性别分化是如何发生的，从而有助于更好地理解人类如何发育成两种不同性别：男性和女性。在发育早期，每个人类胚胎都会经历一段潜在的性别尚未分化期。到妊娠第五周，性腺原基（即将形成性腺的组织）形成一对与

development, every human embryo undergoes a period when it is potentially hermaphroditic. By the fifth week of gestation, gonadal primordia (the tissues that will form the gonad) arise as a pair of **gonadal** (**genital**) **ridges** associated with each embryonic kidney. The embryo is potentially hermaphroditic because at this stage its gonadal phenotype is sexually indifferent—male or female reproductive structures cannot be distinguished, and the gonadal ridge tissue can develop to form male or female gonads. As development progresses, primordial germ cells migrate to these ridges, where an outer cortex and inner medulla form (cortex and medulla are the outer and inner tissues of an organ, respectively). The cortex is capable of developing into an ovary, while the medulla may develop into a testis. In addition, two sets of undifferentiated ducts called the Wolffian and Müllerian ducts exist in each embryo. Wolffian ducts differentiate into other organs of the male reproductive tract, while Müllerian ducts differentiate into structures of the female reproductive tract.

胚胎肾脏相关的**生殖腺嵴（生殖嵴）**。胚胎此时是兼性的，因为此阶段其性腺表型在性别上尚未发生分化——雄性或雌性生殖结构尚无法区分，生殖腺嵴可以发育形成男性或女性性腺。随着发育进行，原始生殖细胞迁移至这些嵴上，并在此形成外部皮质和内部髓质（皮质和髓质分别是器官的外部组织和内部组织）。皮质能够发育成卵巢，而髓质可以发育成睾丸。此外，每个胚胎中还存在两套尚未分化的管道，即沃尔夫管和米勒管。沃尔夫管可分化为男性生殖道的其他器官，而米勒管可分化为女性生殖道的相关结构。

Because gonadal ridges can form either ovaries or testes, they are commonly referred to as **bipotential gonads**. What switch triggers gonadal ridge development into testes or ovaries? The presence or absence of a Y chromosome is the key. If cells of the ridge have an XY constitution, development of the medulla into a testis is initiated around the seventh week. However, in the absence of the Y chromosome, no male development occurs, the cortex of the ridge subsequently forms ovarian tissue, and

由于生殖腺嵴可形成卵巢或者睾丸，因此通常称其为**双潜能性腺**。那么到底是什么引发生殖腺嵴发育成睾丸或卵巢呢？ Y 染色体的存在与否是关键。如果生殖腺嵴上的细胞具有 XY 组成，那么在胚胎发育第 7 周左右髓质便开始发育成睾丸。然而，如果 Y 染色体不存在，雄性发育将不发生，生殖腺嵴的皮质随后便会发育形成卵巢组织，米勒管形成输卵管、子宫、子宫颈和部分阴道。相应的男性或女性生殖道系统将依据所启动的性别发育路径进行平行发

the Müllerian duct forms oviducts (Fallopian tubes), uterus, cervix, and portions of the vagina. Depending on which pathway is initiated, parallel development of the appropriate male or female duct system then occurs, and the other duct system degenerates. If testes differentiation is initiated, the embryonic testicular tissue secretes hormones that are essential for continued male sexual differentiation. As we will discuss in the next section, the presence of a Y chromosome and the development of The testes also inhibit formation of female reproductive organs.

育，而另一条生殖道系统则随之退化。如果启动睾丸分化，则胚胎睾丸组织将分泌男性性别分化所必需的激素。正如我们将在下一节中讨论的内容，Y 染色体存在和睾丸发育将同时抑制女性生殖器官的形成。

In females, as the twelfth week of fetal development approaches, the oogonia within the ovaries begin meiosis, and primary oocytes can be detected. By the twenty-fifth week of gestation, all oocytes become arrested in meiosis and remain dormant until puberty is reached some 10 to 15 years later. In males, on the other hand, primary spermatocytes are not produced until puberty is reached (see Figure 2.9).

在女性中，随着胚胎发育第 12 周的临近，卵巢中的卵原细胞开始进行减数分裂，并且可以检测到初级卵母细胞。到妊娠第 25 周，所有卵母细胞将停滞在减数分裂期并保持休眠状态，直至 10—15 年后青春期的到来。与此不同的是，男性直到青春期才产生初级精母细胞（见图 2.9）。

The Y Chromosome and Male Development

The human Y chromosome, unlike the X, was long thought to be mostly blank genetically. It is now known that this is not true, even though the Y chromosome contains far fewer genes than does the X. Data from the Human Genome Project indicate that the Y chromosome has at least 75 genes, compared to 900—1400 genes on the X. Current analysis of these genes and regions with potential genetic function reveals that some have homologous counterparts on the X chromosome and others do not. For example,

Y 染色体与雄性发育

与 X 染色体不同，人们长期以来一直认为人类 Y 染色体上不含基因信息。现在我们已知并非如此，尽管 Y 染色体上所含基因远比 X 染色体少。人类基因组计划提供的数据表明 Y 染色体上至少含 75 个基因，而 X 染色体上则含 900—1400 个基因。目前对于这些基因及潜在遗传功能区域的分析表明，有些在 X 染色体上有同源区域，而另一些则没有。例如，位于 Y 染色体两端的**假常染色体区**（PARs）与 X 染色体上相应区域具有同源性，并且在减数分裂过

present on both ends of the Y chromosome are so-called **pseudoautosomal regions** (**PARs**) that share homology with regions on the X chromosome and synapse and recombine with it during meiosis. The presence of such a pairing region is critical to segregation of the X and Y chromosomes during male gametogenesis. The remainder of the chromosome, about 95 percent of it, does not synapse or recombine with the X chromosome. As a result, it was originally referred to as the nonrecombining region of the Y (NRY). More recently, researchers have designated this region as the **male-specific region of the Y** (**MSY**). Some portions of the MSY share homology with genes on the X chromosome, and others do not.

程中会进行联会和重组。该配对区域的存在对于雄性配子发生过程中X和Y染色体的准确分离至关重要。Y染色体的其余部分约占95%，不与X染色体进行联会或重组，因此最初被称为Y染色体的非重组区（NRY）。最近，研究人员已将该区段重新命名为**Y染色体雄性特异区**（**MSY**）。该区域中某些部分与X染色体上的基因具有同源性，而其他部分则没有。

The human Y chromosome is diagrammed in Figure 5.2. The MSY is divided about equally between *euchromatic* regions, containing functional genes, and *heterochromatic* regions, lacking genes. Within euchromatin, adjacent to the PAR of the short arm of the Y chromosome, is a critical gene that controls male sexual development, called the ***sex-determining region Y*** (***SRY***). In humans, the absence of a Y chromosome almost always leads to female development; thus, this gene is absent from the X chromosome. At six to eight weeks of development, the *SRY* gene becomes active in XY embryos. *SRY* encodes a protein that causes the undifferentiated gonadal tissue of the embryo to form testes. This protein is called the **testis-determining factor** (**TDF**). *SRY* (or a closely related version) is present in all mammals thus far examined, indicative of its essential function

人类Y染色体如图5.2所示。MSY大致等分为两部分：含功能基因的常染色质区、缺乏基因的异染色质区。在常染色质区内，与Y染色体短臂PAR区相邻的是决定雄性性别发育的关键基因，被称为**Y染色体性别决定区**（***SRY***）。在人类中，Y染色体缺失几乎总是导致雌性发育，因此，该基因不存在于X染色体上。在胚胎发育的6—8周，*SRY*基因在XY胚胎中开始活跃。*SRY*编码的蛋白质会引导胚胎未分化的性腺组织发育成睾丸。该蛋白质被称为**睾丸决定因子**（**TDF**）。迄今为止，人们在所有哺乳动物中均发现了*SRY*基因或功能密切相关的类似基因的存在，这些均表明该基因在整个多元化的动物群体中具有十分重要的功能。

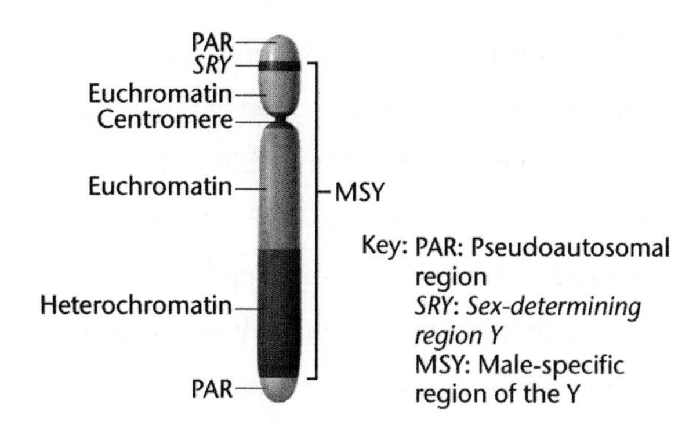

FIGURE 5.2 The regions of the human Y chromosome.

图 5.2 人类 Y 染色体的区域结构。

throughout this diverse group of animals.

Our ability to identify the presence or absence of DNA sequences in rare individuals whose sex-chromosome composition does not correspond to their sexual phenotype has provided evidence that *SRY* is the gene responsible for male sex determination. For example, there are human males who have two X and no Y chromosomes. Often, attached to one of their X chromosomes is the region of the Y that contains *SRY*. There are also females who have one X and one Y chromosome. Their Y is almost always missing the *SRY* gene.These observations argue strongly in favor of the role of *SRY* in providing the primary signal for male development.

在一些罕见个体中，他们的染色体组成与其相应的性别表型并不一致，通过鉴定这些个体的 DNA 序列存在与否为 *SRY* 基因是雄性性别决定基因提供了证据。例如，有些人类男性个体含两条 X 染色体而缺失 Y 染色体，通常他们的一条 X 染色体上附有一段含 *SRY* 基因的 Y 染色体区。还有一些女性个体拥有一条 X 染色体和一条 Y 染色体，但是她们的 Y 染色体几乎总是缺失 *SRY* 基因。这些发现为 *SRY* 基因发挥雄性发育初始信号的作用提供了坚实的证据支持。

Further support of this conclusion involves experiments using **transgenic mice**. These animals are produced from fertilized eggs injected with foreign DNA that is subsequently incorporated into the genetic composition of the developing embryo. In normal mice, a chromosome region designated *Sry* has been identified that is comparable to *SRY* in humans. When mouse DNA containing *Sry* is injected into normal XX mouse eggs, most of the offspring develop into males.

该结论进一步的证据支持来自**转基因小鼠**相关实验。这些动物是通过将外源 DNA 注入受精卵并整合入正在发育的胚胎遗传信息中而产生的。在正常小鼠中已鉴定出与人类 *SRY* 基因相当的染色体区域 *Sry* 基因。当将含 *Sry* 基因的小鼠 DNA 注入正常 XX 小鼠受精卵中时，大多数后代将发育成雄性。

The question of how the product of this gene triggers development of embryonic gonadal tissue into testes rather than ovaries is the key question under investigation. TDF functions as a *transcription factor*, a DNA-binding protein that interacts directly with the regulatory sequences of other genes to stimulate their expression. Thus, TDF behaves as a master switch that controls other genes downstream in the process of sexual differentiation. Interestingly, many identified thus far reside on autosomes, including the human *SOX9* gene located on chromosome 17.

该基因产物如何引发胚胎性腺组织发育成睾丸而不是卵巢是当前研究的关键问题。TDF 的功能是转录因子，即一种能与其他基因的调控序列直接相互作用从而刺激这些基因表达的 DNA 结合蛋白。因此，TDF 是性别分化过程中控制其他下游基因表达的主开关。有趣的是，迄今已鉴定出的许多下游基因都位于常染色体上，包括位于第 17 号染色体上的人类 *SOX9* 基因。

A more recent area of investigation has involved the Y chromosome and paternal age. For many years, it has been known that maternal age is correlated with an elevated rate of offspring with chromosomal defects, including Down syndrome (see Chapter 6). Advanced paternal age has now been associated with an increased risk in offspring of congenital disorders with a genetic basis, including certain cancers, schizophrenia, autism, and other conditions, collectively known as *paternal age effects* (*PAE*). Studies in which the genomes of sperm have been sequenced have demonstrated the presence of specific PAE mutations, including numerous ones on the Y chromosome. Evidence suggests that PAE mutations are positively selected for and result in an enrichment of mutant sperm over time.

近期的研究内容涉及 Y 染色体和父本年龄。多年以来，人们已知产妇年龄与包括唐氏综合征在内的后代染色体缺陷的发生率升高存在关联（请参阅第 6 章）。父本高龄目前也发现与后代先天性遗传疾病的发病风险升高有关，包括某些癌症、精神分裂症、孤独症和其他疾病，这些被称为父亲年龄效应（PAE）。精子基因组的测序研究表明存在特定的 PAE 突变，包括 Y 染色体上存在的大量突变。有证据表明，PAE 突变随着时间推移将被正向选择从而导致突变精子的富集。

ESSENTIAL POINT

The presence or absence of a Y chromosome that contains an intact *SRY* gene is responsible for causing maleness in humans.

基本要点

含完整 *SRY* 基因的 Y 染色体存在与否决定了人类的雄性发育。

5.3 The Ratio of Males to Females in Humans Is Not 1.0

The presence of heteromorphic sex chromosomes in one sex of a species but not the other provides a potential mechanism for producing equal proportions of male and female offspring. This potential depends on the segregation of the X and Y (or Z and W) chromosomes during meiosis, such that half of the gametes of the heterogametic sex receive one of the chromosomes and half receive the other one. As we learned in the previous section, small pseudoautosomal regions of pairing homology do exist at both ends of the human X and Y chromosomes, suggesting that the X and Y chromosomes do synapse and then segregate into different gametes. Provided that both types of gametes are equally successful in fertilization and that the two sexes are equally viable during development, a 1:1 ratio of male and female offspring should result.

The actual proportion of male to female offspring, referred to as the **sex ratio**, has been assessed in two ways. The **primary sex ratio** (**PSR**) reflects the proportion of males to females conceived in a population. The **secondary sex ratio** reflects the proportion of each sex that is born. The secondary sex ratio is much easier to determine but has the disadvantage of not accounting for any disproportionate embryonic or fetal mortality.

When the secondary sex ratio in the human population was determined in 1969 by using

5.3 人类男女性别比并非 1.0

物种中一种性别含异形性染色体组成而另一种性别不含的现象为产生等比例的雄性后代和雌性后代提供了潜在机制。该机制依赖于减数分裂过程中 X 和 Y（或 Z 和 W）染色体的相互分离，这样异配性别产生的配子中一半获得其中一条染色体，另一半则获得另一条染色体。正如我们在上一节中所了解到的，人类 X 和 Y 染色体的两端均存在小的、具有配对同源性的假常染色体区，表明 X、Y 染色体的确可以发生联会，并通过分离进入不同配子中。假设两种不同类型的配子均能同等成功地参与受精，并且在发育过程中两种不同性别具有同等的存活力，那么后代中雄性与雌性的比例将为 1 ∶ 1。

男性和女性后代的实际比例，即**性别比**，可以通过两种方式进行评估。**初级性比**（PSR）反映了群体中所孕育的男性胚胎与女性胚胎的比例。**次级性比**反映了出生个体中两种性别的比例。次级性比的确定相对容易，但其缺陷在于没有考虑任何不成比例的胚胎或胎儿死亡率。

1969 年，人们根据全球人口普查数据确定了人类次级性比，该比例并不等于 1.0。

worldwide census data, it did not equal 1.0. For example, in the Caucasian population in the United States, the secondary ratio was a little less than 1.06, indicating that about 106 males were born for each 100 females. In 1995, this ratio dropped to slightly less than 1.05. In the African-American population in the United States, the ratio was 1.025. In other countries, the excess of male births is even greater than is reflected in these values. For example, in Korea, the secondary sex ratio was 1.15.

例如，在美国白种人人口中，次级性比略低于 1.06，即每 100 名女性出生的同时大约有 106 名男性出生。1995 年，该比率下降至略小于 1.05。在美国非裔人口中，该比率为 1.025。在其他国家中，男性出生过量则比这些数值显示的还要高。例如，在韩国，次级性比为 1.15。

Despite these ratios, it is possible that the PSR is 1.0 and is altered between conception and birth. For the secondary ratio to exceed 1.0, then, prenatal female mortality would have to be greater than prenatal male mortality. However, when this hypothesis was first examined, it was deemed to be false. In a Carnegie Institute study, reported in 1948, the sex of approximately 6000 embryos and fetuses recovered from miscarriages and abortions was determined, and fetal mortality was actually higher in males. On the basis of the data derived from that study, the PSR in U.S. Caucasians was estimated to be 1.079, suggesting that more males than females are conceived in the human population.

尽管存在这样的次级性比，但是初级性比仍可能是 1.0，只是在受孕和出生阶段发生了变化。由于次级性比大于 1.0，所以产前女性胚胎死亡率必定大于产前男性胚胎死亡率。然而，该假设在首次被验证时即被认为是错误的。1948 年报道的一项卡内基研究所的研究确定了大约 6000 名流产胚胎或流产胎儿的性别，实际上男性胚胎的死亡率更高。基于该研究得出的数据，美国白种人的初级性比预计为 1.079，即人群中孕育的男性胚胎数量多于女性胚胎数量。

To explain why, researchers examined the assumptions on which the theoretical ratio is based:

为了解释其中原因，研究人员重新研究了理论性别比所基于的假设：

1. Because of segregation, males produce equal numbers of X- and Y-bearing sperm.

1. 由于染色体分离，男性个体产生等量的含 X 染色体的精子和含 Y 染色体的精子。

2. Each type of sperm has equivalent viability and motility in the female reproductive tract.

2. 每种类型的精子在女性生殖道中都具有同等的存活力和运动性。

3. The egg surface is equally receptive to

3. 卵的表面对于含 X 染色体的精子或

both X- and Y-bearing sperm.

No direct experimental evidence contradicts any of these assumptions.

A PSR favoring male conceptions remained dogma for many decades until, in 2015, a study using an extensive dataset was published that concludes that the PSR is 1.0 – suggesting that equal numbers of males and females are indeed conceived. Among other parameters, the examination of the sex of 3-day-old and 6-day-old embryos conceived using assisted reproductive technology provided the most direct assessment. Following conception, however, mortality was then shown to fluctuate between the sexes, until at birth, more males than females are born. Thus, female mortality during embryonic and fetal development exceeds that of males. Clearly, this is a difficult topic to investigate but one of continued interest. For now, the most recent findings are convincing and contradict the earlier studies.

含 Y 染色体的精子的接受能力相同。

目前尚无直接相关的实验证据与这些假设相抵触。

认为初级性比中男性胚胎比例较高的信条已经存在几十年，直至 2015 年，人们利用广泛的数据进行研究得出初级性比为 1.0 的结论并公开发表，从而表明人类受孕时男性胚胎和女性胚胎的数量确实是相等的。在其他研究数据中，使用辅助生殖技术鉴定 3 日龄和 6 日龄胚胎的性别则提供了最为直接的评估。然而受孕后，不同性别胚胎的死亡率存在波动，直至出生时，男性个体的出生人数多于女性个体。因此，在胚胎和胎儿发育过程中女性死亡率高于男性。显然，这是一个很难进行研究，但人们仍将持续关注的课题。目前，最新的研究发现既有与早期研究结果相一致的，也有与之相悖的。

ESSENTIAL POINT

In humans, the sex ratio at conception and birth remains an active area of research. The most current study shows that equal numbers of males and females are conceived, but that more males than females are born.

基本要点

在人类中，受孕性别比和出生性别比依然是科学研究的活跃领域。最新研究表明，受孕胚胎中男性和女性的数量相等，但是在出生个体中男性多于女性。

5.4 Dosage Compensation Prevents Excessive Expression of X-Linked Genes in Humans and Other Mammals

5.4 剂量补偿避免人类及其他哺乳类动物 X 连锁基因的过量表达

The presence of two X chromosomes in normal human females and only one X in normal

与两性细胞中常染色体数目相等不同，性染色体在两性细胞中的存在方式非常独

human males is unique compared with the equal numbers of autosomes present in the cells of both sexes. On theoretical grounds alone, it is possible to speculate that this disparity should create a "genetic dosage" difference between males and females, with attendant problems, for all X-linked genes. There is the potential for females to produce twice as much of each product of all X-linked genes. The additional X chromosomes in both males and females exhibiting the various syndromes discussed earlier in this chapter are thought to compound this dosage problem. Embryonic development depends on proper timing and precisely regulated levels of gene expression. Otherwise, disease phenotypes or embryonic lethality can occur. In this section, we will describe research findings regarding X-linked gene expression that demonstrate a genetic mechanism of **dosage compensation** that balances the dose of X chromosome gene expression in females and males.

特：正常女性含两条 X 染色体，而正常男性仅含一条 X 染色体。仅从理论出发，我们可以推断这一不同将产生男性与女性间所有 X 连锁基因的"基因剂量"差异，以及随之而来的一些问题。对于 X 连锁基因，女性将产生两倍于男性的"基因剂量"。之前我们所讨论的在男性和女性中由于额外 X 染色体的存在所导致的各种异常性别综合征可能正是由于这种基因产物剂量问题所致。胚胎发育取决于适当的时机和基因表达的精确调控水平。否则，疾病表型或胚胎死亡即会发生。在本节中，我们将介绍 X 连锁基因表达的相关研究发现，这些发现展示了**剂量补偿**的遗传机制，从而实现男性和女性 X 染色体上基因表达剂量的平衡。

Barr Bodies

Murray L. Barr and Ewart G. Bertram's experiments with cats, as well as Keith Moore and Barr's subsequent study in humans, demonstrate a genetic mechanism in mammals that compensates for X chromosome dosage disparities. Barr and Bertram observed a darkly staining body in the interphase nerve cells of female cats that was absent in similar cells of males. In humans, this body can be easily demonstrated in female cells derived from the buccal mucosa (cheek cells) or in fibroblasts (undifferentiated connective tissue cells), but

巴氏小体

默里 · L. 巴尔和尤尔特 · G. 伯特伦对猫的研究实验，以及随后基思 · 穆尔和巴尔对于人类的研究均揭示了哺乳动物中平衡 X 染色体基因剂量差异的遗传机制。巴尔和伯特伦发现在雌性猫的间期神经细胞中有一个深度着色的小体，而在雄性猫的类似细胞中却没有。在人类细胞中，这种小体在女性黏膜细胞（口腔上皮细胞）或成纤维细胞（未分化的结缔组织细胞）中均能很容易地找到，但在类似的男性细胞中却没有。这种高度浓缩的结构，直径约为 1 μm，位于间期细胞的核膜内侧，并且

not in similar male cells. This highly condensed structure, about 1 μ m in diameter, lies against the nuclear envelope of interphase cells, and it stains positively for a number of different DNA-binding dyes.

能与多种 DNA 结合染料结合呈现出阳性着色。

This chromosome structure, called a **sex chromatin body**, or simply a **Barr body**, is an inactivated X chromosome. Susumo Ohno was the first to suggest that the Barr body arises from one of the two X chromosomes. This hypothesis is attractive because it provides a possible mechanism for dosage compensation. If one of the two X chromosomes is inactive in the cells of females, the dosage of genetic information that can be expressed in males and females will be equivalent. Convincing, though indirect, evidence for this hypothesis comes from study of the sex-chromosome syndromes described earlier in this chapter. Regardless of how many X chromosomes a somatic cell possesses, all but one of them appear to be inactivated and can be seen as Barr bodies. For example, no Barr body is seen in the somatic cells of Turner 45,X females; one is seen in Klinefelter 47,XXY males; two in 47,XXX females; three in 48,XXXX females; and so on (Figure 5.3). Therefore, the number of Barr bodies follows an $N-1$ rule, where N is the total number of X chromosomes present.

这种染色体结构被称为**性染色质体**或**巴氏小体**，是一条失活的 X 染色体。大野乾最早提出了巴氏小体由两条 X 染色体中的一条变化而来的假设。这个假设一经提出就非常引人注目，因为它提供了一种可能的基因剂量补偿机制。如果雌性细胞的两条 X 染色体中有一条发生失活，那么雄性细胞和雌性细胞中 X 染色体上遗传信息的表达剂量将相等。我们在本章的开始介绍的性染色体综合征为这一假说提供了间接但极具说服力的证据。不管体细胞中有多少条 X 染色体存在，除一条 X 染色体外，其他均发生失活并以巴氏小体的形态出现。例如，特纳综合征女性个体 45,X 没有巴氏小体；克兰费尔特综合征男性个体 47,XXY 有 1 个巴氏小体；女性个体 47,XXX 有 2 个巴氏小体；以此类推（图 5.3）。因此，巴氏小体的数量遵循 $N-1$ 法则，其中 N 为细胞中存在的 X 染色体总数目。

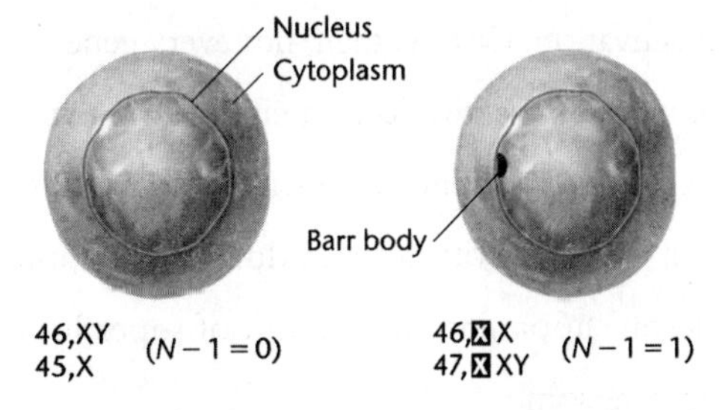

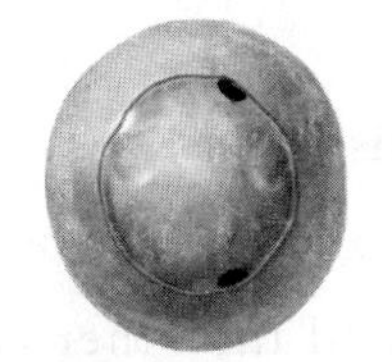

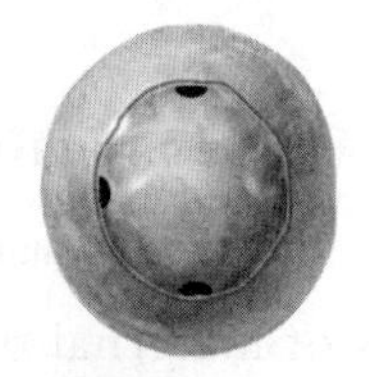

47,⊠⊠X
48,⊠⊠XY
(N − 1 = 2)
48,⊠⊠⊠X
49,⊠⊠⊠XY
(N − 1 = 3)

FIGURE 5.3 Occurrence of Barr bodies in various human karyotypes, where all X chromosomes except one ($N-1$) are inactivated.

图 5.3 不同人类核型中巴氏小体的存在情况，除 1 条 X 染色体外，其他所有的 X 染色体 ($N-1$) 均发生失活。

Although this apparent inactivation of all but one X chromosome increases our understanding of dosage compensation, it further complicates our perception of other matters. For example, because one of the two X chromosomes is inactivated in normal human females, why then is the Turner 45,X individual not entirely normal? Why aren't females with the triplo-X and tetra-X karyotypes (47,XXX and 48,XXXX) completely unaffected by the additional X chromosome? Furthermore, in Klinefelter syndrome (47,XXY), X chromosome inactivation effectively renders the person 46,XY. Why aren't these males unaffected by the extra X chromosome in their nuclei?

尽管除一条X染色体外其他所有X染色体均发生明显失活有助于我们加深对剂量补偿的理解，但也使我们对于其他问题的认识变得更加复杂。例如，由于两条X染色体中有一条发生失活，那么为什么特纳综合征患者（45,X）不能表现出完全正常的性别表型呢？为什么X-三体（47,XXX）或X-四体（48,XXXX）女性个体的性别表型不能完全不受额外存在的X染色体的影响呢？此外，在克兰费尔特综合征(47,XXY)中，X染色体的失活可以有效地使这些个体成为46,XY，但是这些男性患者为什么依然受到细胞核内额外存在的X染色体的影响呢？

One possible explanation is that chromosome inactivation does not normally occur in the very early stages of development of those cells destined to form gonadal tissues. Another possible explanation is that not all genes on each X chromosome forming a Barr body are inactivated. Recent studies have indeed demonstrated that as many as 15 percent of the human X chromosomal genes actually escape inactivation. Clearly, then, not every gene on the X requires inactivation. In either case, excessive expression of certain X-linked genes might still occur at critical times during development despite apparent inactivation of superfluous X chromosomes.

一种可能的解释是：那些注定形成性腺组织的细胞在发育最早期时X染色体失活并未发生。另一种可能的解释是：形成巴氏小体的X染色体上并不是所有基因均发生失活。近期研究表明人类X染色体上多达15%的基因的确发生了失活逃逸。因此，显然X染色体上并非每个基因均发生失活。因此无论如何，虽然细胞中多余的X染色体发生明显失活，但是某些X连锁基因仍可能在发育的关键时刻发生过度表达。

The Lyon Hypothesis

In mammalian females, one X chromosome is of maternal origin, and the other is of paternal origin. Which one is inactivated? Is the

莱昂假说

在雌性哺乳动物中，一条X染色体源于母本，另一条X染色体源于父本。那么到底是哪条X染色体发生失活？X染色体

inactivation random? Is the same chromosome inactive in all somatic cells? In the early 1960s, Mary Lyon, Liane Russell, and Ernest Beutler independently proposed a hypothesis that answers these questions. They postulated that the inac-tivation of X chromosomes occurs randomly in somatic cells at a point early in embryonic development, most likely sometime during the blastocyst stage of development. Once inactivation has occurred, all descendant cells have the same X chromosome inactivated as their initial progenitor cell.

失活是随机发生的吗？在所有体细胞中是同一条 X 染色体发生失活吗？ 20 世纪 60 年代初期，玛丽 · 莱昂、利亚纳 · 拉塞尔和埃内斯特 · 博伊特勒各自独立地提出假设回答了这些问题。他们推测在胚胎发育早期的体细胞中，X 染色体发生随机失活，这一时期十分可能是在胚囊期。一旦 X 染色体发生失活，该初始细胞的所有后代细胞中同一来源的 X 染色体也将随即发生失活。

This explanation, which has come to be called the **Lyon hypothesis**, was initially based on observations of female mice heterozygous for X-linked coat-color genes. The pigmentation of these heterozygous females was mottled, with large patches expressing the color allele on one X and other patches expressing the allele on the other X. This is the phenotypic pattern that would be expected if different X chromosomes were inactive in adjacent patches of cells. Similar mosaic patterns occur in the black and yellow-orange patches of female tortoiseshell and calico cats. Such X-linked coat color patterns do not occur in male cats because all their cells contain the single maternal X chromosome and are therefore hemizygous for only one X-linked coat-color allele.

这一解释最初基于对 X 连锁皮毛颜色决定基因杂合的雌性小鼠的观察，并逐渐形成**莱昂假说**。这些雌性杂合个体的皮毛颜色是斑驳的，一些花色由位于一条 X 染色体上的颜色基因表达决定，而另外一些花色则由另一条 X 染色体上的等位基因决定。这种花纹表型可能是由于相邻细胞中不同 X 染色体发生失活所致。类似的镶嵌花色也出现在黄黑橙相间的雌性玳瑁猫和三色猫中，这种 X 连锁的皮毛花色在雄性猫中不会出现，因为它们所有体细胞中都只含一条来自母本的 X 染色体，因此对于该 X 连锁基因为半合子。

The most direct evidence in support of the Lyon hypothesis comes from studies of gene expression in clones of human fibroblast cells. Individual cells are isolated following biopsy and cultured *in vitro*. A culture of cells derived from a single cell is called a **clone**. The synthesis of the

支持莱昂假说最直接的证据来自人类成纤维细胞克隆中基因表达的研究。单个细胞经活检后进行分离并在体外培养。源自单细胞的细胞体外培养物被称为一个**克隆**。葡萄糖 -6- 磷酸脱氢酶（G6PD）的合成受 X 连锁基因控制。该基因的许多突变

enzyme glucose-6- phosphate dehydrogenase (G6PD) is controlled by an X-linked gene. Numerous mutant alleles of this gene have been detected, and their gene products can be differentiated from the wild-type enzyme by their migration pattern in an electrophoretic field.

型等位基因均已被鉴定，并且它们的基因产物可通过其在电泳场中迁移模式的不同而与野生型葡萄糖-6-磷酸脱氢酶相区分。

Fibroblasts have been taken from females heterozygous for different allelic forms of *G6PD* and studied. The Lyon hypothesis predicts that if inactivation of an X chromosome occurs randomly early in development, and thereafter all progeny cells have the same X chromosome inactivated as their progenitor, such a female should show two types of clones, each containing only one electrophoretic form of *G6PD* product, in approximately equal proportions. This prediction has been confirmed experimentally, and studies involving modern techniques in molecular biology have clearly established that X chromosome inactivation occurs.

研究人员从杂合的、含不同 *G6PD* 等位基因的女性个体中获取了人类成纤维细胞并对其进行了研究。莱昂假说预测，如果 X 染色体失活随机发生在发育早期，随后所有后代细胞中与其祖先细胞来源相同的 X 染色体将发生失活，那么该女性个体应该呈现两种类型的细胞克隆，每种克隆仅含一种 *G6PD* 基因产物的电泳迁移模式，并且比例大致相等。该预测通过实验已得到证实，利用现代分子生物学技术进行的相关研究也已明确确定 X 染色体失活的发生。

One ramification of X-inactivation is that mammalian females are mosaics for all heterozygous X-linked alleles—some areas of the body express only the maternally derived alleles, and others express only the paternally derived alleles. An especially interesting example involves **red-green color blindness**, an X-linked recessive disorder. In humans, hemizygous males are fully color-blind in all retinal cells. However, heterozygous females display mosaic retinas, with patches of defective color perception and surrounding areas with normal color perception. In this example, random inactivation of one or the other X chromosome early in the development of heterozygous females has led to

X 染色体失活的一个结果是雌性哺乳动物对于所有杂合的 X 连锁基因均为镶嵌体——身体的某些部分仅表达母本来源的等位基因，而其他部分则仅表达父本来源的等位基因。一个特别有趣的例子是**红绿色盲**，这是一种 X 连锁隐性遗传疾病。在人类中，半合子男性个体所有的视网膜细胞均为色觉缺陷的，为完全色盲。然而，杂合女性个体的视网膜则呈现马赛克式镶嵌分布，色觉缺陷的视网膜细胞区域成块分布，并且其周围还分布有色觉正常的视网膜细胞区域。在这个例子中，杂合女性个体在胚胎发育早期两条 X 染色体中的任意一条发生随机失活最终导致了这些表型。

these phenotypes.

The Mechanism of Inactivation

The least understood aspect of the Lyon hypothesis is the mechanism of X chromosome inactivation. Somehow, either DNA, the attached histone proteins, or both DNA and histone proteins are chemically modified, silencing most genes that are part of that chromosome. Once silenced, a memory is created that keeps the same homolog inactivated following chromosome replications and cell divisions. Such a process, whereby expression of genes on one homolog, but not the other, is affected, is referred to as **imprinting**. This term also applies to a number of other examples in which genetic information is modified and gene expression is repressed. Collectively, such events are part of the growing field of **epigenetics** (see Chapter 9—Epigenetics).

失活机制

莱昂假说中最鲜为人知的部分是 X 染色体的失活机制。DNA 分子、与 DNA 相结合的组蛋白，或者 DNA 分子与组蛋白一同被化学修饰，从而沉默染色体上大多数基因。这些基因一旦发生沉默，细胞即会形成记忆，并在随后的染色体复制和细胞分裂中使同一来源的染色体继续保持失活状态。这种一条同源染色体上的基因表达受影响而另一条不受影响的过程被称为**印记**。该术语也适用于许多其他遗传信息被修饰和基因表达被抑制的类似情形。总的说来，这些生物学事件都是正在不断发展的**表观遗传学**领域的一部分（请参阅第 9 章——表观遗传学）。

Ongoing investigations are beginning to clarify the mechanism of inactivation. A region of the mammalian X chromosome is the major control unit. This region, located on the proximal end of the p arm in humans, is called the **X-inactivation center** (***Xic***), and its genetic expression *occurs only on the X chromosome that is inactivated*. The *Xic* is about 1 Mb (10^6 base pairs) in length and is known to contain several putative regulatory units and four genes. One of these, ***X-inactive specific transcript*** (***XIST***), is now known to be a critical gene for X-inactivation.

持续开展的研究正在逐步解析 X 染色体失活的分子机制。哺乳动物 X 染色体上存在一个主控单元区，该区域位于 X 染色体 p 臂靠近着丝粒的一端，被称为 **X- 失活中心**（***Xic***），它仅在发生失活的 X 染色体上有基因表达。*Xic* 区全长约 1 Mb（10^6 bp），含多个推定的调控单元和四个基因。**X- 失活特异转录物**（***XIST***）基因是其中一种基因，现已知是X染色体失活过程中的关键基因。

Interesting observations have been made regarding the RNA that is transcribed from the *XIST* gene, many coming from experiments that

人们利用小鼠同源基因（*Xist*）进行的大量实验发现了一些 *XIST* 基因转录得到的 RNA 相关的有趣现象。首先，RNA 转录产

focused on the equivalent gene in the mouse (*Xist*). First, the RNA product is quite large and does not encode a protein, and thus is not translated. The RNA products of *Xist* spread over and coat the X chromosome *bearing the gene that produced them*. Two other noncoding genes at the *Xic* locus, *Tsix* (an antisense partner of *Xist*) and *Xite*, are also believed to play important roles in X-inactivation.

物非常大，而且并不编码蛋白质，因此也不会被翻译。*Xist* 基因转录的 RNA 产物分布并包裹在表达该基因的 X 染色体表面。*Xic* 位点上的另外两个非编码基因 *Tsix*（*Xist* 基因的反义链基因）和 *Xite* 同样被认为在 X 染色体失活中发挥着重要作用。

A second observation is that transcription of *Xist* initially occurs at low levels on all X chromosomes. As the inactivation process begins, however, transcription continues, and is enhanced, only on the X chromosome that becomes inactivated. In 1996, a research group led by Neil Brockdorff and Graeme Penny provided convincing evidence that transcription of *Xist* is the critical event in chromosome inactivation. These researchers introduced a targeted deletion (7 kb) into this gene, disrupting its sequence. As a result, the chromosome bearing the deletion lost its ability to become inactivated.

第二个观察现象是最初在所有 X 染色体上均存在低水平的 *Xist* 基因转录。然而，随着失活进程的启动，转录将仅在失活的 X 染色体上继续增强。1996 年，由尼尔·布罗克多夫和格雷姆·彭尼领导的科研小组提供了令人信服的实验证据证明了 *Xist* 基因转录是染色体失活的关键事件。研究人员在该基因中引入长为 7 kb 的靶向缺失破坏该基因序列，结果这些含缺失突变的染色体丧失了失活的能力。

ESSENTIAL POINT

In mammals, female somatic cells randomly inactivate one of two X chromosomes during early embryonic development, a process important for balancing the expression of X chromosome-linked genes in males and females.

基本要点

在哺乳动物中，雌性体细胞两条 X 染色体中的一条在早期胚胎发育过程中随机发生失活，这一过程对于平衡雄性和雌性个体中 X 染色体连锁基因的表达至关重要。

5.5 The Ratio of X Chromosomes to Sets of Autosomes Can Determine Sex

5.5 X 染色体数目与常染色体组数的比值可以决定性别

We now discuss two interesting cases where

现在我们讨论两个 Y 染色体不参与性

the Y chromosome does not play a role in sex determination. First, in the fruit fly, *Drosophila melanogaster*, even though most males contain a Y chromosome, the Y plays no role. Second, in the roundworm, *Caenorhabditis elegans*, the organism lacks a Y chromosome altogether. In both cases, we shall see that the critical factor is the ratio of X chromosomes to the number of sets of autosomes.

别决定的有趣情形。首先，在黑腹果蝇中，即使大多数雄性个体均含一条 Y 染色体，但是 Y 并不参与性别决定。其次，在秀丽隐杆线虫中，Y 染色体完全缺失。在这两种情况下，我们将看到性别决定的关键因素是 X 染色体数目与常染色体组数之比。

Drosophila melanogaster

Because males and females in *Drosophila melanogaster* (and other *Drosophila* species) have the same general sex-chromosome composition as humans (males are XY and females are XX), we might assume that the Y chromosome also causes maleness in these flies. However, the elegant work of Calvin Bridges in 1921 showed this not to be true. His studies of flies with quite varied chromosome compositions led him to the conclusion that the Y chromosome is not involved in sex determination in this organism. Instead, Bridges proposed that the X chromosomes and autosomes together play a critical role in sex determination.

果蝇

由于黑腹果蝇（包括其他果蝇物种）的雄性个体和雌性个体均具有与人类相同的性染色体组成（雄性为 XY、雌性为 XX），因此我们可能会认为 Y 染色体同样也决定着这些果蝇的雄性发育。但是，1921 年，卡尔文 · 布里奇斯的杰出工作表明并非如此。他研究了染色体组成各异的果蝇并由此得出结论，即 Y 染色体并不参与该生物的性别决定。并且，布里奇斯指出在果蝇性别决定中 X 染色体和常染色体共同扮演着重要角色。

Bridges' work can be divided into two phases: (1) A study of offspring resulting from nondisjunction of the X chromosomes during meiosis in females and (2) subsequent work with progeny of females containing three copies of each chromosome, called triploid (3*n*) females. As we have seen previously in this chapter (and as you will see in Figure 6.1), nondisjunction is the failure of paired chromosomes to segregate

布里奇斯的工作可以分为两部分：(1) 研究雌性个体在减数分裂过程中由于 X 染色体不分离所产生的后代；(2) 随后研究三倍体（3*n*）雌蝇的后代，三倍体雌蝇中每种染色体均含三个拷贝。正如我们在本章前面所见（以及你将在图 6.1 所见），不分离指在第一次减数分裂后期或第二次减数分裂后期配对的染色体分离失败，结果将产生两种类型的异常配子，其中一种包

or separate during the anaphase stage of the first or second meiotic divisions. The result is the production of two types of abnormal gametes, one of which contains an extra chromosome ($n+1$) and the other of which lacks a chromosome ($n-1$). Fertilization of such gametes with a haploid gamete produces ($2n+1$) or ($2n-1$) zygotes. As in humans, if nondisjunction involves the X chromosome, in addition to the normal complement of autosomes, both an XXY and an XO sex-chromosome composition may result. (The "O" signifies that neither a second X nor a Y chromosome is present, as occurs in XO genotypes of individuals with Turner syndrome.)

含一条额外染色体（$n+1$），而另一种则缺少一条染色体（$n-1$）。这些异常配子与单倍体配子受精后则产生（$2n+1$）或（$2n-1$）受精卵。正如在人类中一样，如果X染色体发生不分离，那么将产生常染色体组成正常，性染色体组成分别为XXY和XO的配子类型。（"O"表示不存在第二条X染色体也不存在Y染色体，正如特纳综合征患者的XO基因型。）

Contrary to what was later discovered in humans, Bridges found that the XXY flies were normal females and the XO flies were sterile males. The presence of the Y chromosome in the XXY flies did not cause maleness, and its absence in the XO flies did not produce femaleness. From these data, Bridges concluded that the Y chromosome in *Drosophila* lacks male-determining factors, but since the XO males were sterile, it does contain genetic information essential to male fertility.

与随后在人类中的发现不同，布里奇斯发现XXY是正常雌蝇，而XO则是不育雄蝇。XXY个体中Y染色体的存在并不引发雄性发育，而XO个体中Y染色体的缺失也不会导致雌性发育。布里奇斯从这些数据中得出结论，即果蝇Y染色体缺乏雄性决定因子，但是既然XO雄蝇不育，那么Y染色体上必定含有雄性育性所必需的遗传信息。

Bridges was able to clarify the mode of sex determination in *Drosophila* by studying the progeny of triploid females, which have three copies each of the haploid complement of chromosomes. *Drosophila* has a haploid number of 4, thereby possessing three pairs of autosomes in addition to its pair of sex chromosomes. Triploid females apparently originate from rare diploid eggs fertilized by normal haploid sperm.

通过研究三倍体雌蝇的后代，布里奇斯阐明了果蝇性别决定的模式，三倍体雌蝇含有单倍体中每种染色体的三个拷贝。果蝇的单倍染色体数目为4，因此除一对性染色体外，还含有三对常染色体。三倍体雌蝇显然是由罕见的二倍体卵细胞与正常单倍体精子通过受精产生的。三倍体雌蝇体型大、刚毛粗、眼睛大，并且可育。由于每种染色体所含数目为奇数（3），因此

Triploid females have heavy-set bodies, coarse bristles, and coarse eyes, and they may be fertile. Because of the odd number of each chromosome (3), during meiosis, a variety of different chromosome complements are distributed into gametes that give rise to offspring with a variety of abnormal chromosome constitutions. Correlations between the sexual morphology and chromosome composition, along with Bridges' interpretation, are shown in Figure 5.4.

在减数分裂过程中，各种不同的染色体组合将被分配至配子中，从而产生含有各种异常染色体组成的后代。图 5.4 显示了性别表型与染色体组成之间的相关性以及布里奇斯的解释。

Bridges realized that the critical factor in determining sex is the ratio of X chromosomes to the number of haploid sets of autosomes (A) present. Normal (2X:2A) and triploid (3X:3A) females each have a ratio equal to 1.0, and both are fertile. As the ratio exceeds unity (3X:2A, or 1.5, for example), what was once called a *superfemale* is produced. Because such females are most often inviable, they are now more appropriately called **metafemales**.

布里奇斯意识到决定性别的关键因素是 X 染色体数目与常染色体（A）组数的比值。正常雌蝇（2X ：2A）和三倍体雌蝇（3X ：3A）的比值均等于 1.0，并且均可育。当该比值超过 1 时（例如 3X ：2A，或 1.5），则产生**超雌**个体，这种雌性个体大多不可生存。

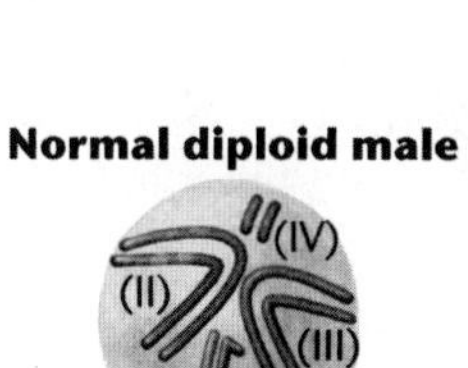

Chromosome formulation	Ratio of X chromosomes to autosome sets	Sexual morphology
3X:2A	1.5	Metafemale
3X:3A	1.0	Female
2X:2A	1.0	Female
3X:4A	0.75	Intersex
2X:3A	0.67	Intersex
X:2A	0.50	Male
XY:2A	0.50	Male
XY:3A	0.33	Metamale

FIGURE 5.4 The ratios of X chromosomes to sets of autosomes (A) and the resultant sexual morphology seen in *Drosophila melanogaster*.

图 5.4 黑腹果蝇中 X 染色体数目与常染色体（A）组数的不同比值及其相应的性别表型。

Normal (XY:2A) and sterile (XO:2A) males each have a ratio of 1:2, or 0.5. When the ratio decreases to 1:3, or 0.33, as in the case of an XY:3A male, infertile **metamales** result. Other flies recovered by Bridges in these studies had an (X:A) ratio intermediate between 0.5 and 1.0. These flies were generally larger, and they exhibited a variety of morphological abnormalities and rudimentary bisexual gonads and genitalia. They were invariably sterile and expressed both male and female morphology, thus being designated as **intersexes**.

正常雄蝇（XY ： 2A）和不育雄蝇（XO ： 2A）的这一比值为 1 ： 2 或 0.5。当比值低至 1 ： 3 或 0.33 时，例如染色体组成为 XY ： 3A 的雄性，则产生不育的**超雄**个体。在这些研究中，布里奇斯还得到了（X ： A）比值介于 0.5 和 1.0 之间的果蝇，这些果蝇通常体型较大，并表现出各种形态异常，两种性腺和外生殖器均发育不全。它们始终是不育的，并且同时表现出雄性和雌性的外部形态特征，因此被称为**兼性**。

Bridges' results indicate that in *Drosophila*, factors that cause a fly to develop into a male are not located on the sex chromosomes but are instead found on the autosomes. Some female-determining factors, however, are located on the X chromosomes. Thus, with respect to primary sex determination, male gametes containing one of each autosome plus a Y chromosome result in male offspring not because of the presence of the Y but because they fail to contribute an X chromosome. This mode of sex determination is explained by the **genic balance theory**. Bridges proposed that a threshold for maleness is reached when the X:A ratio is 1:2 (X:2A), but that the presence of an additional X (XX:2A) alters the balance and results in female differentiation.

布里奇斯的研究结果表明果蝇中决定果蝇发育成雄性的因子并不位于性染色体上，而是位于常染色体上；而一些雌性决定因子则位于 X 染色体上。因此，至于性别决定，含有每种常染色体各一条和一条 Y 染色体的雄性配子能够产生雄性后代的原因不是因为 Y 染色体的存在，而是因为它们无法贡献一条 X 染色体。这种性别决定模式被称为**基因平衡理论**。布里奇斯认为雄性决定的阈值是（X ： A）比值等于 0.5(X ： 2A)；但是当有额外的 X 染色体存在 (XX ： 2A) 时，则会打破平衡，促使个体进行雌性分化。

Numerous genes involved in sex determination in *Drosophila* have been identified. The recessive autosomal gene *transformer* (*tra*), discovered over 50 years ago by Alfred H. Sturtevant, clearly demonstrated that a single autosomal gene could have a profound impact on sex determination. Females

果蝇参与性别决定的许多基因已被鉴定。50 多年前，由艾尔弗雷德 · H. 斯特蒂文特发现的隐性常染色体基因 *transformer*（*tra*）清晰地展示了单一的常染色体基因可能对性别决定产生深远影响。*tra* 基因隐性纯合的雌性将转变为不育雄性，而 *tra* 基因纯合的雄性则不受影响。最近，另一个

homozygous for *tra* are transformed into sterile males, but homozygous males are unaffected. More recently, another gene, *Sex-lethal* (*Sxl*), has been shown to play a critical role, serving as a "master switch" in sex determination. Activation of the X-linked *Sxl* gene, which relies on a ratio of X chromosomes to sets of autosomes that equals 1.0, is essential to female development. In the absence of activation—as when, for example, the X:A ratio is 0.5—male development occurs.

基因 *Sex-lethal*（*Sxl*）也被证明具有至关重要的作用，在性别决定中扮演"主开关"的角色。当 X 染色体数目与常染色体组数比值等于 1.0 时，X 连锁的 *Sxl* 基因被激活，而这对于雌性发育至关重要。在该基因未能被激活的情况下，例如（X ： A）比值为 0.5 时，则发生雄性发育。

Although it is not yet exactly clear how this ratio influences the *Sxl* locus, we do have some insights into the question. The *Sxl* locus is part of a hierarchy of gene expression and exerts control over other genes, including *tra* (discussed in the previous paragraph) and *dsx* (*doublesex*). The wild-type allele of *tra* is activated by the product of *Sxl* only in females and in turn influences the expression of *dsx*. Depending on how the initial RNA transcript of *dsx* is processed (spliced), the resultant dsx protein activates either male- or female-specific genes required for sexual differentiation. Each step in this regulatory cascade requires a form of processing called **RNA splicing**, in which portions of the RNA are removed and the remaining fragments are "spliced" back together prior to translation into a protein. In the case of the *Sxl* gene, the RNA transcript may be spliced in different ways, a phenomenon called **alternative splicing**. A different RNA transcript is produced in females than in males. In potential females, the transcript is active and initiates a cascade of regulatory gene expression, ultimately leading to female

尽管目前我们尚不清楚该比值如何影响 *Sxl* 基因，但是我们确实对该问题已有一些了解。*Sxl* 基因是基因表达层次结构中的一部分，并控制着其他基因，包括上一段中刚讨论的 *tra* 和 *dsx*（*doublesex*）基因。野生型 *tra* 基因仅在雌性个体中被 *Sxl* 基因产物激活，进而影响 *dsx* 基因表达。依据 *dsx* 基因的起始 RNA 转录物进行转录后加工（剪切）的方式，所得 dsx 蛋白将分别激活性别分化所需的雄性特异性基因或雌性特异性基因。该调节级联过程的每一步都需要 **RNA 剪接**这一加工形式，即 RNA 分子中的一部分被移除，剩余的 RNA 片段在翻译成蛋白质之前再次被"连接"在一起。以 *Sxl* 基因为例，其 RNA 转录物可以以多种方式进行剪接，这种现象被称为**选择性剪接**。通过这种方式在雌性个体和雄性个体中可以产生不同的 RNA 转录物。在将发育为雌性的个体中，RNA 转录物有活性，并启动调节基因表达的级联过程，最终导致雌性分化。在将发育为雄性的个体中，RNA 转录物无活性，导致产生不同的基因活性模式，从而发生雄性分化。这种选择性剪接是真核生物基因表达调控众多机制

differentiation. In potential males, the transcript is inactive, leading to a different pattern of gene activity, whereby male differentiation occurs. Alternative splicing is one of the mechanisms involved in the regulation of genetic expression in eukaryotes.

中的一种。

Caenorhabditis elegans

The nematode worm *C. elegans* has become a popular organism in genetic studies, particularly for investigating the genetic control of development. Its usefulness is based on the fact that adults consist of approximately 1000 cells, the precise lineage of which can be traced back to specific embryonic origins. There are two sexual phenotypes in these worms: males, which have only testes, and hermaphrodites, which contain both testes and ovaries. During larval development of hermaphrodites, testes form that produce sperm, which is stored. Ovaries are also produced, but oogenesis does not occur until the adult stage is reached several days later. The eggs that are produced are fertilized by the stored sperm in a process of self-fertilization.

秀丽隐杆线虫

秀丽隐杆线虫已成为遗传研究中的一种常用生物，尤其适用于研究发育的遗传控制。它的优势在于成年线虫由大约 1000 个细胞组成，并且其精确的细胞谱系可以追溯到特定的胚胎起源。线虫有两种性别表型：雄性仅含睾丸，而雌雄同体则同时含睾丸和卵巢。在雌雄同体的幼体发育过程中，睾丸产生精子并将其储存在体内；卵巢也已形成，但直到数天后的成年阶段才进行卵子发生。产生的卵与体内储存的精子受精，该过程称为自体受精。

The outcome of this process is quite interesting (Figure 5.5). The vast majority of organisms that result are hermaphrodites, like the parental worm; less than 1 percent of the offspring are males. As adults, males can mate with hermaphrodites, producing about half male and half hermaphrodite offspring.

这个过程的结果非常有趣（图 5.5）。所产生的绝大多数后代如亲本线虫一样，依然是雌雄同体；仅有不到 1% 的后代是雄性个体。这些个体成年后可与雌雄同体个体进行交配并产生约一半的雄性后代和一半的雌雄同体后代。

The genetic signal that determines maleness in contrast to hermaphroditic development is provided by genes located on both the X chromosome and autosomes. *C. elegans* lacks

线虫中决定雄性发育或雌雄同体发育的遗传信号由位于 X 染色体和常染色体上的基因共同决定。秀丽隐杆线虫完全缺失 Y 染色体——其雌雄同体含两条 X 染色体，

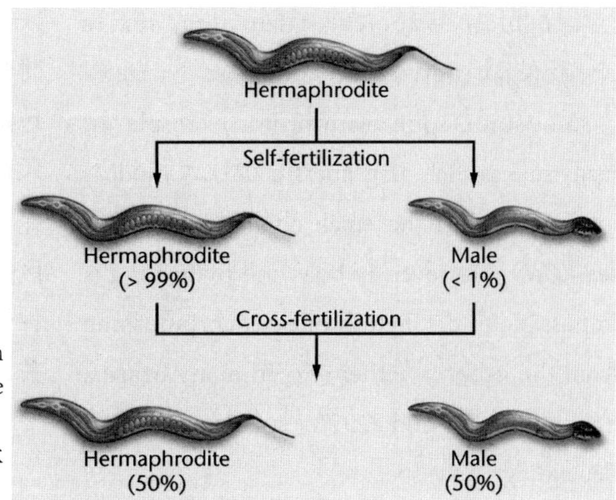

FIGURE 5.5 The outcomes of self-fertilization in a hermaphrodite, and a mating of a hermaphrodite and a male worm.

图 5.5 雌雄同体线虫的自体受精以及雌雄同体线虫与雄性个体的交配所产生的结果。

a Y chromosome altogether—hermaphrodites have two X chromosomes, while males have only one X chromosome. It is believed that, as in *Drosophila*, it is the ratio of X chromosomes to the number of sets of autosomes that ultimately determines the sex of these worms. A ratio of 1.0 (two X chromosomes and two copies of each autosome) results in hermaphrodites, and a ratio of 0.5 results in males. The absence of a heteromorphic Y chromosome is not uncommon in organisms.

而雄性个体仅含一条 X 染色体。人们认为线虫的性别决定机制与果蝇一样，即 X 染色体数目与常染色体组数的比值最终决定线虫性别。1.0 的比值（两条 X 染色体和两套常染色体组）产生雌雄同体，而 0.5 的比值则产生雄性个体。生物体中不存在异形 Y 染色体的情形并不少见。

5.6 Temperature Variation Controls Sex Determination in Reptiles

We conclude this chapter by discussing several cases where the environment—specifically temperature—is the major factor in sex determination. In contrast to situations where sex is determined genetically (as is true of all examples thus far presented in the chapter), the cases that we will now discuss are categorized as **temperature-dependent sex determination** (**TSD**).

5.6 温度变化控制爬行动物的性别决定

在本章的最后，我们将讨论环境，尤其是温度，作为性别决定主要因素的几种情况。本章中迄今为止我们所讨论的所有例子中性别均是由遗传信息所决定的，与此情况不同，我们现在将要讨论的情形都可归为**温度依赖型性别决定（TSD）**。

In many species of reptiles, sex is predetermined at conception by sex-chromosome composition. For example, in many snakes, including vipers, a ZZ/ZW mode is in effect, in which the female is the heterogamous sex (ZW). However, in boas and pythons, it is impossible to distinguish one sex chromosome from the other in either sex. In many lizards, both the XX/XY and ZZ/ZW systems are found, depending on the species.

在许多爬行动物物种中，性别由受孕时性染色体组成决定。例如，许多蛇类，包括毒蛇的性别决定采用 ZZ/ZW 模式，即雌性个体是异配性别（ZW）。然而，在蟒蛇和巨蟒中我们很难将不同性染色体加以区分。在许多蜥蜴中，XX/XY 和 ZZ/ZW 两种性别决定模式均存在，这与具体物种有关。

In still other reptilian species, however, TSD is the norm, including all crocodiles, most turtles, and some lizards, where sex determination is achieved according to the incubation temperature of eggs during a critical period of embryonic development. Three distinct patterns of TSD emerge (cases in Figure 5.6). In case I, low temperatures yield 100 percent females, and high temperatures yield 100 percent males. Just the opposite occurs in case II. In case III, low and high temperatures yield 100 percent females,

然而在其他爬行动物物种中，温度依赖型性别决定十分常见，包括所有鳄类、大多数龟类和一些蜥蜴类，它们的性别由胚胎发育关键时期受精卵的孵化温度决定。存在三种截然不同的温度依赖型性别决定模式（图 5.6）。在情况 I 中，低温产生 100％的雌性后代，而高温则产生 100％的雄性后代。情况 II 恰恰与 I 相反。在情况 III 中，低温和高温均产生 100％的雌性后代，而在中间温度段则随温度不同得到不同比例的雄性后代。第三种情形存在于鳄类、

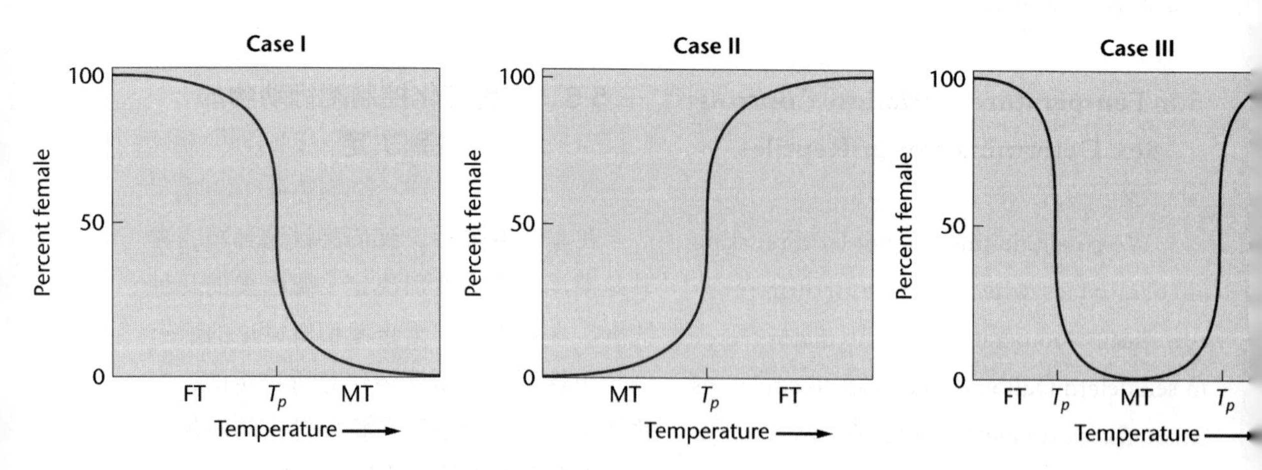

FIGURE 5.6 Three different patterns of temperature-dependent sex determination (TSD) in reptiles, as described in the text. The relative pivotal temperature T_p is crucial to sex determination during a critical point during embryonic development. FT = Female-determining temperature; MT = male-determining temperature.

图 5.6 爬行动物中三种不同的温度依赖型性别决定（TSD）模式。在胚胎发育的关键节点，中枢温度 T_p 对于性别决定至关重要。FT：雌性决定温度；MT：雄性决定温度。

while intermediate temperatures yield various proportions of males. The third pattern is seen in various species of crocodiles, turtles, and lizards, although other members of these groups are known to exhibit the other patterns.

龟类和蜥蜴类的不同种类中，尽管已知这些群体的其他成员也采用其他模式。

Two observations are noteworthy. First, in all three patterns, certain temperatures result in both male and female offspring; second, this pivotal temperature T_P range is fairly narrow, usually spanning less than 5℃, and sometimes only 1℃. The central question raised by these observations is: What are the metabolic or physiological parameters affected by temperature that lead to the differentiation of one sex or the other?

有两个现象值得一提：首先，在这三种模式中，某些温度下会同时产生雄性后代和雌性后代；其次，中枢温度T_P范围很窄，通常跨度小于5℃，有时仅为1℃。由这些观察引发的中心问题是：受温度影响导致性别分化的代谢参数或生理参数到底是什么？

The answer is thought to involve steroids (mainly estrogens) and the enzymes involved in their synthesis. Studies clearly demonstrate that the effects of temperature on estrogens, androgens, and inhibitors of the enzymes controlling their synthesis are involved in the sexual differentiation of ovaries and testes. One enzyme in particular, **aromatase**, converts androgens (male hormones such as testosterone) to estrogens (female hormones such as estradiol). The activity of this enzyme is correlated with the pathway of reactions that occurs during gonadal differentiation activity and is high in developing ovaries and low in developing testes. Researchers in this field, including Claude Pieau and colleagues, have proposed that a thermosensitive factor mediates the transcription of the reptilian aromatase gene, leading to temperature-dependent sex determination. Several other genes are likely to be involved in this mediation.

人们认为与类固醇（主要是雌激素）及参与其合成的酶类有关。已有的研究清楚地表明温度影响雌激素、雄激素以及控制它们合成的相关酶类的抑制因子，这些均与卵巢和睾丸的性别分化有关。特别是可以将雄激素（如睾酮）转变为雌激素（如雌二醇）的**芳香化酶**，该酶活性与性腺分化过程中的反应路径有关，在卵巢发育过程中活性高，而在睾丸发育过程中活性低。包括克劳德·皮奥及其同事在内的该领域研究人员指出，爬行类动物芳香化酶基因的转录受一种温度敏感因子调控，导致温度依赖型性别决定，并且可能还存在其他几种基因也参与该过程。

The involvement of sex steroids in gonadal differentiation has also been documented in birds, fishes, and amphibians. Thus, sex-determining mechanisms involving estrogens seem to be characteristic of numerous nonmammalian vertebrates.

鸟类、鱼类和两栖动物中也有性类固醇参与性腺分化的相关报道。因此，雌激素参与的性别决定机制似乎是许多非哺乳类脊椎动物的共同特征。

ESSENTIAL POINT

Although chromosome composition determines the sex of some reptiles, many others show that temperature-dependent effects during egg incubation are critical for sex determination.

基本要点

尽管染色体组成决定了某些爬行动物的性别，但受精卵孵化过程中的温度依赖效应对于许多其他爬行动物的性别决定十分重要。

6 Chromosome Mutations: Variation in Number and Arrangement

6 染色体突变：染色体数目与结构的变异

CHAPTER CONCEPTS

■ The failure of chromosomes to properly separate during meiosis results in variation in the chromosome content of gametes and subsequently in offspring arising from such gametes.

■ Plants often tolerate an abnormal genetic content, but, as a result, they often manifest unique phenotypes. Such genetic variation has been an important factor in the evolution of plants.

■ In animals, genetic information is in a delicate equilibrium whereby the gain or loss of a chromosome, or part of a chromosome, in an otherwise diploid organism often leads to lethality or to an abnormal phenotype.

■ The rearrangement of genetic information within the genome of a diploid organism may be tolerated by that organism but may affect the viability of gametes and the phenotypes of organisms arising from those gametes.

■ Chromosomes in humans contain fragile sites—regions susceptible to breakage, which lead to abnormal phenotypes.

本章速览

■ 减数分裂过程中染色体的分离失败将导致产生的配子中染色体组成发生变化，由这些配子发育而来的后代个体也将出现表型变化。

■ 植物通常可以耐受异常的遗传信息组成，但是也常明显呈现出独特的表型。这些遗传变异已成为植物进化中的重要因素之一。

■ 在动物中，遗传信息保持着精妙的平衡，因此二倍体生物中整条染色体或部分染色体的额外获得或缺失通常会导致个体死亡或异常表型。

■ 二倍体生物可能可以耐受基因组内遗传信息的重新排列，但是重新排列的遗传信息可能影响该个体所产生的配子存活率以及由这些配子产生的后代表型。

■ 人类染色体含有一些脆性位点，这些区域易于断裂并会导致异常表型。

In previous chapters, we have emphasized

在前面的章节中，我们已经重点讨论

how mutations and the resulting alleles affect an organism's phenotype and how traits are passed from parents to offspring according to Mendelian principles. In this chapter, we look at phenotypic variation that results from more substantial changes than alterations of individual genes—modifications at the level of the chromosome.

了突变及其产生的等位基因如何影响生物的表型，以及性状如何按照孟德尔定律从父母遗传给后代。在本章，我们将着眼的表型变异并不是由单基因突变所致，而是由更为重大的改变所引起——染色体水平的变化。

Although most members of diploid species normally contain precisely two haploid chromosome sets, many known cases vary from this pattern. Modifications include a change in the total number of chromosomes, the deletion or duplication of genes or segments of a chromosome, and rearrangements of the genetic material either within or among chromosomes. Taken together, such changes are called **chromosome mutations** or **chromosome aberrations** in order to distinguish them from gene mutations. Because the chromosome is the unit of genetic transmission, according to Mendelian laws, chromosome aberrations are passed to offspring in a predictable manner, resulting in many unique genetic outcomes.

尽管二倍体物种的大多数成员通常情况下精确地含有两套单倍染色体组，但已知存在许多与此不同的情形，包括染色体总数目变化、染色体上基因或片段的缺失或重复，以及染色体内或染色体间遗传物质的重排。总的来说，这些变化被称为**染色体突变**或**染色体畸变**，从而将其与基因突变相区别。由于染色体是遗传物质传递的单元，根据孟德尔定律，染色体畸变以可预测的方式传递给后代，从而产生许多独特的遗传结果。

Because the genetic component of an organism is delicately balanced, even minor alterations of either content or location of genetic information within the genome can result in some form of phenotypic variation. More substantial changes may be lethal, particularly in animals. Throughout this chapter, we consider many types of chromosomal aberrations, the phenotypic consequences for the organism that harbors an aberration, and the impact of the aberration on the offspring of an affected

由于生物的遗传组成处于微妙的平衡状态，因此即使基因组内遗传信息无论是组成还是位置发生极其微小的变化，也能导致某种形式的表型变异。更为重大的变化可能是致命的，尤其对于动物而言。贯穿本章，我们将讨论多种类型的染色体畸变、含染色体畸变的生物表型后果以及畸变对个体后代的影响。我们还将讨论染色体畸变在进化过程中的地位。

individual. We will also discuss the role of chromosome aberrations in the evolutionary process.

6.1 Variation in Chromosome Number: Terminology and Origin

Variation in chromosome number ranges from the addition or loss of one or more chromosomes to the addition of one or more haploid sets of chromosomes. Before we embark on our discussion, it is useful to clarify the terminology that describes such changes. In the general condition known as **aneuploidy**, an organism gains or loses one or more chromosomes but not a complete set. The loss of a single chromosome from an otherwise diploid genome is called *monosomy*. The gain of one chromosome results in *trisomy*. These changes are contrasted with the condition of **euploidy**, where complete haploid sets of chromosomes are present. If more than two sets are present, the term **polyploidy** applies. Organisms with three sets are specifically *triploid*, those with four sets are *tetraploid*, and so on. Table 6.1 provides an organizational framework for you to follow as we discuss each of these categories of aneuploid and euploid variation and the subsets within them.

As we consider cases that include the gain or loss of chromosomes, it is useful to examine how such aberrations originate. For instance, how do the syndromes arise where the number of sex-determining chromosomes in humans is altered, as described in Chapter 5? As you may recall, the gain (47,XXY) or loss (45,X) of

6.1 染色体数目的变异：专业术语与起源

染色体数目变异从一条或多条染色体的增加或缺失到一组或多组单倍染色体组的增加都有。在我们开始讨论之前，清楚地了解描述这些变化的专业术语十分必要。在**非整倍性**的情况下，生物体获得或丢失一条或多条染色体，而不是完整的成组染色体。从二倍体基因组中丢失一条染色体的情形称为单体。额外获得一条染色体的情形则称为三体。这些变异与**整倍性**的情形形成对比，整倍性中存在完整的单倍染色体组。如果存在两组以上的完整单倍染色体组，则使用专业术语**“多倍性”**一词。含有三组染色体的生物体被特定地称为三倍体，含有四组的生物体称为四倍体，以此类推。表 6.1 提供了一个组织架构，可在讨论这些非整倍体和整倍体变异种类及其亚类时方便对照。

当我们讨论染色体增加或缺失的情况时，考察此类畸变产生的缘由十分重要。例如第 5 章中所描述的，人类性别决定染色体数目发生变化时，相关的综合征是如何发生的？如前所述，二倍体基因组中 X 染色体的增加（47,XXY）或缺失（45,X）均影响表型，并分别导致**克兰费尔特综合**

an X chromosome from an otherwise diploid genome affects the phenotype, resulting in **Klinefelter syndrome** or **Turner syndrome**, respectively. Human females may contain extra X chromosomes (e.g., 47,XXX, 48,XXXX), and some males contain an extra Y chromosome (47,XYY).

征和**特纳综合征**。人类女性个体还可能含额外的多条 X 染色体（例如 47,XXX，48,XXXX），而一些男性个体则含额外的 Y 染色体（47,XYY）。

TABLE 6.1 Terminology for Variation in Chromosome Numbers
表 6.1 不同染色体数目变异的命名

Term	Explanation
Aneuploidy	$2n \pm x$ chromosomes
Monosomy	$2n - 1$
Disomy	$2n$
Trisomy	$2n + 1$
Tetrasomy, pentasomy, etc.	$2n + 2$, $2n + 3$, etc.
Euploidy	Multiples of n
Diploidy	$2n$
Polyploidy	$3n$, $4n$, $5n$, . . .
Triploidy	$3n$
Tetraploidy, pentaploidy, etc.	$4n$, $5n$, etc.
Autopolyploidy	Multiples of the same genome
Allopolyploidy (amphidiploidy)	Multiples of closely related genomes

Such chromosomal variation originates as a random error during the production of gametes, a phenomenon referred to as **nondisjunction**, whereby paired homologs fail to disjoin during segregation. This process disrupts the normal distribution of chromosomes into gametes. The results of nondisjunction during meiosis I and meiosis II for a single chromosome of a diploid organism are shown in Figure 6.1.

这种染色体变异源自配子形成过程中发生的随机错误，一种被称为“**不分离**”的现象，即成对存在的同源染色体在分离过程中无法正确分开。这一过程阻止了染色体正常分配进入配子。二倍体生物中单条染色体在减数分裂 I 期和减数分裂 II 期发生不分离的结果如图 6.1 所示。

As you can see, abnormal gametes can form that contain either two members of the affected chromosome or none at all. Fertilizing these with a normal haploid gamete produces a zygote with either three members (trisomy) or only one member (monosomy) of this chromosome.

如你所见，同时含同源染色体的两条或不含同源染色体中任意一条的异常配子均可能产生。当这些异常配子与正常的单倍体配子受精形成合子时，则合子将同时含该同源染色体的三个成员（三体）或仅一个成员（单体）。不分离导致了人类或

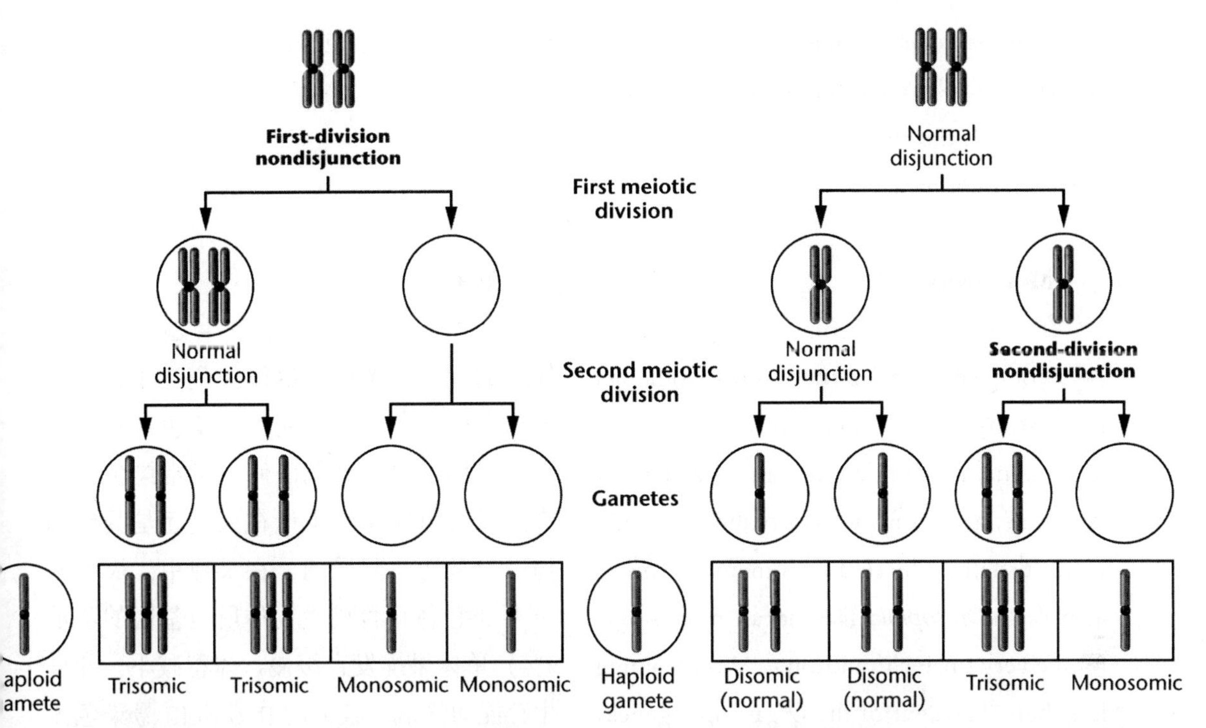

FIGURE 6.1 Nondisjunction during the first and second meiotic divisions. In both cases, some of the gametes that are formed either contain two members of a specific chromosome or lack that chromosome. After fertilization by a gamete with normal haploid content, monosomic, disomic (normal), or trisomic zygotes are produced.

图 6.1 在减数分裂 I 期和减数分裂 II 期发生的不分离。在这两种情况下，形成的一些配子中或者同时含某特定同源染色体的两条，或者不含该同源染色体的任意一条。这些异常配子与正常单倍体配子进行受精，则会产生单体合子、二体（正常）合子或者三体合子。

Nondisjunction leads to a variety of aneuploid conditions in humans and other organisms.

其他物种中非整倍体情形的产生。

ESSENTIAL POINT

Alterations of the precise diploid content of chromosomes are referred to as chromosomal aberrations or chromosomal mutations.

基本要点

染色体精确二倍体组成的变化称为染色体畸变或染色体突变。

6.2 Monosomy and Trisomy Result in a Variety of Phenotypic Effects

6.2 单体和三体导致不同的表型效应

We turn now to a consideration of variations in the number of autosomes and the genetic consequence of such changes. The most common examples of aneuploidy, where an

现在我们转向讨论常染色体数目变异及这种变化的遗传结果。非整倍性即生物体所含染色体数目并不是单倍染色体组的整数倍，其中最常见的情形是正常二倍染

organism has a chromosome number other than an exact multiple of the haploid set, are cases in which a single chromosome is either added to, or lost from, a normal diploid set.

色体组中额外增加或缺失一条染色体。

Monosomy

The loss of one chromosome produces a $2n-1$ complement called **monosomy**. Although monosomy for the X chromosome occurs in humans, as we have seen in 45,X Turner syndrome, monosomy for any of the autosomes is not usually tolerated in humans or other animals. In *Drosophila*, flies that are monosomic for the very small chromosome IV (containing less than 5 percent of the organism's genes) develop mores lowly, exhibit reduced body size, and have impaired viability. Monosomy for the larger chromosomes II and III is apparently lethal because such flies have never been recovered.

单体

一条染色体的缺失会产生 $2n-1$ 的染色体组成，称为**单体**。尽管我们在 45,X（特纳综合征）中看到了人类 X 染色体缺失产生的单体，但是其他任何一条常染色体缺失产生的单体通常不论在人类还是其他动物中都是无法耐受的。在果蝇中，最小的 IV 号染色体（含该生物体所有基因的不足 5%）单体果蝇发育缓慢、体型减小，并且生存能力降低。较大的 II 号和 III 号染色体的单体显然是致命的，因为这样的果蝇从未被发现。

The failure of monosomic individuals to survive is at first quite puzzling, since at least a single copy of every gene is present in the remaining homolog. However, one explanation is that if just one of those genes is represented by a lethal allele, monosomy unmasks the recessive lethal allele that is tolerated in heterozygotes carrying the corresponding wild-type allele, leading to the death of the organism. In other cases, a single copy of a recessive gene due to monosomy may be insufficient to provide life-sustaining function for the organism, a phenomenon called **haploinsufficiency**.

起初单体个体难以存活的现象十分令人费解，因为保留下来的同源染色体上至少每种基因尚存一个拷贝。但是，一种解释是，如果这些基因拷贝中恰好有一个是致死等位基因，那么这样的单体无法掩盖在携带野生型等位基因的杂合个体中可以耐受的隐性致死基因效应，从而导致生物体死亡。在其他情形中，单体中的一个隐性基因拷贝也许不足以为生物体提供维持生命所需的功能，这种现象被称为**单倍剂量不足**。

Aneuploidy is better tolerated in the plant kingdom. Monosomy for autosomal chromosomes has been observed in maize,

植物界则对非整倍性的耐受能力相对较好。在玉米、烟草、月见草和曼陀罗等植物中均已发现常染色体的单体。然而，

tobacco, the evening primrose (*Oenothera*), and the jimson weed (*Datura*), among many other plants. Nevertheless, such monosomic plants are usually less viable than their diploid derivatives. Haploid pollen grains, which undergo extensive development before participating in fertilization, are particularly sensitive to the lack of one chromosome and are seldom viable.

这些单体植株的生存力通常比其二倍体亲缘植株要差。单倍体花粉粒在参与受精前将进行进一步的发育，它们对于单条染色体的缺失尤其敏感，并且几乎不可存活。

Trisomy

In general, the effects of **trisomy** ($2n+1$) parallel those of monosomy. However, the addition of an extra chromosome produces somewhat more viable individuals in both animal and plant species than does the loss of a chromosome. In animals, this is often true, provided that the chromosome involved is relatively small. However, the addition of a large autosome to the diploid complement in both *Drosophila* and humans has severe effects and is usually lethal during development.

三体

通常，**三体**（$2n+1$）的遗传效应与单体类似。然而，增加一条额外染色体比缺失一条染色体在动物和植物物种中拥有更多的存活个体。在动物中，如果所涉及的染色体相对较小，则更是如此。但是，在果蝇和人的二倍体组成中额外多出一条较大的常染色体则会产生严重后果，并且通常在发育过程中致死。

In plants, trisomic individuals are viable, but their phenotype may be altered. A classical example involves the jimson weed, *Datura*, whose diploid number is 24. Twelve primary trisomic conditions are possible, and examples of each one have been recovered. Each trisomy alters the phenotype of the plant's capsule sufficiently to produce a unique phenotype. These capsule phenotypes were first thought to be caused by mutations in one or more genes.

在植物中，三体个体可以存活，但其表型可能发生变化。一个经典例子是曼陀罗，其二倍染色体数目是 24。它有 12 种可能的三体情形，并且每种三体均已被发现。每种三体的蒴果表型均发生改变并产生独特性状，这些蒴果表型曾被认为是由单基因或多基因突变所致。

Still another example is seen in the rice plant (*Oryza sativa*), which has a haploid number of 12. Trisomic strains for each chromosome have been isolated and studied—the plants of 11

另一个例子是水稻，其单倍染色体数目是 12。每种染色体相应的三体植株均已得到分离和研究。其中 11 种三体植株可彼此区分，并且也可与野生型植株相区分。

strains can be distinguished from one another and from wild-type plants. Trisomics for the longer chromosomes are the most distinctive, and the plants grow more slowly. This is in keeping with the belief that larger chromosomes cause greater genetic imbalance than smaller ones. Leaf structure, foliage, stems, grain morphology, and plant height also vary among the various trisomies.

最长染色体所对应的三体最为独特，该植株生长较为缓慢，这与“较大染色体比较小染色体引起的遗传失衡更大”这一观点相一致。此外，不同三体的叶片结构、茎、谷粒形态和植株高度也各有差异。

Down Syndrome: Trisomy 21

The only human autosomal trisomy in which a significant number of individuals survive longer than a year past birth was discovered in 1866 by John Langdon Down. The condition is now known to result from trisomy of chromosome 21, one of the G group, and is called **Down syndrome** or simply **trisomy 21** (designated 47,21+). This trisomy is found in approximately 1 infant in every 800 live births. While this might seem to be a rare, improbable event, there are approximately 4000—5000 such births annually in the United States, and there are currently over 250,000 individuals with Down syndrome.

唐氏综合征：21 三体

约翰·兰登·唐于 1866 年发现了人类唯一出生后存活一年以上并且个体数量众多的常染色体三体。现在已知该综合征由 G 组中 21 号染色体的三体引起，被称为**唐氏综合征**或简称 **21 三体**（表示为 47,21+）。这种三体的发病率为每 800 个出生存活婴儿中出现 1 例。尽管看似罕见、发生概率低，但在美国，每年有 4000—5000 例患儿出生，并且目前唐氏综合征患者数量已达到 25 万以上。

Typical of other conditions classified as syndromes, many phenotypic characteristics *may* be present in trisomy 21, but any single affected individual usually exhibits only a subset of these. In the case of Down syndrome, there are 12 to 14 such characteristics, with each individual, on average, expressing 6 to 8 of them. Nevertheless, the outward appearance of these individuals is very similar, and they bear a striking resemblance to one another. This is, for the most part, due to

与其他典型综合征类似，21 三体可能存在许多表型特征，但是任何一位患者通常仅表现出其中部分特征。唐氏综合征共有 12—14 种表型特征，每个个体平均表现其中的 6—8 种。然而，这些患者的外在特征非常类似，而且彼此之间有着惊人的相似之处。这主要是由于患者眼睛都有明显的内眦赘皮，并且通常面部平坦、头小而圆。此外，唐氏综合征患者身材矮小，舌头弯曲突出（从而导致嘴巴保持部分张开），

a prominent epicanthic fold in each eye and the typically flat face and round head. People with Down syndrome are also characteristically short and may have a protruding, furrowed tongue (which causes the mouth to remain partially open) and short, broad hands with characteristic palm and fingerprint patterns. Physical, psychomotor, and cognitive disabilities are evident, and poor muscle tone is characteristic. While life expectancy is shortened to an average of about 50 years, individuals are known to survive into their 60s.

手掌短而宽、具有特征性的掌纹和指纹，身体、心理和认知发育存在明显障碍，并伴有肌张力不良。虽然患者的平均寿命为50岁左右，但目前已知有些个体可以存活至60多岁。

Children with Down syndrome are prone to respiratory disease and heart malformations, and they show an incidence of leukemia approximately 20 times higher than that of the normal population. However, careful medical scrutiny and treatment throughout their lives can extend their survival significantly. A striking observation is that death in older adults with Down syndrome is frequently due to Alzheimer disease. The onset of this disease occurs at a much earlier age than in the normal population.

唐氏综合征患儿易患呼吸系统疾病和心脏畸形，并且白血病发病率比正常人群高约20倍。但是，终身精细的医疗护理可以明显延长其生存时间。一个惊人发现是年长的唐氏综合征患者的死亡通常由阿尔茨海默病引起，该疾病在唐氏综合征患者中的发病年龄比正常人群要早得多。

Because Down syndrome is common in our population, a comprehensive understanding of the underlying genetic basis has long been a research goal. Investigations have given rise to the idea that a critical region of chromosome 21 contains the genes that are dosage sensitive in this trisomy and responsible for the many phenotypes associated with the syndrome. This hypothetical portion of the chromosome has been called the **Down syndrome critical region (DSCR)**. A mouse model was created in 2004 that is trisomic for the DSCR, although

由于唐氏综合征在人群中较为常见，因此深入了解其潜在遗传基础一直是科学研究的目标。研究表明三体21号染色体上一个关键区域所含基因具有剂量敏感性，并可引发该综合征相关的许多表型。这一染色体假设区域被称为**唐氏综合征关键区（DSCR）**。2004年，人们建立了DSCR三体小鼠模型，但是其中一些小鼠并未表现出该综合征相关特征。尽管如此，这依然是一种重要的研究途径。

some mice do not exhibit the characteristics of the syndrome. Nevertheless, this remains an important investigative approach.

Current studies of the DSCR region in both humans and mice have led to several interesting findings. We now believe that the three copies of the genes present in this region are necessary, but themselves not sufficient, for the cognitive deficiencies characteristic of the syndrome. Another finding involves the important observation that Down syndrome individuals have a decreased risk of developing a number of cancers involving solid tumors, including lung cancer and melanoma. This health benefit has been correlated with the presence of an extra copy of the *DSCR1* gene, which encodes a protein that suppresses *vascular endothelial growth factor* (*VEGF*). This suppression, in turn, blocks the process of angiogenesis. As a result, the overexpression of this gene inhibits tumors from forming proper vascularization, diminishing their growth. A 14-year study published in 2002 involving 17,800 Down syndrome individuals revealed an approximate 10 percent reduction in cancer mortality in contrast to a control population.

目前对人和小鼠 DSCR 区的研究均获得了一些有趣的发现。我们现在可以确信，位于该区域的基因存在三个拷贝对于该综合征的认知缺陷特征是必要的，但仅有这些基因并不足够。另一个源自患者观察的重要发现是唐氏综合征患者罹患包括实体瘤在内的癌症（如肺癌和黑色素瘤）的风险会降低。这种健康裨益与 *DSCR1* 基因额外存在的拷贝相关，该基因编码的蛋白质可以抑制血管内皮生长因子（VEGF），这种抑制作用可以阻止血管生成。结果便是该基因的过表达抑制了肿瘤血管形成，从而降低了它们的生长。2002 年发表的一项为期 14 年、涉及 17 800 名唐氏综合征患者的研究结果表明，与对照人群相比，唐氏综合征患者的癌症死亡率降低约 10%。

The Origin of the Extra 21st Chromosome in Down Syndrome

Most frequently, this trisomic condition occurs through nondisjunction of chromosome 21 during meiosis. Failure of paired homologs to disjoin during either anaphase I or II may lead to gametes with the $n+1$ chromosome composition. About 75 percent of these errors

唐氏综合征中额外的 21 号染色体来源

在最常见的情况下，这种三体是由减数分裂过程中 21 号染色体不分离而导致的。在后期 I 或后期 II，配对的同源染色体分离失败均可导致具有 $n+1$ 条染色体组成的配子类型产生。在导致唐氏综合征的错误中，约 75% 可归因于第一次减数分裂期间发生

leading to Down syndrome are attributed to nondisjunction during the first meiotic division. Subsequent fertilization with a normal gamete creates the trisomic condition.

的不分离。随后这些组成异常的配子与正常配子受精则会产生三体。

Chromosome analysis has shown that, while the additional chromosome may be derived from either the mother or father, the ovum is the source in about 95 percent of 47,21+ trisomy cases. Before the development of techniques using polymorphic markers to distinguish paternal from maternal homologs, this conclusion was supported by the more indirect evidence derived from studies of the age of mothers giving birth to infants with Down syndrome. Figure 6.2 shows the relationship between the incidence of children born with Down syndrome and maternal age, illustrating the dramatic increase as the age of the mother increases. While the frequency is about 1 in 1000 at maternal age 30, a tenfold increase to a frequency of 1 in 100 is noted at age 40. The frequency increases still further to about 1 in 30 at age 45. A very alarming statistic is that as the age of childbearing women exceeds 45, the probability of a child born with Down syndrome continues to increase substantially. In spite of this high probability, substantially more than half of such births occur to women younger than 35 years, because the overwhelming proportion of

染色体分析表明，尽管额外多出的染色体可能来自母亲或者父亲，但在约 95% 的 47,21+ 三体病例中，卵子是其来源。在使用多态性标记区分父本与母本同源染色体的技术出现之前，这一结论也得到了很多来自唐氏综合征患儿母亲年龄研究所得间接证据的支持。图 6.2 显示了唐氏综合征患儿出生率与母亲年龄之间的关系，随着母亲年龄增加，患儿出生率急剧增加。孕产妇在 30 岁时，患儿发生率约为 1/1000；在 40 岁时，发生率则增长至 10 倍，高达 1/100；在 45 岁时，发生率进一步上升至 1/30。非常令人震惊的统计数字是当育龄妇女超过 45 岁时，孕育唐氏综合征患儿的可能性还将继续大幅增加。尽管概率如此之高，但实际上目前超过一半的唐氏综合征患儿都是由 35 岁以下女性孕育所生的，因为在人群中，绝大多数的孕育涉及该年龄以下的女性。

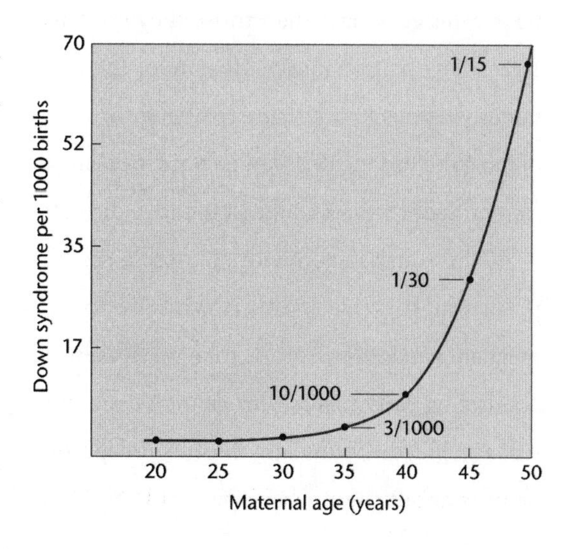

FIGURE 6.2 Incidence of children born with Down syndrome related to maternal age.

图 6.2 唐氏综合征患儿出生率与母亲年龄之间的关系。

pregnancies in the general population involve women under that age.

Although the nondisjunctional event that produces Down syndrome seems more likely to occur during oogenesis in women over the age of 35, we do not know with certainty why this is so. However, one observation may be relevant. Meiosis is initiated in all the eggs of a human female when she is still a fetus, until the point where the homologs synapse and recombination has begun. Then oocyte development is arrested in meiosis I. Thus, all primary oocytes have been formed by birth. When ovulation begins at puberty, meiosis is reinitiated in one egg during each ovulatory cycle and continues into meiosis II. The process is once again arrested after ovulation and is not completed unless fertilization occurs.

尽管导致唐氏综合征的不分离事件似乎更易于发生在 35 岁以上女性的卵子发生过程中，但我们尚不确定其缘由。然而，一种现象可能与此相关。当人类女性尚在胎儿期时，所有卵细胞均已开始减数分裂过程，直至同源染色体间发生联会和重组。然后，卵母细胞的发育停滞于减数分裂 I 期。因此，所有初级卵母细胞在出生时即已形成。当青春期开始排卵时，在每一个排卵周期将有一个卵母细胞重新启动减数分裂，并继续进入减数分裂 II 期。如果没有发生受精作用，那么排卵后该过程将再次被终止，不能完成完整的减数分裂过程。

The end result of this progression is that each ovum that is released has been arrested in meiosis I for about a month longer than the one released during the preceding cycle. As a consequence, women 30 or 40 years old produce ova that are significantly older and that have been arrested longer than those they ovulated 10 or 20 years previously. In spite of the logic underlying this hypothesis explaining the cause of the increased incidence of Down syndrome as women age, it remains difficult to prove directly.

这一进程的最终结果是，每个被释放的卵子在减数分裂 I 期中被阻滞的时间总比前一周期释放的卵子长一个月左右。因此，30 或 40 岁女性所产生的卵子必定比其 10 或 20 年前所排的卵经历阻滞的时间更长。尽管该假设的逻辑能够解释随着育龄女性年龄的增加，唐氏综合征患儿发病率升高的原因，但是仍然难以直接进行证明。

These statistics obviously pose a serious problem for the woman who becomes pregnant late in her reproductive years. Genetic counseling early in such pregnancies is highly recommended. Counseling informs prospective parents about the probability that their child

这些统计数据显然给在生育期较后阶段受孕的女性提出了一个十分严肃的问题。强烈建议这些孕妇在妊娠早期进行遗传咨询。咨询将告知这些准父母其子女患病的可能性，并对他们进行有关唐氏综合征的教育。尽管部分唐氏综合征患者患有中度

will be affected and educates them about Down syndrome. Although some individuals with Down syndrome experience moderate to severe cognitive delays, most experience only mild to moderate delays. These individuals are increasingly integrated into society, including school, the work force, and social and recreational activities. A genetic counselor may also recommend a prenatal diagnostic technique in which fetal cells are isolated and cultured.

至重度的认知发育迟缓，但是大多数患者仅为轻度至中度认知发育迟缓。这些个体越来越多地融入社会，包括学校、劳动力市场、社交和娱乐活动。遗传咨询师也建议采用产前诊断技术，从而分离并培养胎儿细胞用于分析鉴定。

In **amniocentesis** and **chorionic villus sampling** (**CVS**), the two most familiar approaches, fetal cells are obtained from the amniotic fluid or the chorion of the placenta, respectively. In a newer approach, fetal cells and DNA are derived directly from the maternal circulation, a technique referred to as **noninvasive prenatal genetic diagnosis** (**NIPGD**). Requiring only a 10-mL maternal blood sample, this procedure will become increasingly more common because it poses no risk to the fetus. After fetal cells are obtained and cultured, the karyotype can be determined by cytogenetic analysis. If the fetus is diagnosed as being affected, further counseling for parents will be offered regarding the options open to them, one of which is abortion of the fetus. Obviously, this is a difficult decision involving both religious and ethical issues.

羊膜腔穿刺术和**绒毛膜绒毛吸取术**（CVS）是两种最为熟悉的方法，分别从羊水或胎盘绒毛膜中获得胎儿细胞。在一种新型的**无创产前遗传诊断**（NIPGD）技术中，胎儿细胞和 DNA 可以直接取自母体循环系统。由于只需要 10 mL 的孕妇血液样本，对胎儿没有风险，因此该方法将越来越普遍。获得胎儿细胞并进行细胞培养后，通过细胞遗传学分析鉴定核型。如果诊断结果显示胎儿患有唐氏综合征，准父母将得到进一步的咨询服务，提供相关的选择，其中之一便是进行胎儿流产。显然，这是一个涉及宗教和伦理问题的艰难选择。

Since Down syndrome is caused by a random error—nondisjunction of chromosome 21 during maternal or paternal meiosis—the occurrence of the disorder is *not* expected to be inherited. Nevertheless, Down syndrome occasionally runs in families. These instances,

由于唐氏综合征由随机错误引起，即母本或父本减数分裂过程中 21 号染色体不分离，因此该疾病的发生理论上不遗传。然而，唐氏综合征偶尔在一些家族中世代发生。这些情形称为家族性唐氏综合征，与 21 号染色体易位相关，属于另一种染色

referred to as familial Down syndrome, involve a translocation of chromosome 21, another type of chromosomal aberration, which we will discuss later in the chapter.

体畸变，我们将在本章稍后进行讨论。

Human Aneuploidy

Besides Down syndrome, only two human trisomies, and no autosomal monosomies, survive to term: **Patau** and **Edwards syndromes** (47, 13+ and 47, 18+, respectively). Even so, these individuals manifest severe malformations and early lethality.

The preceding observation leads us to ask whether many other aneuploid conditions arise but that the affected fetuses do not survive to term. That this is the case has been confirmed by karyotypic analysis of spontaneously aborted fetuses. These studies reveal two striking statistics: (1) Approximately 20 percent of all conceptions terminate in spontaneous abortion (some estimates are considerably higher); and (2) about 30 percent of all spontaneously aborted fetuses demonstrate some form of chromosomal imbalance. This suggests that at least 6 percent (0.20 × 0.30) of conceptions contain an abnormal chromosome complement. A large percentage of fetuses demonstrating chromosomal abnormalities are aneuploids.

An extensive review of this subject by David H. Carr has revealed that a significant percentage of aborted fetuses are trisomic for one of the chromosome groups. Trisomies for every human chromosome have been recovered. Interestingly, the monosomy with the highest incidence among abortuses is the

人类非整倍性

除唐氏综合征外，还有两种人类三体可以存活，**帕托综合征**和**爱德华综合征**（分别为47, 13+和47, 18+），并且没有常染色体单体。即便如此，这些个体仍表现出严重畸形和早期夭折。

已有的观察使我们会问是否还会产生许多其他非整倍性染色体异常，但患儿不能存活的情形。通过对自然流产胎儿进行核型分析，可以证实的确存在这种情况。这些研究揭示了两个惊人的统计数据：（1）所有受孕中约有20%会因自然流产而终止（有些预计会更高）；（2）所有自然流产的胎儿中约有30%存在某种形式的染色体不平衡。这表明至少6%（0.20 × 0.30）的胎儿含有异常染色体组成。显示染色体异常的胎儿大部分是非整倍体。

戴维 · H. 卡尔对该课题进行了更为深入的研究并且揭示了流产胎儿中很大一部分是某染色体的三体。每种人类染色体的三体个体均已被发现。有趣的是，流产中发生率最高的单体是45, X，如果胎儿能够存活至足月出生，即为特纳综合征婴儿。然而，常染色体单体很少被发现，虽然不

45, X condition, which produces an infant with Turner syndrome if the fetus survives to term. Autosomal monosomies are seldom found, however, even though nondisjunction should produce $n-1$ gametes with a frequency equal to $n+1$ gametes. This finding suggests that gametes lacking a single chromosome are functionally impaired to a serious degree or that the embryo dies so early in its development that recovery occurs infrequently. We discussed the potential causes of monosomic lethality earlier in this chapter. Carr's study also found various forms of polyploidy and other miscellaneous chromosomal anomalies.

分离产生的（$n-1$）配子类型与（$n+1$）配子类型的频率相当。该发现表明缺乏单条染色体的配子在功能上严重受损，或者胚胎在其发育早期即死亡，以至于很难被发现。我们在本章的前面讨论了单体致死的潜在原因。卡尔的研究还发现了多种形式的多倍性和其他各种染色体异常。

These observations support the hypothesis that normal embryonic development requires a precise diploid complement of chromosomes to maintain the delicate equilibrium in the expression of genetic information. The prenatal mortality of most aneuploids provides a barrier against the introduction of these genetic anomalies into the human population.

这些观察结果支持了以下假设：正常的胚胎发育需要精确的二倍染色体组成以便维持遗传信息表达过程中的精妙平衡。大多数非整倍体的产前死亡为避免将这些遗传异常引入人类群体提供了一道屏障。

ESSENTIAL POINT

Studies of monosomic and trisomic disorders are increasing our understanding of the delicate genetic balance that is essential for normal development.

基本要点

单体和三体疾病的研究正逐步加深人们理解精妙的遗传平衡对于正常发育必不可少的重要性。

6.3 Polyploidy, in Which More Than Two Haploid Sets of Chromosomes Are Present, Is Prevalent in Plants

6.3 含两套以上单倍染色体组成的多倍体在植物界中广泛存在

The term *polyploidy* describes instances in which more than two multiples of the haploid

“多倍性”用以描述两个以上单倍染色体组同时存在的情况。多倍体的命名基

chromosome set are found. The naming of polyploids is based on the number of sets of chromosomes found: A triploid has $3n$ chromosomes; a tetraploid has $4n$; a pentaploid, $5n$; and so forth (Table 6.1). Several general statements can be made about polyploidy. This condition is relatively infrequent in many animal species but is well known in lizards, amphibians, and fish, and is much more common in plant species. Usually, odd numbers of chromosome sets are not reliably maintained from generation to generation because a polyploid organism with an uneven number of homologs often does not produce genetically balanced gametes. For this reason, triploids, pentaploids, and so on, are not usually found in plant species that depend solely on sexual reproduction for propagation.

于存在的染色体组数目：三倍体含 $3n$ 条染色体；四倍体含 $4n$ 条染色体；五倍体含 $5n$ 条染色体，以此类推（表 6.1）。关于多倍性还有几点需要说明：多倍性在许多动物物种中相对较为少见，但在蜥蜴、两栖类动物和鱼类中广泛存在，在植物物种中则更为常见。通常，含奇数组数的染色体组很难在世代交替过程中被稳定保持，因为同源染色体数目为奇数的多倍体生物通常难以产生遗传均衡的配子。正因如此，在仅依赖有性生殖进行繁衍的植物物种中，三倍体、五倍体等并不多见。

Polyploidy originates in two ways: (1) The addition of one or more extra sets of chromosomes, identical to the normal haploid complement of the same species, resulting in **autopolyploidy**; or (2) the combination of chromosome sets from different species occurring as a consequence of hybridization, resulting in **allopolyploidy** (from the Greek word *allo*, meaning "other" or "different"). The distinction between auto-and allopolyploidy is based on the genetic origin of the extra chromosome sets, as shown in Figure 6.3.

多倍性通过两种方式产生：（1）增加一组或多组与同一物种的正常单倍染色体组成完全相同的染色体，形成**同源多倍性**；（2）通过杂交将来自不同物种的染色体组进行组合，形成**异源多倍性**（“*allo*”源于希腊语，意为“其他的”或“不同的”）。同源多倍性和异源多倍性的区别在于额外增加的成组染色体组的遗传起源，如图 6.3 所示。

In our discussion of polyploidy, we use the following symbols to clarify the origin of additional chromosome sets. For example, if A represents the haploid set of chromosomes of any organism, then

$$A = a_1 + a_2 + a_3 + a_4 + \cdots + a_n$$

在多倍性的讨论中，我们将使用以下符号表明额外增加的染色体组的遗传来源，例如，如果 A 代表某生物的单倍染色体组，则

$$A = a_1 + a_2 + a_3 + a_4 + \cdots + a_n$$

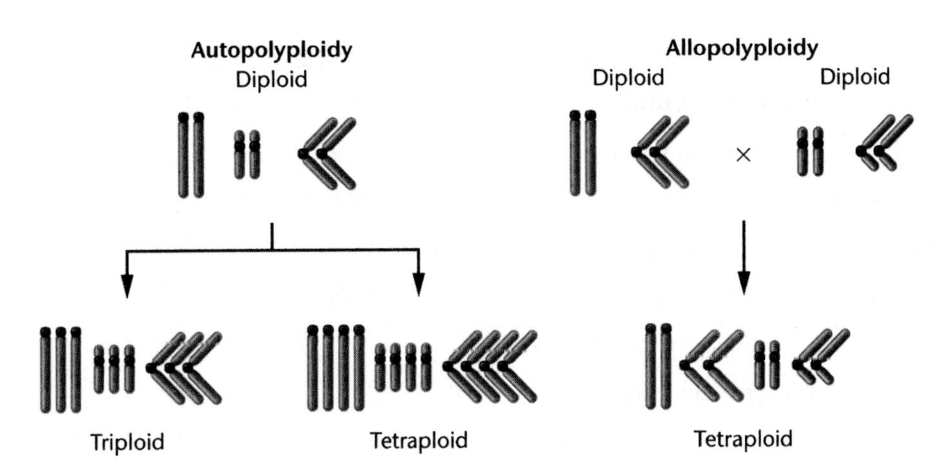

FIGURE 6.3 Contrasting chromosome origins of an autopolyploid versus an allopolyploid karyotype.

图 6.3 比较同源多倍体核型与异源多倍体核型中的染色体来源。

where a_1, a_2, and so on, are individual chromosomes and n is the haploid number. A normal diploid organism is represented simply as *AA*.

其中 a_1、a_2 等表示每一条染色体，n 则是物种单倍染色体的总数目。正常二倍体生物可以简单地表示为 *AA*。

Autopolyploidy

In autopolyploidy, each additional set of chromosomes is identical to the parent species. Therefore, triploids are represented as *AAA*, tetraploids are *AAAA*, and so forth.

Autotriploids arise in several ways. A failure of all chromosomes to segregate during meiotic divisions can produce a diploid gamete. If such a gamete is fertilized by a haploid gamete, a zygote with three sets of chromosomes is produced. Or, rarely, two sperm may fertilize an ovum, resulting in a triploid zygote. Triploids are also produced under experimental conditions by crossing diploids with tetraploids. Diploid organisms produce gametes with n chromosomes, while tetraploids produce $2n$ gametes. Upon fertilization, the desired triploid is produced.

同源多倍性

在同源多倍性中，每组额外增加的染色体均与亲本物种完全相同。因此，三倍体表示为 *AAA*，四倍体表示为 *AAAA*，以此类推。

同源三倍体可由多种途径产生。减数分裂过程中所有染色体分离失败将产生二倍体配子。如果该配子与单倍体配子受精，则产生含有三组单倍染色体组成的合子。或者，在罕见情况下，两个精子同时与一颗卵子发生受精作用，也将产生三倍体合子。三倍体还可在实验条件下通过二倍体与四倍体进行杂交产生。二倍体生物产生含 n 条染色体的配子，四倍体生物则产生含 $2n$ 条染色体的配子，二者受精后即得到预期的三倍体个体。

Because they have an even number of chromosomes, **autotetraploids** ($4n$) are theoretically more likely to be found in nature than are autotriploids. Unlike triploids, which often produce genetically unbalanced gametes with odd numbers of chromosomes, tetraploids are more likely to produce balanced gametes when involved in sexual reproduction.

由于**同源四倍体**（$4n$）所含染色体数目为偶数，因此理论上在自然界中比同源三倍体更容易被发现。与三倍体通常产生含奇数数目染色体、遗传不平衡的配子不同，四倍体更可能产生遗传平衡的配子，从而可以参与有性生殖过程。

How polyploidy arises naturally is of great interest to geneticists. In theory, if chromosomes have replicated, but the parent cell never divides and instead reenters interphase, the chromosome number will be doubled. That this very likely occurs is supported by the observation that tetraploid cells can be produced experimentally from diploid cells. This is accomplished by applying cold or heat shock to meiotic cells or by applying colchicine to somatic cells undergoing mitosis. Colchicine, an alkaloid derived from the autumn crocus, interferes with spindle formation, and thus replicated chromosomes cannot separate at anaphase and do not migrate to the poles. When colchicine is removed, the cell can reenter interphase. When the paired sister chromatids separate and uncoil, the nucleus contains twice the diploid number of chromosomes and is therefore $4n$. This process is shown in Figure 6.4.

多倍性在自然界中的产生机制引起遗传学家极大的兴趣。理论上，如果染色体已经复制，但是亲代细胞不发生分裂，而是重新进入间期，那么染色体数目便会加倍。在实验条件下，观察由二倍体细胞产生四倍体细胞的结果即可支持这种可能性的发生。通过对减数分裂细胞施加冷休克或热休克，或者对正进行有丝分裂的体细胞使用秋水仙碱处理，均可实现这一实验结果。秋水仙碱是一种来自秋水仙的生物碱，能够干扰纺锤体形成，因此已完成复制的染色体在有丝分裂后期无法正确分离，也无法迁移至细胞两极。当秋水仙碱被移除后，细胞重新进入间期，配对的姐妹染色单体相互分开并解螺旋，此时细胞核内将含有两倍的二倍染色体数目，即 $4n$。该过程如图 6.4 所示。

In general, autopolyploids are larger than their diploid relatives. This increase seems to be due to larger cell size rather than greater cell number. Although autopolyploids do not contain new or unique information compared with their diploid relatives, the flower and fruit of plants are often increased in size, making such

通常，同源多倍体比其相应的二倍体亲缘个体更大。这种个体的增大似乎是由于细胞体积增大而不是因为细胞数量增多。尽管同源多倍体与其二倍体亲缘个体相比并不含有新的或特有的遗传信息，但是植物的花朵和果实通常有所增大，从而使这些品种具有更高的园艺价值或商业价值。

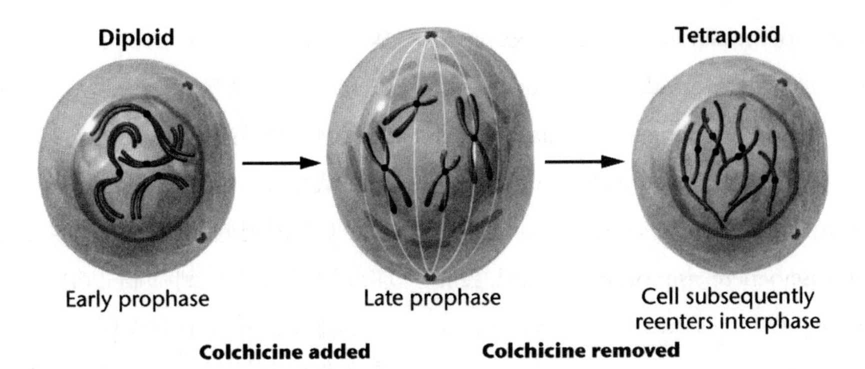

FIGURE 6.4 The potential involvement of colchicine in doubling the chromosome number. Two pairs of homologous chromosomes are shown. While each chromosome had replicated its DNA earlier during interphase, the chromosomes do not appear as double structures until late prophase. When anaphase fails to occur normally, the chromosome number doubles if the cell reenters interphase.

图 6.4 秋水仙碱导致染色体数目加倍。图中显示两对同源染色体。每条染色体在间期进行 DNA 复制，直至前期末染色体呈现出二倍结构。当后期无法正常进行时，如果细胞再次进入间期，那么染色体数目将加倍。

varieties of greater horticultural or commercial value. Economically important triploid plants include several potato species of the genus *Solanum*, Winesap apples, commercial bananas, seedless watermelons, and the cultivated tiger lily *Lilium tigrinum*. These plants are propagated asexually. Diploid bananas contain hard seeds, but the commercial, triploid, “seedless” variety has edible seeds. Tetraploid alfalfa, coffee, peanuts, and McIntosh apples are also of economic value because they are either larger or grow more vigorously than do their diploid or triploid counterparts. Many of the most popular varieties of hosta plant are tetraploid. In each case, leaves are thicker and larger, the foliage is more vivid, and the plant grows more vigorously. The commercial strawberry is an octoploid.

具有重要经济价值的三倍体植物包括茄属的几种马铃薯品种、晚熟苹果、商品化香蕉、无籽西瓜和栽培种虎皮百合。这些物种均为无性繁殖。二倍体香蕉含有坚硬的种子，但是商品化的三倍体“无籽”品种的种子则可食用。四倍体紫花苜蓿、咖啡、花生和麦金托什苹果也同样具有经济价值，因为它们比其相应的二倍体或三倍体品种个体更大或者生长更旺盛。许多倍受欢迎的玉簪品种也是四倍体，每种品种的叶片都更厚、更大，叶片颜色更鲜艳，并且植株生长也更旺盛。商品化草莓是八倍体。

How cells with increased ploidy values express different phenotypes from their diploid counterparts has been investigated. Gerald Fink and his colleagues created strains of the yeast *Saccharomyces cerevisiae* with one, two, three,

细胞染色体倍性增加与其相应二倍体物种具有表型差异的原因已获解析。杰拉尔德·芬克及其同事获得了酿酒酵母的单倍体、二倍体、三倍体和四倍体菌株并监测细胞周期中所有基因的表达水平。使用

or four copies of the genome and then examined the expression levels of all genes during the cell cycle. Using the stringent standards of at least a tenfold increase or decrease of gene expression, Fink and coworkers identified numerous cases where, as ploidy increased, gene expression either increased or decreased at least tenfold. Among these cases are two genes that encode **G1 cyclins**, which are repressed when ploidy increases. G1 cyclins facilitate the cell's movement through G1 of the cell cycle, which is thus delayed when expression of these genes is repressed. The polyploid cell stays in the G1 phase longer and, on average, grows to a larger size before it moves beyond the G1 stage of the cell cycle, providing a clue as to how other polyploids demonstrate increased cell size.

基因表达水平上升至少 10 倍范围或降低至少 1/10 的严格标准，芬克与其同事鉴定了随细胞倍性增加，基因表达增加至少 10 倍或降低至少 1/10 的许多情形。其中，两种编码 **G1 细胞周期蛋白**的基因在细胞倍性增加过程中表达受到抑制。G1 细胞周期蛋白可促进细胞通过细胞周期的 G1 期，因此如果这两种基因的表达受到抑制，那么细胞将滞留于 G1 期。多倍体细胞在 G1 期滞留的时间越长，其跨过细胞周期 G1 阶段的过程中体积会变得越大，从而为了解多倍体生物呈现更大细胞体积的原因提供了线索。

Allopolyploidy

Polyploidy can also result from hybridizing two closely related species. If a haploid ovum from a species with chromosome sets *AA* is fertilized by sperm from a species with sets *BB*, the resulting hybrid is *AB*, where $A = a_1, a_2, a_3, \cdots, a_n$ and $B = b_1, b_2, b_3, \cdots, b_n$. The hybrid organism may be sterile because of its inability to produce viable gametes. Most often, this occurs when some or all of the *a* and *b* chromosomes are not homologous and therefore cannot synapse in meiosis. As a result, unbalanced genetic conditions result. If, however, the new *AB* genetic combination undergoes a natural or an induced chromosomal doubling, two copies of all *a* chromosomes and two copies of all *b* chromosomes will be present, and they will

异源多倍性

多倍体也可通过两种亲缘关系相近的物种进行杂交而产生。如果来自染色体组成 *AA* 物种的单倍体卵子与来自染色体组成 *BB* 物种的精子进行受精，则所得杂交后代为 *AB*，其中 $A = a_1, a_2, a_3, \cdots, a_n$，$B = b_1, b_2, b_3, \cdots, b_n$。杂交后代可能由于不能产生可存活配子而不育。最常见的情形是，部分或全部的 *a* 和 *b* 染色体相互之间不同源，因此无法在减数分裂中进行联会，结果导致产生不均衡的染色体遗传组成。但是，如果新产生的 *AB* 遗传组合经历自然的或人工诱导的染色体加倍，则所有 *a* 染色体和所有 *b* 染色体均有两个拷贝，那么在减数分裂过程中将可以进行染色体配对，最终产生可育的 *AABB* 四倍体。这些过程如图 6.5 所示。由于该多倍体中含有相当于来自不同物种

pair during meiosis. As a result, a fertile *AABB* tetraploid is produced. These events are shown in Figure 6.5. Since this polyploid contains the equivalent of four haploid genomes derived from separate species, such an organism is called an **allotetraploid**. When both original species are known, an equivalent term, **amphidiploid**, is preferred in describing the allotetraploid.

的四组单倍体基因组，因此这种生物被称为**异源四倍体**。当两种原始来源物种均已知时，另一种同义专业术语“**双二倍体**”则被推荐用于描述该类异源四倍体。

Amphidiploid plants are often found in nature. Their reproductive success is based on their potential for forming balanced gametes. Since two homologs of each specific chromosome are present, meiosis occurs normally (Figure 6.5) and fertilization successfully propagates the plant sexually. This discussion assumes the simplest situation, where none of the chromosomes in set *A* are

双二倍体植物常见于自然界中。它们能成功进行繁衍是基于其具有形成均衡配子的潜力。由于每种特定染色体均存在两条同源染色体，所以减数分裂可以正常发生（图 6.5），并通过成功受精实现植物的有性生殖。这里的讨论假设的是最简单的情况，即 *A* 组染色体与 *B* 组染色体彼此均不同源。如果由亲缘关系相近的物种形成双二倍体，那么一些 *a* 染色体和 *b* 染色体之

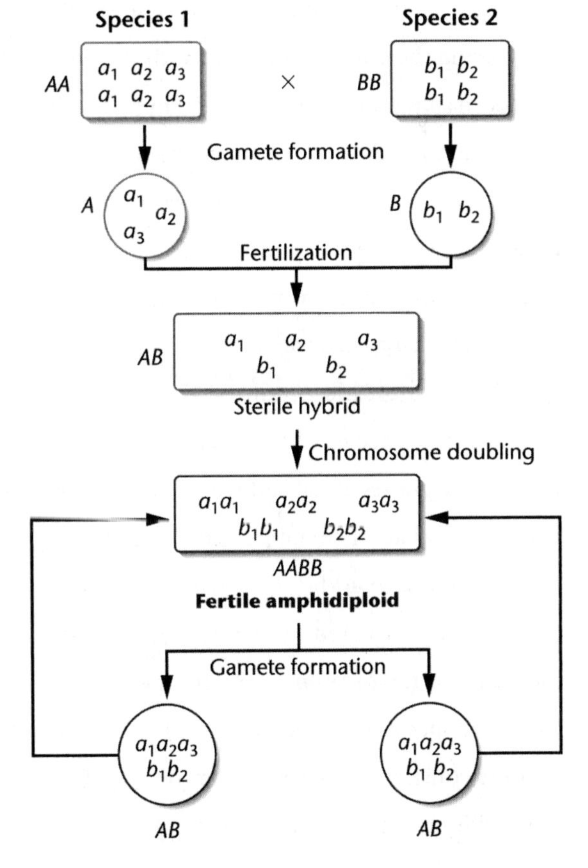

FIGURE 6.5 The origin and propagation of an amphidiploid. Species 1 contains genome *A* consisting of three distinct chromosomes, a_1, a_2, and a_3. Species 2 contains genome *B* consisting of two distinct chromosomes, b_1 and b_2. Following fertilization between members of the two species and chromosome doubling, a fertile amphidiploid containing two complete diploid genomes (*AABB*) is formed.

图 6.5 双二倍体的起源与繁殖。物种 1 的基因组 *A* 含 3 种不同染色体 a_1、a_2 和 a_3。物种 2 的基因组 *B* 含 2 种不同染色体 b_1 和 b_2。两种物种经受精和染色体加倍后获得含两套完整二倍体基因组的可育双二倍体 (*AABB*)。

homologous to those in set B. In amphidiploids formed from closely related species, some homology between a and b chromosomes is likely. Allopolyploids are rare in most animals because mating behavior is most often species-specific, and thus the initial step in hybridization is unlikely to occur.

间可能存在同源。异源多倍体在大多数动物中并不常见，因为动物的交配行为大多具有物种特异性，因此杂交的第一步就不可能发生。

A classical example of amphidiploidy in plants is the cultivated species of American cotton, *Gossypium*. This species has 26 pairs of chromosomes: 13 are large and 13 are much smaller. When it was discovered that Old World cotton had only 13 pairs of large chromosomes, allopolyploidy was suspected. After an examination of wild American cotton revealed 13 pairs of small chromosomes, this speculation was strengthened. J. O. Beasley reconstructed the origin of cultivated cotton experimentally by crossing the Old World strain with the wild American strain and then treating the hybrid with colchicine to double the chromosome number. The result of these treatments was a fertile amphidiploid variety of cotton. It contained 26 pairs of chromosomes as well as characteristics similar to the cultivated variety.

植物双二倍体的一个经典例子是栽培种美棉。该物种有 26 对染色体：13 对大染色体和 13 对小染色体。当人们发现亚洲棉仅含 13 对大染色体时，就推测栽培种美棉可能是异源多倍体。在发现野生美洲棉含 13 对小染色体时，人们更坚定了这种推测的可能性。J. O. 比斯利通过将亚洲棉与野生美洲棉进行杂交，然后用秋水仙碱处理杂交后代使其染色体数目加倍，在实验条件下重现了栽培种美棉的起源过程。经过这些操作得到的可育双二倍体棉花品种含 26 对染色体，并与栽培种美棉特征相似。

Amphidiploids often exhibit traits of both parental species. An interesting example involves the grasses wheat and rye. Wheat (genus *Triticum*) has a basic haploid genome of seven chromosomes. In addition to normal diploids ($2n = 14$), cultivated autopolyploids exist, including tetraploid ($4n = 28$) and hexaploid ($6n = 42$) species. Rye (genus *Secale*) also has a genome consisting of seven chromosomes. The only cultivated species is the diploid plant ($2n =$

双二倍体通常同时表现出两种亲本物种的特征。一个有趣的例子涉及小麦和黑麦。小麦（小麦属）单倍体基因组含 7 条染色体，除了正常二倍体（$2n$=14）外，还有栽培种同源多倍体存在，包括四倍体（$4n$=28）和六倍体（$6n$=42）。黑麦（黑麦属）同样含由 7 条染色体组成的单倍体基因组，并且栽培品种仅有二倍体（$2n$=14）。

14).

Using the technique outlined in Figure 6.5, geneticists have produced various hybrids. When tetraploid wheat is crossed with diploid rye and the F_1 plants are treated with colchicine, a hexaploid variety ($6n = 42$) is obtained; the hybrid, designated *Triticale*, represents a new genus. Other *Triticale* varieties have been created. These hybrid plants demonstrate characteristics of both wheat and rye. For example, they combine the high-protein content of wheat with rye's high content of the amino acid lysine, which is low in wheat and thus is a limiting nutritional factor. Wheat is considered to be a high-yielding grain, whereas rye is noted for its versatility of growth in unfavorable environments. Triticale species that combine both traits have the potential of significantly increasing grain production. This and similar programs designed to improve crops through hybridization have long been under way in several developing countries.

利用图 6.5 中简述的技术，遗传学家已经创造出各种不同的杂交品种。用四倍体小麦与二倍体黑麦进行杂交得到 F_1 代，然后用秋水仙碱处理 F_1 代后可获得六倍体品种（$6n=42$），该杂交品种被称为小黑麦，是一个新的属。其他小黑麦品种也已被创建，这些杂交品种同时表现出小麦和黑麦的特性。例如，它们具有小麦的高蛋白含量和黑麦的高赖氨酸含量。赖氨酸在小麦中含量很低，这也是小麦的限制性营养因子。小麦是高产谷物，而黑麦则因其在恶劣环境下的强生存能力而闻名。同时具有这两种性状的小黑麦具有显著提高谷物产量的潜力。通过杂交设计改良农作物的这项计划及类似项目在若干发展中国家已开展了很长时间。

ESSENTIAL POINT

When complete sets of chromosomes are added to the diploid genome, these sets can have an identical or a diverse genetic origin, creating either autopolyploidy or allopolyploidy, respectively.

基本要点

当完整的成组染色体增加至二倍体基因组中时，这些成组染色体可以具有相同的遗传起源，也可以具有不同的遗传起源，从而分别产生同源多倍性或异源多倍性。

6.4 Variation Occurs in the Composition and Arrangement of Chromosomes

6.4 染色体结构和排列顺序的变异

The second general class of chromosome aberrations includes changes that delete, add,

染色体畸变的第二大类包括缺失、添加或者重新排列一条或多条染色体的部分

or rearrange substantial portions of one or more chromosomes. Included in this broad category are deletions and duplications of genes or part of a chromosome and rearrangements of genetic material in which a chromosome segment is inverted, exchanged with a segment of a nonhomologous chromosome, or merely transferred to another chromosome. Exchanges and transfers are called translocations, in which the locations of genes are altered within the genome. These types of chromosome alterations are illustrated in Figure 6.6.

片段。包括在这一大类中的有基因或部分染色体的缺失和重复；遗传物质重排，如染色体片段倒位、非同源染色体片段间发生交换或仅转移至另一条染色体上。交换或转移被称为易位，即基因位置在基因组内发生变化。这些染色体畸变类型如图 6.6 所示。

In most instances, these structural

在大多情况下，这些结构改变是沿染

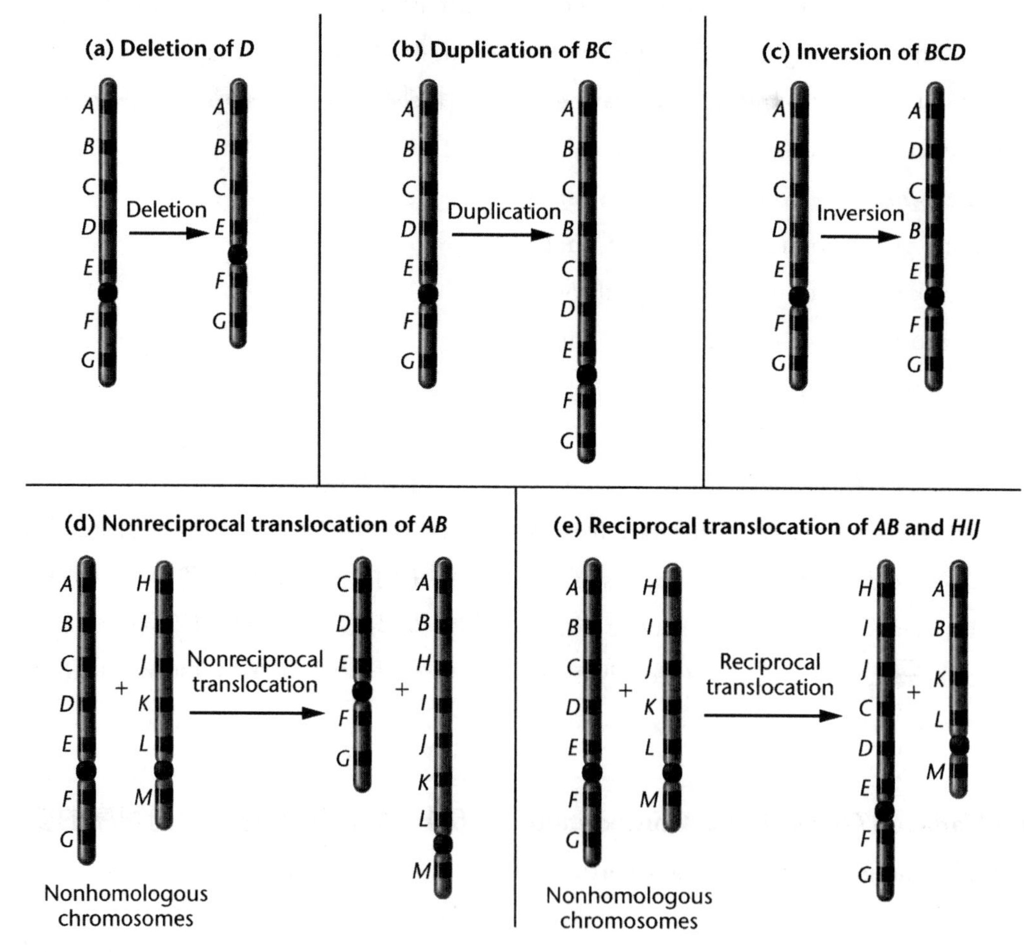

FIGURE 6.6 An overview of the five different types of gain, loss, or rearrangement of chromosome segments.

图 6.6 5 种不同的染色体畸变类型，包括染色体片段的添加、缺失和重排。

changes are due to one or more breaks along the axis of a chromosome, followed by either the loss or rearrangement of genetic material. Chromosomes can break spontaneously, but the rate of breakage may increase in cells exposed to chemicals or radiation. The ends produced at points of breakage are "sticky" and can rejoin other broken ends. If breakage and rejoining do not reestablish the original relationship and if the alteration occurs in germ plasm, the gametes will contain the structural rearrangement, which is heritable.

色体发生一处或多处断裂所致，随后发生遗传物质的丢失或重排。染色体会自发发生断裂，但是当细胞暴露于化学物质或辐射时，染色体发生自发断裂的频率可能会增加。断裂位点产生的末端具有"黏性"，并且可与其他断裂末端重新黏合。如果断裂和重新黏合不能恢复原有的染色体结构，或者该变化发生于种质中，那么配子将含有这些结构重排并具有可遗传性。

If the aberration is found in one homolog but not the other, the individual is said to be *heterozygous for the aberration*. In such cases, unusual but characteristic pairing configurations are formed during meiotic synapsis. These patterns are useful in identifying the type of change that has occurred. If no loss or gain of genetic material occurs, individuals bearing the aberration "heterozygously" are likely to be unaffected phenotypically. However, the unusual pairing arrangements often lead to gametes that are duplicated or deficient for some chromosomal regions. When this occurs, the offspring of "carriers" of certain aberrations have an increased probability of demonstrating phenotypic changes.

如果二倍体个体的一条同源染色体中含染色体畸变，而另一条同源染色体不含，那么该个体被称为染色体畸变杂合子。在这些个体中，减数分裂联会时将形成异常的、具有特征性的配对构象。这些构象有助于鉴定已发生的染色体畸变类型。如果遗传物质没有发生缺失或增加，畸变杂合子个体可能在表型上不受影响。然而，染色体的异常配对通常会导致一些染色体区发生缺失或重复。当这些情况发生时，染色体畸变携带者的后代将有很大概率表现出表型变化。

6.5 A Deletion Is a Missing Region of a Chromosome

6.5 缺失是染色体上发生丢失的一段区域

When a chromosome breaks in one or more places and a portion of it is lost, the missing piece is called a **deletion** (or a **deficiency**).

当染色体在一处或多处发生断裂而导致部分片段丢失时，该发生丢失的片段称为**缺失**。缺失可以发生在染色体端部或染

The deletion can occur either near one end or within the interior of the chromosome. These are **terminal** and **intercalary deletions**, respectively [Figure 6.7(a) and (b)]. The portion of the chromosome that retains the centromere region is usually maintained when the cell divides, whereas the segment without the centromere is eventually lost in progeny cells following mitosis or meiosis. For synapsis to occur between a chromosome with a large intercalary deletion and a normal homolog, the unpaired region of the normal homolog must "buckle out" into a **deletion**, or **compensation, loop** [Figure 6.7(c)].

色体内部，分别为**末端缺失**和**中间缺失**［图 6.7（a）和（b）］。含有着丝粒区的部分染色体在细胞分裂时通常得以保留，而不含着丝粒的染色体部分在有丝分裂或减数分裂的子代细胞中最终会丢失。为了在具有较大中间缺失的染色体和正常同源染色体间进行联会配对，正常同源染色体中不能被配对的区域必须"环出"，形成**缺失环**或**补偿环**［图 6.7（c）］。

If only a small part of a chromosome is deleted, the organism might survive. However, a deletion of a portion of a chromosome need not be very great before the effects become severe.

如果仅缺失小部分染色体，则该生物可能可以存活。但是，染色体发生缺失的部分不能过多，否则将产生十分严重的影响。我们将在随后讨论的人类猫叫综合征

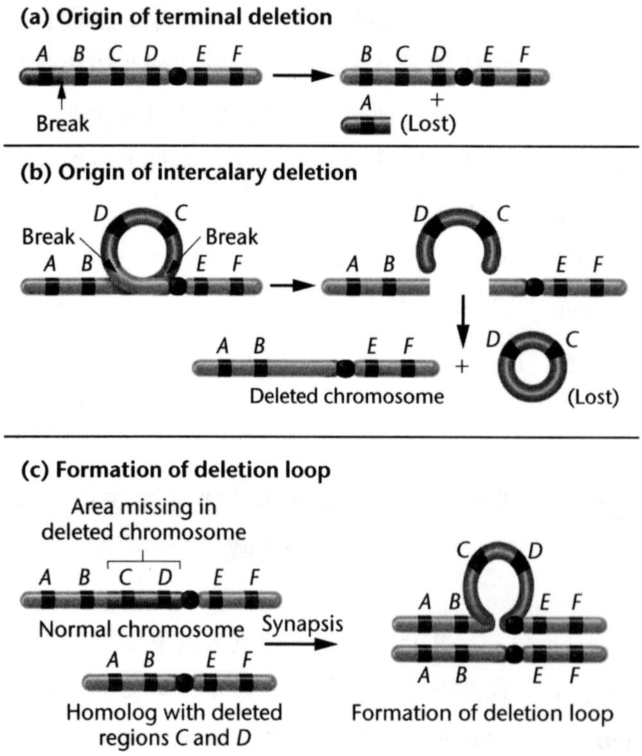

FIGURE 6.7 Origins of (a) a terminal and (b) an intercalary deletion. In (c), pairing occurs between a normal chromosome and one with an intercalary deletion by looping out the undeleted portion to form a deletion (or compensation) loop.

图 6.7 末端缺失（a）和中间缺失（b）的产生。在（c）中，正常染色体与中间缺失染色体通过将未缺失部分环出形成缺失环（或补偿环）而进行联会配对。

We see an example of this in the following discussion of the cri du chat syndrome in humans. If even more genetic information is lost as a result of a deletion, the aberration is often lethal, in which case the chromosome mutation never becomes available for study.

例子中看到这种情况。如果由于染色体缺失而丢失过多的遗传信息，那么通常会致死，这样的染色体畸变当然也无法获得并进行研究。

Cri du Chat Syndrome in Humans

In humans, the **cri du chat syndrome** results from the deletion of a small terminal portion of chromosome 5. It might be considered a case of *partial monosomy*, but since the region that is missing is so small, it is better referred to as a *segmental deletion*. This syndrome was first reported by Jérôme Lejeune in 1963, when he described the clinical symptoms, including an eerie cry similar to the meowing of a cat, after which the syndrome is named. This syndrome is associated with the loss of a small, variable part of the short arm of chromosome 5 . Thus, the genetic constitution may be designated as 46,5p−, meaning that the individual has all 46 chromosomes but that some or all of the p arm (the petite, or short, arm) of one member of the chromosome 5 pair is missing.

人类猫叫综合征

人类**猫叫综合征**是由人类 5 号染色体末端发生小片段缺失所引起的，也可以被认为是部分单体，但由于缺失区域的片段非常小，因此将其作为节段缺失更合适。热罗姆·勒热纳于 1963 年首次报道了这种综合征，并描述了该综合征的临床症状，包括患者哭声类似于猫叫声并按该特征对其进行了命名。该综合征与 5 号染色体短臂端小且长度可变的片段丢失有关，因此，该个体的遗传组成可以表示为 46,5p−，意为该个体含有全部 46 条人类染色体，但是其中一条 5 号染色体的短臂发生部分或全部缺失。

Infants with this syndrome exhibit intellectual disability, delayed development, small head size, and distinctive facial features in addition to abnormalities in the glottis and larynx, leading to the characteristic crying sound.

患有该综合征的婴儿除了声门和喉结结构异常导致特征性哭声外，还表现出智力障碍、发育迟缓、头部较小、面部特征独特等特征。

Since 1963, hundreds of cases of cri du chat syndrome have been reported worldwide. An incidence of 1 in 20,000—50,000 live births has been estimated. Most often, the condition is not inherited but instead results from the sporadic

自 1963 年以来，全世界已报告了数百例猫叫综合征病例，预计新生儿发病率为 1/50 000—1/20 000。在大多数情况下，这种疾病并不遗传，由配子中染色体的零星丢失所致。染色体短臂中发生缺失的片段长

loss of chromosomal material in gametes. The length of the short arm that is deleted varies somewhat; longer deletions tend to result in more severe intellectual disability and developmental delay. Although the effects of the syndrome are severe, most individuals achieve motor and language skills. The deletion of several genes, including the telomerase reverse transcriptase (TERT) gene, has been implicated in various phenotypic changes in cri du chat syndrome.

度各有不同，较长片段缺失会导致更严重的智力障碍和发育迟缓。尽管该综合征的影响很严重，但大多数患者可以获得运动和语言技能。研究表明缺失的多种基因，如端粒酶逆转录酶（TERT）基因与猫叫综合征的各种表型变化有关。

6.6 A Duplication Is a Repeated Segment of a Chromosome

6.6 重复是多次出现的染色体片段

When any part of the genetic material—a single locus or a large piece of a chromosome—is present more than once in the genome, it is called a **duplication**. As in deletions, pairing in heterozygotes can produce a compensation loop. Duplications may arise as the result of unequal crossing over between synapsed chromosomes during meiosis (Figure 6.8) or through a replication error prior to meiosis. In the former case, both a duplication and a deletion are produced.

当遗传物质的任何部分（单个基因座或一段染色体大片段）在基因组中不止一次出现时，即称为**重复**。与缺失类似，杂合子中同源染色体配对将产生补偿环。减数分裂过程中，联会的染色体间发生不等交换可能导致重复（图 6.8），或者在减数分裂之前由于复制错误也可能导致重复。在前一种情况下，重复和缺失将同时产生。

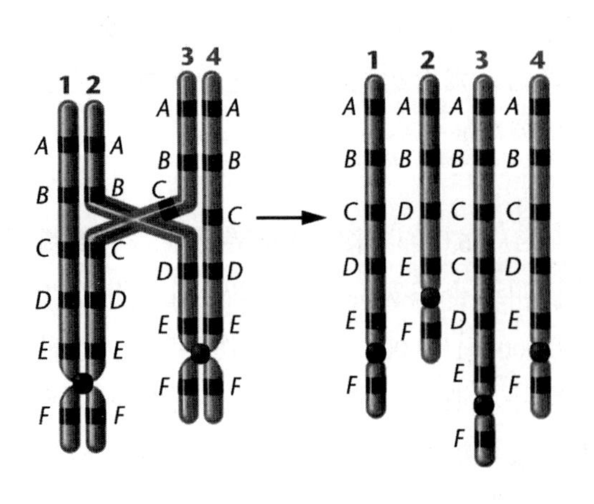

FIGURE 6.8 The origin of duplicated and deficient regions of chromosomes as a result of unequal crossing over. The tetrad on the left is mispaired during synapsis. A single crossover between chromatids 2 and 3 results in the deficient (chromosome 2) and duplicated (chromosome 3) chromosomal regions shown on the right. The two chromosomes uninvolved in the crossover event remain normal in gene sequence and content.

图 6.8 不等交换的一个结果是产生染色体片段的重复或缺失。图左侧的四联体在联会中发生配对错误，染色单体 2 与染色单体 3 间的单交换产生了图右侧的染色体片段缺失（染色体 2）和重复（染色体 3）。不参与交换事件的两条染色体则在基因序列和组成上保持不变。

We consider three interesting aspects of duplications. First, they may result in gene redundancy. Second, as with deletions, duplications may produce phenotypic variation. Third, according to one convincing theory, duplications have also been an important source of genetic variability during evolution.

我们将讨论有关重复的三个方面的有趣内容：第一，重复导致基因冗余；第二，与缺失一样，重复产生表型变异；第三，根据已确定的理论，重复也是进化过程中遗传变异的重要来源。

Gene Redundancy—Ribosomal RNA Genes

Although many gene products are not needed in every cell of an organism, other gene products are known to be essential components of all cells. For example, ribosomal RNA must be present in abundance to support protein synthesis. The more metabolically active a cell is, the higher the demand for this molecule. We might hypothesize that a single copy of the gene encoding rRNA is inadequate in many cells. Studies using the technique of molecular hybridization, which enables us to determine the percentage of the genome that codes for specific RNA sequences, show that our hypothesis is correct. Indeed, multiple copies of genes code for rRNA. Such DNA is called **rDNA**, and the general phenomenon is referred to as **gene redundancy**. For example, in the common intestinal bacterium *Escherichia coli* (*E. coli*), about 0.7 percent of the haploid genome consists of rDNA—the equivalent of seven copies of the gene. In *Drosophila melanogaster*, 0.3 percent of the haploid genome, equivalent to 130 gene copies, consists of rDNA. Although the presence of multiple copies of the same gene is not restricted to those coding for rRNA, we will

基因冗余——核糖体 RNA 基因

尽管许多基因产物并非在生物体每个细胞中均需要，但已知一些基因产物是所有细胞的必需组分。例如，核糖体 RNA 必须大量存在以支持蛋白质合成，并且细胞的代谢活性越强，对该分子的需求就越高。我们可以假设，单一拷贝编码 rRNA 的基因在许多细胞中都是不能满足需求的。使用分子杂交技术可以确定编码特定 RNA 序列的基因所占基因组的百分比，从而表明我们的假设是正确的。事实上，编码 rRNA 的基因具有多个拷贝，这些 DNA 被称为 **rDNA**，并且这种常见现象被称为**基因冗余**。例如，在常见的肠道细菌——大肠杆菌中，单倍体基因组中约 0.7% 由 rDNA 组成，相当于该基因有 7 个拷贝。在果蝇中，rDNA 占单倍体基因组的 0.3%，相当于有 130 个 rDNA 基因拷贝。尽管同种基因存在多个拷贝并不局限于 rRNA 编码基因，但是在本节中我们将重点介绍它们。

focus on them in this section.

In some cells, particularly oocytes, even the normal amplification of rDNA is insufficient to provide adequate amounts of rRNA needed to construct ribosomes. For example, in the amphibian *Xenopus laevis*, 400 copies of rDNA are present per haploid genome. These genes are all found in a single area of the chromosome known as the **nucleolar organizer region** (**NOR**). In *Xenopus* oocytes, the NOR is selectively replicated to further increase rDNA copies, and each new set of genes is released from its template. Each set forms a small nucleolus, and as many as 1500 of these "micronucleoli" have been observed in a single oocyte. If we multiply the number of micronucleoli (1500) by the number of gene copies in each NOR (400), we see that amplification in *Xenopus* oocytes can result in over half a million gene copies! If each copy is transcribed only 20 times during the maturation of the oocyte, in theory, sufficient copies of rRNA are produced to result in well over 12 million ribosomes.

在某些细胞，尤其是卵母细胞中，即使rDNA正常扩增，也不足以提供足够数量的rRNA用于构建核糖体。例如，在两栖类动物非洲爪蟾中，单倍体基因组中存在400个rDNA拷贝。这些基因全部位于染色体的一个区域，即**核仁组织区（NOR）**。在非洲爪蟾卵母细胞中，NOR被选择性复制以进一步增加rDNA基因拷贝，并且每组新基因均被从其模板中释放出来，形成一个小核仁。在单个卵母细胞中已观察到多达1500个类似的“微核仁”。如果我们将微核仁的数量（1500）乘以每个NOR中的基因拷贝数（400），那么我们将发现非洲爪蟾卵母细胞中的基因扩增可产生超过50万个基因拷贝！理论上，在卵母细胞的成熟过程中，如果每个基因拷贝被转录20次，那么合成的rRNA拷贝将足以构建1200万个以上的核糖体。

The *Bar* Mutation in *Drosophila*

Duplications can cause phenotypic variation that might at first appear to be caused by a simple gene mutation. The *Bar*-eye phenotype in *Drosophila* is a classic example. Instead of the normal oval-eye shape, Bar-eyed flies have narrow, slit-like eyes. This phenotype is inherited in the same way as a dominant X-linked mutation.

果蝇的棒眼突变

重复可以导致表型变异，并且这些变异看似是由简单的基因突变引起的。果蝇的棒眼表型即是一个经典例子。与正常卵圆形眼睛不同，棒眼果蝇的眼睛呈窄狭缝状。这种表型的遗传方式与X连锁显性突变相同。

In the early 1920s, Alfred H. Sturtevant and Thomas H. Morgan discovered and investigated

在20世纪20年代初期，艾尔弗雷德·H. 斯特蒂文特和托马斯·H. 摩尔根发现并研

this "mutation." Normal wild-type females (B^+/B^+) have about 800 facets in each eye. Heterozygous females (B/B^+) have about 350 facets,while homozygous females (B/B) average only about 70 facets. Females were occasionally recovered with even fewer facets and were designated as *double Bar* (B^D/B^+).

究了这种"突变"。正常野生型雌蝇（B^+/B^+）每只眼睛约含800只小眼面。杂合雌蝇(B/B^+)约含350只小眼面，而纯合雌蝇(B/B)平均只含约70只小眼面。偶尔还发现一些小眼面数量更少的雌蝇，称为超棒眼（B^D/B^+）。

About 10 years later, Calvin Bridges and Herman J. Muller compared the polytene X chromosome banding pattern of the *Bar* fly with that of the wild-type fly. These chromosomes contain specific banding patterns that have been well categorized into regions. Their studies revealed that one copy of the region designated as 16A is present on both X chromosomes of wild-type flies but that this region was duplicated in *Bar* flies and triplicated in *double Bar* flies. These observations provided evidence that the *Bar* phenotype is not the result of a simple chemical change in the gene but is instead a duplication.

大约10年后，卡尔文·布里奇斯和赫尔曼·J. 马勒对棒眼果蝇与野生型果蝇的多线X染色体的带型进行了比较。这些染色体含有已明确区域化的特定染色体带型。他们的研究揭示了野生型果蝇的两条X染色体上均存在一个拷贝的16A区，但该区域在棒眼果蝇中发生重复且在超棒眼果蝇中存在三个拷贝。这些发现为棒眼表型并非由基因简单的化学变化所致，而是由染色体重复所致提供了证据。

The Role of Gene Duplication in Evolution

During the study of evolution, it is intriguing to speculate on the possible mechanisms of genetic variation. The origin of unique gene products present in more recently evolved organisms but absent in ancestral forms is a topic of particular interest. In other words, how do "new" genes arise?

基因重复在进化中的作用

在进化研究中，推测遗传变异的可能机制令人着迷。高度进化的生物中存在的独特基因产物在其祖先物种中并不存在，因此这些独特基因产物的起源尤其令人感兴趣。换句话说，"新"基因是如何产生的？

In 1970, Susumo Ohno published a provocative monograph, *Evolution by Gene Duplication*, in which he suggested that gene duplication is essential to the origin of new genes during evolution. Ohno's thesis is based

1970年，大野乾出版了一部引人入胜的专著《基因重复造成的进化》。在书中，他指出基因重复对于进化过程中新基因的起源至关重要。大野的论点所基于的假设是许多基因在基因组中仅存在一个拷贝，

on the supposition that the gene products of many genes, present as only a single copy in the genome, are indispensable to the survival of members of any species during evolution. Therefore, unique genes are not free to accumulate mutations sufficient to alter their primary function and give rise to new genes.

在进化过程中它们的基因产物对于任何物种成员的生存都必不可少。因此，唯一拷贝的基因不可能自由积累到足以改变其主要功能的突变并由此衍生新基因。

However, if an essential gene is duplicated in the germ line, major mutational changes in this extra copy will be tolerated in future generations because the original gene provides the genetic information for its essential function. The duplicated copy will be free to acquire many mutational changes over extended periods of time. Over short intervals, the new genetic information may be of no practical advantage. However, over long evolutionary periods, the duplicated gene may change sufficiently so that its product assumes a divergent role in the cell. The new function may impart an "adaptive" advantage to organisms, enhancing their fitness. Ohno has outlined a mechanism through which sustained genetic variability may have originated.

但是，如果必需基因在生殖细胞系中发生重复，则该额外基因拷贝的主要突变将在后代中被容忍和保留，因为原始基因为其基本功能提供了遗传信息的保障。随着时间推移，重复的基因拷贝将自由积累许多基因突变。短时间内，新遗传信息可能并没有实际价值。但是，在漫长的进化过程中，重复基因可能发生足够的遗传变化，从而使其基因产物在细胞中发挥不同作用。新功能可以为物种赋予“适应性”优势，增强其适应性。大野已经概述了持续遗传变异可能的产生机制。

Ohno's thesis is supported by the discovery of genes that have a substantial amount of their organization and DNA sequence in common, but whose gene products are distinct. For example, trypsin and chymotrypsin fit this description, as do myoglobin and the various forms of hemoglobin. The DNA sequence is so similar (homologous) in each case that we may conclude that members of each pair of genes arose from a common ancestral gene through duplication. During evolution, the related genes diverged sufficiently that their products became

大野的论点得到了一些基因发现的支持，这些基因的组织含量大，具有共同的 DNA 序列，但它们的基因产物却截然不同。例如，胰蛋白酶和胰凝乳蛋白酶就是这种情形，肌红蛋白和各种形式的血红蛋白也是如此。在这些例子中，DNA 序列十分相似（同源），因此我们可以得出结论，每对基因的成员都是通过重复由共同的祖先基因产生的。在进化过程中，相关基因产生了充分的差异，因此它们的基因产物变得独特。

unique.

Other support includes the presence of **multigene families**—groups of contiguous genes whose products perform the same, or very similar functions. Again, members of a family show DNA sequence homology sufficient to conclude that they share a common origin and arose through the process of gene duplication. One of the most interesting supporting examples is the case of the *SRGAP2* gene in primates. This gene is known to be involved in the development of the brain. Humans have at least four similar copies of the gene, while all nonhuman primates have only a single copy. Several duplication events can be traced back to 3.4 million years ago, to 2.4 million years ago, and finally to 1 million years ago, resulting in distinct forms of *SRGAP2* labeled A—D. These evolutionary periods coincide with the emergence of the human lineage in primates. The function of these genes has now been related to the regulation and formation of dendritic spines in the brain, which is believed to contribute to the evolution of expanded brain function in humans, including the development of language and social cognition.

其他的支持证据包括**多基因家族**，即基因产物执行相同功能或相似功能的多种基因。同样，基因家族成员的DNA序列同源性也足以推断出它们具有共同的起源，并通过基因重复产生。最有趣的支持性实例之一是灵长类动物的*SRGAP2*基因。已知该基因参与大脑发育。人类至少含四个类似的该基因拷贝，而所有非人类的灵长类动物仅含一个基因拷贝。基因重复事件可以追溯到340万年前、240万年前和100万年前，最终产生了不同形式、标记为A—D的*SRGAP2*基因。这些进化阶段与灵长类动物中人类谱系的出现相吻合。这些基因的功能已表明与大脑中树突棘的调节和形成有关，人们认为这些有助于人类大脑功能的扩展进化，包括语言和社会认知的发展。

Other examples of gene families arising from duplication during evolution include the various types of human hemoglobin polypeptide chains, as well as the immunologically important T-cell receptors and antigens encoded by the major histocompatibility complex.

进化过程中因重复形成基因家族的其他例子包括各种类型的人类血红蛋白多肽链，以及免疫学上十分重要的由主要组织相容性复合体编码的T细胞受体和抗原。

Duplications at the Molecular Level: Copy Number Variations (CNVs)

分子水平的重复：拷贝数变异（CNVs）

As we entered the era of genomics and

当进入基因组学时代并能够对整个基

became capable of sequencing entire genomes, we quickly realized that duplications of *portions of genes*, most often involving thousands of base pairs, occur on a regular basis. When individuals in the same species are compared, the number of copies of any given duplicated sequence within a given gene is found to differ—sometimes there are larger and sometimes smaller numbers of copies, a condition described as **copy number variation** (**CNV**). Such duplications are found in both coding and noncoding regions of the genome.

因组进行测序时，我们很快意识到基因部分的重复现象时常发生，并且通常涉及数千个碱基对。当对同一物种的不同个体进行比较时，指定基因内任一重复序列的拷贝数目都不同——有时拷贝数多，有时少，这种情况被称为**拷贝数变异**（CNV）。这种重复在基因组编码区和非编码区中均有发现。

CNVs are of major interest in genetics because they are now believed to play crucial roles in the expression of many of our individual traits, in both normal and diseased individuals. Currently, when CNVs of sizes ranging from 50 bp to 3 Mb are considered, it is estimated that they occupy between 5-10 percent of the human genome. Current studies have focused on finding associations with human diseases. CNVs appear to have both positive and negative associations with many diseases in which the genetic basis is not yet fully understood. For example, pathogenic CNVs have been associated with autism and other neurological disorders, and with cancer. Additionally, CNVs are suspected to be associated with Type I diabetes and cardiovascular disease.

CNVs 在遗传学研究中引起人们的极大兴趣，因为人们认为它们在许多人类个体特征表达中起着至关重要的作用，无论是正常个体还是患病个体。目前据估计，片段大小为 50 bp—3 Mb 的 CNVs 占人类基因组的 5%—10%。当前的研究主要聚焦于发现 CNVs 与人类疾病的关系。CNVs 似乎与许多遗传学基础尚未完全了解的人类疾病显示出正向和负向关联。例如，病原性 CNVs 与孤独症、其他神经系统疾病以及癌症有关。此外，CNVs 与 I 型糖尿病和心血管疾病也可能有关。

In some cases, entire gene sequences are duplicated and impact individuals. For example, a higher-than-average copy number of the gene *CCL3L1* imparts an HIV-suppressive effect during viral infection, diminishing the progression to AIDS. Another finding has

在一些情况下，完整的基因序列会发生重复并对个体产生影响。例如，在病毒感染期间，拷贝数高于平均数的基因 *CCL3L1* 具有 HIV 抑制效应，可以减缓疾病向艾滋病发展的进程。另一个发现则将特定的突变 CNV 位点与肺癌的某些亚型患者人群相

associated specific mutant CNV sites with certain subset populations of individuals with lung cancer—the greater number of copies of the *EGFR* (*Epidermal Growth Factor Receptor*) gene, the more responsive are patients with non-small-cell lung cancer to treatment. Finally, the greater the reduction in the copy number of the gene designated *DEFB*, the greater the risk of developing Crohn's disease, a condition affecting the colon. Relevant to this chapter, these findings reveal that duplications and deletions are no longer restricted to textbook examples of these chromosomal mutations.

关联——*EGFR* 表皮生长因子受体基因拷贝数越多，非小细胞肺癌患者对治疗的响应越强。最后，*DEFB* 基因的拷贝数减少越多，患结肠相关的克罗恩病的风险就越大。这些与本章相关的发现均表明重复和缺失不再局限于教科书中的这些染色体畸变实例。

ESSENTIAL POINT

Deletions or duplications of segments of a gene or a chromosome may be the source of mutant phenotypes such as cri du chat syndrome in humans and *Bar* eyes in *Drosophila*, while duplications can be particularly important as a source of redundant or new genes.

基本要点

基因或染色体片段的缺失或重复可能是突变表型，如人类猫叫综合征和果蝇棒眼形成的原因，同时，重复作为冗余基因或新基因的产生源泉尤其重要。

6.7 Inversions Rearrange the Linear Gene Sequence

The **inversion**, another class of structural variation, is a type of chromosomal aberration in which a segment of a chromosome is turned around 180 degrees within a chromosome. An inversion does not involve a loss of genetic information but simply rearranges the linear gene sequence. An inversion requires breaks at two points along the length of the chromosome and subsequent reinsertion of the inverted segment. Figure 6.9 illustrates how an inversion might

6.7 倒位将线性基因序列进行重排

另一类结构变异——**倒位**是一种染色体片段在染色体内发生 180°旋转的染色体畸变。倒位不涉及遗传信息丢失，只是重新排列了线性基因顺序。倒位需要沿着染色体在两处发生断裂，随后片段发生 180°旋转后重新插入染色体中。图 6.9 展示了倒位发生的可能机制：在染色体发生断裂之前形成染色体环，新产生的“黏性末端”靠近在一起并重新发生连接。

arise. By forming a chromosomal loop prior to breakage, the newly created "sticky ends" are brought close together and rejoined.

The inverted segment may be short or quite long and may or may not include the centromere. If the centromere is not part of the rearranged chromosome segment, it is a **paracentric inversion**, which is the type shown in Figure 6.9. If the centromere is part of the inverted segment, it is described as a **pericentric inversion**.

倒位片段可以短也可以相当长，可以包含着丝粒也可以不包含着丝粒。如果重排的染色体片段不含着丝粒，则为**臂内倒位**；如果着丝粒位于倒位片段内，则为**臂间倒位**，如图 6.9 所示。

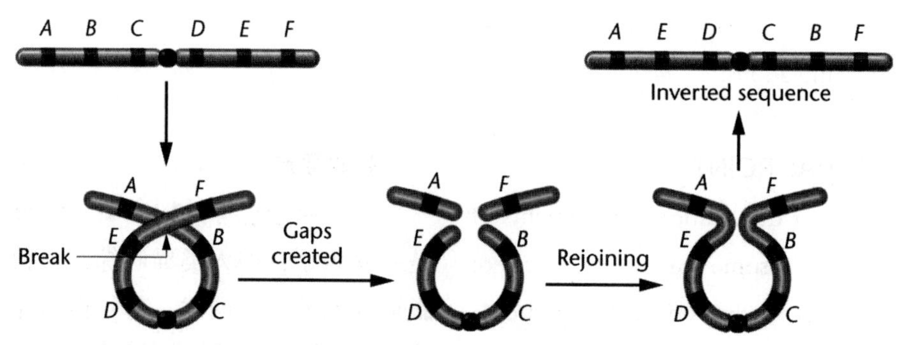

FIGURE 6.9 One possible origin of a pericentric inversion.

图 6.9 臂间倒位的一种可能发生机制。

Consequences of Inversions during Gamete Formation

倒位在配子形成过程中的影响

If only one member of a homologous pair of chromosomes has an inverted segment, normal *linear synapsis* during meiosis is not possible. Organisms with one inverted chromosome and one noninverted homolog are called **inversion heterozygotes**. Pairing between two such chromosomes in meiosis is accomplished only if they form an inversion loop (Figure 6.10).

如果同源染色体对中仅有一条含倒位片段，那么两条同源染色体在减数分裂过程中将不可能进行正常的线性联会。同时含一条倒位染色体和一条正常染色体的个体被称为**倒位杂合子**。这样的两条染色体在减数分裂过程中只有通过形成**倒位环**才能完成配对（图 6.10）。

If crossing over does not occur within the inverted segment of the inversion loop, the homologs will segregate, which results in two

如果倒位环的倒位片段内不发生交换，则同源染色体分开，产生两条正常的染色单体和两条含倒位片段的染色单体，并分

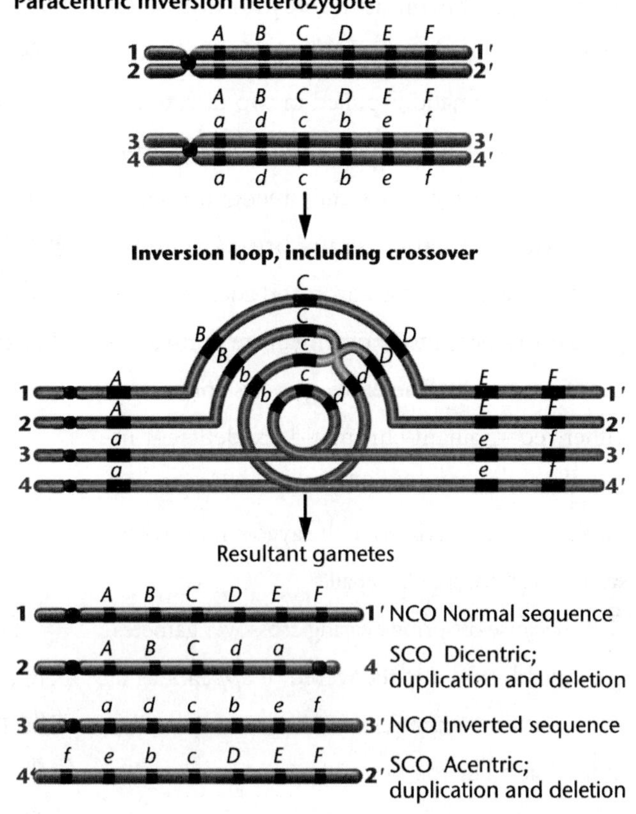

FIGURE 6.10 The effects of a single crossover (SCO) within an inversion loop in a paracentric inversion heterozygote, where two altered chromosomes are produced, one acentric and one dicentric. Both chromosomes also contain duplicated and deficient regions.

图 6.10 臂内倒位杂合子倒位环内的单交换(SCO)效应，产生了两条结构变化的染色体，一条无着丝粒，一条含双着丝粒。这两条染色体均包含重复区域和缺失区域。

normal and two inverted chromatids that are distributed into gametes. However, if crossing over does occur within the inversion loop, abnormal chromatids are produced. The effect of a single crossover (SCO) event within a paracentric inversion is diagrammed in Figure 6.10.

配至配子中。但是，如果在倒位环内发生交换，则会产生异常的染色单体。图 6.10 示意了臂内倒位中发生单交换（SCO）事件的影响。

In any meiotic tetrad, a single crossover between nonsister chromatids produces two parental chromatids and two recombinant chromatids. When the crossover occurs within a paracentric inversion, however, one recombinant **dicentric chromatid** (two centromeres) and one recombinant **acentric chromatid** (lacking a centromere) are produced. Both contain duplications and deletions of chromosome segments as well. During anaphase,

在任何一个减数分裂四联体中，非姐妹染色单体间的单交换均会产生两个亲本型染色单体和两个重组型染色单体。但是，当交换发生在臂内倒位中时，一个重组型**双着丝粒染色单体**和一个重组型**无着丝粒染色单体**将会产生，二者均包含染色体片段的重复和缺失。在减数分裂后期，无着丝粒染色单体随机移动到细胞的任意一极，或者可能发生丢失，而含双着丝粒的染色单体则被同时向细胞两极拉动。这种极化

an acentric chromatid moves randomly to one pole or the other or may be lost, while a dicentric chromatid is pulled in two directions. This polarized movement produces *dicentric bridges* that are cytologically recognizable. A dicentric chromatid usually breaks at some point so that part of the chromatid goes into one gamete and part into another gamete during the meiotic divisions. Therefore, gametes containing either recombinant chromatid are deficient in genetic material. In animals, when such a gamete participates in fertilization, the zygote most often develops abnormally, if at all.

运动将产生细胞学上可识别的双着丝粒桥。双着丝粒染色单体通常会发生断裂，因此在减数分裂过程中，部分染色单体会进入一个配子，而另一部分则进入另一配子。因此，含任何一种重组型染色单体的配子遗传信息均存在缺失。在动物中，这样的配子即使参与受精，合子在大多数情况下也会发育异常。

Because offspring bearing crossover gametes are inviable and not recovered, it *appears* as if the inversion suppresses crossing over. Actually, in inversion heterozygotes, the inversion has the effect of *suppressing the recovery of crossover products* when chromosome exchange occurs within the inverted region. Moreover, up to one-half of the viable gametes have the inverted chromosome, and the inversion will be perpetuated within the species. The cycle will be repeated continuously during meiosis in future generations.

因为含交换型配子的后代不育且无法获得，所以宏观上似乎是倒位抑制了交换的发生。实际上，在倒位杂合子中，当倒位染色体区域内发生染色体交换时，倒位具有抑制交换型后代存活的效应。而且，多达一半的可存活配子含倒位染色体，并将倒位持续在物种内遗传保存。这样的循环往复在后代减数分裂过程中将不断重复。

Evolutionary Advantages of Inversions

Because recovery of crossover products is suppressed in inversion heterozygotes, groups of specific alleles at adjacent loci within inversions may be preserved from generation to generation. If the alleles of the involved genes confer a survival advantage on the organisms maintaining them, the inversion is beneficial to the evolutionary survival of the species. For example,

倒位的进化意义

因为倒位杂合子的交换型后代存活被抑制，所以倒位片段内相邻基因座上特定的等位基因组合可以世代保存。如果所涉及基因的等位基因形式可以为生物体赋予生存优势，那么倒位将有利于该物种的进化生存。例如，如果等位基因组合 *ABcDef* 比 *AbCdeF* 或 *abcdEF* 更具有生存适应性，则有效配子将包含这组有利的基因组合，

if a set of alleles *ABcDef* is more adaptive than sets *AbCdeF* or *abcdEF*, effective gametes will contain this favorable set of genes, undisrupted by crossing over.

且不会由于交换而发生改变。

In laboratory studies, the same principle is applied using **balancer chromosomes**, which contain inversions. When an organism is heterozygous for a balancer chromosome, desired sequences of alleles are preserved during experimental work.

在实验室研究中，研究人员使用含倒位的**平衡染色体**正是应用了这一原理。当生物体是平衡染色体杂合子时，在实验工作中理想的等位基因序列将得以保留。

6.8 Translocations Alter the Location of Chromosomal Segments in the Genome

6.8 易位改变了基因组中染色体片段的位置

Translocation, as the name implies, is the movement of a chromosomal segment to a new location in the genome. Reciprocal translocation, for example, involves the exchange of segments between two nonhomologous chromosomes. The least complex way for this event to occur is for two nonhomologous chromosome arms to come close to each other so that an exchange is facilitated. Figure 6.11(a) shows a simple reciprocal translocation in which only two breaks are required. If the exchange includes internal chromosome segments, four breaks are required, two on each chromosome.

易位，顾名思义，指染色体片段移动至基因组内其他新的位置。例如，相互易位涉及两条非同源染色体间的片段交换，这种情形发生的最简单方式是两条非同源染色体的臂彼此靠近，从而有利于交换发生。图 6.11（a）显示了一个简单的相互易位过程，只需两个断裂位点。如果交换涉及染色体内部片段，则需要四个断裂位点，每条染色体上分别有两个。

The genetic consequences of reciprocal translocations are, in several instances, similar to those of inversions. For example, genetic information is not lost or gained. Rather, there is only a rearrangement of genetic material. The presence of a translocation does not, therefore, directly alter the viability of individuals bearing

在一些情况下，相互易位的遗传效应与倒位类似。例如，遗传信息既没有丢失也没有增加，而只有遗传物质的重新线性排列。因此，易位的存在并不直接改变易位携带者的生存能力。

it.

Homologs that are heterozygous for a reciprocal translocation undergo unorthodox synapsis during meiosis. As shown in Figure 6.11(b), pairing results in a cross-like configuration. As with inversions, genetically unbalanced gametes are also produced as a result of this unusual alignment during meiosis. In the case of translocations, however, aberrant gametes are not necessarily the result of crossing over.

相互易位杂合子的同源染色体在减数分裂过程中进行非正常联会。如图 6.11 (b) 所示，配对产生“十”字形构象。与倒位一样，减数分裂过程中这种异常联会将导致遗传不均衡配子的产生。但是，在易位中，异常配子不一定是交换发生的结果。为了看清不均衡配子的产生方式，请关注图 6.11 (b) 和图 6.11 (c) 中的同源着丝粒。根据自由组合定律，在第一次减数分裂的后期，

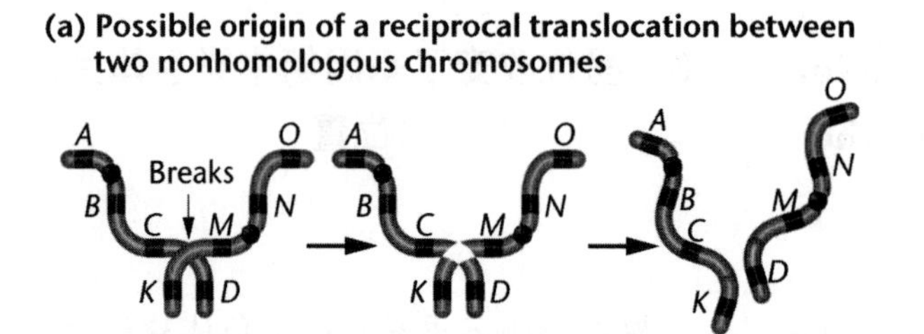

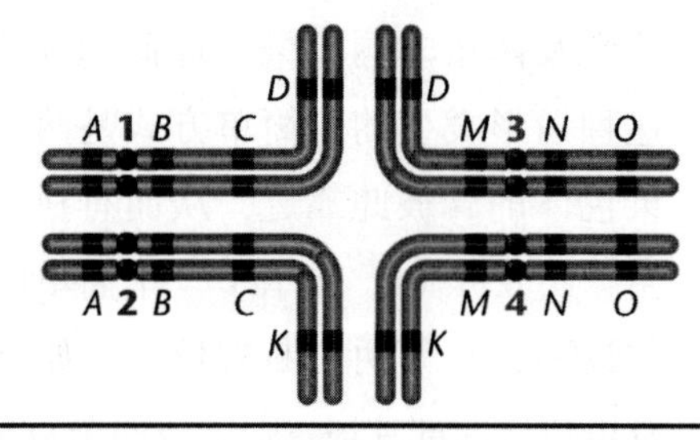

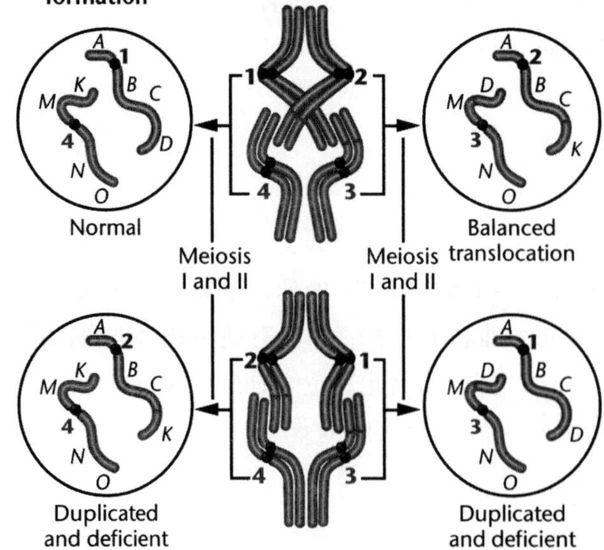

FIGURE 6.11 (a) Possible origin of a reciprocal translocation. (b) Synaptic configuration formed during meiosis in an individual that is heterozygous for the translocation. (c) Two possible segregation patterns, one of which leads to a normal and a balanced gamete (called alternate segregation) and one that leads to gametes containing duplications and deficiencies (called adjacent segregation).

图 6.11 (a) 相互易位可能的发生机制。(b) 易位杂合子个体在减数分裂过程中的联会构象。(c) 两种可能的同源染色体分离模式，一种得到正常配子和均衡配子（交互分离），另一种则得到含重复和缺失的配子（邻近分离）。

To see how unbalanced gametes are produced, focus on the homologous centromeres in Figure 6.11(b) and Figure 6.11(c). According to the principle of independent assortment, the chromosome containing centromere 1 migrates randomly toward one pole of the spindle during the first meiotic anaphase; it travels along with *either* the chromosome having centromere 3 *or* the chromosome having centromere 4. The chromosome with centromere 2 moves to the other pole along with the chromosome containing *either* centromere 3 *or* centromere 4. This results in four potential meiotic products. The 1,4 combination contains chromosomes that are not involved in the translocation. The 2,3 combination, however, contains translocated chromosomes. These contain a complete complement of genetic information and are balanced. The other two potential products, the 1,3 and 2,4 combinations, contain chromosomes displaying duplicated and deleted segments. To simplify matters, crossover exchanges are ignored here.

含着丝粒 1 的染色体随机向纺锤体的一极移动；它与含着丝粒 3 的染色体或含着丝粒 4 的染色体同时移动。含着丝粒 2 的染色体与含着丝粒 3 或含着丝粒 4 的染色体同时向另一极移动，这将产生四种减数分裂产物。1、4 组合中不含易位染色体，2、3 组合中含易位染色体，这些均含完整的遗传信息组成，并且是均衡的。另外两种产物，即 1、3 和 2、4 组合，所含染色体均存在重复和缺失片段。为简化讨论，此处没有涉及交叉互换。

When incorporated into gametes, the resultant meiotic products are genetically unbalanced. If they participate in fertilization, lethality often results. As few as 50 percent of the progeny of parents that are heterozygous for a reciprocal translocation survive. This condition, called **semisterility**, has an impact on the reproductive fitness of organisms, thus playing a role in evolution. Furthermore, in humans, such an unbalanced condition results in partial monosomy or trisomy, leading to a variety of physical or biochemical abnormalities at birth.

当进入配子时，所得减数分裂产物在遗传上是不均衡的。如果它们参与受精，通常会致死。相互易位杂合子的后代中仅有 50%可以存活，这种情况称为**半不育**，影响生物体的生殖适应性，因此在进化中发挥作用。此外，在人类中，这种不均衡状况将导致部分单体或三体的产生，个体出生时即表现出各种生理、生化异常。

Translocations in Humans: Familial Down Syndrome

Research conducted since 1959 has revealed numerous translocations in members of the human population. One common type of translocation involves breaks at the extreme ends of the short arms of two nonhomologous acrocentric chromosomes. These small segments are lost, and the larger segments fuse at their centromeric region. This type of translocation produces a new, large submetacentric or metacentric chromosome, often called a **Robertsonian translocation**.

One such translocation accounts for cases in which Down syndrome is familial (inherited). Earlier in this chapter, we pointed out that most instances of Down syndrome are due to trisomy 21. This chromosome composition results from nondisjunction during meiosis in one parent. Trisomy accounts for over 95 percent of all cases of Down syndrome. In such instances, the chance of the same parents producing a second affected child is extremely low. However, in the remaining families with a child with Down syndrome, it occurs with a much higher frequency over several generations; that is, it "runs in families.".

Cytogenetic studies of the parents and their offspring from these unusual cases explain the cause of **familial Down syndrome**. Analysis reveals that one of the parents contains a 14/21, D/G translocation (Figure 6.12). That is, one parent has the majority of the G-group chromosome 21 translocated to one end of the

人类群体中的易位：家族性唐氏综合征

自1959年以来的研究表明，人类群体中存在大量易位。一种常见易位类型是两条非同源的近端着丝粒染色体的短臂末端发生断裂。这些小片段发生丢失，而较大片段则在其着丝粒区域发生融合。这种易位将产生一条新的、长的近中着丝粒染色体或中着丝粒染色体，通常称为**罗伯逊易位**。

家族遗传性唐氏综合征的发病机理即是该种易位。在本章的前面，我们指出唐氏综合征的大多数情况是由21三体导致的。这种染色体组成由一位亲本在减数分裂过程中同源染色体不分离所致。三体占唐氏综合征所有病例的95%以上。在这种情况下，同一对父母生下第二个患病孩子的概率非常低。但是，在一些唐氏综合征患儿家庭中，该症状在世代中的发生概率较高，也就是说，它在家族中世代遗传。

对这些异常病例的父母及其后代进行的细胞遗传学研究揭示了**家族性唐氏综合征**的病因。分析表明，其中一个亲本含14/21，D/G易位（图6.12）。也就是说，一个亲本G组21号染色体的大部分易位到了D组14号染色体的一端。即使该个体只含45条染色体，但是其表型正常。在减数

D-group chromosome 14. This individual is phenotypically normal, even though he or she has only 45 chromosomes. During meiosis, one-fourth of the individual's gametes have two copies of chromosome 21: a normal chromosome and a second copy translocated to chromosome 14. When such a gamete is fertilized by a standard haploid gamete, the resulting zygote has 46 chromosomes but three copies of chromosome 21. These individuals exhibit Down syndrome. Other potential surviving offspring contain either the standard diploid genome (without a translocation) or the balanced translocation like the parent. Both cases result in normal individuals. Although not illustrated in Figure 6.12, two other gametes may be formed, though rarely. Such gametes are unbalanced, and upon fertilization, lethality occurs.

分裂过程中，该个体产生的配子中，1/4 含两个 21 号染色体拷贝：一个为正常的 21 号染色体，另一个拷贝则易位到 14 号染色体上。当这种配子与正常的单倍体配子受精时，合子含 46 条染色体，但 21 号染色体有 3 个拷贝。这些个体将表现出唐氏综合征。其他可能存活的后代或者是不含易位染色体的正常二倍体个体，或者是与亲本一样的平衡易位携带者，这两种情形均表现为正常个体。尽管图 6.12 中未显示，但还可能以极低概率产生另外两种配子，这些配子遗传上不均衡，即使参与受精也会发生致死效应。

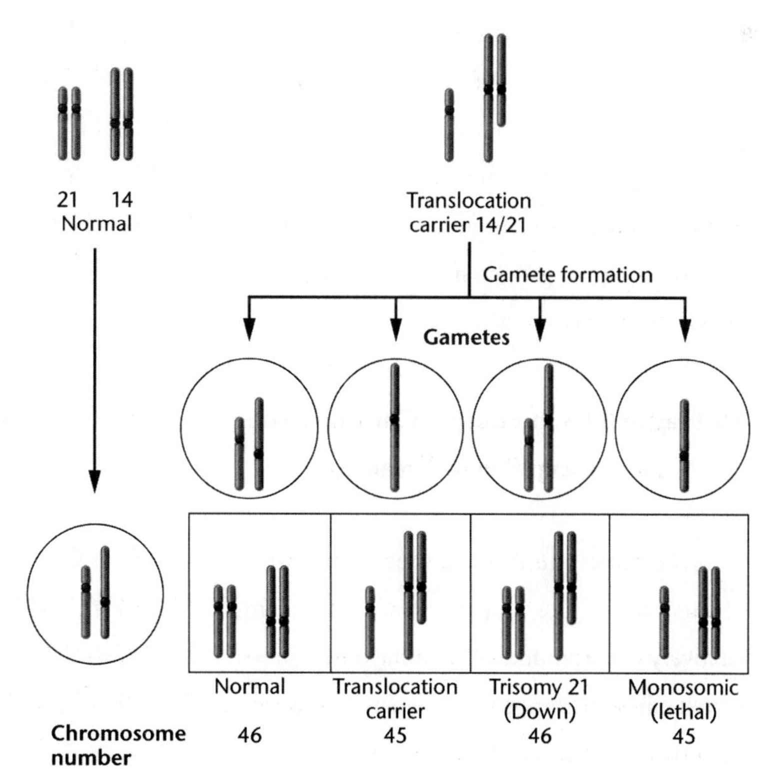

FIGURE 6.12 Chromosomal involvement and translocation in familial Down syndrome.

图 6.12 家族性唐氏综合征所涉及的染色体及易位。

The above findings have allowed geneticists to resolve the seeming paradox of an inherited trisomic phenotype in an individual with an apparent diploid number of chromosomes. It is also unique that the "carrier," who has 45 chromosomes and exhibits a normal phenotype, does not contain the *complete* diploid amount of genetic material. A small region is lost from both chromosomes 14 and 21 during the translocation event. This occurs because the ends of both chromosomes have broken off prior to their fusion. These specific regions are known to be two of many chromosomal locations housing multiple copies of the genes encoding rRNA, the major component of ribosomes. Despite the loss of up to 20 percent of these genes, the carrier is unaffected.

上述发现使遗传学家能够解释具有明显二倍染色体数目的个体却表现出可遗传的三体表型这一看似矛盾的情况。另外，同样独特的是含 45 条染色体且表型正常的"携带者"并不含完整的二倍体数量的遗传物质。在易位过程中，14 号染色体和 21 号染色体均有小区域遗传信息丢失。这是由于两条染色体的末端在融合之前已经发生断裂。已知这些特定区域是诸多富含多拷贝 rRNA 编码基因的染色体区域中的两个，该基因产物是核糖体的主要成分。尽管易位携带者有多达 20% 的核糖体 rRNA 基因发生丢失，但其表型不受影响。

ESSENTIAL POINT

Inversions and translocations may initially cause little or no loss of genetic information or deleterious effects. However, heterozygous combinations of the involved chromosome segments may result in genetically abnormal gametes following meiosis, with lethality or inviability often ensuing.

基本要点

倒位和易位可能几乎不会或根本不会造成遗传信息的损失或有害影响。但是，含染色体畸变片段的杂合个体可能在减数分裂后产生遗传异常的配子，这些配子通常会致死或不能存活。

6.9 Fragile Sites in Human Chromosomes Are Susceptible to Breakage

6.9 人类染色体的脆性位点

We conclude this chapter with a brief discussion of the results of an intriguing discovery made around 1970 during observations of metaphase chromosomes prepared following human cell culture. In cells derived from

在本章的最后，我们简要讨论 1970 年左右培养人类细胞并观察中期染色体时发现的有趣现象。源自某些个体的细胞，沿着一条染色体的某些特定区域，染色体无法被染色，从而表现出缺口。在染色体也

certain individuals, a specific area along one of the chromosomes failed to stain, giving the appearance of a gap. In other individuals whose chromosomes displayed such morphology, the gaps appeared at other positions within the set of chromosomes. Such areas eventually became known as **fragile sites**, since they appeared to be susceptible to chromosome breakage when cultured in the absence of certain chemicals such as folic acid, which is normally present in the culture medium.

呈现类似情形的其他个体中，缺口则出现在染色体组内的其他位置。这些区域最终被称为**脆性位点**，因为细胞培养缺乏某些化学物质，如叶酸时，染色体似乎易于发生断裂的情形，而培养基中通常含有叶酸。

Because they represent points along the chromosome that are susceptible to breakage, these sites may indicate regions where the chromatin is not tightly coiled. Note that even though almost all studies of fragile sites have been carried out *in vitro* using mitotically dividing cells, clear associations have been established between several of these sites and the corresponding altered phenotype, including intellectual disability and cancer.

因为这些脆性位点代表了沿着染色体存在的易于发生断裂的位点，所以这些位点可能表示染色质未能进行高度折叠的区域。值得注意的是，尽管几乎所有的脆性位点研究都是在体外利用有丝分裂细胞进行的，但是研究结果显示，其中一些位点与相应的表型变化，包括智力障碍、癌症等，均有十分明确的关联。

Fragile-X Syndrome

Most fragile sites do not appear to be associated with any clinical syndrome. However, individuals bearing a folate-sensitive site on the X chromosome may exhibit the **fragile-X syndrome** (**FXS**), the most common form of inherited intellectual disability. This syndrome affects about 1 in 4000 males and 1 in 8000 females. All males bearing this X chromosome exhibit the syndrome, while about 60 percent of females bearing one affected chromosome exhibit the syndrome. In addition to intellectual disability, affected males and females have

脆性 X 染色体综合征

大多数脆性位点似乎与临床综合征无关，但是，X 染色体上含叶酸敏感位点的个体可能表现出**脆性 X 染色体综合征**（FXS），这是最为常见的一种遗传性智力障碍。该综合征的发病率在男性中为 1/4000，女性中则为 1/8000。所有携带脆性 X 染色体的男性个体均表现出该综合征的相应症状，而大约 60％携带一条脆性 X 染色体的女性表现出该综合征的相应症状。除智力障碍外，患病男性和女性还具有特征性的长而狭窄的脸形、突出的下巴及大耳朵。

characteristic long, narrow faces with protruding chins and enlarged ears.

A gene that spans the fragile site is now known to be responsible for this syndrome. Named *FMR1*, it is one of a growing number of genes that have been discovered in which a sequence of three nucleotides is repeated many times, expanding the size of the gene. Such **trinucleotide repeats** are also recognized in other human disorders, including Huntington disease and myotonic dystrophy. In *FMR1*, the trinucleotide sequence CGG is repeated in an untranslated area adjacent to the coding sequence of the gene (called the "upstream" region). The number of repeats varies immensely within the human population, and a high number correlates directly with expression of fragile-X syndrome. Normal individuals have between 6 and 54 repeats, whereas those with 55 to 230 repeats are considered carriers of the disorder. More than 230 repeats lead to expression of the syndrome.

现在已知跨越该脆性位点的基因是导致这种综合征的原因。该基因为 *FMR1*，是越来越多被发现的含三个核苷酸重复序列的基因之一，重复序列会增大基因的大小。类似的**三核苷酸重复**在其他人类疾病中也有被发现，包括亨廷顿病和强直性肌营养不良。在 *FMR1* 中，三核苷酸序列 CGG 在邻近基因编码序列的非翻译区，即基因的"上游"区发生重复。重复次数在人群中变化很大，并且重复次数与脆性 X 染色体综合征的表达直接相关。正常个体含 6—54 次重复，疾病携带者含 55—230 次重复。超过 230 次重复将导致该综合征的表达。

It is thought that, once the gene contains this increased number of repeats, it becomes chemically modified so that the bases within and around the repeats are methylated, an epigenetic process that inactivates the gene. The normal product of the *FMR1* gene is an RNA-binding protein, FMRP, known to be produced in the brain. Evidence is now accumulating that directly links the absence of the protein in the brain with the cognitive defects associated with the syndrome.

人们认为，一旦基因包含这种数量增加的重复，基因就会被化学修饰，从而使重复序列内部和周围的碱基发生甲基化，通过表观遗传过程使基因失活。*FMR1* 基因的正常产物是已知合成于大脑中的 RNA 结合蛋白 FMRP。目前越来越多的证据表明，大脑中该蛋白质的缺失与综合征相关的认知缺陷直接相关。

From a genetic standpoint, an interesting aspect of fragile-X syndrome is the instability

从遗传学角度来看，脆性 X 染色体综合征的有趣之处在于 CGG 重复序列数目的

of the number of CGG repeats. An individual with 6 to 54 repeats transmits a gene containing the same number of copies to his or her offspring. However, carrier individuals with 55 to 230 repeats, though not at risk to develop the syndrome, may transmit to their offspring a gene with an increased number of repeats. This number increases in future generations, demonstrating the phenomenon known as **genetic anticipation**. Once the threshold of 230 repeats is exceeded, retardation becomes more severe in each successive generation as the number of trinucleotide repeats increases. Interestingly, expansion from the carrier status (55 to 230 repeats) to the syndrome status (over 230 repeats) occurs only during the transmission of the gene by the maternal parent, not by the paternal parent. Thus, a carrier male may transmit a stable chromosome to his daughter, who may subsequently transmit an unstable chromosome with an increased number of repeats to her offspring. Their grandfather was the source of the original chromosome.

不稳定性。含 6—54 次重复序列的个体将含同样数目的基因拷贝传递给后代，然而含 55—230 次重复序列的携带者尽管本身没有患病风险，但可能将增加了重复次数的基因拷贝传递给后代。重复的拷贝数在后代中增加，呈现出**遗传早现**现象。一旦超过 230 次重复次数的阈值，则随着三核苷酸重复次数的增加，随后的每个连续世代中智力发育迟缓症状就会变得更加严重。有趣的是，从携带者状态（55—230 次重复）到综合征患病状态（超过 230 次重复），重复拷贝数扩增的现象仅发生在母本基因传递过程中，与父本无关。因此，男性携带者传递给女儿的是重复拷贝数目稳定的染色体，而其女儿随后会将重复拷贝数增加的不稳定染色体传递给她的后代，即外祖父是原始染色体的来源。

The Link between Fragile Sites and Cancer

While the study of the fragile-X syndrome first brought unstable chromosome regions to the attention of geneticists, a link between an autosomal fragile site and lung cancer was reported in 1996 by Carlo Croce, Kay Huebner, and their colleagues. They have subsequently postulated that the defect is associated with the formation of a variety of different tumor types. Croce and Huebner first showed that the *FHIT* gene (standing for *f*ragile *hi*stidine *t*riad), located within the well-defined

脆性位点与癌症的关系

对脆性 X 染色体综合征的研究首次引起了遗传学家对不稳定染色体区的关注，与此同时，在 1996 年，卡洛 · 克罗斯、凯 · 许布纳及其同事报道了常染色体脆性位点与肺癌之间的联系。随后他们又推测，该缺陷与多种类型癌症的形成有关。克罗斯和许布纳首次指出位于 3 号染色体短臂上 *FRA3B* 脆性位点内的 *FHIT* 基因（意为“脆性三联组氨酸”），在来自肺癌患者的癌细胞中常发生改变或缺失。现在，更广泛

fragile site designated as *FRA3B* on the p arm of chromosome 3, is often altered or missing in cells taken from tumors of individuals with lung cancer. More extensive studies have now revealed that the normal protein product of this gene is absent in cells of many other cancers, including those of the esophagus, breast, cervix, liver, kidney, pancreas, colon, and stomach. Genes such as *FHIT* that are located within fragile regions undoubtedly have an increased susceptibility to mutations and deletions.

的研究揭示了该基因的正常蛋白质产物在许多其他癌细胞中同样不存在，包括食管癌、乳腺癌、子宫颈癌、肝癌、肾癌、胰腺癌、结肠癌和胃癌。毫无疑问，位于脆性位点区域内的基因如 *FHIT* 对突变和缺失的敏感性有所提升。

The study of this and still other fragile sites is but one example of how chromosomal abnormalities of many sorts are linked to cancer.

这一研究以及其他脆性位点的相关研究只是诸多染色体异常与癌症之间相关联的例子之一。

ESSENTIAL POINT

Fragile sites in human mitotic chromosomes have sparked research interest because one such site on the X chromosome is associated with the most common form of inherited mental retardation, while other autosomal sites have been linked to various forms of cancer.

基本要点

人类有丝分裂染色体上的脆性位点引起了人们的研究兴趣，因为 X 染色体上的一个类似位点与一种最为常见的遗传性智力低下相关，同时其他常染色体脆性位点则与多种形式的癌症相关。

7 Linkage and Chromosome Mapping in Eukaryotes

7 真核生物的连锁与染色体作图

CHAPTER CONCEPTS

■ Chromosomes in eukaryotes contain many genes whose locations are fixed along the length of the chromosomes.

■ Unless separated by crossing over, alleles present on a chromosome segregate as a unit during gamete formation.

■ Crossing over between homologs is a process of genetic recombination during meiosis that creates gametes with new combinations of alleles that enhance genetic variation within species.

■ Crossing over between homologs serves as the basis for the construction of chromosome maps.

■ While exchange occurs between sister chromatids during mitosis, no new recombinant chromatids are created.

本章速览

■ 真核生物染色体上包含很多基因，它们在染色体上的位置固定并且沿着染色体排列。

■ 染色体上的等位基因在配子形成过程中如果不随着交换彼此分开，那么它们将作为一个单元进行分离。

■ 在减数分裂中，同源染色体交换是亲本遗传信息重组的过程，这样产生的配子含有新的等位基因组合，大大增加了物种的遗传变异。

■ 同源染色体交换是构建染色体图的基础。

■ 有丝分裂过程中姐妹染色单体间也会发生交换，但并不产生重组型染色单体。

Walter Sutton, along with Theodor Boveri, was instrumental in uniting the fields of cytology and genetics. As early as 1903, Sutton pointed out the likelihood that there must be many more “unit factors” than chromosomes in most organisms. Soon thereafter, genetics investigations revealed that certain genes segregate as if they were somehow joined or linked together. Further investigations showed

沃尔特·萨顿和特奥多尔·博韦里在将细胞学和遗传学进行融合的过程中发挥了重要作用。萨顿早在1903年就指出，在大多数生物体中，“单位因子”的数量必定比染色体更多。此后不久，遗传学研究发现某些基因在分离时似乎以某种方式彼此联系或捆绑在一起。进一步的研究表明，这些基因位于同一染色体上，并且确实可能作为一个单元进行世代传递。现在我们

that such genes are part of the same chromosome and may indeed be transmitted as a single unit. We now know that most chromosomes contain a very large number of genes. Those that are part of the same chromosome are said to be *linked* and to demonstrate **linkage** in genetic crosses.

知道大多数染色体上都含有大量数目的基因。那些位于同一染色体上的基因被认为是连锁的，并在遗传杂交过程中表现出**连锁**。

Because the chromosome, not the gene, is the unit of transmission during meiosis, linked genes are not free to undergo independent assortment. Instead, the alleles at all loci of one chromosome should, in theory, be transmitted as a unit during gamete formation. However, in many instances this does not occur. During the first meiotic prophase, when homologs are paired or synapsed, a reciprocal exchange of chromosome segments can take place. This **crossing over** event results in the reshuffling, or **recombination**, of the alleles between homologs, and it always occurs during the tetrad stage.

因为染色体是减数分裂过程中的传递单元，而不是基因，所以连锁的基因彼此间不能进行自由组合，相反，在配子形成过程中，一条染色体所有基因座上的等位基因理论上应该作为一个单元进行传递。然而，在许多情况下并非如此。在第一次减数分裂前期，同源染色体进行配对或者联会时，染色体节段间会发生相互交换。这种**交换**会导致同源染色体上等位基因重新排序或**重组**，并且总是发生在四联体阶段。

The frequency of crossing over between any two loci on a single chromosome is proportional to the distance between them. Therefore, depending on which loci are being studied, the percentage of recombinant gametes varies. This correlation allows us to construct chromosome maps, which give the relative locations of genes on chromosomes.

单条染色体上任意两个基因座间发生交换的频率与它们之间的距离成正比。因此，研究的基因座不同，重组型配子的百分比也随之不同。这种相关性使我们能够构建染色体图，从而给出基因在染色体上的相对位置。

In this chapter, we will discuss linkage, crossing over, and chromosome mapping in more detail.

在本章中，我们将详细讨论连锁、交换和染色体作图。

7.1 Genes Linked on the Same Chromosome Segregate Together

7.1 同一染色体上的连锁基因相伴分离

A simplified overview of the major theme of

本章主题的简要概述请见图 7.1。该图

this chapter is given in Figure 7.1, which contrasts the meiotic consequences of (a) independent assortment, (b) linkage *without* crossing over, and (c) linkage *with* crossing over. In Figure 7.1(a), we see the results of independent assortment of two pairs of chromosomes, each containing one heterozygous gene pair. No linkage is exhibited. When a large number of meiotic events are observed, four genetically different gametes are formed in equal proportions, and each contains a different combination of alleles of the two genes.

对 (a) 自由组合，(b) 连锁不交换和 (c) 连锁且交换三种情形的减数分裂结果进行了对比。在图 7.1 (a) 中，我们可以看到两对染色体发生自由组合的结果，每个子代细胞含一对杂合基因对，不发生连锁。当考察大量减数分裂事件时，四种遗传上不同的配子等比例形成，每种配子所含两种基因的等位基因组合不同。

Now let's compare these results with what occurs if the same genes are linked on the same chromosome. If no crossing over occurs between the two genes [Figure 7.1(b)], only two genetically different gametes are formed. Each gamete receives the alleles present on one homolog or the other, which is transmitted intact as the result of segregation. This case demonstrates *complete linkage*, which produces only **parental** or **noncrossover gametes**. The two parental gametes are formed in equal proportions. Though complete linkage between two genes seldom occurs, it is useful to consider the theoretical consequences of this concept.

现在，将这些结果与位于同一染色体上且存在连锁的相同基因进行比较。如果两基因间不发生交换 [图 7.1 (b)]，则仅产生两种遗传上不同的配子类型。每种配子仅接受来自一条或另一条同源染色体上存在的等位基因，它们在分离时作为一个整体被捆绑传递。这种情况是完全连锁，仅产生**亲本型**或**非交换型配子**，并且两种亲本型配子等比例形成。尽管两基因间完全连锁的情形十分少见，但这对于理解连锁这一概念的理论结果十分有用。

Figure 7.1(c) shows the results of crossing over between two linked genes. As you can see, this crossover involves only two nonsister chromatids of the four chromatids present in the tetrad. This exchange generates two new allele combinations, called **recombinant** or **crossover gametes**. The two chromatids not involved in the exchange result in noncrossover gametes, like those in Figure 7.1(b). The frequency with

图 7.1 (c) 显示了两个连锁基因之间发生交换的结果。如你所见，这种交换仅涉及四联体中两条非姐妹染色单体。这种交换会产生两种新的等位基因组合，被称为**重组型**或**交换型配子**。不参与交换的两条染色单体产生非交换型配子，如图 7.1 (b) 所示。任何两个连锁基因间发生交换的频率通常与二者沿染色体线性排列相隔的距离成正比。理论上，两个随机选择的基因

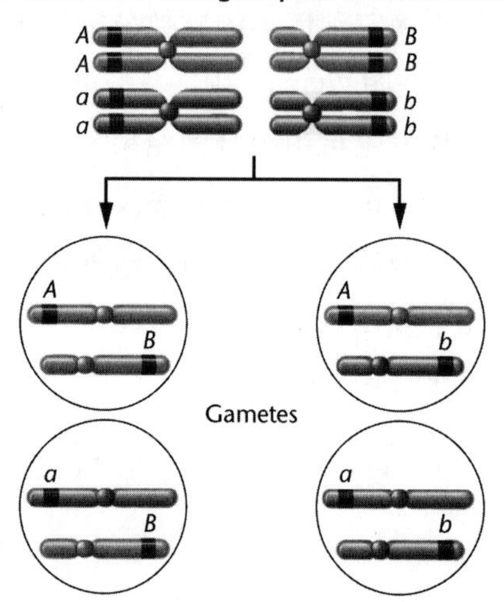

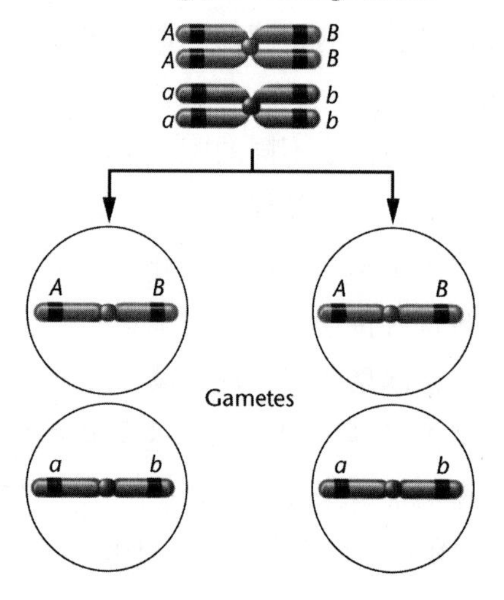

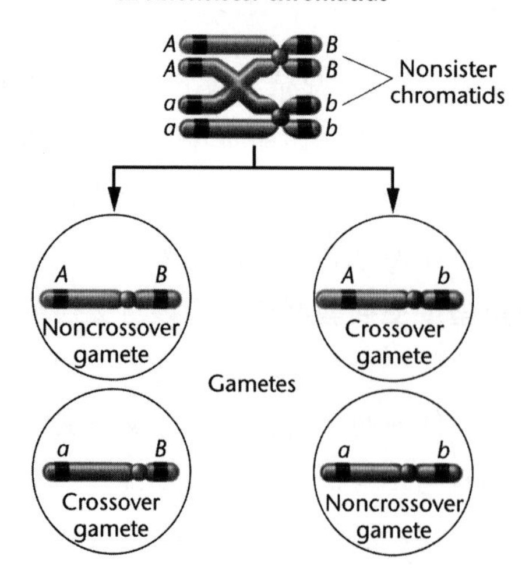

FIGURE 7.1 Results of gamete formation when two heterozygous genes are (a) on two different pairs of chromosomes; (b) on the same pair of homologs, but with no exchange occurring between them; and (c) on the same pair of homologs, but with an exchange occurring between two nonsister chromatids. Note in this and the following figures that members of homologous pairs of chromosomes are shown in two different colors. This convention was established in Chapter 2 (see, for example, Figure 2.9).

图 7.1 两对杂合基因配子形成的结果：（a）两对基因位于两对不同染色体上；（b）两对基因位于一对同源染色体上，但基因间不发生交换；（c）两对基因位于一对同源染色体上，而且两条非姐妹染色单体之间发生一次交换。注意在本图及随后的示意图中同源染色体的两个成员将用不同颜色进行表示。该惯例建立于第 2 章（如图 2.9）。

which crossing over occurs between any two linked genes is generally proportional to the distance separating the respective loci along the chromosome. In theory, two randomly selected genes can be so close to each other that crossover events are too infrequent to be detected easily. As shown in Figure 7.1(b), this complete linkage produces only parental gametes. On the other hand, if a small but distinct distance separates two genes, few recombinant and many parental gametes will be formed. As the distance between the two genes increases, the proportion of recombinant gametes increases and that of the parental gametes decreases.

可能非常接近，以至于很少发生交换，因此也很难被检测到。如图 7.1（b）所示，这种完全连锁仅产生亲本型配子。此外，如果两基因间距离很近但十分明显，则会产生少量重组型配子和大量亲本型配子。随着两基因间距离增加，重组型配子的比例会增加，而亲本型配子的比例则会减少。

As we will discuss later in this chapter, when the loci of two linked genes are far apart, the number of recombinant gametes approaches, but does not exceed, 50 percent. If 50 percent recombinants occur, the result is a 1:1:1:1 ratio of the four types (two parental and two recombinant gametes). In this case, transmission of two linked genes is indistinguishable from that of two unlinked, independently assorting genes. That is, the proportion of the four possible genotypes is identical, as shown in Figure 7.1(a) and (c).

正如我们在本章稍后将要讨论的那样，当两个连锁基因的基因座相距很远时，重组型配子的数量将无限接近但不超过 50%。如果产生 50% 的重组型配子，那么四种类型（两种亲本型和两种重组型配子）的比率是 1 ∶ 1 ∶ 1 ∶ 1。在这种情况下，两个连锁基因的传递与发生自由组合的不连锁基因的传递将无法区分，也就是说，如图 7.1（a）和（c）所示，四种可能的基因型将等比例产生。

The Linkage Ratio

If complete linkage exists between two genes because of their close proximity, and organisms heterozygous at both loci are mated, a unique F_2 phenotypic ratio results, which we designate the **linkage ratio**. To illustrate this ratio, let's consider a cross involving the closely linked, recessive, mutant genes *heavy* wing vein (*hv*)

连锁比

如果由于两基因间位置紧密而存在完全连锁，并且参与杂交的两个生物体在这两个基因座上的等位基因均为杂合，那么将产生独特的 F_2 代表型比例，我们称之为**连锁比**。为了说明这一比例，我们考虑一个果蝇杂交实验（图 7.2）。该实验涉及两个紧密连锁的隐性突变基因，分别是重翅

and *brown* eye (*bw*) in *Drosophila melanogaster* (Figure 7.2). The normal, wild-type alleles hv+ and bw^+ are both dominant and result in thin wing veins and red eyes, respectively.

脉基因（*hv*）和褐眼基因（*bw*），相对应的正常野生型等位基因分别是细翅脉基因（hv^+）和红眼基因（bw^+），均为显性。

In this cross, flies with normal thin wing veins and mutant brown eyes are mated to flies with mutant heavy wing veins and normal red eyes. In more concise terms, brown-eyed flies are crossed with heavy-veined flies. If we extend the system of genetic symbols established in Chapter 4, linked genes are represented by placing their allele designations above and below a single or double horizontal line. Those above the line are located at loci on one homolog, and those below are located at the homologous loci on the other homolog. Thus,we represent the P_1 generation as follows:

在这个杂交实验中，正常细翅脉、褐眼果蝇与重翅脉、正常红眼果蝇进行交配。简言之，即为褐眼果蝇与重翅脉果蝇杂交。如果我们使用第4章中建立的遗传符号系统，那么即可通过将等位基因符号置于单/双水平线的上方和下方来表示连锁基因。线上方的基因位于一条同源染色体的基因座上，而线下方的基因则位于另一条同源染色体相对应的同源基因座上。因此，P_1代可以表示如下：

$$P_1:\ \frac{hv^+bw}{hv^+bw} \times \frac{hv\,bw^+}{hv\,bw^+}$$

thin, brown　　heavy, red

These genes are located on an autosome, so no distinction between males and females is necessary.

这些基因均位于常染色体上，因此无须区分雄性和雌性。

In the F_1 generation, each fly receives one chromosome of each pair from each parent. All flies are heterozygous for both gene pairs and exhibit the dominant traits of thin wing veins and red eyes:

在F_1代中，每只果蝇均从每个亲本的每对染色体中获得一条。所有果蝇在这两个基因座上的等位基因均为杂合，表现出细翅脉和红眼的显性性状：

$$F_1:\ \frac{hv^+bw}{hvbw^+}$$

thin, red

As shown in Figure 7.2(a), when the F_1 generation is interbred, each F_1 individual forms only parental gametes because of complete linkage. After fertilization, the F_2 generation is produced in a 1:2:1 phenotypic and genotypic ratio. One-fourth of this generation shows thin

如图7.2(a)所示，当F_1代间进行杂交时，由于完全连锁，F_1代个体仅产生亲本型配子。在受精后，F_2代的表型和基因型比例均为1 ∶ 2 ∶ 1。该世代中1/4表型为细翅脉和褐眼；1/2为两种野生型性状，即细翅脉和红眼；1/4为重翅脉和红眼。简言之，

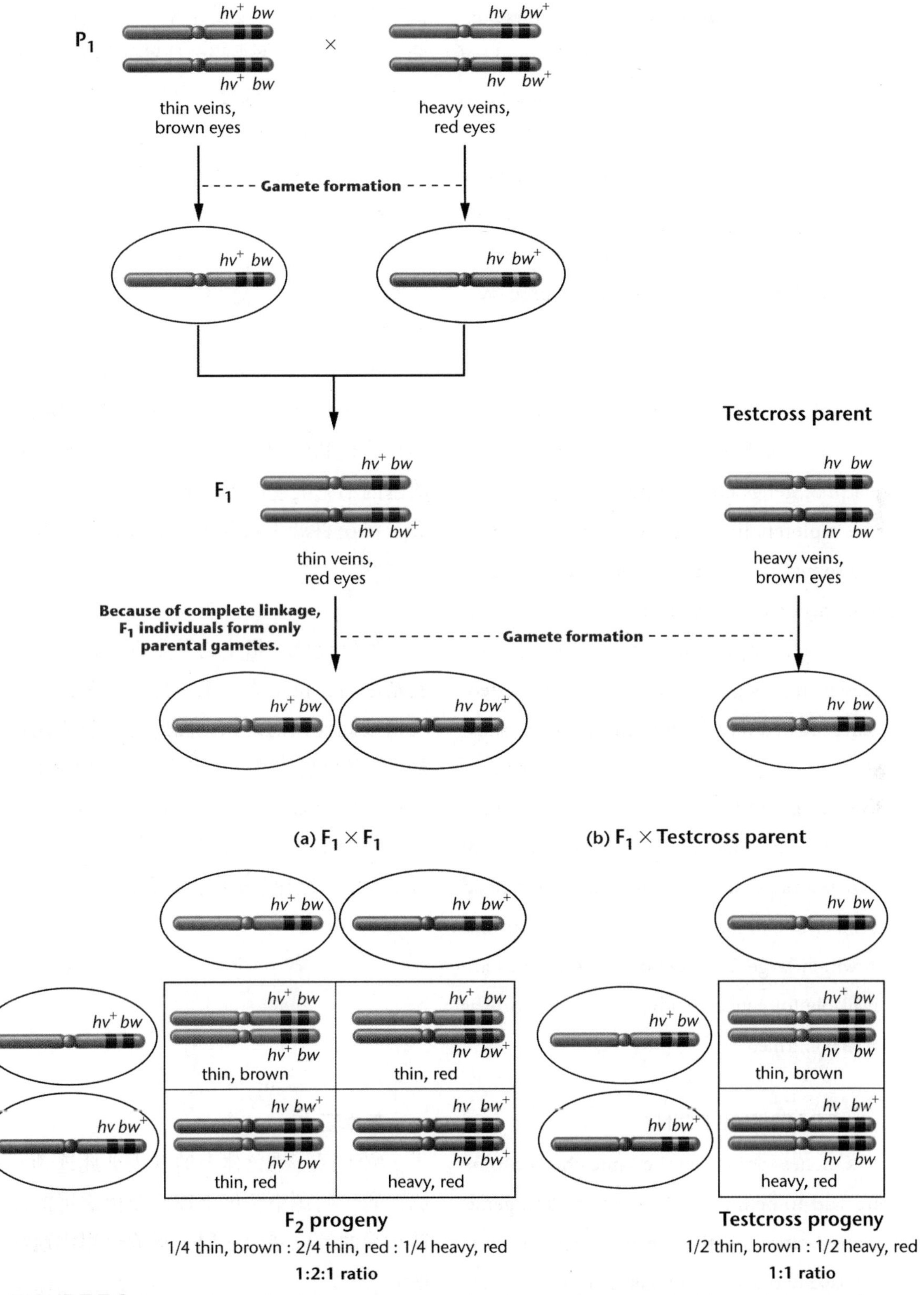

FIGURE 7.2 Results of a cross involving two genes located on the same chromosome and demonstrating complete linkage. (a) The F_2 results of the cross. (b) The results of a testcross involving the F_1 progeny.

图 7.2 涉及位于同一染色体上的两个完全连锁基因间的杂交结果。(a) 杂交的 F_2 代结果。(b) F_1 代的测交结果。

wing veins and brown eyes; one-half shows both wild-type traits, namely, thin wing veins and red eyes; and one-fourth shows heavy wing veins and red eyes. In more concise terms, the ratio is 1 heavy:2 wild:1 brown. Such a 1:2:1 ratio is characteristic of complete linkage. Complete linkage is usually observed only when genes are very close together and the number of progeny is relatively small.

比率为 1 重翅脉 ： 2 野生型 ： 1 褐眼。这样的 1 ： 2 ： 1 比例是完全连锁的特征。通常只有在基因距离很近且子代数量相对较少的情况下，才能观察到完全连锁。

Figure 7.2(b) demonstrates the results of a testcross with the F_1 flies. Such a cross produces a 1:1 ratio of thin, brown and heavy, red flies. Had the genes controlling these traits been incompletely linked or located on separate autosomes, the testcross would have produced four phenotypes rather than two.

图 7.2（b）展示了 F_1 代果蝇测交实验的结果，该杂交产生 1 ： 1 的细翅脉褐眼果蝇、重翅脉红眼果蝇。如果决定这些性状的基因不完全连锁或位于不同常染色体上，那么测交将产生四种表型后代，而不是仅有两种。

When large numbers of mutant genes present in any given species are investigated, genes located on the same chromosome show evidence of linkage to one another. As a result, **linkage groups** can be established, one for each chromosome. In theory, the number of linkage groups should correspond to the haploid number of chromosomes. In diploid organisms in which large numbers of mutant genes are available for genetic study, this correlation has been confirmed.

当研究任一指定物种中存在的大量突变基因时，位于同一染色体上的基因就呈现出彼此连锁的证据。因此可以建立**连锁群**，每条染色体就是一个连锁群。理论上，连锁群的数量与单倍染色体数量相等。在含大量突变基因、可用于遗传研究的二倍体生物中，这种相关性已得到证实。

ESSENTIAL POINT

Genes located on the same chromosome are said to be linked. Alleles of linked genes located close together on the same homolog are usually transmitted together during gamete formation.

基本要点

位于同一染色体上的基因彼此连锁。位于同一同源染色体上彼此距离靠近的连锁基因的等位基因在配子形成过程中通常捆绑遗传。

7.2 Crossing Over Serves as the Basis of Determining the Distance between Genes during Mapping

It is highly improbable that two randomly selected genes linked on the same chromosome will be so close to one another along the chromosome that they demonstrate complete linkage. Instead, crosses involving two such genes almost always produce a percentage of offspring resulting from recombinant gametes. This percentage is variable and depends on the distance between the two genes along the chromosome. This phenomenon was first explained around 1910 by two *Drosophila* geneticists, Thomas H. Morgan and his undergraduate student, Alfred H. Sturtevant.

7.2 交换是染色体作图中确定基因间距离的基础

同一染色体上随机选择的两个连锁基因沿着染色体彼此间距离十分靠近，以至于发生完全连锁，这种情形的可能性极小。事实上，涉及两个基因的杂交实验几乎总会产生一定比例的由重组型配子形成的后代。该百分比可变，并取决于沿染色体排列的两基因间的距离。这一现象最早是 1910 年左右由果蝇遗传学家托马斯 · H. 摩尔根和他的学生艾尔弗雷德 · H. 斯特蒂文特首先发现并阐明的。

Morgan, Sturtevant, and Crossing Over

In his studies, Morgan investigated numerous *Drosophila* mutations located on the X chromosome. When he analyzed crosses involving only one trait, he deduced the mode of X-linked inheritance. However, when he made crosses involving two X-linked genes, his results were initially puzzling. For example, female flies expressing the mutant *yellow* body (*y*) and *white* eyes (*w*) alleles were crossed with wild-type males (gray bodies and red eyes). The F_1 females were wild type, while the F_1 males expressed both mutant traits. In the F_2 generation, the vast majority of the offspring showed the expected parental phenotypes—either yellow-bodied, white-eyed flies or wild-type flies (gray-bodied, red-eyed). However, the remaining flies, less

摩尔根、斯特蒂文特和交换

摩尔根在他的工作中研究了大量位于 X 染色体上的果蝇突变。当分析仅涉及一种性状的杂交实验时，他发现了 X 连锁遗传模式。然而，当研究涉及两种 X 连锁基因的杂交实验时，最初得到的结果却令人费解。例如，突变型黄体（*y*）白眼（*w*）的雌蝇与野生型雄蝇（灰体红眼）杂交，F_1 代雌蝇均为野生型，而 F_1 代雄蝇则同时表达两种突变性状。在 F_2 代中，绝大多数后代个体表现出预期的亲本表型——黄体白眼或者野生型灰体红眼。但是，剩下不到 1.0% 的果蝇则或者是黄体红眼或者是灰体白眼。似乎在 F_1 代雌蝇的配子形成过程中，两种突变型等位基因彼此之间以某种方式发生了分离。图 7.3 的 cross A 展示了该杂交实验的情况，所用数据来自之后斯特蒂

than 1.0 percent, were either yellow-bodied with red eyes or gray-bodied with white eyes. It was as if the two mutant alleles had somehow separated from each other on the homolog during gamete formation in the F_1 female flies. This cross is illustrated in cross A of Figure 7.3, using data later compiled by Sturtevant.

文特收集所得。

When Morgan studied other X-linked genes, the same basic pattern was observed, but the proportion of the unexpected F_2 phenotypes differed. For example, in a cross involving the mutant *white* eye (w), *miniature* wing (m) alleles, the majority of the F_2 again showed the parental phenotypes, but a much higher proportion of the offspring appeared as if the mutant genes had separated during gamete formation. This is illustrated in cross B of Figure 7.3, again using data subsequently compiled by Sturtevant.

当摩尔根研究其他 X 连锁基因时，观察到了相同的基本模式，但是意想不到的是 F_2 代表型比例有所不同。例如，在涉及突变型白眼（w）、小翅（m）等位基因的杂交实验中，大多数 F_2 代个体再次呈现出亲本表型，但两种突变型等位基因在配子形成过程中发生分离产生的后代比例更高。图 7.3 中 cross B 展示了该杂交实验的情况，同样使用了斯特蒂文特随后收集的数据。

Morgan was faced with two questions: (1) What was the source of gene separation, and (2) why did the frequency of the apparent separation vary depending on the genes being studied? The answer he proposed for the first question was based on his knowledge of earlier cytological observations made by F.A. Janssens and others. Janssens had observed that synapsed homologous chromosomes in meiosis wrapped around each other, creating **chiasmata** (sing., *chiasma*) where points of overlap are evident. Morgan proposed that chiasmata could represent points of genetic exchange.

摩尔根面临着两个需要回答的问题：（1）基因分离的原因是什么？（2）为什么基因分离频率随所研究基因的不同而变化？对于第一个问题，他基于对 F. A. 让森斯等人所做的早期细胞生物学观察结果的了解给出了自己的观点。让森斯观察到，减数分裂过程中联会的同源染色体彼此缠绕，形成明显的彼此重叠的**交叉**。摩尔根指出，交叉即代表了遗传交换发生的位点。

In the crosses shown in Figure 7.3, Morgan postulated that if an exchange occurs during gamete formation between the mutant genes on the two X chromosomes of the F_1 females, the

在图 7.3 所示的杂交实验中，摩尔根推测，如果 F_1 代雌性个体在配子形成过程中位于两条 X 染色体上的突变型基因间发生交换，则将出现独特表型。他认为，与未

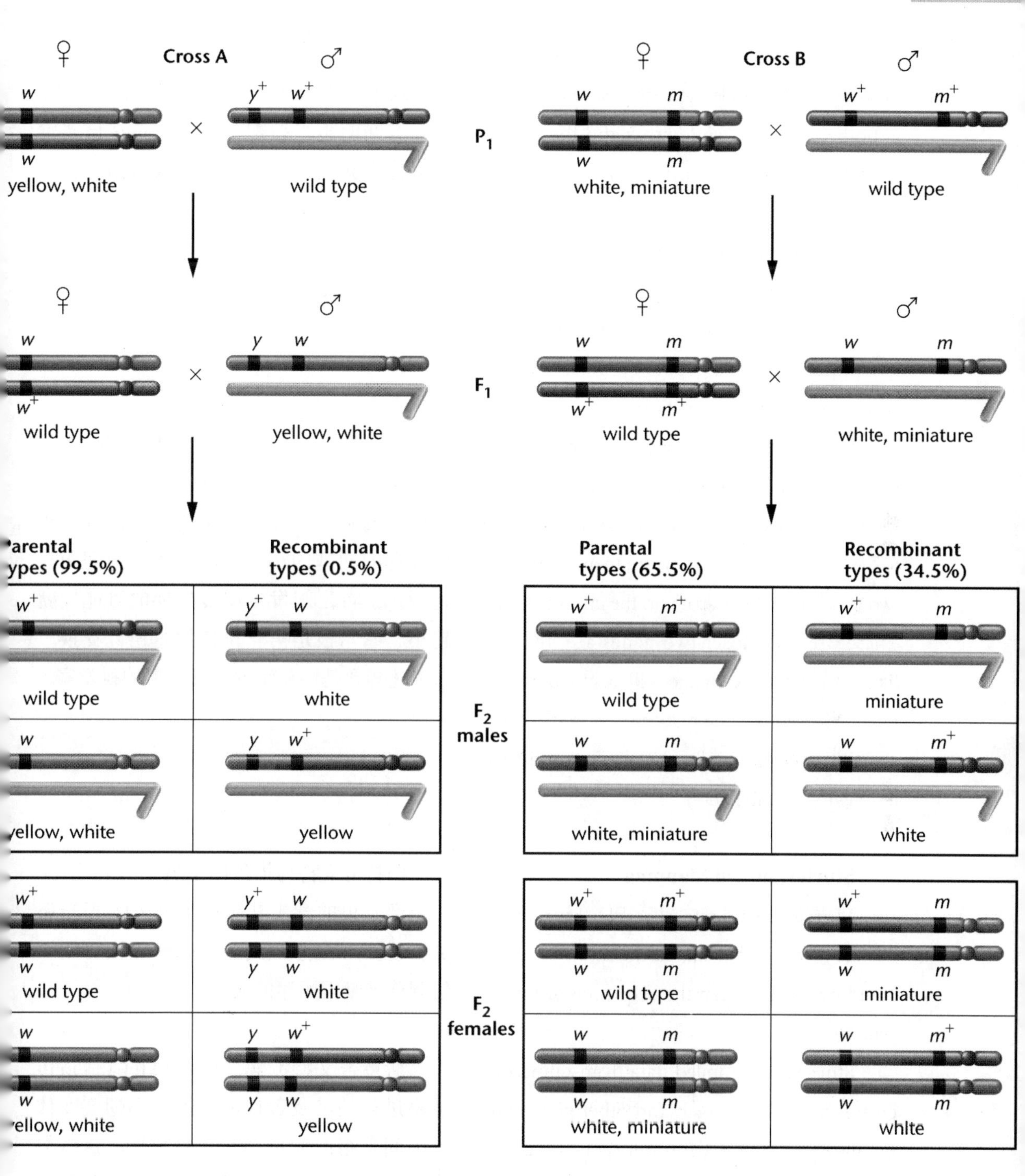

FIGURE 7.3 The F_1 and F_2 results of crosses involving the *yellow* (y), *white* (w) mutations (cross A), and the *white*, *miniature* (m) mutations (cross B), as compiled by Sturtevant. In cross A, 0.5 percent of the F_2 flies (males and females) demonstrate recombinant phenotypes, which express either *white* or *yellow*. In cross B, 34.5 percent of the F_2 flies (males and females) demonstrate recombinant phenotypes, which are either *miniature* or *white* mutants.

图 7.3 涉及黄体（y）白眼（w）突变的杂交 A 和白眼（w）小翅（m）突变的杂交 B 的 F_1 代和 F_2 代实验结果（根据斯特蒂文特收集的数据）。在杂交 A 中，F_2 代果蝇（雄蝇和雌蝇）中 0.5% 表现为重组表型，即或者是白眼或者是黄体。在杂交 B 中，F_2 代果蝇（雄蝇和雌蝇）中 34.5% 表现为重组表型，即或者是小翅或者是白眼。

unique phenotypes will occur. He suggested that such exchanges led to *recombinant gametes* in both the *yellow-white* cross and the *white-miniature* cross, in contrast to the parental gametes that have undergone no exchange. On the basis of this and other experiments, Morgan concluded that linked genes exist in a linear order along the chromosome and that a variable amount of exchange occurs between any two genes during gamete formation.

发生交换的亲本型配子不同，这种交换同时导致了黄体 - 白眼基因座间和白眼 - 小翅基因座间的重组型配子的产生。在该实验及其他实验的基础上，摩尔根得出结论，即连锁基因沿染色体呈线性排列，并且在配子形成过程中任意两基因之间将发生数量不等的交换。

In answer to the second question, Morgan proposed that two genes located relatively close to each other along a chromosome are less likely to have a chiasma form between them than if the two genes are farther apart on the chromosome. Therefore, the closer two genes are, the less likely a genetic exchange will occur between them. Morgan was the first to propose the term **crossing over** to describe the physical exchange leading to recombination.

在回答第二个问题时，摩尔根推测，沿着染色体线性排列的基因中，彼此距离较近的两基因间形成交叉的可能性要比相距较远的两基因小。因此，两基因间距离越近，二者之间发生遗传交换的可能性就越小。摩尔根是第一位使用术语“**交换**”来描述导致基因重组的物理交换的科学家。

Sturtevant and Mapping

斯特蒂文特与染色体作图

Morgan's student, Alfred H. Sturtevant, was the first to realize that his mentor's proposal could be used to map the sequence of linked genes.

摩尔根的学生艾尔弗雷德 · H. 斯特蒂文特是第一位意识到其导师的提议可用于绘制连锁基因顺序的人。

Sturtevant compiled data from numerous crosses made by Morgan and other geneticists involving recombination between the genes represented by the *yellow*, *white*, and *miniature* mutants. These data are shown in Figure 7.3. The following recombination between each pair of these three genes, published in Sturtevant's paper in 1913, is as follows:

斯特蒂文特汇集了摩尔根和其他遗传学家进行的涉及黄体、白眼和小翅突变体基因间重组的大量杂交实验数据。这些数据如图 7.3 所示。这三个基因两两之间的重组情况发表在斯特蒂文特 1913 年的论文中，如下所示：

(1) *yellow* – *white* 0.5%

（1）黄体 - 白眼：0.5%

(2) *white* – *miniature* 34.5%

(3) *yellow* – *miniature* 35.4%

Because the sum of (1) and (2) approximately equals (3), Sturtevant suggested that the recombination frequencies between linked genes are additive. On this basis, he predicted that the order of the genes on the X chromosome is *yellow-white-miniature*. In arriving at this conclusion, he reasoned as follows: the *yellow* and *white* genes are apparently close to each other because the recombination frequency is low. However, both of these genes are much farther apart from *miniature* genes because the *white-miniature* and *yellow-miniature* combinations show larger recombination frequencies. Because *miniature* shows more recombination with *yellow* than with *white* (35.4 percent versus 34.5 percent), it follows that *white* is located between the other two genes, not outside of them.

Sturtevant knew from Morgan's work that the frequency of exchange could be used as an estimate of the distance between two genes or loci along the chromosome. He constructed a **chromosome map** of the three genes on the X chromosome, setting 1 map unit (mu) equal to 1 percent recombination between two genes. In the preceding example, the distance between *yellow* and *white* is thus 0.5 mu, and the distance between *yellow* and *miniature* is 35.4 mu. It follows that the distance between *white* and *miniature* should be 35.4 mu − 0.5 mu = 34.9 mu. This estimate is close to the actual frequency of recombination between *white* and *miniature* (34.5 percent). The map for these three genes is

(2) 白眼 - 小翅：34.5%

(3) 黄体 - 小翅：35.4%

由于 (1) 和 (2) 的加和大约等于 (3)，所以斯特蒂文特推测连锁基因间的重组频率是可加和的。在此基础上，他推测 X 染色体上的基因顺序是黄体 - 白眼 - 小翅。在得出这一结论的过程中，他给出的理由如下：黄体和白眼基因间显然彼此靠近，因为重组频率低。但是，这两个基因均与小翅基因相距较远，因为白眼 - 小翅和黄体 - 小翅杂交组合均显示出更大的重组频率。因为小翅与黄体间的基因重组频率比与白眼间的重组频率大（35.4%大于 34.5%），所以白眼基因位于另外两基因之间，而不是位于它们的外侧。

斯特蒂文特从摩尔根的工作中发现交换频率可用于估计沿着染色体顺序排列的两基因间或基因座间的距离。他绘制了 X 染色体上三个基因的**染色体图**，1 个图距（mu）等于两基因间 1% 的重组频率。因此在前面的示例中，黄体和白眼基因间的距离为 0.5 mu，黄体和小翅基因间的距离是 35.4 mu。并且，白眼和小翅基因间的距离是 35.4 mu − 0.5 mu = 34.9 mu。该估值接近白眼与小翅基因间的实际重组频率（34.5%）。这三个基因的染色体图如图 7.4 所示。这些数据之间不能精确进行加和的事实源于作图相关实验的不精确性，尤其随着基因间距离的增加，不精确性越突出。

shown in Figure 7.4. The fact that these do not add up perfectly is due to the imprecision of mapping experiments, particularly as the distance between genes increases.

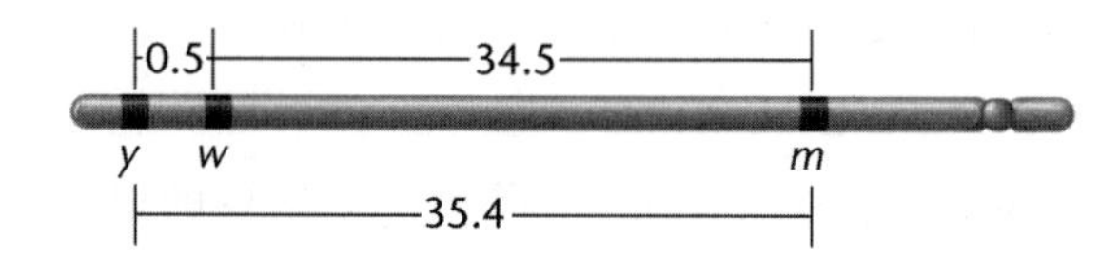

FIGURE 7.4 A map of the *yellow* (*y*), *white* (*w*), and *miniature* (*m*) genes on the X chromosome of *Drosophila melanogaster*. Each number represents the percentage of recombinant offspring produced in one of three crosses, each involving two different genes.

图 7.4 黑腹果蝇 X 染色体上三个基因黄体（*y*）、白眼（*w*）、小翅（*m*）的染色体图。数字代表涉及其中两个不同基因的杂交实验所得重组型后代的百分比。

In addition to these three genes, Sturtevant considered two other genes on the X chromosome and produced a more extensive map that included all five genes. He and a colleague, Calvin Bridges, soon began a search for autosomal linkage in *Drosophila*. By 1923, they had clearly shown that linkage and crossing over are not restricted to X-linked genes but can also be demonstrated with autosomes. During this work, they made another interesting observation. Crossing over in *Drosophila* was shown to occur only in females. The fact that no crossing over occurs in males made genetic mapping much less complex to analyze in *Drosophila*. However, crossing over does occur in both sexes in most other organisms.

除了这三个基因外，斯特蒂文特还研究了 X 染色体上的另外两个基因，并绘制了一个扩大的、涵盖所有五个基因的染色体图。他和他的同事卡尔文 · 布里奇斯很快开始在果蝇中寻找常染色体连锁现象。至 1923 年，他们已经清楚地发现连锁和交叉并不仅限于 X 连锁基因，同样也适用于常染色体上的基因。在这项工作中，他们还发现了另一个有趣的现象，即果蝇的基因交换仅在雌性个体中发生。雄性个体中不发生基因交换的事实使得果蝇的遗传图谱分析得以简化。然而，在大多数其他生物中，两种性别确实均发生基因交换。

Although many refinements in chromosome mapping have been developed since Sturtevant's initial work, his basic principles are considered to be correct. These principles are used to produce detailed chromosome maps of organisms for which large numbers of linked mutant genes are known. Sturtevant's findings

尽管自斯特蒂文特的最初工作以来，染色体作图已经得到了许多改进，但他提出的基本原理仍被认为是正确的。这些原理被用于绘制已知有大量连锁突变基因的生物体的详细染色体图。斯特蒂文特的发现对于遗传学中更为广阔的领域也具有重要历史意义。1910 年，**遗传的染色体学说**

are also historically significant to the broader field of genetics. In 1910, the **chromosomal theory of inheritance** was still widely disputed—even Morgan was skeptical of this theory before he conducted his experiments. Research has now firmly established that chromosomes contain genes in a linear order and that these genes are the equivalent of Mendel's unit factors.

仍然受到广泛争议，甚至摩尔根在进行实验之前对该理论也持怀疑态度。现在的研究已确凿地表明染色体含线性顺序排列的基因，并且这些基因等同于孟德尔所说的单位因子。

Single Crossovers

Why should the relative distance between two loci influence the amount of recombination and crossing over observed between them? During meiosis, a limited number of crossover events occur in each tetrad. These recombinant events occur randomly along the length of the tetrad. Therefore, the closer two loci reside along the axis of the chromosome, the less likely any single-crossover event will occur between them. The same reasoning suggests that the farther apart two linked loci are, the more likely a random crossover event will occur between them.

单交换

为什么两基因座间的相对距离会影响它们之间可观察的重组和交换数量呢？在减数分裂期间，每个四联体中发生有限数量的交换。这些重组事件按四联体的长度随机发生。因此，两个基因座沿着染色体长轴靠得越近，它们之间发生任何单交换的可能性就越小。相同的推理表明，两个连锁的基因座相距越远，它们之间发生随机交换的可能性就越大。

In Figure 7.5(a), a **single crossover** occurs between two nonsister chromatids but not between the two loci; therefore, the crossover is not detected because no recombinant gametes are produced. In Figure 7.5(b), where two loci are quite far apart, a crossover does occur between them, yielding gametes in which the traits of interest are recombined.

在图 7.5（a）中，两条非姐妹染色单体之间发生了一次**单交换**，但并未发生在两个基因座之间。由于未产生重组型配子，因此该交换无法被检测到。在图 7.5（b）中，两个基因座相距很远，二者之间确实发生了交换，产生了所研究性状发生了重组的配子。

When a single crossover occurs between two nonsister chromatids, the other two chromatids of the tetrad are not involved in this exchange and enter the gamete unchanged. Even if a single crossover occurs 100 percent of the

当两条非姐妹染色单体之间发生单交换时，四联体的另外两条染色单体不参与该交换并保持不变地进入配子。即使两个连锁基因之间 100%发生了单交换，随后也仅观察到 50%的潜在配子形成了重组，这

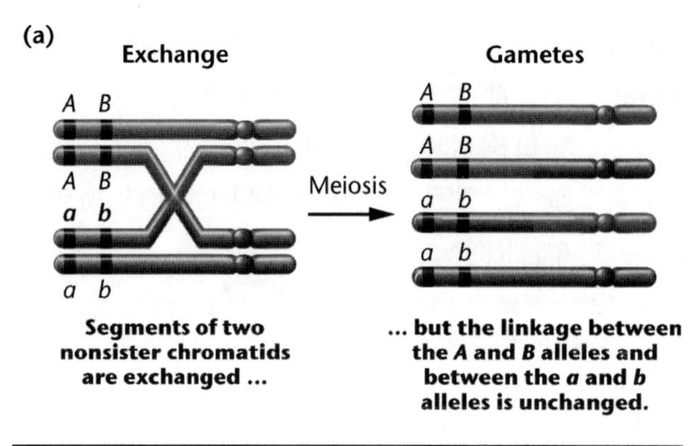

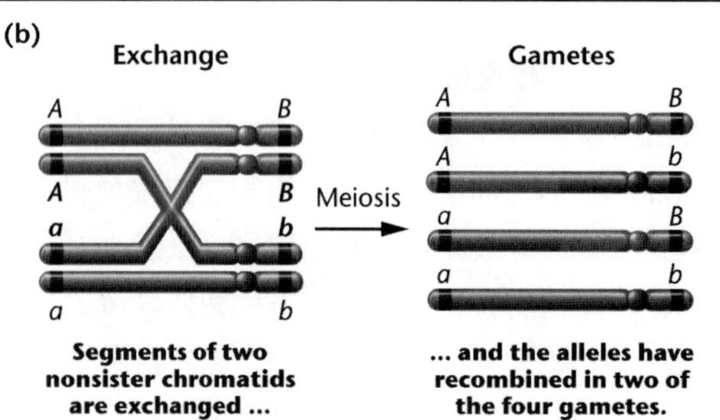

FIGURE 7.5 Two examples of a single crossover between two nonsister chromatids and the gametes subsequently produced. In (a) the exchange does not alter the linkage arrangement between the alleles of the two genes, only parental gametes are formed, and the exchange goes undetected. In (b) the exchange separates the alleles, resulting in recombinant gametes, which are detectable.

图 7.5 两条非姐妹染色单体间发生单交换及其随后所产生配子类型的两个示例。在 (a) 中，交换没有改变两个基因等位基因间的连锁排列方式，即只有亲本型配子产生，该交换无法被检测到。在 (b) 中，交换将等位基因分离，产生重组型配子，从而可以被检测到。

time between two linked genes, recombination is subsequently observed in only 50 percent of the potential gametes formed. This concept is diagrammed in Figure 7.6. Theoretically, if we consider only single exchanges and observe 20 percent recombinant gametes, crossing over

一概念如图 7.6 所示。理论上，如果仅考虑单交换并观察到 20%的重组型配子，则实际上 40%的四联体发生了交换。在这些条件下，总的规律是参与两基因间交换的四联体百分比是产生重组型配子百分比的两倍。因此，由于交换而观察到的重组比例

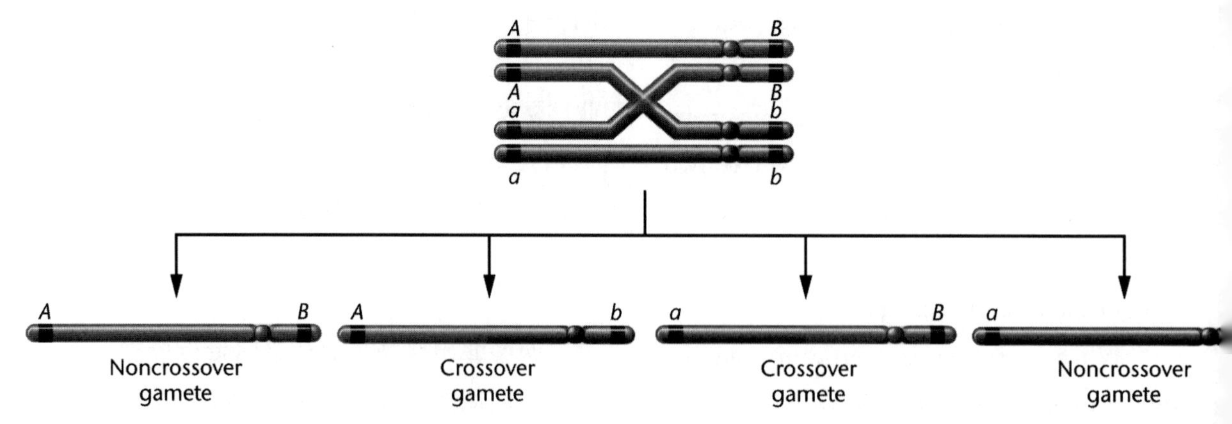

FIGURE 7.6 The consequences of a single exchange between two nonsister chromatids occurring in the tetrad stage. Two noncrossover (parental) and two crossover (recombinant) gametes are produced.

图 7.6 在四联体阶段两条非姐妹染色单体间发生单交换的结果。产生了两种非交换型（亲本型）配子和两种交换型（重组型）配子。

actually occurred in 40 percent of the tetrads. Under these conditions, the general rule is that the percentage of tetrads involved in an exchange between two genes is twice the percentage of recombinant gametes produced. Therefore, the theoretical limit of observed recombination due to crossing over is 50 percent.

理论极限值为 50%。

When two linked genes are more than 50 mu apart, a crossover can theoretically be expected to occur between them in 100 percent of the tetrads. If this prediction were achieved, each tetrad would yield equal proportions of the four gametes shown in Figure 7.6, just as if the genes were on different chromosomes and assorting independently. However, this theoretical limit is seldom achieved.

当两连锁基因相距超过 50 mu 时，理论上可以预计在 100%四联体中两基因间均发生交换。如果这一预测实现了，则每个四联体将等比例地产生图 7.6 所示的四种配子类型，就如同这些基因位于不同染色体上并且进行了自由组合。但是，这一理论极限比例很少实现。

ESSENTIAL POINT

Crossover frequency between linked genes during gamete formation is proportional to the distance between genes, providing the experimental basis for mapping the location of genes relative to one another along the chromosome.

基本要点

在配子形成过程中，连锁基因间的交换频率与基因间距离成正比，从而为确定沿染色体排列的各基因间相对位置提供了实验依据。

7.3 Determining the Gene Sequence during Mapping Requires the Analysis of Multiple Crossovers

7.3 染色体作图中确定基因顺序需要分析多交换

The study of single crossovers between two linked genes provides the basis of determining the distance between them. However, when many linked genes are studied, their sequence along the chromosome is more difficult to determine. Fortunately, the discovery

两个连锁基因间的单交换研究为确定二者之间的距离奠定了基础。但是，当研究许多连锁基因时，它们沿染色体排列的顺序则较难确定。幸运的是，四联体中染色单体间多交换的发现为绘制更为复杂的染色体图提供了便利。正如我们接下来将

that multiple exchanges occur between the chromatids of a tetrad has facilitated the process of producing more extensive chromosome maps. As we shall see next, when three or more linked genes are investigated simultaneously, it is possible to determine first the sequence of the genes and then the distances between them.

要看到的那样，当同时研究三个或三个以上连锁基因时，可以先确定基因的顺序，然后再确定它们之间的距离。

Multiple Crossovers

It is possible that in a single tetrad, two, three, or more exchanges will occur between nonsister chromatids as a result of several crossover events. Double exchanges of genetic material result from **double crossovers** (**DCOs**), as shown in Figure 7.7. For a double exchange to be studied, three gene pairs must be investigated, each heterozygous for two alleles. Before we determine the frequency of recombination among all three loci, let's review some simple probability calculations.

多交换

在单个四联体中，由于存在多次交换事件，非姐妹染色单体之间可能发生两次交换、三次交换或多次交换。遗传物质的两次交换由**双交换**（DCOs）产生，如图 7.7 所示。在研究双交换时，必须考察三组基因对，并且每组等位基因必须是杂合的。在确定所有三个基因座间的重组频率前，让我们回顾一些简单的概率计算。

As we have seen, the probability of a single exchange occurring between the *A* and *B* or the *B* and *C* genes relates directly to the distance between the respective loci. The closer *A* is to *B* and *B* is to *C*, the less likely a single exchange will occur between either of the two sets of loci. In the case of a double crossover, two separate and independent events or exchanges must occur simultaneously. The mathematical probability of two independent events occurring simultaneously is equal to the product of the individual probabilities (the **product law**).

如前所述，在基因 *A* 和 *B*、基因 *B* 和 *C* 之间发生单交换的概率直接与相应基因座间的距离有关。*A* 与 *B*、*B* 与 *C* 间的距离越近，这两组基因座中相应一组发生单交换的概率就越小。在双交换发生时，两个相互独立的单交换必定同时发生。两个独立事件同时发生的概率在数学上等于这两个事件单独发生概率的乘积（**相乘定律**）。

Suppose that crossover gametes resulting from single exchanges are recovered 20 percent of the time ($p = 0.20$) between *A* and *B*, and 30

假设 *A*、*B* 间由单交换产生的交换型配子比例为 20%（$p = 0.20$），*B*、*C* 间由单交换产生的交换型配子比例为 30%（$p = 0.30$）。

percent of the time ($p = 0.30$) between B and C. The probability of recovering a double-crossover gamete arising from two exchanges (between A and B, and between B and C) is predicted to be $(0.20)\ (0.30) = 0.06$, or 6 percent. It is apparent from this calculation that the frequency of double-crossover gametes is always expected to be much lower than that of either single-crossover class of gametes.

由以上两种单交换产生的 A 与 B、B 与 C 间的双交换型配子比例预计为 $(0.20)(0.30)=0.06$，即 6%。通过该计算可以明显看出双交换型配子比例总是比任何一种单交换型配子比例要低很多。

If three genes are relatively close together along one chromosome, the expected frequency of double-crossover gametes is extremely low. For example, suppose the A—B distance in Figure 7.7 is 3 mu and the B—C distance is 2 mu. The expected double-crossover frequency is $(0.03)(0.02) = 0.0006$, or 0.06 percent. This translates to only 6 events in 10,000. Thus, in a mapping experiment where closely linked genes are involved, very large numbers of offspring are required to detect double-crossover events. In this example, it is unlikely that a double crossover will be observed even if 1000 offspring are examined. Thus, it is evident that if four or five genes are being mapped, even fewer triple and quadruple crossovers can be expected to occur.

如果三个基因沿一条染色体排列的相对距离较近时，那么双交换型配子的预期频率将非常低。例如，图 7.7 中假设 A—B 距离为 3 mu，而 B—C 距离为 2 mu，则预期双交换频率为 $(0.03)(0.02)=0.0006$，即 0.06%。换句话说，每 10 000 个事件中只出现 6 个。因此，在涉及紧密连锁基因的作图实验中，需要大量后代才能检测到双交换。在这个例子中，即使考察 1000 个后代也不太可能观察到双交换型后代。因此，很明显，如果定位四个或五个基因，那么三交换型后代、四交换型后代的检出可能性更低。

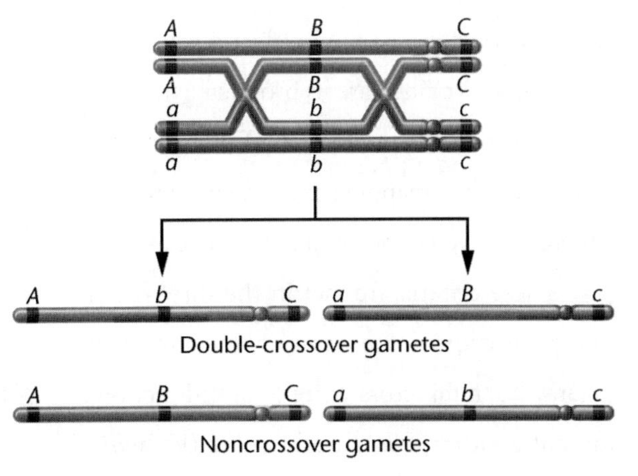

FIGURE 7.7 Consequences of a double exchange occurring between two nonsister chromatids. Because the exchanges involve only two chromatids, two noncrossover gametes and two double-crossover gametes are produced.

图 7.7 两条非姐妹染色单体间发生双交换的结果。由于交换仅涉及两条单体，因此产生了两种非交换型配子和两种双交换型配子。

ESSENTIAL POINT

Determining the sequence of genes in a three-point mapping experiment requires analysis of the double-crossover gametes, as reflected in the phenotype of the offspring receiving those gametes.

基本要点

三点作图实验中确定基因顺序需要分析双交换型配子，这些配子的后代表型可以反映相应的配子类型。

Three-Point Mapping in *Drosophila*

The information in the preceding section enables us to map three or more linked genes in a single cross. To illustrate the mapping process in its entirety, we examine two situations involving three linked genes in two quite different organisms.

果蝇的三点作图

上一节中所介绍的内容让我们能够使用单因子杂交定位三个或三个以上的连锁基因。为了完整地展现基因作图过程，我们考察在两种截然不同的生物中定位三个连锁基因的情形。

To execute a successful mapping cross, three criteria must be met:

1. The genotype of the organism producing the crossover gametes must be heterozygous at all loci under consideration.

2. The cross must be constructed so that genotypes of all gametes can be determined accurately by observing the phenotypes of the resulting offspring. This is necessary because the gametes and their genotypes can never be observed directly. To overcome this problem, each phenotypic class must reflect the genotype of the gametes of the parents producing it.

3. A sufficient number of offspring must be produced in the mapping experiment to recover a representative sample of all crossover classes.

为了实施成功的基因作图杂交实验，必须满足三个条件：

1. 产生交换型配子的生物体基因型在所有待研究的基因座上必须是杂合的。

2. 构建杂交实验体系时，所有配子的基因型可以通过观察所得后代的表型精确确定。这十分必要，因为我们永远无法直接观察到配子及其基因型。为了解决这一问题，每种表型类别必须能反映其父母所产生的相应配子的基因型。

3. 在基因作图实验中必须产生足够数量的后代，以获得具有代表性的所有交换型类别样本。

These criteria are met in the three-point mapping cross from *Drosophila* shown in Figure 7.8. In this cross, three X-linked recessive mutant genes—*yellow* body color (*y*), *white*

图7.8所示的果蝇三点作图杂交实验满足这些条件。在该杂交实验中，考察三个X连锁隐性突变基因——黄体（*y*）、白眼（*w*）和棘眼（*ec*）。为了图示该杂交实验，我们

eye color (*w*), and *echinus* eye shape (*ec*)—are considered. To diagram the cross, we must assume some theoretical sequence, even though we do not yet know if it is correct. In Figure 7.8, we initially assume the sequence of the three genes to be *y*—*w*—*ec*. If this assumption is incorrect, our analysis will demonstrate this and reveal the correct sequence.

必须假设一些可能的基因顺序，即使我们尚不清楚该假设是否正确。图 7.8 中，我们最初假设这三个基因的顺序为 *y*—*w*—*ec*。如果该假设不正确，我们的分析将有所显示并最终发现正确顺序。

In the P_1 generation, males hemizygous for all three wild-type alleles are crossed to females that are homozygous for all three recessive mutant alleles. Therefore, the P_1 males are wild type with respect to body color, eye color, and eye shape. They are said to have a *wild-type phenotype*. The females, on the other hand, exhibit the three mutant traits—yellow body color, white eyes, and echinus eye shape.

在 P_1 代中，含所有三个野生型等位基因的雄性半合子果蝇与所有三个基因均为隐性突变纯合的雌蝇杂交。因此，就体色、眼睛颜色和眼睛形状而言，P_1 代雄蝇是野生型而雌蝇则表现出三种突变性状：黄体、白眼和棘眼。

This cross produces an F_1 generation consisting of females that are hete-rozygous at all three loci and males that, because of the Y chromosome, are hemizygous for the three mutant alleles. Phenotypically, all F_1 females are wild type, while all F_1 males are yellow, white, and echinus. The genotype of the F_1 females fulfills the first criterion for mapping; that is, it is heterozygous at the three loci and can serve as the source of recombinant gametes generated by crossing over. Note that because of the genotypes of the P_1 parents, all three mutant alleles in the F_1 female are on one homolog and all three wild-type alleles are on the other homolog. With other females, other arrangements are possible that could produce a heterozygous genotype. For example, a heterozygous female could have the *y* and *ec* mutant alleles on one homolog and

该杂交产生的 F_1 代个体，雌蝇在所有三个基因座上均为杂合的，雄蝇则由于 Y 染色体的存在，三个基因座均为突变型半合子。从表型上看，所有 F_1 代雌蝇均为野生型，所有 F_1 代雄蝇均为黄体、白眼和棘眼。F_1 代雌蝇基因型符合基因作图的第一个条件：即三个基因座均为杂合的，可以作为通过杂交产生重组型配子的来源。请注意，由于 P_1 代亲本的基因型，F_1 代雌蝇中的所有三个突变型等位基因均位于一条同源染色体上，而所有三个野生型等位基因则位于另一条同源染色体上。当然，产生杂合基因型的雌蝇也可能具有其他方式的基因排布。例如，杂合雌蝇可能在一条同源染色体上含 *y* 和 *ec* 突变型等位基因，而在另一条同源染色体上含 *w* 突变型等位基因。如果是这种情况，那么在 P_1 代杂交中，一个亲本表型是黄体、棘眼；而另一个亲本

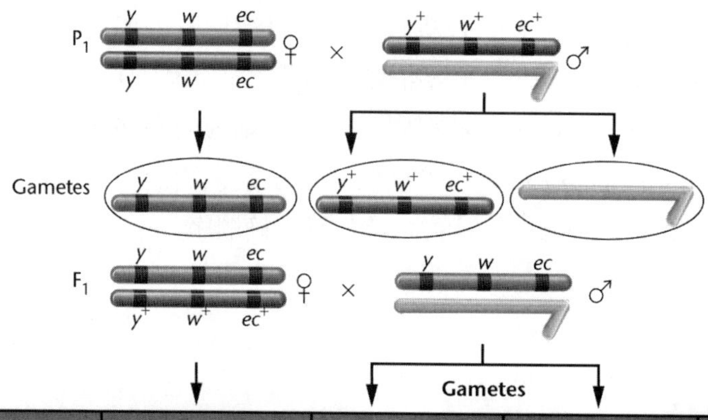

Origin of female gametes	Gametes	Male gamete: *y w ec*	Male gamete: Y	F_2 phenotype	Observed number	Category, total, and percentage
NCO (*y w ec* / y^+ w^+ ec^+)	1) *y w ec*	*y w ec* / *y w ec*	*y w ec*	*y w ec*	4685	Non-crossover
	2) y^+ w^+ ec^+	y^+ w^+ ec^+ / *y w ec*	y^+ w^+ ec^+	y^+ w^+ ec^+	4759	9444 94.44%
SCO (*y w ec* / y^+ w^+ ec^+)	3) *y* w^+ ec^+	*y* w^+ ec^+ / *y w ec*	*y* w^+ ec^+	*y* w^+ ec^+	80	Single crossover between *y* and *w*
	4) y^+ *w ec*	y^+ *w ec* / *y w ec*	y^+ *w ec*	y^+ *w ec*	70	150 1.50%
SCO (*y w ec* / y^+ w^+ ec^+)	5) *y w* ec^+	*y w* ec^+ / *y w ec*	*y w* ec^+	*y w* ec^+	193	Single crossover between *w* and *ec*
	6) y^+ w^+ *ec*	y^+ w^+ *ec* / *y w ec*	y^+ w^+ *ec*	y^+ w^+ *ec*	207	400 4.00%
DCO (*y w ec* / y^+ w^+ ec^+)	7) *y* w^+ *ec*	*y* w^+ *ec* / *y w ec*	*y* w^+ *ec*	*y* w^+ *ec*	3	Double crossover between *y* and *w* and between *w* and *ec*
	8) y^+ *w* ec^+	y^+ *w* ec^+ / *y w ec*	y^+ *w* ec^+	y^+ *w* ec^+	3	6 0.06%

Map of *y*, *w*, and *ec* loci: *y* —1.56— *w* —4.06— *ec*

FIGURE 7.8 A three-point mapping cross involving the *yellow* (*y* or y^+), *white* (*w* or w^+), and *echinus* (*ec* or ec^+) genes in *Drosophila melanogaster*. NCO, SCO, and DCO refer to noncrossover, single-crossover, and double-crossover groups, respectively. Centromeres are not drawn on the chromosomes, and only two nonsister chromatids are initially shown in the left-hand column.

图 7.8 涉及黄体（*y* 或 y^+）、白眼（*w* 或 w^+）和棘眼（*ec* 或 ec^+）三个基因的黑腹果蝇三点作图杂交实验。NCO、SCO 和 DCO 分别指非交换组、单交换组和双交换组。染色体着丝粒并未示出，左侧栏中仅显示两条非姐妹染色单体。

the *w* allele on the other. This would occur if, in the P_1 cross, one parent was yellow, echinus and the other parent was white.

表型是白眼。

In our cross, the second criterion is met by virtue of the gametes formed by the F_1 males. Every gamete contains either an X chromosome bearing the three mutant alleles or a Y chromosome, which is genetically inert for the three loci being considered. Whichever type participates in fertilization, the genotype of the gamete produced by the F_1 female will be expressed phenotypically in the F_2 male and female offspring derived from it. Thus, all F_1 noncrossover and crossover gametes can be detected by observing the F_2 phenotypes.

在我们的杂交实验中，第二个条件由 F_1 代雄蝇形成的配子满足。每个配子或含一条携带三个突变型等位基因的 X 染色体或含一条 Y 染色体，这对所研究的三个基因座在遗传上均没有贡献。无论哪种类型的配子参与受精，由 F_1 代雌蝇产生的配子基因型均将在由其产生的 F_2 代雄性和雌性后代表型中体现出来。因此，通过观察 F_2 代个体表型就可以考察所有的 F_1 代非交换型配子和交换型配子。

With these two criteria met, we can now construct a chromosome map from the crosses shown in Figure 7.8. First, we determine which F_2 phenotypes correspond to the various noncrossover and crossover categories. To determine the noncrossover F_2 phenotypes, we must identify individuals derived from the parental gametes formed by the F_1 female. Each such gamete contains an X chromosome *unaffected by crossing over*. As a result of segregation, approximately equal proportions of the two types of gametes and, subsequently, the F_2 phenotypes, are produced. Because they derive from a heterozygote, the genotypes of the two parental gametes and the resultant F_2 phenotypes complement one another. For example, if one is wild type, the other is completely mutant. This is the case in the cross being considered. In other situations, if one chromosome shows one mutant allele, the second chromosome shows

满足这两个条件后，我们可以根据图 7.8 所示的杂交实验构建染色体图。首先，我们需要确定 F_2 代表型分别对应的各种非交换型和交换型类别。为了确定非交换型 F_2 代表型，我们必须鉴定出由 F_1 代雌蝇产生的亲本型配子形成的个体。这些配子均含一条未参与交换的 X 染色体。作为分离的结果，产生比例近似相等的两种类型的配子以及随后的两种 F_2 代表型。因为它们来自杂合子，所以所产生的两种亲本型配子的基因型和相应产生的两种 F_2 代表型彼此互补。例如，如果一种是野生型，那么另一种则是完全突变型。这就是当前我们正在考察的杂交实验情形。在其他情况下，如果一条染色体显示一个突变型等位基因，则另一条染色体会显示另外两个突变型等位基因，以此类推。因此，它们被称为配子和表型的互补组。

the other two mutant alleles, and so on. They are therefore called reciprocal classes of gametes and phenotypes.

The two noncrossover phenotypes are most easily recognized because they exist in the greatest proportion. Figure 7.8 shows that gametes 1 and 2 are present in the greatest numbers. Therefore, flies that express yellow, white, and echinus phenotypes and flies that are normal (or wild type) for all three characters constitute the noncrossover category and represent 94.44 percent of the F_2 offspring.

两种非交换型表型最容易被识别，因为它们存在的数量比例最大。图 7.8 显示配子 1 和 2 的数量最多。因此表型为黄体、白眼、棘眼的果蝇以及三种性状均正常（或野生型）的果蝇构成非交换型，占 F_2 代个体的 94.44%。

The second category that can be easily detected is represented by the double-crossover phenotypes. Because of their low probability of occurrence, they must be present in the least numbers. Remember that this group represents two independent but simultaneous single-crossover events. Two reciprocal phenotypes can be identified: gamete 7, which shows the mutant traits yellow, echinus but normal eye color; and gamete 8, which shows the mutant trait white but normal body color and eye shape. Together these double-crossover phenotypes constitute only 0.06 percent of the F_2 offspring.

第二类容易被检测到的是双交换型表型。由于它们出现的概率低，因此必定以最少的数量出现。请记住，该组代表两个彼此独立但同时发生的单交换。两种互补型表型可被鉴定出：配子 7，表现出突变性状黄体、棘眼，但眼睛颜色正常；配子 8，表现出突变性状白眼，但体色和眼睛形状正常。这些双交换型表型合在一起仅占 F_2 代个体的 0.06%。

The remaining four phenotypic classes represent two categories resulting from single crossovers. Gametes 3 and 4, reciprocal phenotypes produced by single-crossover events occurring between the *yellow* and *white* loci, are equal to 1.50 percent of the F_2 offspring; gametes 5 and 6, constituting 4.00 percent of the F_2 offspring, represent the reciprocal phenotypes resulting from single-crossover events occurring between the *white* and *echinus* loci.

剩下的四种表型代表了由单交换产生的两种类别。配子 3 和 4 是由黄体和白眼基因座间发生的单交换产生的互补型表型，占 F_2 代个体的 1.50%。配子 5 和 6 占 F_2 代个体的 4.00%，是由白眼和棘眼基因座间发生单交换产生的互补型表型。

The map distances separating the three loci can now be calculated. The distance between *y* and *w* or between *w* and *ec* is equal to the percentage of all detectable exchanges occurring between them. For any two genes under consideration, this includes all appropriate single crossovers as well as all double crossovers. The latter are included because they represent two simultaneous single crossovers. For the *y* and *w* genes, this includes gametes 3, 4, 7, and 8, totaling 1.50%+ 0.06%, or 1.56 mu. Similarly, the distance between *w* and *ec* is equal to the percentage of offspring resulting from an exchange between these two loci: gametes 5, 6, 7, and 8, totaling 4.00%+ 0.06%, or 4.06 mu. The map of these three loci on the X chromosome is shown at the bottom of Figure 7.8.

现在可以计算三个基因座间的图谱距离。*y* 和 *w* 间，*w* 和 *ec* 间的距离等于它们之间发生的所有可检测交换所占的百分比。对于研究的任意两个基因，包括所有相关的单交换以及所有双交换。包括双交换是因为它们代表了两个同时发生的单交换。对于 *y* 和 *w* 基因，这包括配子 3、4、7 和 8，总计 1.50% + 0.06%或 1.56 mu。同样，*w* 和 *ec* 间的距离等于这两个基因座间所有的交换型后代百分比：配子 5、6、7 和 8，总计 4.00% + 0.06%或 4.06 mu。X 染色体上这三个基因座的基因图显示在图 7.8 的底部。

Determining the Gene Sequence

In the preceding example, the sequence (or order) of the three genes along the chromosome was assumed to be *y—w—ec*. Our analysis shows this sequence to be consistent with the data. However, in most mapping experiments the gene sequence is not known, and this constitutes another variable in the analysis. In our example, had the gene sequence been unknown, it could have been determined using a straight-forward method.

确定基因顺序

在前面的示例中，假设沿着染色体排列的三个基因的序列（或顺序）为 *y—w—ec*。我们的分析表明，该序列与所得数据相吻合。但是，在大多数基因作图实验中，基因序列未知，从而成为实验分析中的另一个变量。在我们的示例中，如果基因序列未知，则可以使用一种直接方法进行确定。

This method is based on the fact that there are only three possible arrangements, each containing one of the three genes between the other two:

(I) *w—y—ec* (*y* in the middle)

(II) *y—ec—w* (*ec* in the middle)

此方法基于以下事实，只有三种可能的排列方式，每种排列分别对应三个基因中任意一个位于中间的情况：

(I) *w—y—ec* (*y* 位于中间)

(II) *y—ec—w* (*ec* 位于中间)

(III) y—w—ec (w in the middle)

Use the following steps during your analysis to determine the gene order:

1. Assuming any one of the three orders, first determine the arrangement of alleles along each homolog of the heterozygous parent giving rise to noncrossover and crossover gametes (the F_1 female in our example).

2. Determine whether a double-crossover event occurring within that arrangement will produce the observed double-crossover phenotypes. Remember that these phenotypes occur least frequently and are easily identified.

3. If this order does not produce the predicted phenotypes, try each of the other two orders. One must work!

In Figure 7.9, the above steps are applied to each of the three possible arrangements (I, II, and III above). A full analysis can proceed as follows:

(III) y—w—ec (w 位于中间)

在分析过程中，请按照以下步骤确定基因顺序：

1. 假设基因序列是以上三种中的任意一种，先确定产生非交换型配子和交换型配子的杂合亲本等位基因沿每条染色体排列的顺序（本例中是 F_1 代雌蝇）。

2. 确定所选的基因排列顺序所发生的双交换是否能产生实际观察到的双交换型后代表型。请记住：这些后代发生频率最低并且易于识别。

3. 如果该基因排列顺序不能产生预期表型，则继续尝试另两种基因排列顺序。其中必定有一种符合！

在图 7.9 中，上述步骤一一被应用于三种可能的排列方式（以上的 I、II 和 III）。完整的分析过程如下：

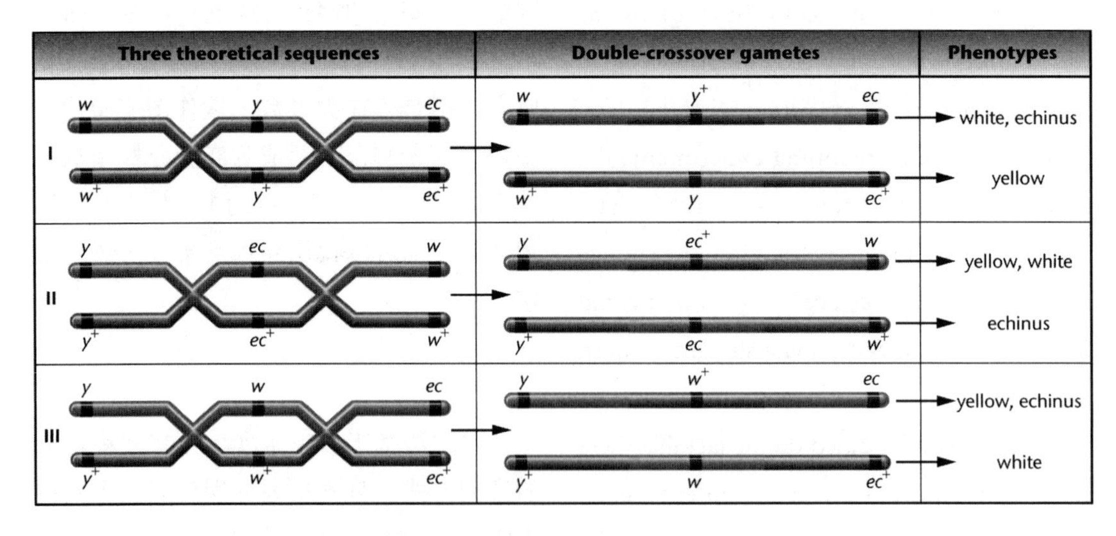

FIGURE 7.9 The three possible sequences of the *white*, *yellow*, and *echinus* genes, the results of a double-crossover in each case, and the resulting phenotypes produced in a testcross. For simplicity, the two noncrossover chromatids of each tetrad are omitted.

图 7.9 白眼、黄体和棘眼基因可能的三种排列顺序，以及每种排列顺序在测交实验中发生双交换及所产生的表型结果。为了简便起见，每个四联体中的两条非交换型染色单体被省略，未示出。

1. Assuming that y is between w and ec, arrangement I of alleles along the homologs of the F_1 heterozygote is

$$\frac{w \quad y \quad ec}{w^+ \quad y^+ \quad ec^+}$$

We know this because of the way in which the P_1 generation was crossed: The P_1 female contributes an X chromosome bearing the w, y, and ec alleles, while the P_1 male contributes an X chromosome bearing the w^+, y^+, and ec^+ alleles.

2. A double-crossover within that arrangement yields the following gametes

$\underline{w \quad y^+ \quad ec}$ and $\underline{w^+ \quad y \quad ec^+}$

Following fertilization, if y is in the middle, the F_2 double-crossover phenotypes will correspond to these gametic genotypes, yielding offspring that express the white, echinus phenotype and offspring that express the yellow phenotype. Instead, determination of the actual double-crossover phenotypes reveals them to be yellow, echinus flies and white flies. Therefore, our assumed order is incorrect.

3. If we consider arrangement II with the ec/ec^+ alleles in the middle or arrangement III with the w/w^+ alleles in the middle

$$\text{(II)} \frac{y \quad ec \quad w}{y^+ \quad ec^+ \quad w^+} \quad \text{or} \quad \text{(III)} \frac{y \quad w \quad ec}{y^+ \quad w^+ \quad ec^+}$$

we see that arrangement II again provides predicted double-crossover phenotypes that do not correspond to the actual (observed) double-crossover phenotypes. The predicted phenotypes are yellow, white flies and echinus flies in the F2 generation. Therefore, this order is also incorrect. However, arrangement III produces the observed phenotypes—yellow, echinus flies

1. 假设y位于w和ec之间（排列方式I），则F_1代杂合子同源染色体上等位基因的排列方式是：

$$\frac{w \quad y \quad ec}{w^+ \quad y^+ \quad ec^+}$$

这可由杂交亲本P_1代可知：P_1代雌蝇贡献一条含w、y、ec等位基因的X染色体，而P_1代雄蝇贡献一条含w^+、y^+、ec^+等位基因的X染色体。

2. 在该基因排列顺序下的双交换产生以下配子类型：

$\underline{w \quad y^+ \quad ec}$ 和 $\underline{w^+ \quad y \quad ec^+}$

如果y位于中间，则受精后F_2代双交换型表型将与这些配子基因型相对应，即后代表型分别为白眼棘眼果蝇，或者为黄体果蝇。但是实际产生的双交换型表型为黄体棘眼果蝇和白眼果蝇。因此，我们之前假定的基因排列顺序不正确。

3. 如果我们考虑等位基因ec/ec^+位于中间的排列方式II或等位基因w/w^+位于中间的排列方式III：

$$\text{(II)} \frac{y \quad ec \quad w}{y^+ \quad ec^+ \quad w^+} \quad \text{或} \quad \text{(III)} \frac{y \quad w \quad ec}{y^+ \quad w^+ \quad ec^+}$$

我们看到由排列方式II预测产生的双交换型表型再次与实际观测到的双交换型后代表型不相符。预测产生的F_2代表型分别是黄体白眼果蝇和棘眼果蝇。因此，排列方式II的基因顺序也不正确。然而，排列方式III可以产生观测到的双交换型后代——黄体棘眼果蝇和白眼果蝇。因此，w基因位于中间的这种基因排列顺序是正确

and white flies. Therefore, this arrangement, with the *w* gene in the middle, is correct.

To summarize, first determine the arrangement of alleles on the homologs of the heterozygote yielding the crossover gametes by locating the reciprocal noncrossover phenotypes. Then, test each of three possible orders to determine which yields the observed double-crossover phenotypes—the one that does so represents the correct order.

的。

总而言之，首先通过互补非交换型表型确定产生交换型配子的杂合亲本同源染色体上的等位基因排列方式。然后，一一检验三种可能的基因排列顺序，从而确定可以获得实际观测到的双交换型后代的表型——能够符合的即为正确的基因排列顺序。

Solving an Autosomal Mapping Problem

Having established the basic principles of chromosome mapping, we will now consider a related problem in maize (corn). This analysis differs from the preceding example in two ways. First, the previous mapping cross involved X-linked genes. Here, we consider autosomal genes. Second, in the discussion of this cross we have changed our use of symbols, as first suggested in Chapter 4. Instead of using the gene symbols and superscripts (e.g., bm^+, v^+, and pr^+), we simply use + to denote each wild-type allele. This system is easier to manipulate but requires a better understanding of mapping procedures.

解决常染色体作图中的问题

建立了染色体作图的基本原理后，我们现在考虑玉米中的一个相关问题。该分析与前面的示例有两处不同：第一，先前的基因作图杂交实验涉及X连锁基因。在这里，我们研究常染色体上的基因。第二，在讨论本杂交实验时，我们将不使用第4章中首次建议的基因符号表示方法，即不再使用基因符号和上角标（例如bm^+、v^+和pr^+），而是简单地使用“+”表示相应的野生型等位基因。该系统更为简化，但是需要对于基因作图有更深入的理解。

When we look at three autosomally linked genes in maize, the experimental cross must still meet the same three criteria we established for the X-linked genes in *Drosophila*: (1) One parent must be heterozygous for all traits under consideration; (2) the gametic genotypes produced by the heterozygote must be apparent from observing the phenotypes of the offspring; and (3) a sufficient sample size must be available for complete analysis.

当我们研究玉米的三个常染色体上连锁基因时，杂交实验仍然必须符合果蝇X连锁基因研究中建立的三条标准：（1）一个亲本对于所有研究性状必须是杂合的；（2）由杂合亲本所产生的配子基因型必须可以通过观察后代表型清晰地确定；（3）必须有足够数目的样本量以进行完整分析。

In maize, the recessive mutant genes *brown* midrib (*bm*), *virescent* seedling (*v*), and *purple* aleurone (*pr*) are linked on chromosome 5. Assume that a female plant is known to be heterozygous for all three traits, but we do not know (1) the arrangement of the mutant alleles on the maternal and paternal homologs of this heterozygote, (2) the sequence of genes, or (3) the map distances between the genes. What genotype must the male plant have to allow successful mapping? To meet the second criterion, the male must be homozygous for all three recessive mutant alleles. Otherwise, offspring of this cross showing a given phenotype might represent more than one genotype, making accurate mapping impossible.

在玉米中，隐性突变基因棕色中肋（*bm*）、绿化幼苗（*v*）和紫色糊粉（*pr*）位于5号染色体上并且连锁。假定雌株所有三种性状的基因型均为杂合，但我们并不知道（1）突变型等位基因在该杂合子的母本和父本同源染色体上的排列方式；（2）基因排列顺序；（3）基因间的图距。那么雄株的基因型必须是哪种才能进行成功的基因作图呢？为了满足第二条标准，雄株的三个等位基因必须是隐性纯合的。否则，杂交后代的一种指定表型可能同时代表多种基因型，从而无法进行精确作图。

Figure 7.10 diagrams this cross. As shown, we known either the arrangement of alleles nor the sequence of loci in the heterozygous female. Several possibilities are shown, but we have yet to determine which is correct. We don't know the sequence in the testcross male parent either, and so we must designate it randomly. Note that we have initially placed *v* in the middle. This may or may not be correct.

图7.10展示了该杂交实验。如图所示，我们既不知道杂合雌株等位基因的排列方式，也不清楚基因座顺序。图中显示了多种可能性，但是我们尚不能确定哪种正确。我们也不知道测交雄性亲本的基因序列，因此我们必须进行随机指定。请注意，我们最初已将*v*放在中间。这一假设可能正确也可能不正确。

The offspring are arranged in groups of two for each pair of reciprocal phenotypic classes. The two members of each reciprocal class are derived from no crossing over (NCO), one of two possible single-crossover events (SCO), or a double-crossover (DCO).

后代按照成对存在的互补表型类别，两两为一组进行排列。每个互补组的两个成员或者来自非交换（NCO），或者是两种可能的单交换（SCO）之一，或者来自双交换（DCO）。

To solve this problem, refer to Figures 7.10 and 7.11 as you consider the following questions.

为了解决该问题，请在考虑以下问题时参考图7.10和7.11。

1. *What is the correct heterozygous arrangement of alleles in the female parent?*

1. 雌性亲本中正确的杂合等位基因排布方式是什么？

(a) Some possible allele arrangements and gene sequences in a heterozygous female

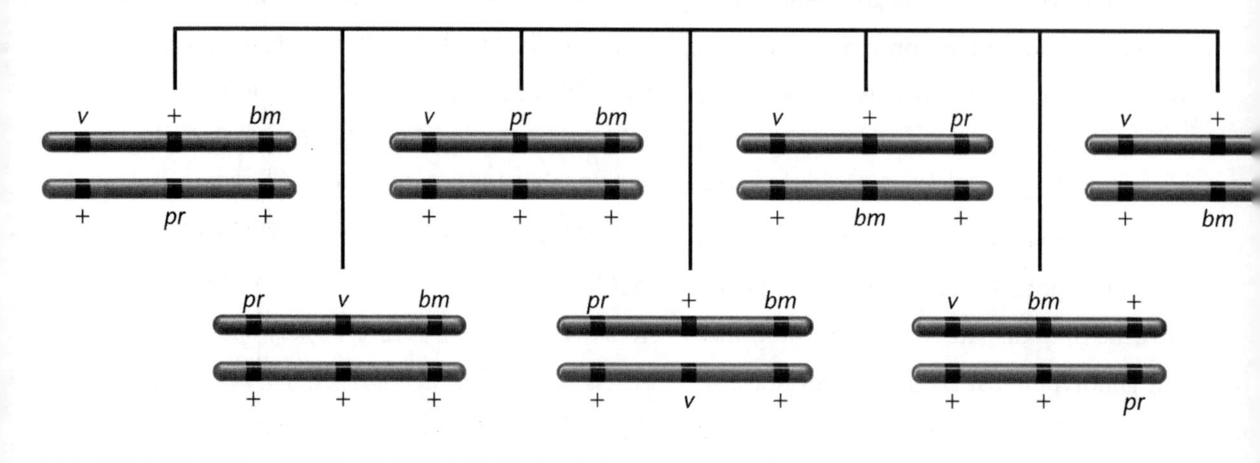

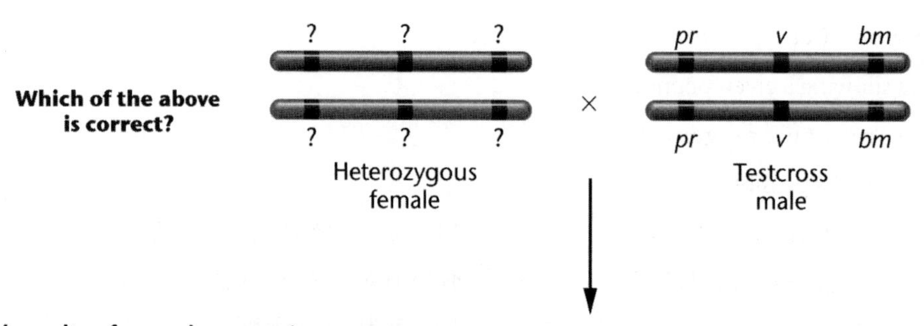

(b) Actual results of mapping cross*

Phenotypes of offspring	Number	Total and percentage	Exchange classification
+ *v* *bm*	230	467 42.1%	Noncrossover (NCO)
pr + +	237		
+ + *bm*	82	161 14.5%	Single crossover (SCO)
pr *v* +	79		
+ *v* +	200	395 35.6%	Single crossover (SCO)
pr + *bm*	195		
pr *v* *bm*	44	86 7.8%	Double crossover (DCO)
+ + +	42		

* The sequence *pr* – *v* – *bm* may or may not be correct.

FIGURE 7.10 (a) Some possible allele arrangements and gene sequences in a heterozygous female. The data from a three-point mapping cross, depicted in (b), where the female is testcrossed, provide the basis for determining which combination of arrangement and sequence is correct. [See Figure 7.11(d).]

图 7.10 (a) 杂合雌性个体可能的等位基因排列方式和可能的基因座顺序。三点作图杂交实验数据如 (b) 所示，测交雌性个体为确定哪种等位基因排列方式和基因座顺序正确提供了基础。[见图 7.11(d)]

Determine the two noncrossover classes, those that occur with the highest frequency. In this case, they are $\underline{+\ v\ bm}$ and $\underline{pr\ +\ +}$. Therefore, the alleles on the homologs of the female parent must be arranged as shown in Figure 7.11(a). These homologs segregate into gametes, unaffected by any recombination event. Any other arrangement of alleles will not yield the observed noncrossover classes. (Remember that $\underline{+\ v\ bm}$ is equivalent to $\underline{pr^+ v\ bm}$ and that $\underline{pr\ +\ +}$ is equivalent to $\underline{pr\ v^+ bm^+}$.)

确定两种非交换型类别，它们的发生频率最高。在本例中，它们是 $\underline{+\ v\ bm}$ 和 $\underline{pr\ +\ +}$。因此，母本同源染色体上的等位基因必定如图 7.11（a）所示进行排列。这些同源染色体发生分离进入配子中，未受任何重组事件的影响。任何其他方式的等位基因排列均不能得到所观察的非交换型后代。（请记住，$\underline{+\ v\ bm}$ 等同于 $\underline{pr^+ v\ bm}$，而 $\underline{pr\ +\ +}$ 等同于 $\underline{pr\ v^+ bm^+}$。）

2. *What is the correct sequence of genes?*

We know that the arrangement of alleles is

$$\frac{+\quad v\quad bm}{pr\quad +\quad +}$$

2. 基因的正确排列顺序是什么？

我们知道等位基因的排布方式是：

$$\frac{+\quad v\quad bm}{pr\quad +\quad +}$$

But is the gene sequence correct? That is, will a double-crossover event yield the observed double-crossover phenotypes after fertilization? Observation shows that it will not [Figure 7.11(b)]. Now try the other two orders [Figure 7.11(c) and (d)] maintaining the same arrangement of alleles:

但是这样的基因顺序正确吗？也就是说，受精后发生双交换是否会产生所观察的双交换型表型？观察结果表明该种基因排列不可以［图 7.11（b）］。现在尝试另外两种基因排列顺序［图 7.11（c）和（d）］同时保持同样的等位基因排布方式：

$$\frac{+\quad bm\quad v}{pr\quad +\quad +}\quad \text{or}\quad \frac{v\quad +\quad bm}{+\quad pr\quad +}$$

$$\frac{+\quad bm\quad v}{pr\quad +\quad +}\quad \text{或}\quad \frac{v\quad +\quad bm}{+\quad pr\quad +}$$

Only the order on the right yields the observed double-crossover gametes [Figure 7.11(d)]. Therefore, the *pr* gene is in the middle. From this point on, work the problem using this arrangement and sequence, with the *pr* locus in the middle.

只有右侧的基因排列顺序才能产生所观察的双交换型配子［图 7.11（d）］。因此，*pr* 基因位于中间。从现在开始，使用 *pr* 基因座位于中间的这种排布方式和基因顺序来回答待解决问题。

3. *What is the distance between each pair of genes?*

Having established the sequence of loci as *v*—*pr*—*bm*, we can determine the distance between *v* and *pr* and between *pr* and *bm*.

3. 每对基因之间的距离是多少？

因为已将基因座的序列确定为 *v*—*pr*—*bm*，我们可以确定 *v* 与 *pr* 之间以及 *pr* 与 *bm* 之间的距离。请记住，两基因间的图距

Remember that the map distance between two genes is calculated on the basis of all detectable recombination events occurring between them. This includes both single- and double-crossover events.

根据它们之间发生的所有可检出的重组进行计算，既包括单交换，也包括双交换。

Figure 7.11(e) shows that the phenotypes *v pr +* and *+ + bm* result from single crossovers between the *v* and *pr* loci, accounting for 14.5 percent of the offspring [according to data in Figure 7.10(b)]. By adding the percentage of double crossovers (7.8 percent) to the number obtained for single crossovers, the total distance between the *v* and *pr* loci is calculated to be 22.3 mu.

图 7.11（e）显示表型 *v pr +* 和 *+ + bm* 由 *v* 和 *pr* 间的单交换产生，占后代总数的 14.5% [根据图 7.10（b）中数据可得]。通过加上双交换的百分比（7.8%），即可计算出 *v* 和 *pr* 间的基因图距为 22.3 mu。

Figure 7.11(f) shows that the phenotypes *v + +* and *+ pr bm* result from single crossovers between the *pr* and *bm* loci, totaling 35.6 percent. Added to the double crossovers (7.8 percent), the distance between *pr* and *bm* is calculated to be 43.4 mu. The final map for all three genes in this example is shown in Figure 7.11(g).

图 7.11（f）显示表型 *v + +* 和 *+ pr bm* 由 *pr* 和 *bm* 间的单交换产生，总计 35.6%。加上双交换频率（7.8%），计算得出 *pr* 和 *bm* 间的图距为 43.4 mu。此示例中所有三个基因的最终遗传图如图 7.11（g）所示。

7.4 As the Distance between Two Genes Increases, Mapping Estimates Become More Inaccurate

7.4 随着基因间距离的增加，染色体作图的精确性将随之下降

So far, we have assumed that crossover frequencies are directly proportional to the distance between any two loci along the chromosome. However, it is not always possible to detect all crossover events. A case in point is a double exchange that occurs between the two loci in question. As shown in Figure 7.12(a), if a double exchange occurs, the original arrangement of alleles on each nonsister

到目前为止，我们假设交换频率与沿染色体排列的任何两基因座间的距离成正比。但是，并非所有交换都能被检测到。一个恰当的例子是两个待研究的基因座之间发生的双交换。如图 7.12（a）所示，如果发生双交换，每条非姐妹染色单体上等位基因恢复其原始排布顺序，那么，即使发生了交换，也无法被检测到。这种现象对于两个基因座间所有的偶数次交换均是

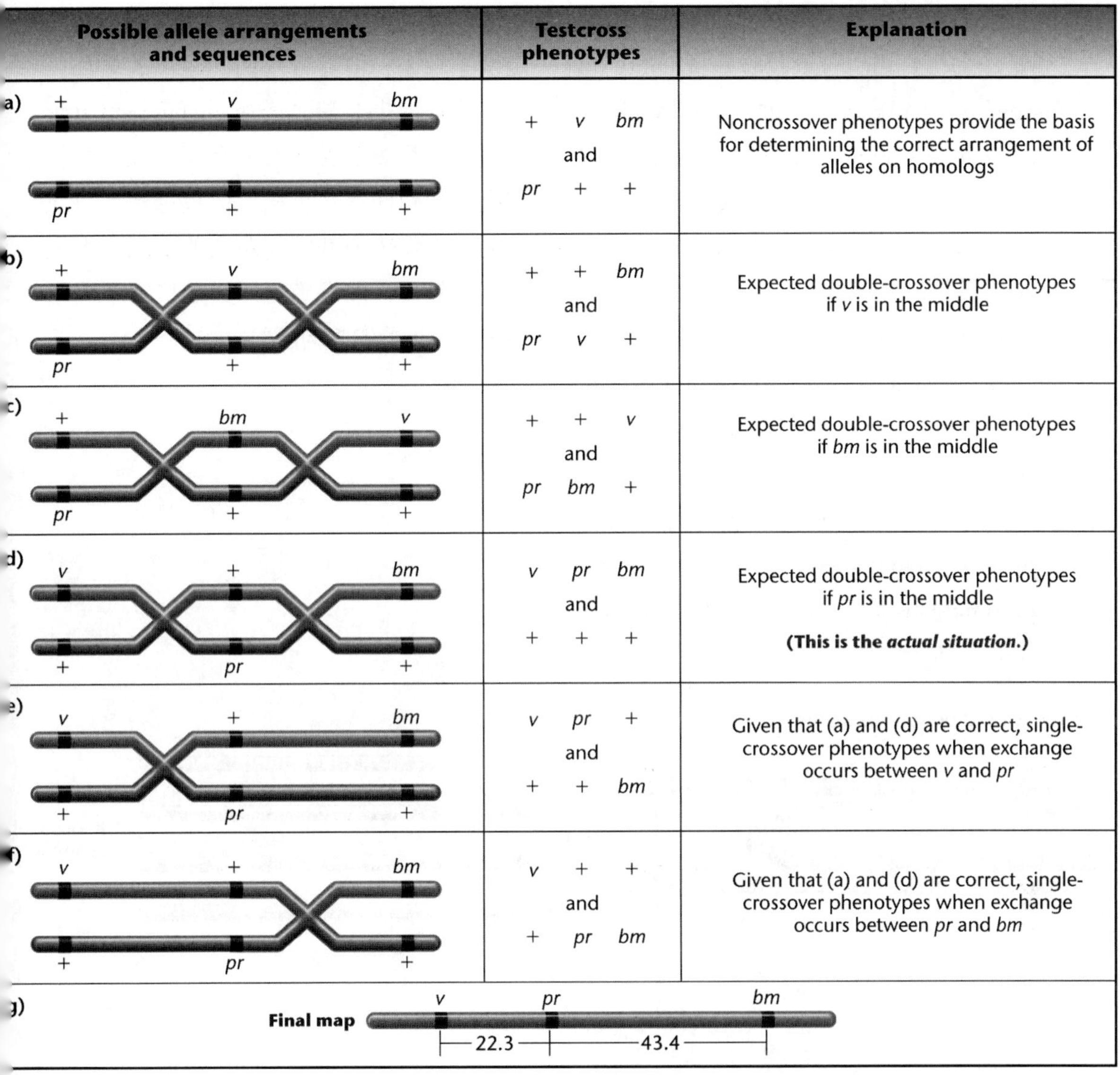

	Possible allele arrangements and sequences	Testcross phenotypes	Explanation
(a)	+ *v* *bm* / *pr* + +	+ *v* *bm* and *pr* + +	Noncrossover phenotypes provide the basis for determining the correct arrangement of alleles on homologs
(b)	+ *v* *bm* / *pr* + +	+ + *bm* and *pr* *v* +	Expected double-crossover phenotypes if *v* is in the middle
(c)	+ *bm* *v* / *pr* + +	+ + *v* and *pr* *bm* +	Expected double-crossover phenotypes if *bm* is in the middle
(d)	*v* + *bm* / + *pr* +	*v* *pr* *bm* and + + +	Expected double-crossover phenotypes if *pr* is in the middle **(This is the *actual situation*.)**
(e)	*v* + *bm* / + *pr* +	*v* *pr* + and + + *bm*	Given that (a) and (d) are correct, single-crossover phenotypes when exchange occurs between *v* and *pr*
(f)	*v* + *bm* / + *pr* +	*v* + + and + *pr* *bm*	Given that (a) and (d) are correct, single-crossover phenotypes when exchange occurs between *pr* and *bm*
(g)	**Final map** *v* — 22.3 — *pr* — 43.4 — *bm*		

FIGURE 7.11 Steps utilized in producing a map of the three genes in the cross in Figure 7.10, where neither the arrangement of alleles nor the sequence of genes in the heterozygous female parent is known.

图 7.11 制作图 7.10 所示杂交实验中三种基因染色体图的步骤，在该杂交实验中，杂合雌性亲本个体的等位基因排列方式和基因座顺序均未知。

homolog is recovered. Therefore, even though crossing over has occurred, it is impossible to detect. This phenomenon is true for all even-numbered exchanges between two loci.

Furthermore, as a result of complications posed by *multiple-strand exchanges*, mapping determinations usually underestimate the actual

如此。

此外，由于多链交换因素的存在，遗传作图通常会低估两基因间的实际距离。两个基因距离越远，发生交换未被检测的

distance between two genes. The farther apart two genes are, the greater the probability that undetected crossovers will occur. While the discrepancy is minimal for two genes relatively close together, the degree of inaccuracy increases as the distance increases, as shown in the graph of map distance versus recombination frequency in Figure 7.12(b). There, the theoretical frequency where a direct correlation between recombination and map distance exists is contrasted with the actual frequency observed as the distance between two genes increases. The most accurate maps are constructed from experiments where genes are relatively close together.

可能性就越大。相对靠近的两个基因间测定偏差较小，但随着基因间距离的增加，不准确程度也会随之增加，图距与重组频率间的关系如图 7.12（b）所示。在图中，重组和图距间存在正相关的理论频率与观察到的用作表征基因间距离的实际频率之间形成对比。最准确的图谱是根据基因距离相对靠近的实验构建的。

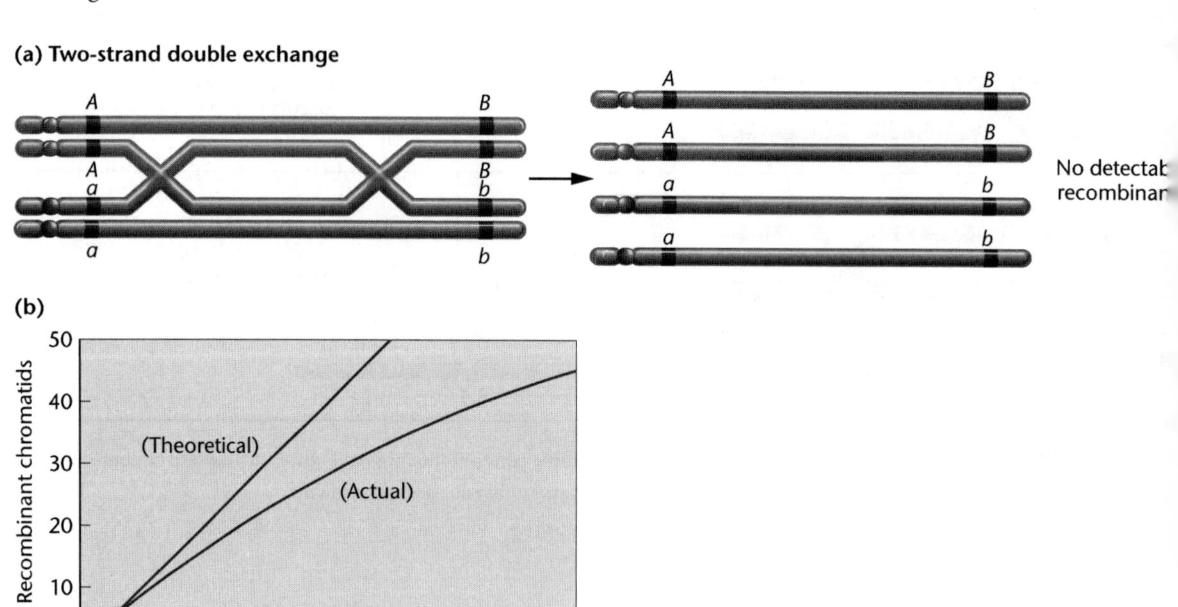

FIGURE 7.12 (a) A double crossover is undetected because no rearrangement of alleles occurs. (b) The theoretical and actual percentage of recombinant chromatids versus map distance. The straight line shows the theoretical relationship if a direct correlation between recombination and map distance exists. The curved line is the actual relationship derived from studies of *Drosophila*, *Neurospora*, and *Zea mays*.

图 7.12 （a）由于未发生等位基因重排列，所以该双交换无法被检出。（b）重组染色单体的理论百分比和实际百分比与图距的关系。直线显示了重组与图距间呈正相关的理论关系。曲线显示了由果蝇、链孢霉和玉米研究中所得的实际关系。

Interference and the Coefficient of Coincidence

As shown in our maize example, we can predict the expected frequency of multiple exchanges, such as double crossovers, once the distance between genes is established. For example, in the maize cross, the distance between *v* and *pr* is 22.3 mu, and the distance between *pr* and *bm* is 43.4 mu. If the two single crossovers that make up a double crossover occur independently of one another, we can calculate the expected frequency of double crossovers (DCO_{exp}):

$$DCO_{exp} = (0.223) \times (0.434) = 0.097 = 9.7\%$$

Often in mapping experiments, the observed DCO frequency is less than the expected number of DCOs. In the maize cross, for example, only 7.8 percent DCOs are observed when 9.7 percent are expected. **Interference** (***I***), the phenomenon through which a crossover event in one region of the chromosome inhibits a second event in nearby regions, causes this reduction.

To quantify the disparities that result from interference, we calculate the **coefficient of coincidence** (***C***):

$$C = \frac{\text{Observed DCO}}{\text{Expected DCO}}$$

In the maize cross, we have

$$C = \frac{0.078}{0.097} = 0.804$$

Once we have found *C*, we can quantify interference using the simple equation

$$I = 1 - C$$

干涉与并发系数

在我们所举的玉米示例中，一旦确定了基因间的距离，我们即可预测多交换例如双交换的预期发生频率。例如，在玉米杂交实验中，*v* 和 *pr* 间的距离为 22.3 mu，*pr* 和 *bm* 间的距离为 43.4 mu。如果构成双交换的两个单交换彼此独立，那么可以计算出双交换的预期频率（DCO_{exp}）：

$$DCO_{exp} = (0.223) \times (0.434) = 0.097 = 9.7\%$$

通常在作图实验中，观察到的 DCO 频率会低于预期发生的双交换数量。例如，在玉米杂交实验中，双交换发生频率预期为 9.7%，但实际仅观察到 7.8%的双交换。**干涉**（***I***）导致了这种减少。干涉是染色体一个区域中的交换抑制附近区域第二个交换发生的现象。

为了量化由干涉引起的偏差，我们计算**并发系数**（***C***）：

$$C = \frac{\text{观察到的 DCO}}{\text{预期的 DCO}}$$

在玉米杂交实验中，我们得到：

$$C = \frac{0.078}{0.097} = 0.804$$

一旦得到并发系数 *C*，我们即可使用以下的简单计算量化干涉：

$$I = 1 - C$$

In the maize cross, we have

$$I = 1.000 - 0.804 = 0.196$$

If interference is complete and no double crossovers occur, then $I = 1.0$. If fewer DCOs than expected occur, I is a positive number and positive interference has occurred. If more DCOs than expected occur, I is a negative number and negative interference has occurred. In the maize example, I is a positive number (0.196), indicating that 19.6 percent fewer double crossovers occurred than expected.

Positive interference is most often the rule in eukaryotic systems. In general, the closer genes are to one another along the chromosome, the more positive interference occurs. In fact, interference in *Drosophila* is often complete within a distance of 10 mu, and no multiple crossovers are recovered. This observation suggests that physical constraints preventing the formation of closely aligned chiasmata contribute to interference. This interpretation is consistent with the finding that interference decreases as the genes in question are located farther apart. In the maize cross in Figures 7.10 and 7.11, the three genes are relatively far apart, and 80 percent of the expected double crossovers are observed.

ESSENTIAL POINT

Interference describes the extent to which a crossover in one region of a chromosome influences the occurrence of a crossover in an adjacent region of the chromosome and is quantified by calculating the coefficient of coincidence (C).

在玉米杂交实验中，我们得到

$$I = 1.000 - 0.804 = 0.196$$

如果完全干涉则无双交换发生，即 $I = 1.0$。如果双交换的发生数量少于预期，则 I 为正值，即发生正干涉；如果双交换的发生数量超过预期，则 I 为负值，即发生负干涉。在玉米的例子中，I 为正值（0.196），表明发生的双交换比预期少 19.6%。

正干涉通常在真核生物系统中大量存在。一般而言，沿着染色体排列的基因彼此间距离越近，发生正干涉的程度就越大。实际上，果蝇通常在图距 10 mu 之内发生完全干涉，无任何多交换现象发生。此观察结果表明，物理限制因素阻止了紧密排列的交叉发生，从而造成干涉。这一解释与干涉随着所讨论基因间距离变远而减少的发现相一致。在图 7.10 和 7.11 的玉米杂交实验中，所研究的三个基因相距较远，因此可以观察到 80%的预期双交换。

基本要点

干涉描述染色体一个区域中的交换影响其附近区域发生其他交换的程度，并可通过计算并发系数（C）进行量化。

7.5 Chromosome Mapping Is Now Possible Using DNA Markers and Annotated Computer Databases

Although traditional methods based on recombination analysis have produced detailed chromosomal maps in several organisms, such maps in other organisms (including humans) that do not lend themselves to such studies are greatly limited. Fortunately, the development of technology allowing direct analysis of DNA has greatly enhanced mapping in those organisms. We will address this topic using humans as an example.

Progress has initially relied on the discovery of **DNA markers** that have been identified during recombinant DNA and genomic studies. These markers are short segments of DNA whose sequence and location are known, making them useful *landmarks* for mapping purposes. The analysis of human genes in relation to these markers has extended our knowledge of the location within the genome of countless genes, which is the ultimate goal of mapping.

The earliest examples are the DNA markers referred to as **restriction fragment length polymorphisms (RFLPs)** (see Chapter 10—Genetic Testing) and **microsatellites**. RFLPs are polymorphic sites generated when specific DNA sequences are recognized and cut by restriction enzymes. Microsatellites are short repetitive sequences that are found throughout the genome, and they vary in the number of repeats at any given site. For example, the two-nucleotide sequence CA is repeated 5—50

7.5 当今，使用 DNA 标记和计算机注释数据库进行染色体作图已成为可能

尽管使用基于重组分析的传统方法已构建了多种生物体详细的染色体图，但在其他生物（包括人类）中此类作图研究方法受到极大限制，并不适用。幸运的是，直接分析 DNA 的相关技术不断发展，极大地推动了这些生物的染色体作图工作。我们将以人类为例讨论这一话题。

最初的进展依赖于 DNA 重组和基因组研究过程中 **DNA 标记**的鉴定和发现。这些标记是序列和位置均已知的短 DNA 片段，从而成为可用于染色体作图的有用地标。分析人类基因与这些标记的关系拓展了我们对于基因组中无数基因定位的了解，这也正是染色体作图的最终目标。

最早的 DNA 标记是**限制性片段长度多态性（RFLPs）**（请参阅第 10 章——遗传检测）和**微卫星**。RFLPs 是特定 DNA 序列被限制性内切酶识别并剪切后产生的多态性位点。微卫星是在整个基因组中发现的短重复序列，微卫星在任何给定位点的重复数目不同。例如，二核苷酸序列 CA 在每一位点重复 5—50 次 [$(CA)_n$]，并且在整个基因组中平均大约每 10 000 个碱基出现一个位点。微卫星不仅可通过重复次数进行鉴定，而且还可通过侧翼 DNA 序列进行鉴

times per site [$(CA)_n$] and appears throughout the genome approximately every 10,000 bases, on average. Microsatellites may be identified not only by the number of repeats but by the DNA sequences that flank them. More recently, variation in single nucleotides (called **single-nucleotide polymorphisms** or **SNPs**) has been utilized. Found throughout the genome, up to several million of these variations may be screened for an association with a disease or trait of interest, thus providing geneticists with a means to identify and locate related genes.

定。最近，人们已经利用单核苷酸的变异（**单核苷酸多态性**或 SNPs）。在整个基因组中与所关注的疾病或性状相关联的可被筛选的此类变异多达数百万个，从而为遗传学家提供了鉴定和定位相关基因的手段。

Cystic fibrosis offers an early example of a gene located by using DNA markers. It is a life-shortening autosomal recessive exocrine disorder resulting in excessive, thick mucus that impedes the function of organs such as the lung and pancreas. After scientists established that the gene causing this disorder is located on chromosome 7, they were then able to pinpoint its exact location on the long arm (the q arm) of that chromosome.

囊性纤维化提供了使用 DNA 标记定位基因的早期示例。囊性纤维化是一种导致寿命缩短的常染色体隐性遗传外分泌疾病，由于黏液分泌过多、浓稠，会阻碍肺和胰腺等器官的功能。在科学家确定导致该疾病的基因位于 7 号染色体后，他们便能够将其精确定位于该染色体的长臂（q 臂）上。

In 2007, using SNPs as DNA markers, associations between 24 genomic locations were established with seven common human diseases: Type 1 (insulin dependent) and Type 2 diabetes, Crohn disease (inflammatory bowel disease), hypertension, coronary artery disease, bipolar disorder, and rheumatoid arthritis. In each case, an inherited susceptibility effect was mapped to a specific location on a specific chromosome within the genome. In some cases, this either confirmed or led to the identification of a specific gene involved in the cause of the disease.

2007 年，使用 SNPs 作为 DNA 标记建立了 24 个基因组定位与 7 种常见人类疾病之间的关联：1 型（胰岛素依赖型）糖尿病、2 型糖尿病、克罗恩病（炎症性肠病）、高血压、冠心病、双相情感障碍和类风湿性关节炎。在每种疾病中，遗传的易感性效应均被定位至基因组中特定染色体上的特定位置。在某些疾病中，这可以证实或有助于鉴定致病原因相关的特定基因。

During the past 15 years or so, dramatic

在过去 15 年左右的时间里，DNA 测序

improvements in DNA sequencing technology have resulted in a proliferation of **sequence maps** for humans and many other species. Sequence maps provide the finest level of mapping detail because they pinpoint the nucleotide sequence of genes (and noncoding sequences) on a chromosome. The Human Genome Project resulted in sequence maps for all human chromosomes, providing an incredible level of detail about human gene sequences, the specific location of genes on a chromosome, and the proximity of genes and noncoding sequences to each other, among other details. For instance, when human chromosome sequences were analyzed by software programs, an approach called **bioinformatics**, geneticists could utilize such data to map possible protein-coding sequences in the genome. This led to the identification of thousands of potential genes that were previously unknown.

技术的显著进步使得人类和许多其他物种的**序列图**激增。序列图提供了最精细的序列细节，因为它们精确定位了染色体上基因的核苷酸序列（以及非编码序列）。人类基因组计划获得了所有人类染色体的序列图，提供了令人难以置信的人类基因序列细节、基因在染色体上的特定位置、基因和非编码序列彼此之间的接近程度以及其他细节。例如，当使用**生物信息学**方法，即利用软件程序分析人类染色体序列时，遗传学家可以利用这些数据在基因组中定位可能的蛋白质编码序列。这将促使之前未知的数以千计的潜在基因得以被鉴定。

The many Human Genome Project databases that are now available make it possible to map genes along a human chromosome in base-pair distances rather than recombination frequency. This distinguishes what is referred to as a physical map of the genome from the genetic maps described above. When the genome sequence of a species is available, mapping by linkage or other genetic mapping approaches becomes obsolete.

现在可供使用的许多人类基因组计划数据库使得沿着人类染色体以碱基对为距离而不以重组频率定位基因成为可能。这被称为基因组的物理图谱，从而与上述遗传图谱相区分。当物种的基因组序列已知时，就不会再使用连锁或其他遗传作图方法进行作图了。

ESSENTIAL POINT

Human linkage studies have been enhanced by the use of newly discovered molecular DNA markers.

基本要点

新发现的分子 DNA 标记的使用促进了人类基因连锁的研究。

7.6 Other Aspects of Genetic Exchange

7.6 关于遗传交换的几点补充

Careful analysis of crossing over during gamete formation allows us to construct chromosome maps in many organisms. However, we should not lose sight of the real biological significance of crossing over, which is to generate genetic variation in gametes and, subsequently, in the offspring derived from the resultant eggs and sperm. Many unanswered questions remain, which we consider next.

仔细分析配子形成过程中的交换让我们能够构建许多生物的染色体图。但是，我们不应该忽视交换真正的生物学意义，即产生了配子及随后由这些卵和精子发育所得后代的遗传变异。许多尚未解决的问题仍然存在，我们接下来将对此进行讨论。

Crossing Over—A Physical Exchange between Chromatids

交换——染色单体间的物理交换

Once genetic mapping was understood, it was of great interest to investigate the relationship between chiasmata observed in meiotic prophase I and crossing over. Are chiasmata visible manifestations of crossover events? If so, then crossing over in higher organisms appears to result from an actual physical exchange between homologous chromosomes. That this is the case was demonstrated independently in the 1930s by Harriet Creighton and Barbara McClintock in *Zea mays* and by Curt Stern in *Drosophila*.

一旦理解了遗传作图，研究减数分裂前期 I 观察到的交叉与交换之间的关系便引起了人们极大的兴趣。交叉是交换的明显表征吗？如果是这样，那么高等生物中的交换似乎源自同源染色体之间实际发生的物理交换。20 世纪 30 年代哈丽雅特 · 克赖顿和芭芭拉 · 麦克林托克在玉米中的研究以及柯特 · 斯特恩在果蝇中的研究分别独立地对该推测进行了证明。

Since the experiments are similar, we will consider only the work with maize. Creighton and McClintock studied two linked genes on chromosome 9. At one locus, the alleles *colorless* (*c*) and *colored* (*C*) control endosperm coloration. At the other locus, the alleles *starchy* (*Wx*) and *waxy* (*wx*) control the carbohydrate characteristics of the endosperm. The maize

由于实验相似，因此我们仅介绍玉米相关的研究工作。克赖顿和麦克林托克研究位于玉米 9 号染色体上的两个连锁基因。在其中一个基因座上，等位基因无色（*c*）和有色（*C*）决定胚乳着色。在另一个基因座上，等位基因糯质（*Wx*）和非糯（*wx*）决定胚乳碳水化合物的特性。研究所用的玉米植株在两个基因座上都是杂合的。该

plant studied is heterozygous at both loci. The key to this experiment is that one of the homologs contains two unique cytological markers. The markers consist of a densely stained knob at one end of the chromosome and a translocated piece of another chromosome (8) at the other end. The arrangements of alleles and cytological markers can be detected cytologically and are shown in Figure 7.13.

实验的关键在于其中一条同源染色体含两种独特的细胞学标记。染色体一端所含标记是显著染色的结节，另一端则是来自另一条染色体（8 号染色体）的易位片段。等位基因和细胞标记的排布情况可利用细胞学方法进行检测，如图 7.13 所示。

Creighton and McClintock crossed this plant to one homozygous for the *colored* allele (*c*) and heterozygous for the endosperm alleles. They obtained a variety of different phenotypes in the offspring, but they were most interested in a crossover result involving the chromosome with the unique cytological markers. They examined the chromosomes of this plant with the colorless, waxy phenotype (Case I in Figure 7.13) for the presence of the cytological markers. If physical exchange between homologs accompanies genetic crossing over, the translocated chromosome will still be present, but the knob will not—this is exactly

克赖顿和麦克林托克将该植株与胚乳着色等位基因（*c*）纯合、胚乳特性等位基因杂合的植株进行杂交。他们获得了多种表型的后代，但其中最令人感兴趣的则是含独特细胞学标记的染色体参与交换所产生的后代。他们在表型为无色、非糯的植株染色体（图 7.13 中的情况 I）中检查细胞学标记的存在情况。如果同源染色体间的物理交换伴随着遗传交换，那么易位染色体片段将仍然存在，同时结节将不存在——事实上结果正是如此。在第二种表型的植株（情况 II）中，有色、糯质可能由非重组型配子产生，也可能由遗传交换产生。这些植株的一些个体染色体中应该含着色显

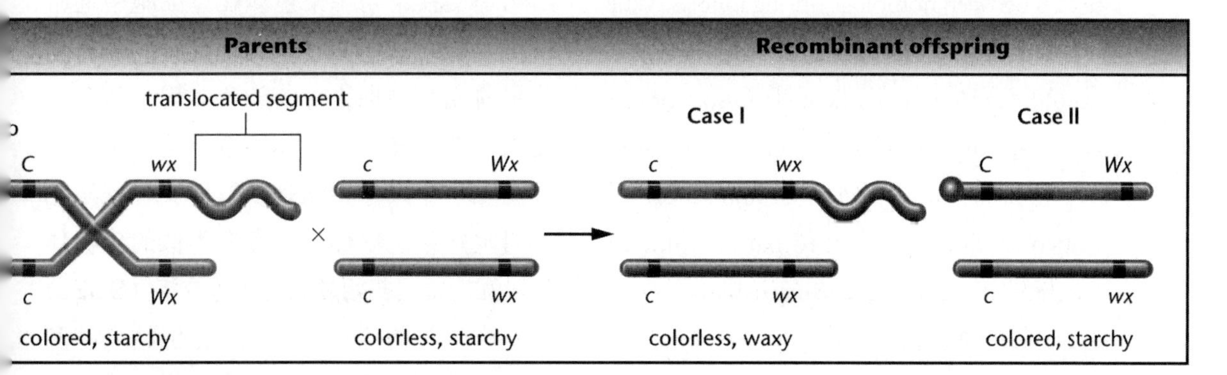

FIGURE 7.13 The phenotypes and chromosome compositions of parents and recombinant offspring in Creighton and McClintock's experiment in maize. The knob and translocated segment served as cytological markers, which established that crossing over involves an actual exchange of chromosome arms.

图 7.13 在克赖顿和麦克林托克的玉米实验中亲本和重组型后代的表型和染色体组成。结节和易位片段作为细胞标记，用于确定在染色体臂间实际发生的交换。

what happened. In a second plant (Case II), the phenotype colored, starchy should result from either nonrecombinant gametes or crossing over. Some of the plants then ought to contain chromosomes with the dense knob but not the translocated chromosome. This condition was also found, and the conclusion that a physical exchange takes place was again supported. Along with Stern's findings with *Drosophila*, this work clearly established that crossing over has a cytological basis.

著的结节，而不含易位染色体片段。这种情形也同样被发现，因此再次支持了物理交换发生的结论。该研究工作与斯特恩在果蝇研究中的发现一同清晰地证明了交换具有细胞生物学理论基础。

ESSENTIAL POINT

Cytological investigations of both maize and *Drosophila* reveal that crossing over involves a physical exchange of segments between nonsister chromatids.

基本要点

玉米和果蝇中的细胞生物学研究表明，交换是涉及非姐妹染色单体间染色体片段发生物理交换的过程。

Sister Chromatid Exchanges between Mitotic Chromosomes

Considering that crossing over occurs between synapsed homologs in meiosis, we might ask whether a similar physical exchange occurs between homologs during mitosis. While homologous chromosomes do not usually pair up or synapse in somatic cells (*Drosophila* is an exception), each individual chromosome in prophase and metaphase of mitosis consists of two identical sister chromatids, joined at a common centromere. Surprisingly, several experimental approaches have demonstrated that reciprocal exchanges similar to crossing over occur between sister chromatids. These **sister chromatid exchanges (SCEs)** do not produce new allelic combinations, but evidence

有丝分裂染色体间的姐妹染色单体交换

由于交换发生在减数分裂过程中联会的同源染色体间，所以我们可能会问有丝分裂过程中同源染色体间是否也发生类似的物理交换。同源染色体在体细胞中通常不发生配对或联会（果蝇除外），在有丝分裂前期和中期，每条染色体均由两条完全相同的姐妹染色单体组成，并共用一个着丝粒。令人惊讶的是，多种实验方法已证明姐妹染色单体之间存在类似的染色体片段互换。这些**姐妹染色单体交换（SCEs）**并不产生新的等位基因组合，但越来越多的证据表明这些事件具有重要的生物学意义。

is accumulating that attaches significance to these events.

Identification and study of SCEs are facilitated by several modern staining techniques. In one technique, cells replicate for two generations in the presence of the thymidine analog **bromodeoxyuridine** (**BrdU**). Following two rounds of replication, each pair of sister chromatids has one member with one strand of DNA labeled with BrdU and the other member with both strands labeled with BrdU. Using a differential stain, chromatids with the analog in both strands stain differently than chromatids with BrdU in only one strand. As a result, SCEs are readily detectable if they occur. These sister chromatids are sometimes referred to as **harlequin chromosomes** because of their patchlike appearance.

多种现代染色技术促进了 SCEs 的鉴定和研究。在其中一种技术中，细胞在胸苷类似物**溴脱氧尿苷**（**BrdU**）存在下，会复制两代。经过两轮 DNA 复制，每对姐妹染色单体中一个成员 DNA 的一条链被 BrdU 标记，而另一个成员 DNA 的两条链则均被 BrdU 标记。使用差异染色技术，DNA 双链均含类似物的染色单体与 DNA 中仅一条链含类似物的染色单体的染色情形会不同。因此，如果发生姐妹染色单体交换，则很容易被检测到。这些姐妹染色单体由于它们的拼接状外观有时也被称为**花斑染色体**。

The significance of SCEs is still uncertain, but several observations have generated great interest in this phenomenon. We know, for example, that agents that induce chromosome damage (viruses, X rays, ultraviolet light, and certain chemical mutagens) increase the frequency of SCEs. The frequency of SCEs is also elevated in **Bloom syndrome**, a human disorder caused by a mutation in the *BLM* gene on chromosome 15. This rare, recessively inherited disease is characterized by prenatal and postnatal delays in growth, a great sensitivity of the facial skin to the sun, immune deficiency, a predisposition to malignant and benign tumors, and abnormal behavior patterns. The chromosomes from cultured leukocytes, bone marrow cells, and fibroblasts

SCEs 的生物学意义尚不明确，但一些观察结果已经引发人们对该现象的极大兴趣，例如，我们已知的诱发染色体损伤的因子（病毒、X 射线、紫外线和某些化学诱变剂）会增加 SCEs 的发生频率。在**布卢姆综合征**（一种由位于 15 号染色体上的 *BLM* 基因突变引起的人类疾病）中，SCEs 的发生频率会升高。这种罕见的隐性遗传疾病的特征是个体在产前和产后生长迟缓、面部皮肤对阳光非常敏感、免疫缺陷、具恶性和良性肿瘤易感性以及存在异常的行为模式。与纯合或杂合的正常个体细胞相比，源自人工培养后的纯合个体的白细胞、骨髓细胞和成纤维细胞染色体非常脆弱且不稳定。除存在大量的姐妹染色单体交换外，非同源染色体间的断裂和重排情况也有所增加。詹姆斯 · 杰曼及其同事的研究工作

derived from homozygotes are very fragile and unstable compared to those of homozygous and heterozygous normal individuals. Increased breaks and rearrangements between nonhomologous chromosomes are observed in addition to excessive amounts of sister chromatid exchanges. Work by James German and colleagues suggests that the *BLM* gene encodes an enzyme called **DNA helicase**, which is best known for its role in DNA replication.

表明 *BLM* 基因编码的 **DNA 解旋酶**在 DNA 复制中具有十分重要的作用。

ESSENTIAL POINT

Recombination events between sister chromatids in mitosis, referred to as sister chromatid exchanges (SCEs), occur at an elevated frequency in the human disorder, Bloom syndrome.

基本要点

有丝分裂中姐妹染色单体间的重组被称为姐妹染色单体交换（SCEs），它在人类疾病布卢姆综合征中发生频率较高。

8 Genetic Analysis and Mapping in Bacteria and Bacteriophages

8 细菌和噬菌体的遗传分析与染色体作图

CHAPTER CONCEPTS

■ Bacterial genomes are most often contained in a single circular chromosome.

■ Bacteria have developed numerous ways in which they can exchange and recombine genetic information between individual cells, including conjugation, transformation, and transduction.

■ The ability to undergo conjugation and to transfer a portion or all of the bacterial chromosome from one cell to another is governed by the presence of genetic information contained in the DNA of a "fertility," or F factor.

■ The F factor can exist autonomously in the bacterial cytoplasm as a plasmid, or it can integrate into the bacterial chromosome, where it facilitates the transfer of the host chromosome to the recipient cell, leading to genetic recombination.

■ Genetic recombination during conjugation provides a means of mapping bacterial genes.

■ Bacteriophages are viruses that have bacteria as their hosts. During infection of the bacterial host, bacteriophage DNA is injected into the host cell, where it is replicated and

本章速览

■ 细菌基因组在大多情况下存在于一条环形染色体上。

■ 细菌采用多种方式进行细胞间遗传信息的交换和重组，这些方式包括接合、转化和转导。

■ 细菌的部分或全部染色体通过接合的方式从一个细胞向另一个细胞进行转移的能力由位于“致育因子”或 F 因子上的遗传信息决定。

■ F 因子作为质粒既可独立自主地存在于细菌细胞质中，也可整合到细菌染色体上，从而促进宿主染色体向受体细胞转移并引发遗传重组。

■ 接合过程中产生的遗传重组为细菌基因作图提供了方法。

■ 噬菌体是以细菌为宿主的病毒。当噬菌体侵染细菌宿主时，噬菌体 DNA 会进入宿主细胞内，并在宿主细胞内进行复制，指导噬菌体增殖。

directs the reproduction of the bacteriophage.

■ Rarely, following infection, bacteriophage DNA integrates into the host chromosome, becoming a prophage, where it is replicated along with the bacterial DNA.

■ 在极少数情况下，噬菌体在侵染细菌宿主后，其DNA会整合入宿主染色体中变为原噬菌体，并且随着细菌DNA的复制而复制。

In this chapter, we shift from consideration of mapping genetic information in eukaryotes to discussion of the analysis and mapping of genes in **bacteria** (prokaryotes) and **bacteriophages**, viruses that use bacteria as their hosts. The study of bacteria and bacteriophages has been essential to the accumulation of knowledge in many areas of genetic study. For example, much of what we know about molecular genetics, recombinational phenomena, and gene structure was initially derived from experimental work with them. Furthermore, our extensive knowledge of bacteria and their resident plasmids has led to their widespread use in DNA cloning and other recombinant DNA studies.

在本章，我们将把目光从真核生物遗传信息的作图转向原核生物**细菌**和以细菌为宿主的**噬菌体**的基因分析和作图上。细菌和噬菌体的研究对于遗传学许多领域的知识积累都至关重要。例如，我们对分子遗传学、重组现象和基因结构的许多了解最初都是从利用它们作为研究对象的实验工作中获得的。此外，我们对细菌及其质粒的深入了解已使得它们在DNA克隆和其他重组DNA研究中获得了广泛应用。

Bacteria and their viruses are especially useful research organisms in genetics for several reasons. They have extremely short reproductive cycles—literally hundreds of generations, giving rise to billions of genetically identical bacteria or phages, can be produced in short periods of time. Furthermore, they can be studied in pure cultures. That is, a single species or mutant strain of bacteria or one type of virus can be isolated and investigated independently of other similar organisms.

出于多种原因，细菌及病毒成为遗传学研究中特别有用的物种。它们的繁殖周期极短，可以在短时间内繁殖数百个世代，产生数以亿计遗传上完全相同的细菌或噬菌体。此外，人们可以获得这些物种的纯培养物进行研究，即可以分离获得单一物种或单一突变体的细菌菌株或单一类型的病毒株，从而独立于其他类似生物体进行单独研究。

In this chapter, we focus on genetic recombination and chromosome mapping. Complex processes have evolved in bacteria

在本章中，我们将重点介绍遗传重组和染色体作图。细菌和噬菌体经过进化已经形成许多促进群体内单细胞间遗传信息

and bacteriophages that facilitate the transfer of genetic information between individual cells within populations. As we shall see, these processes are the basis for the chromosome mapping analysis that forms the cornerstone of molecular genetic investigations of bacteria and the viruses that invade them.

转移的复杂过程。正如我们将要看到的那样，这些过程是染色体作图分析的基础，同时也成为对细菌及其入侵者病毒开展分子遗传学研究的奠基石。

8.1 Bacteria Mutate Spontaneously and Are Easily Cultured

8.1 细菌能够自发突变并易于培养

It has long been known that pure cultures of bacteria give rise to cells that exhibit heritable variation, particularly with respect to growth under unique environmental conditions. Mutant cells that arise spontaneously in otherwise pure cultures can be isolated and established independently from the parent strain by using established selection techniques. As a result, mutations for almost any desired characteristic can now be isolated. Because bacteria and viruses usually contain only a single chromosome and are therefore haploid, all mutations are expressed directly in the descendants of mutant cells, adding to the ease with which these microorganisms can be studied.

人们很早就知道，细菌纯培养物可以产生具有可遗传变异的细胞，尤其在独特环境条件下进行培养时。纯培养物中自发产生的突变细胞可以使用已有的选择技术从其亲本菌株中分离获得，并进行独立培养和鉴定。因此，如今几乎所有预期性状的突变体均可分离得到。由于细菌和病毒通常仅含一条染色体，即为单倍体，因此所有突变均可在突变细胞后代中直接表达，这使得这些微生物的研究变得更为简便。

Bacteria are grown in a liquid culture medium or in a petri dish on a semisolid agar surface. If the nutrient components of the growth medium are simple and consist only of an organic carbon source (such as glucose or lactose) and a variety of ions, including Na^+, K^+, Mg^{2+}, Ca^{2+}, and NH_4^+, present as inorganic salts, it is called **minimal medium**. To grow on such a medium, a bacterium must be able to

细菌可以在液体培养基中或在培养皿的半固体琼脂表面上进行生长。如果培养基中营养组成简单，仅由有机碳源（例如葡萄糖或乳糖）和包括 Na^+、K^+、Mg^{2+}、Ca^{2+} 与 NH_4^+ 在内的以无机盐形式存在的各种离子组成，则这种培养基被称为**基本培养基**。为了在基本培养基上生长，细菌必须能够合成所有必需的有机化合物，例如氨基酸、嘌呤、嘧啶、维生素和脂肪酸。

synthesize all essential organic compounds (e.g., amino acids, purines, pyrimidines, vitamins, and fatty acids). A bacterium that can accomplish this remarkable biosynthetic feat—one that we ourselves cannot duplicate—is a **prototroph**. It is said to be wild-type for all growth requirements. On the other hand, if a bacterium loses the ability to synthesize one or more organic components through mutation, it is an **auxotroph**. For example, if a bacterium loses the ability to make histidine, then this amino acid must be added as a supplement to the minimal medium for growth to occur. The resulting bacterium is designated as an his^- auxotroph, in contrast to its prototrophic his^+ counterpart.

能够完成这一非凡的、人类不可能完成的生物合成壮举的细菌是**原养型**细菌，它们对于所有的生长需求均为野生型。此外，如果细菌由于突变失去合成其中一种或多种有机组分的能力，即为**营养缺陷型**细菌。例如，如果细菌无法合成组氨酸，那么该氨基酸则必须作为添加剂补充至基本培养基中，从而有助于该细菌的生长。该细菌表示为 his^- 缺陷型菌株，与之相对应的原养型细菌则表示为 his^+。

To study bacterial growth quantitatively, an inoculum of bacteria is placed in liquid culture medium. Cells grown in liquid medium can be quantified by transferring them to a semisolid medium in a petri dish. Following incubation and many divisions, each cell gives rise to a visible colony on the surface of the medium. If the number of colonies is too great to count, then a series of successive dilutions (a technique called **serial dilution**) of the original liquid culture is made and plated, until the colony number is reduced to the point where it can be counted (Figure 8.1). This technique allows the number of bacteria present in the original culture to be calculated.

为了定量研究细菌生长，人们将细菌接种物置于液体培养基中培养。液体培养基中生长的细胞可通过将其转移至培养皿中的半固体培养基上来进行定量分析。经过培养和多次细胞分裂，每个细胞在培养基表面都会形成肉眼可见的菌落。如果菌落数量太多无法进行计数，则可通过对原始液体培养物做一系列的次第稀释（这一技术被称为**连续稀释**），然后涂布平板，直至菌落数量减少至可以计数为止(图8.1)。利用这种技术即可计算出原始培养物中所存在的细菌数量。

As an example, let's assume that the three dishes in Figure 8.1 represent serial dilutions of 10^{-3}, 10^{-4}, and 10^{-5} (from left to right). We need only select the dish in which the number of colonies can be counted accurately. Assuming

例如，假设图 8.1 中的三个培养皿分别代表 10^{-3}、10^{-4} 和 10^{-5}（从左到右）系列稀释培养物。我们只需选择菌落数可以准确计数的培养皿即可。假设所使用的样品为 1 mL，由于每个菌落来自一个细菌细胞，

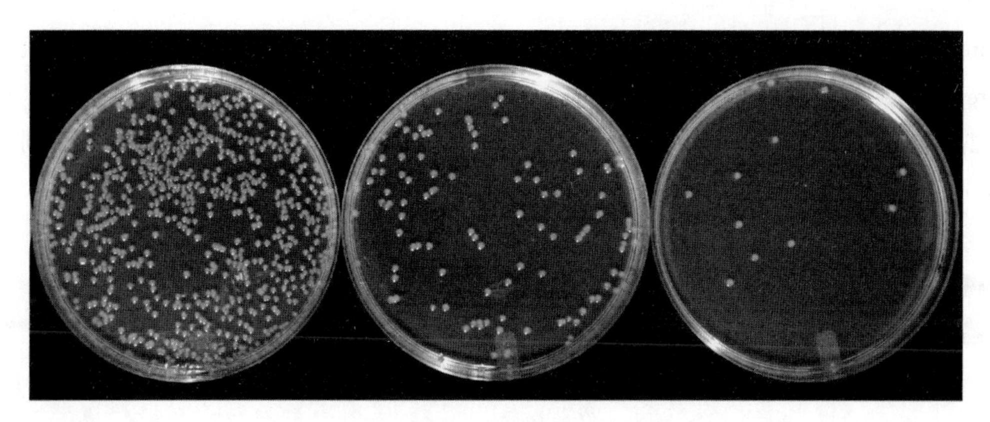

FIGURE 8.1 Results of the serial dilution technique and subsequent culture of bacteria. Each dilution varies by a factor of 10. Each colony is derived from a single bacterial cell.

图 8.1 系列连续稀释及随后的细菌培养。每个不同稀释度的因子都是 10。每个克隆都源自单一的细菌细胞。

that a 1 mL sample was used, and because each colony arose from a single bacterium, the number of colonies multiplied by the dilution factor represents the number of bacteria in each milliliter of the initial inoculum used to start the serial dilutions. In Figure 8.1, the rightmost dish has 12 colonies.The dilution factor for a 10^{-5} dilution is 10^5. Therefore, the initial number of bacteria was 12×10^5 per mL.

因此菌落数乘以稀释倍数就是每毫升用于连续稀释的初始接种物中的细菌细胞数量。在图 8.1 中，最右边的培养皿中有 12 个菌落，其稀释度为 10^{-5}，即稀释倍数为 10^5。因此，细菌的初始数目为 12×10^5 个 /mL。

8.2 Genetic Recombination Occurs in Bacteria

8.2 细菌的基因重组

Development of techniques that allowed the identification and study of bacterial mutations led to detailed investigations of the transfer of genetic information between individual organisms. As we shall see, as with meiotic crossing over in eukaryotes, the process of genetic recombination in bacteria provided the basis for the development of chromosome mapping methodology. It is important to note at the outset of our discussion that the term genetic

用于细菌突变鉴定和研究的相关技术的发展使得人们能够对个体细胞间遗传信息的转移开展详细的研究。正如我们将要看到的，与真核生物减数分裂过程中的交换一样，细菌的遗传重组过程也为染色体作图方法的发展奠定了基础。在开始讨论前，需要提醒的是：术语“遗传重组”应用于细菌时指一个细胞染色体上存在的一个或多个基因被遗传上不同的另一个细胞染色体上的相应基因所替代。尽管这与我

recombination, as applied to bacteria, refers to the replacement of one or more genes present in the chromosome of one cell with those from the chromosome of a genetically distinct cell. While this is somewhat different from our use of the term in eukaryotes—where it describes crossing over resulting in a reciprocal exchange—the overall effect is the same: Genetic information is transferred, and it results in an altered genotype.

们在真核生物中所使用的“遗传重组”有所不同——在真核生物中指染色体片段间发生相互交换——但总体效果是相同的：遗传信息被转移，并产生基因型改变。

We will discuss three processes that result in the transfer of genetic information from one bacterium to another: *conjugation, transformation, and transduction*. Collectively, knowledge of these processes has helped us understand the origin of genetic variation between members of the same bacterial species, and in some cases, between members of different species. When transfer of genetic information occurs between generations of the same species, the term **vertical gene transfer** applies. When transfer occurs between unrelated cells, the term horizontal gene transfer is used. The **horizontal gene transfer** process has played a significant role in the evolution of bacteria. Often, the genes discovered to be involved in horizontal transfer are those that also confer survival advantages to the recipient species. For example, one species may transfer antibiotic resistance genes to another species. Or genes conferring enhanced pathogenicity may be transferred. Thus, the potential for such transfer is a major concern in the medical community. In addition, horizontal gene transfer has been a major factor in the process of speciation in bacteria. Many, if not most, bacterial species have been the recipient of

我们将讨论导致遗传信息从一个细菌细胞转移到另一个细菌细胞的三种不同过程：接合、转化和转导。总的来说，对于这些过程的了解有助于我们理解同一细菌物种不同个体间，甚至不同物种的个体间所存在的遗传变异的起源。当遗传信息在同一物种的世代交替间发生转移时，称为**垂直基因转移**。当遗传信息在无亲缘关系的细胞间发生转移时，称为**水平基因转移**。水平基因转移过程在细菌进化中扮演着十分重要的作用。通常来讲，发现的水平基因转移相关基因也是那些同时赋予受体物种生存优势的基因。例如，一个物种可以将抗生素抗性基因转移至另一物种中，或者被转移的基因可以增强致病性。因此，这种基因转移在医学界是潜在的重要关注点。此外，水平基因转移也是细菌物种形成的一个主要因素。许多细菌物种，即使不是绝大多数，均可成为其他物种基因的受体。

genes from other species.

Conjugation in Bacteria: The Discovery of F^+ and F^- Strains

Studies of bacterial recombination began in 1946, when Joshua Lederberg and Edward Tatum showed that bacteria undergo **conjugation**, a process by which genetic information from one bacterium is transferred to and recombined with that of another bacterium. Their initial experiments were performed with two multiple auxotrophs (nutritional mutants) of *E. coli* strain K12. As shown in Figure 8.2, strain A required methionine (met) and biotin (bio) in order to grow, whereas strain B required threonine (thr), leucine (leu), and thiamine (thi). Neither strain would grow on minimal medium. The two strains were first grown separately in supplemented media, and then cells from both were mixed and grown together for several more generations. They were then plated on minimal medium. Any cells that grew on minimal medium were prototrophs. It is highly improbable that any of the cells containing two or three mutant genes would undergo spontaneous mutation simultaneously at two or three independent locations to become wild-type cells. Therefore, the researchers assumed that any prototrophs recovered must have arisen as a result of some form of genetic exchange and recombination between the two mutant strains.

In this experiment, prototrophs were recovered at a rate of $1/10^7$ (or 10^{-7}) cells plated. The controls for this experiment involved separate plating of cells from strains A and B

细菌间的接合：F^+ 和 F^- 菌株的发现

细菌重组研究始于 1946 年，乔舒亚 · 莱德伯格和爱德华 · 塔特姆指出细菌间可以发生**接合**。在这一过程中，一个细菌细胞中的遗传信息会转移至另一个细菌细胞中并发生遗传重组。他们最初的实验使用了两种大肠杆菌 K12 菌株的多基因营养缺陷型突变体。如图 8.2 所示，菌株 A 的生长需要甲硫氨酸（met）和生物素（bio），而菌株 B 则需要苏氨酸（thr）、亮氨酸（leu）和硫胺素（thi），因此二者均不能在基本培养基上生长。在实验中，他们首先将这两种菌株在补充培养基中分别培养，然后将两种细胞混合继续培养数代，最后将培养的菌液涂布在基本培养基上。能够在基本培养基上生长的任何细胞均为原养型细胞。由于任何含两个或三个突变基因的细胞在这些基因座上几乎不可能同时发生自发突变成为野生型细胞，因此研究人员认为所获得的任何原养型细胞一定是由两个突变细菌菌株间发生了某种形式的遗传交换和重组导致的。

在该实验中，培养细胞中原养型细胞的出现概率是 $1/10^7$（或 10^{-7}）。该实验的对照组包括在基本培养基上分别培养的来自菌株 A 和 B 的细胞，均没有发现原养型

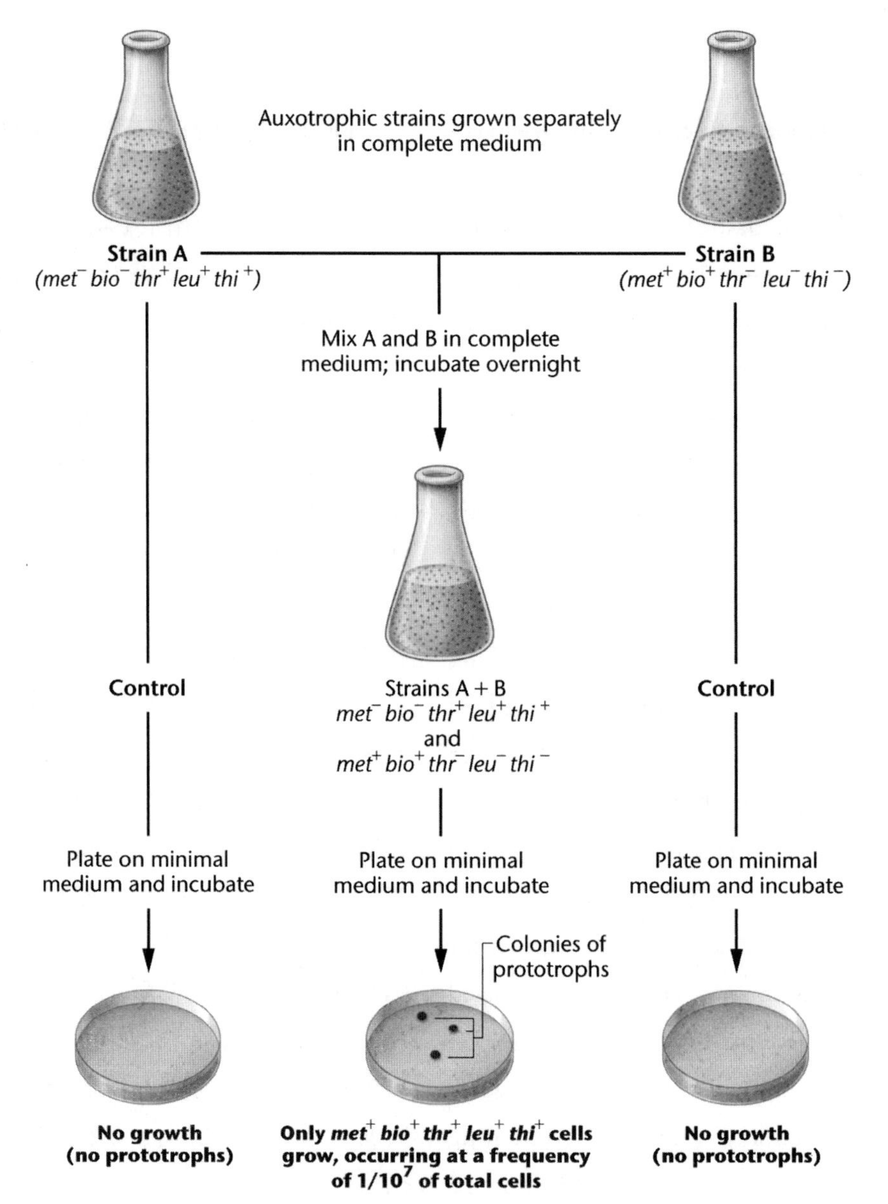

FIGURE 8.2 Genetic recombination of two auxotrophic strains producing prototrophs. Neither auxotroph grows on minimal medium, but prototrophs do, suggesting that genetic recombination has occurred.

图 8.2 两种缺陷型菌株产生原养型菌株的遗传重组。任何一种缺陷型菌株均无法在基本培养基上生长，而原养型菌株可以，表明发生了遗传重组。

on minimal medium. No prototrophs were recovered. Based on these observations, Lederberg and Tatum proposed that genetic exchange had occurred. Lederberg and Tatum's findings were soon followed by numerous experiments that elucidated the genetic basis

细胞。基于这些观察，莱德伯格和塔特姆猜测发生了遗传交换。不久之后，大量实验继莱德伯格和塔特姆的发现阐明了接合的遗传学基础。很快人们便清楚地发现不同的细菌菌株参与了遗传物质的单向转移。当细胞作为部分染色体的供体时，它们被

of conjugation. It quickly became evident that different strains of bacteria are involved in a unidirectional transfer of genetic material. When cells serve as donors of parts of their chromosomes, they are designated as **F^+ cells** (F for "fertility"). Recipient bacteria receive the donor chromosome material (now known to be DNA), and recombine it with part of their own chromosome. They are designated as **F^- cells**.

称为 **F^+ 细胞**（F 表示育性）；受体细菌则接收供体细胞的染色体（即 DNA），并将其与自身的部分染色体进行重组，它们被称为 **F^- 细胞**。

Experimentation subsequently established that cell contact is essential for chromosome transfer to occur. Support for this concept was provided by Bernard Davis, who designed the Davis U-tube for growing F^+ and F^- cells shown in Figure 8.3. At the base of the tube is a sintered glass filter with a pore size that allows passage of the liquid medium but that is too small to allow the passage of bacteria. The F^+ cells are placed on one side of the filter, and F^- cells on

随后的实验证明细胞接触对于染色体转移的发生是必需的。伯纳德·戴维斯为这一理论提供了证据支持，他设计了戴维斯 U 形管用于培养 F^+ 和 F^- 细胞，如图 8.3 所示。在 U 形管的底部是玻璃过滤板，其孔径尺寸可以允许液体培养基通过，但无法允许细菌细胞通过。F^+ 细胞位于过滤板的一侧，F^- 细胞则位于另一侧。培养基在过滤板的两侧来回流动，因此两侧细胞在生长过程中分享着同样的培养基。当戴维

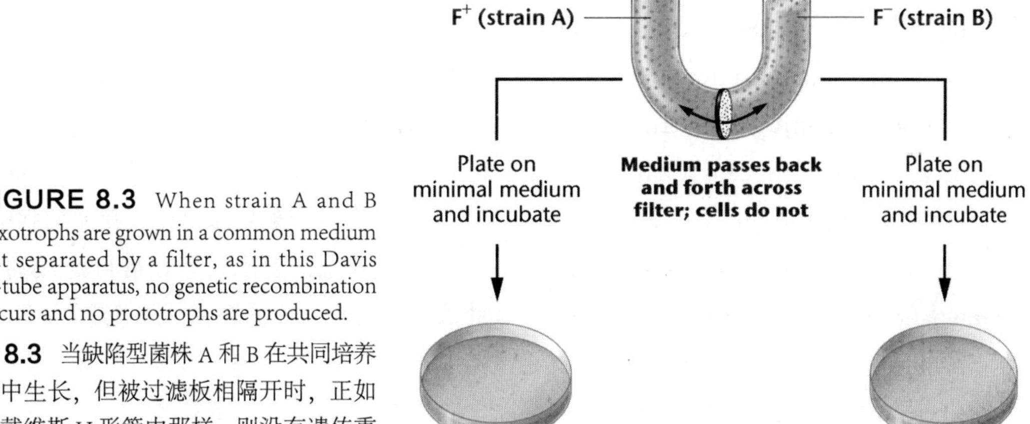

FIGURE 8.3 When strain A and B auxotrophs are grown in a common medium but separated by a filter, as in this Davis U-tube apparatus, no genetic recombination occurs and no prototrophs are produced.

图 8.3 当缺陷型菌株 A 和 B 在共同培养基中生长，但被过滤板相隔开时，正如在戴维斯 U 形管中那样，则没有遗传重组发生，也不会产生原养型细菌。

the other side. The medium is moved back and forth across the filter so that the cells share a common medium during bacterial incubation. When Davis plated samples from both sides of the tube on minimal medium, no prototrophs were found, and he logically concluded that *physical contact between cells of the two strains is essential to genetic recombination*. We now know that this physical interaction is the initial step in the process of conjugation established by a structure called the **F pilus** (or **sex pilus**; pl. pili). Bacteria often have many pili, which are tubular extensions of the cell. After contact is initiated between mating pairs, chromosome transfer is then possible.

斯分别从试管的两侧取样并在基本培养基上进行培养时，没有发现原养型细胞的产生，因此他通过逻辑推断得出结论：两种菌株细胞间的物理接触对于遗传重组是必需的。现在我们知道这种物理相互作用是接合发生的第一步，由称为**F菌毛**（或**性菌毛**）的结构实现。细菌通常含有许多菌毛，它们是位于细胞表面的管状延伸。在进行接合的两个细胞个体间开始接触后，染色体转移成为可能。

Later evidence established that F^+ cells contain a **fertility factor** (**F factor**) that confers the ability to donate part of their chromosome during conjugation. Experiments by Joshua and Esther Lederberg and by William Hayes and Luca Cavalli-Sforza showed that certain conditions eliminate the F factor in otherwise fertile cells. However, if these "infertile" cells are then grown with fertile donor cells, the F factor is regained.

随后的证据表明，F^+细胞含**致育因子**（**F因子**），从而赋予细胞在接合过程中提供自身部分染色体的能力。乔舒亚和埃丝特·莱德伯格以及威廉·海斯和卢卡·卡瓦利-斯福尔扎的实验表明，某些条件可造成原本有育性的细胞丢失F因子。但是，如果这些"不育的"细胞与有育性的供体细胞共培养，则可以重新获得F因子。

The conclusion that the F factor is a mobile element is further supported by the observation that, following conjugation and genetic recombination, recipient cells always become F^+. Thus, in addition to the *rare* cases of gene transfer from the bacterial chromosome (genetic recombination), the F factor itself is passed to *all* recipient cells. On this basis, the initial cross of Lederberg and Tatum (see Figure 8.2) can be interpreted as follows:

在接合和遗传重组后，受体细胞总会变为F^+，这一观察结果进一步支持了F因子是可移动元件的结论。因此，除罕见情况下细菌染色体的基因转移产生遗传重组外，F因子本身会被传递至所有受体细胞中。基于这一基础，莱德伯格和塔特姆最初设计的杂交实验（见图8.2）可以解释如下：

Strain A		**Strain B**
F^+	×	F^-
Donor		Recipient

Characterization of the F factor confirmed these conclusions. Like the bacterial chromosome, though distinct from it, the F factor has been shown to consist of a circular, double-stranded DNA molecule, equivalent to about 2 percent of the bacterial chromosome (about 100,000 nucleotide pairs). There are 19 genes contained within the F factor whose products are involved in the transfer of genetic information, excluding those involved in the formation of the sex pilus.

F 因子的特征证实了这些结论。与细菌染色体一样，尽管与之不同，F 因子是环状双链 DNA 分子，大小约为细菌染色体的 2%（约 100 000 个核苷酸对）。F 因子中含 19 个基因，其产物除了与性菌毛形成有关外，均与遗传信息转移有关。

Geneticists believe that transfer of the F factor during conjugation involves separation of the two strands of its double helix and movement of one of the two strands into the recipient cell. Both strands, one moving across the conjugation tube and one remaining in the donor cell, are replicated. The result is that both the donor *and* the recipient cells become F^+. This process is diagrammed in Figure 8.4.

遗传学家认为，接合过程中 F 因子的转移涉及双螺旋两条链的分离和双链中的一条进入受体细胞。这两条链中一条通过接合管，另一条继续保留在供体细胞中，并且两条链均会被复制。结果是供体细胞和受体细胞都变为 F^+。该过程如图 8.4 所示。

To summarize, an *E. coli* cell may or may not contain the F factor. When this factor is present, the cell is able to form a sex pilus and potentially serve as a donor of genetic information. During conjugation, a copy of the F factor is almost always transferred from the F^+ cell to the F^- recipient, converting the recipient to the F^+ state. The question remained as to exactly why such a low proportion of cells involved in these matings (10^{-7}) also results in genetic recombination. The answer awaited further experimentation.

总而言之，大肠杆菌细胞可能含 F 因子，也可能不含 F 因子。当存在该因子时，该细胞能够形成性菌毛，并可成为遗传信息的供体。在接合过程中，F 因子的拷贝几乎总是从 F^+ 细胞转移至 F^- 受体，从而将受体细胞变为 F^+ 细胞。问题在于，参与接合过程的细胞中发生遗传重组的概率为何如此之低（10^{-7}）。答案尚待进一步实验研究的揭晓。

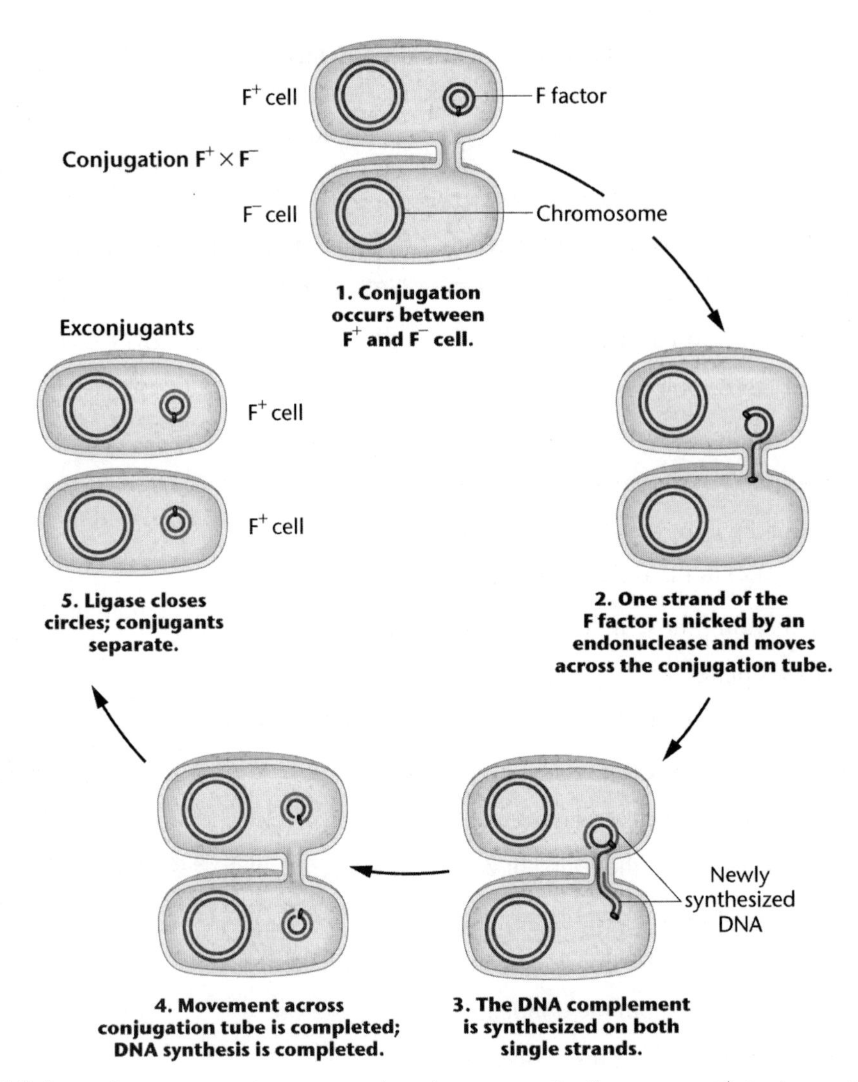

FIGURE 8.4 An $F^+ \times F^-$ mating demonstrating how the recipient F^- cell converts to F^+. During conjugation, one strand of the F factor is transferred to the recipient cell, converting it to F^+. Single strands in both donor and recipient cells are replicated. Newly replicated DNA is depicted by a lighter shade of gray as the F factor is transferred.

图 8.4 本图显示了 $F^+ \times F^-$ 接合过程中受体 F^- 细胞如何变为 F^+ 细胞。在接合过程中，F 因子的一条 DNA 链转移至受体细胞中，将其变为 F^+ 细胞。供体细胞和受体细胞的单链均被复制。在 F 因子转移过程中新复制的 DNA 链用浅灰色表示。

As you soon shall see, the F factor is in reality an autonomous genetic unit called a *plasmid*. However, in our historical coverage of its discovery, we will continue to refer to it as a factor.

下边你将看到，F 因子实际上是一种独立自治的遗传单元，被称为质粒。然而，在对质粒发现的历史叙述中，我们将继续称其为因子。

ESSENTIAL POINT

Conjugation may be initiated by a

基本要点

接合由细胞质中存在 F 因子这一质粒

bacterium housing a plasmid called the F factor in its cytoplasm, making it a donor (F^+) cell. Following conjugation, the recipient (F^-) cell receives a copy of the F factor and is converted to the F^+ status.

从而成为供体（F^+）细胞的细菌引发。发生接合后，受体细胞（F^-）将获得F因子的一个拷贝，并成为 F^+ 细胞。

Hfr Bacteria and Chromosome Mapping

Subsequent discoveries not only clarified how genetic recombination occurs but also defined a mechanism by which the *E. coli* chromosome could be mapped. Let's address chromosome mapping first.

In 1950, Cavalli-Sforza treated an F^+ strain of *E. coli* K12 with nitrogen mustard, a chemical known to induce mutations. From these treated cells, he recovered a genetically altered strain of donor bacteria that underwent recombination at a rate of $1/10^4$ (or 10^{-4}), 1000 times more frequently than the original F^+ strains. In 1953, Hayes isolated another strain that demonstrated an elevated frequency. Both strains were designated **Hfr**, for **high-frequency recombination**. Because Hfr cells behave as donors, they are a special class of F^+ cells.

Another important difference was noted between Hfr strains and the original F^+ strains. If the donor is from an Hfr strain, recipient cells, though sometimes displaying genetic recombination, almost never become Hfr; that is, they remain F^-. In comparison, then,

$F^+ \times F^- \rightarrow F^+$ (low rate of recombination)

$Hfr \times F^- \rightarrow F^-$ (higher rate of recombination)

Perhaps the most significant characteristic of Hfr strains is the *nature of recombination*. In any given strain, certain genes are more

Hfr 菌株与染色体作图

随后的发现不仅阐明了遗传重组是如何发生的，而且还明确了可用于大肠杆菌染色体作图的机制。我们首先来谈染色体作图。

1950年，卡瓦利-斯福尔扎使用可诱发突变的化学试剂氮芥处理大肠杆菌K12的 F^+ 菌株。在这些被处理的细胞中，他发现了遗传改造后重组频率可达 $1/10^4$（或 10^{-4}）的供体细菌菌株，该重组频率是原始 F^+ 菌株的1000倍。1953年，海斯分离得到另一株高重组频率的菌株。这两种菌株均被命名为 **Hfr**，意思是**高频重组**。由于Hfr细胞是供体细胞，因此它们是 F^+ 细胞的一种特殊类别。

在Hfr菌株和原始 F^+ 菌株之间的另一个重要区别是：如果供体细胞是Hfr菌株，那么受体细胞尽管有时表现出遗传重组，但几乎永远不会变成Hfr。也就是说，它们依然是 F^- 菌株。两相比较，可见：

$F^+ \times F^- \rightarrow F^+$（低重组频率）

$Hfr \times F^- \rightarrow F^-$（高重组频率）

Hfr菌株最显著的特征可能就是其重组特性。在任何指定菌株中，某些基因比其他基因的重组频率更高，而有些基因则根

frequently recombined than others, and some not at all. This *nonrandom* pattern was shown to vary between Hfr strains. Although these results were puzzling, Hayes interpreted them to mean that some physiological alteration of the F factor had occurred, resulting in the production of Hfr strains of *E. coli*.

本不发生重组。这种非随机模式表明 Hfr 菌株间存在不同。尽管这些结果令人费解，但海斯指出这意味着 F 因子发生了某些生理变化，从而导致大肠杆菌 Hfr 菌株的产生。

In the mid-1950s, experimentation by Elie Wollman and François Jacob elucidated the difference between Hfr and F^+ strains and showed how Hfr strains allow genetic mapping of the *E. coli* chromosome. In their experiments, Hfr and F^- strains with suitable marker genes were mixed, and recombination of specific genes was assayed at different times. To accomplish this, a culture containing a mixture of an Hfr and an F^- strain was first incubated, and samples were removed at various intervals and placed in a blender. The shear forces in the blender separated conjugating bacteria so that the transfer of the chromosome was terminated. The cells were then assayed for genetic recombination.

20 世纪 50 年代中期，埃利 · 沃尔曼和弗朗索瓦 · 雅各布通过实验阐明了 Hfr 和 F^+ 菌株之间的差异，并显示了如何使用 Hfr 菌株进行大肠杆菌染色体的遗传作图。在他们的实验中，将具有合适标记基因的 Hfr 和 F^- 菌株混合培养，并在不同时间检测特定基因的重组情况。为此，首先要培养含 Hfr 和 F^- 菌株的混合培养物；然后在不同的时间间隔取样并置于搅拌器中，利用搅拌器的剪切力把进行接合的细菌分开，从而终止染色体转移；最后分析细胞的遗传重组。

This process, called the **interrupted mating technique**, demonstrated that specific genes of a given Hfr strain were transferred and recombined sooner than others. The graph in Figure 8.5 illustrates this point. During the first 8 minutes after the two strains were mixed, no genetic recombination was detected. At about 10 minutes, recombination of the azi^R gene was detected, but no transfer of the ton^s, lac^+, or gal^+ genes was noted. By 15 minutes, 50 percent of the recombinants were azi^R, and 15 percent were ton^s; but none were lac^+ or gal^+. Within 20 minutes, the lac^+ was found among the

这一过程被称为**中断杂交技术**，可以表明指定的 Hfr 菌株中特定基因比其他基因转移和重组得更早。图 8.5 对此进行了展示。在两种菌株混合培养的前 8 min 内，未检测到遗传重组。在大约 10 min 时，检测到 azi^R 基因的重组，但未发现 ton^s、lac^+ 或 gal^+ 基因的转移；到 15 min 时，50%的重组子是 azi^R，15%是 ton^s，但没有发现 lac^+ 或 gal^+ 重组子；在 20 min 内，重组子中发现了 lac^+；25 min 内，发现 gal^+ 也发生了转移。沃尔曼和雅各布证明了基因的有序转移与接合时间的长短相关。

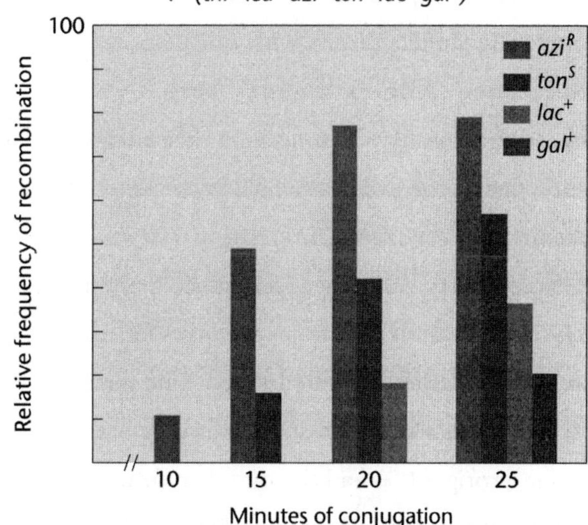

FIGURE 8.5 The progressive transfer during conjugation of various genes from a specific Hfr strain of *E. coli* to an F^- strain. Certain genes (*azi* and *ton*) transfer more quickly than others and recombine more frequently. Others (*lac* and *gal*) take longer to transfer, and recombinants are found at a lower frequency.

图 8.5 特定大肠杆菌 Hfr 菌株与 F^- 菌株在接合中不同基因连续转移的过程。某些基因（*azi* 和 *ton*）比其他基因转移更快，重组频率更高。另一些基因 (*lac* 和 *gal*) 则需要更长时间才能转移，并且重组频率相对较低。

recombinants; and within 25 minutes, gal^+ was also being transferred. Wollman and Jacob had demonstrated *an ordered transfer of genes* that correlated with the length of time conjugation proceeded.

It appeared that the chromosome of the Hfr bacterium was transferred linearly and that the gene order and distance between genes, as measured in minutes, could be predicted from such experiments (Figure 8.6). This process, sometimes referred to as **time mapping**, served as the basis for the first genetic map of the *E. coli* chromosome. Minutes in bacterial mapping are similar to map units in eukaryotes.

实验显示 Hfr 细菌的染色体会发生线性转移，并且基因顺序和基因间距离可通过该实验以分钟为单位进行预测（图 8.6）。这一过程有时也被称为**时间作图**，它是绘制第一张大肠杆菌染色体遗传图的基础。细菌作图中所用的单位“分钟”与真核生物遗传作图中的单位“图距”类似。

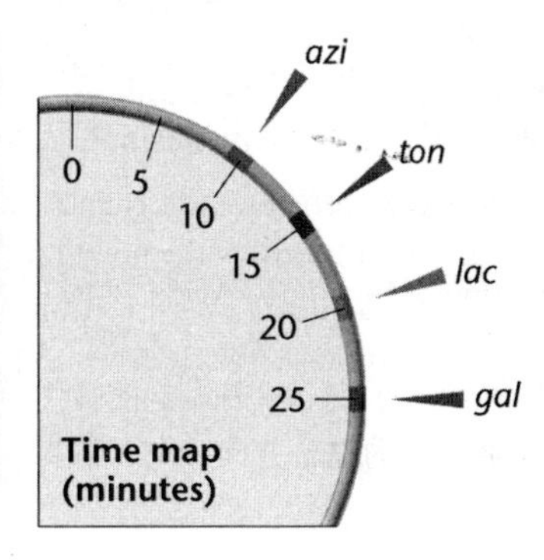

FIGURE 8.6 A time map of the genes studied in the experiment depicted in Figure 8.5.

图 8.6 图 8.5 所示实验中所研究基因以时间为单位的染色体遗传图。

Wollman and Jacob then repeated the same type of experiment with other Hfr strains, obtaining similar results with one important difference. Although genes were always transferred linearly with time, as in their original experiment, the order in which genes entered seemed to vary from Hfr strain to Hfr strain [Figure 8.7(a)]. When they reexamined the entry rate of genes, and thus the genetic maps for each strain, a definite pattern emerged. The major difference between each strain was simply the point of origin (*O*) and the direction in which entry proceeded from that point [Figure 8.7(b)].

随后，沃尔曼和雅各布使用其他 Hfr 菌株重复了该实验，获得了相似的实验结果，但存在一个重要区别。尽管与最初的实验一样，基因总是随时间线性转移，但是基因转移的顺序因 Hfr 菌株的不同而不同 [图 8.7（a）]。当他们重新检查基因转移的速度及每种菌株的遗传图时，发现了明确的基因转移模式。菌株之间的主要区别仅在于起点（*O*）的位置和从起点开始进行基因转移的方向 [图 8.7（b）]。

To explain these results, Wollman and Jacob postulated that the *E. coli* chromosome

为了解释这些结果，沃尔曼和雅各布推测大肠杆菌染色体是环形的（一个封闭

(a)

Hfr strain	(earliest)			Order of transfer				(latest)
H	*thr* –	*leu* –	*azi* –	*ton* –	*pro* –	*lac* –	*gal* –	*thi*
1	*leu* –	*thr* –	*thi* –	*gal* –	*lac* –	*pro* –	*ton* –	*azi*
2	*pro* –	*ton* –	*azi* –	*leu* –	*thr* –	*thi* –	*gal* –	*lac*
7	*ton* –	*azi* –	*leu* –	*thr* –	*thi* –	*gal* –	*lac* –	*pro*

(b)

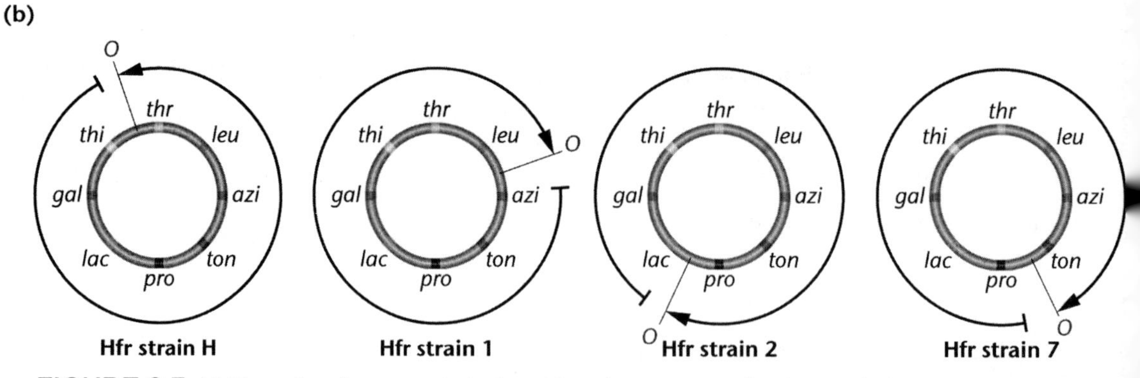

FIGURE 8.7 (a) The order of gene transfer in four Hfr strains, suggesting that the *E. coli* chromosome is circular. (b) The point where transfer originates (*O*) is identified in each strain. The origin is determined by the point of integration into the chromosome of the F factor, and the direction of transfer is determined by the orientation of the F factor as it integrates. The arrowheads indicate the points of initial transfer.

图 8.7 （a）四种 Hfr 菌株的基因转移顺序，表明大肠杆菌染色体是环形的。（b）每种菌株确定的转移起始点（*O*）。起始位点由 F 因子整合入染色体的位点决定，转移方向由 F 因子的整合方向决定。箭头表明了起始转移位点。

is circular (a closed circle, with no free ends). If the point of origin (*O*) varies from strain to strain, a different sequence of genes will be transferred in each case. But what determines *O*? They proposed that *in various Hfr strains, the F factor integrates into the chromosome at different points and that its position determines the site of O*. A case of integration is shown in step 1 of Figure 8.8. During conjugation between an Hfr and an F⁻ cell, the position of the F factor determines the initial point of transfer (steps 2 and 3). Those genes adjacent to *O* are transferred first, and the F factor becomes the last part that can be transferred (step 4). However, conjugation rarely, if ever, lasts long enough to allow the entire chromosome to pass across the conjugation tube (step 5). *This proposal explains why most recipient cells, when mated with Hfr cells, remain F⁻.*

的环形，没有自由末端）。如果复制起点（*O*）因菌株而异，那么不同菌株的基因转移顺序也将不同。但是 *O* 的位置由什么因素决定呢？他们指出，在不同的 Hfr 菌株中，F 因子整合入大肠杆菌染色体的位置不同，而正是该位置决定了 *O* 的所在位点。图 8.8 的步骤 1 展示了 F 因子整合的情况。在 Hfr 和 F⁻ 细胞进行接合的过程中，F 因子的位置决定了基因转移的起始点（步骤 2 和 3）。与 *O* 相邻的基因首先被转移，而 F 因子则成为可被转移的最后部分（步骤 4）。但是，接合即使存在，也几乎很少可以持续足够长的时间以保证整个大肠杆菌染色体都能经过接合管（步骤 5）。该假设解释了为什么大多数受体细胞与 Hfr 细胞接合后仍保持为 F⁻ 细胞。

Figure 8.8 also depicts the way in which the two strands making up a DNA molecule behave during transfer, allowing for the entry of one strand of DNA into the recipient (see step 3). Following replication, the entering DNA now has the potential to recombine with its homologous region of the host chromosome. The DNA strand that remains in the donor also undergoes replication.

图 8.8 还描述了组成 DNA 分子的两条链在转移过程中的行为方式，即允许一条 DNA 链进入受体(请参见步骤 3)。在复制后，进入的 DNA 现在有可能与其宿主染色体的同源区进行重组。保留在供体中的 DNA 链也会进行复制。

Use of the interrupted mating technique with different Hfr strains allowed researchers to map the entire *E. coli* chromosome. Mapped in time units, strain K12 (or *E. coli* K12) was shown to be 100 minutes long. While modern genome analysis of the *E. coli* chromosome has now established the presence of just over

使用中断杂交技术和不同 Hfr 菌株使研究人员可以绘制整个大肠杆菌的染色体图。以时间为单位进行作图，K12 菌株（或大肠杆菌 K12）的染色体长度为 100 min。对大肠杆菌染色体的现代基因组分析目前已经确定了 4000 多种蛋白质编码序列的存在，而这种最初的基因定位作图方法确定了大

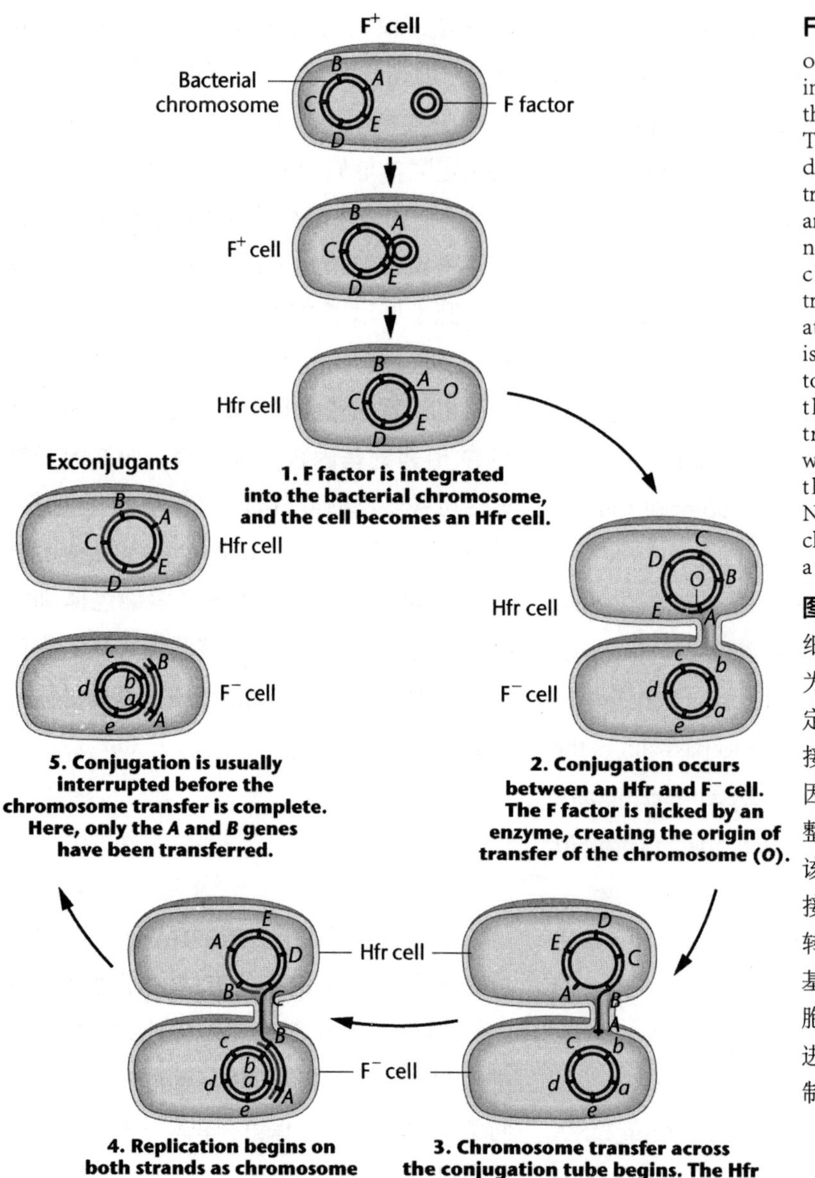

FIGURE 8.8 Conversion of F^+ to an Hfr state occurs by integrating the F factor into the bacterial chromosome. The point of integration determines the origin (*O*) of transfer. During conjugation, an enzyme nicks the F factor, now integrated into the host chromosome, initiating transfer of the chromosome at that point. Conjugation is usually interrupted prior to complete transfer. Only the *A* and *B* genes are transferred to the F^- cell, which may recombine with the host chromosome. Newly replicated DNA of the chromosome is depicted by a lighter shade of gray.

图 8.8 通过 F 因子整合入细菌染色体将 F^+ 细胞转变为 Hfr 状态。整合位点决定了转移起点（*O*）。在接合过程中，一种酶将 F 因子剪切，从而产生切口，整合入宿主染色体中，从该切点起始染色体转移。接合通常在染色体完全被转移前被打断，因此仅有基因 *A* 和 *B* 被转移入 F^- 细胞，从而能与宿主染色体进行重组。染色体中新复制的 DNA 用浅灰色表示。

4000 protein-coding sequences, this original mapping procedure established the location of approximately 1000 genes.

约 1000 个基因的位置。

ESSENTIAL POINT

When the F factor is integrated into the donor cell chromosome (making it Hfr), the donor chromosome moves unidirectionally

基本要点

当 F 因子整合入供体细胞染色体中（使其成为 Hfr）时，供体染色体单向移动至受体细胞中，并启动重组，为以时间为单位

into the recipient, initiating recombination and providing the basis for time mapping of the bacterial chromosome.

进行细菌染色体作图奠定了基础。

Recombination in $F^+ \times F^-$ Matings: A Reexamination

The preceding experiment helped geneticists better understand how genetic recombination occurs during $F^+ \times F^-$ matings. Recall that recombination occurs much less frequently than in Hfr$\times F^-$ matings and that random gene transfer is involved. The current belief is that when F^+ and F^- cells are mixed, conjugation occurs readily and each F^- cell involved in conjugation with an F^+ cell receives a copy of the F factor, *but no genetic recombination occurs*. However, at an extremely low frequency in a population of F^+ cells, the F factor integrates spontaneously from the cytoplasm to a random point in the bacterial chromosome, converting the F^+ cell to the Hfr state, as we saw in Figure 8.8. Therefore, in $F^+ \times F^-$ matings, the extremely low frequency of genetic recombination (10^{-7}) is attributed to the rare, newly formed Hfr cells, which then undergo conjugation with F^- cells. Because the point of integration of the F factor is random, the gene or genes that are transferred by any newly formed Hfr donor *will also appear to be random within the larger F^+/F^- population*. The recipient bacterium will appear as a recombinant but will remain F^-. If it subsequently undergoes conjugation with an F^+ cell, it will then be converted to F^+.

$F^+ \times F^-$ 重组：再验证

先前的实验帮助遗传学家更好地了解了$F^+ \times F^-$交配过程中遗传重组发生的过程。回想一下，$F^+ \times F^-$重组发生频率比Hfr$\times F^-$低得多，并且涉及随机的基因转移。目前的观点是，当F^+与F^-细胞混合培养时，容易发生接合，并且参与接合的每个F^-细胞均从F^+细胞获得一个F因子的拷贝，但是不发生遗传重组。然而，在F^+细胞群体中，细胞质中的F因子以极低的发生频率自发整合到细菌染色体的随机位点上，从而将F^+细胞转化为Hfr状态，如图8.8所示。因此，$F^+ \times F^-$交配中极低的遗传重组频率（10^{-7}）应归因于罕见的、新形成的Hfr细胞与F^-细胞的接合。因为F因子的整合位点是随机的，所以任何新生成的Hfr供体细胞转移一个或多个基因在较大的F^+/F^-群体中也将呈现出随机特点。受体细菌虽为重组子，但仍保持F^-状态。如果随后它与F^+细胞接合，则将变为F^+。

The F' State and Merozygotes

In 1959, during experiments with Hfr strains of *E. coli*, Edward Adelberg discovered that the F factor could lose its integrated status, causing the cell to revert to the F^+ state (Figure 8.9, Step 1). When this occurs, the F factor frequently carries several adjacent bacterial genes along with it (Step 2). Adelberg labeled this

F' 状态与局部杂合子

1959 年，在使用大肠杆菌 Hfr 菌株进行实验时，爱德华 · 阿德尔贝格发现 F 因子可以失去其整合状态，从而使细胞恢复 F^+ 状态（图 8.9，步骤 1）。在发生这一过程时，F 因子通常会携带其相邻的细菌染色体上的几个基因（步骤 2）。阿德尔贝格将这种状态表示为 F'，以将其与 F^+ 和 Hfr 相

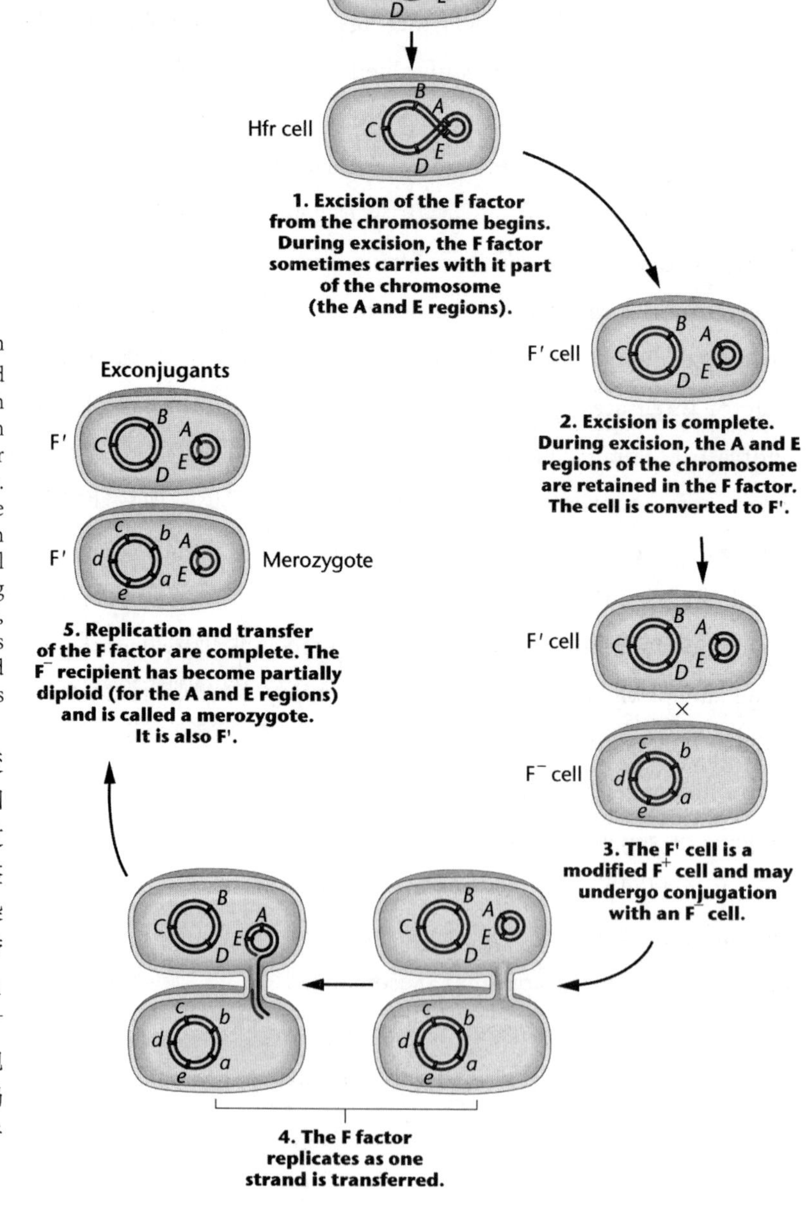

FIGURE 8.9 Conversion of an Hfr bacterium to F' and its subsequent mating with an F^- cell. The conversion occurs when the F factor loses its integrated status. During excision from the chromosome, it carries with it one or more chromosomal genes (*A* and *E*). Following conjugation with an F^- cell, the recipient cell becomes partially diploid and is called a merozygote; it also behaves as an F^+ donor cell.

图 8.9 由 Hfr 细菌转变为 F' 细菌及随后与 F^- 细胞进行的接合。该转变过程发生在 F 因子丢失其整合状态时。当 F 因子从染色体上脱落时，它会携带一个或多个染色体基因（*A* 和 *E*）同时脱落。随后与 F^- 细胞发生接合，受体细胞变为部分二倍体，称为局部杂合子；其行为类似于 F^+ 供体细胞。

condition F' to distinguish it from F^+ and Hfr. F', like Hfr, is thus another special case of F^+, but this conversion is from Hfr to F'.

区别。F' 与 Hfr 一样，也是 F^+ 的另一种特殊存在状态，但是这种转换是从 Hfr 到 F' 的。

The presence of bacterial genes within a cytoplasmic F factor creates an interesting situation. An F' bacterium behaves like an F^+ cell by initiating conjugation with F^- cells (Figure 8.9, Step 3). When this occurs, the F factor, containing chromosomal genes, is transferred to the F^- cell (Step 4). As a result, whatever chromosomal genes are part of the F factor are now present as duplicates in the recipient cell (Step 5) because the recipient still has a complete chromosome. This creates a partially diploid cell called a **merozygote**. Pure cultures of F' merozygotes can be established. They have been extremely useful in the study of bacterial genetics, particularly in genetic regulation.

细胞质 F 因子中细菌基因的存在将产生一个有趣的情况：F' 细菌可以像 F^+ 细胞一样启动与 F^- 细胞的接合（图 8.9，步骤 3）。当这种接合发生时，包含染色体基因的 F 因子会被转移至 F^- 细胞中（步骤 4）。结果便是，由于受体细胞内含有完整的染色体，所以不论作为 F 因子部分所携带的染色体基因是哪些，此时均会以二倍体形式存在于受体细胞中（步骤 5）。如此产生的部分二倍体细胞被称为**局部杂合子**。纯的 F' 局部杂合子培养物可以通过分离获得，并且它们在细菌遗传学，特别是遗传调控研究中非常有用。

8.3 The F Factor Is an Example of a Plasmid

8.3 F 因子是一种质粒

The preceding sections introduced the extrachromosomal heredity unit required for conjugation called the F factor. When it exists autonomously in the bacterial cytoplasm, the F factor is composed of a double-stranded closed circle of DNA. These characteristics place the F factor in the more general category of genetic structures called **plasmids**. These structures contain one or more genes and often, quite a few. Their replication depends on the same enzymes that replicate the chromosome of the host cell, and they are distributed to daughter cells along with the host chromosome during cell division.

前面的部分介绍了接合所需的染色体外遗传单元——F 因子。F 因子由闭合环形 DNA 双链组成，并自主地存在于细菌细胞质中。这些特征使 F 因子可以归入一类更为普遍的遗传结构类别——**质粒**中。这些结构包含一个或多个基因，通常是很多基因。它们的复制依赖于与宿主细胞染色体复制相同的酶类，并且在细胞分裂过程中，会与宿主染色体一同分配到子细胞中。

Plasmids are generally classified according to the genetic information specified by their DNA. The F factor plasmid confers fertility and contains the genes essential for sex pilus formation, on which genetic recombination depends. Other examples of plasmids include the R and Col plasmids.

质粒通常根据其 DNA 所含的遗传信息进行分类。F 因子质粒赋予细胞育性，并且含遗传重组所需性菌毛形成的决定基因。质粒的其他示例还有 R 质粒和 Col 质粒。

Most **R plasmids** consist of two components: the **resistance transfer factor** (**RTF**) and one or more **r-determinants** (Figure 8.10). The RTF encodes genetic information essential to transferring the plasmid between bacteria, and the r-determinants are genes that confer resistance to antibiotics or mercury. While RTFs are similar in a variety of plasmids from different bacterial species, r-determinants are specific for resistance to one class of antibiotic and vary widely. Resistance to tetracycline, streptomycin, ampicillin, sulfonamide, kanamycin, or chloramphenicol is most frequently encountered. Sometimes several r-determinants occur in a single plasmid, conferring multiple resistance to several antibiotics (Figure 8.10). Bacteria bearing these plasmids are of great medical significance not only because of their multiple resistance but because of the ease with which the plasmids can be transferred to other bacteria.

大多数 **R 质粒**由两部分组成：**抗药性转移因子**（RTF）和一个或多个**抗性决定因子**（图 8.10）。RTF 编码细菌间转移质粒所必需的遗传信息，而抗性决定因子则是赋予细胞抵抗抗生素或汞的能力的基因。RTFs 在不同细菌物种的多种质粒中都较相似，但是抗性决定因子则是对一类抗生素具有特异性抗性，并且差异很大。最常见的是对四环素、链霉素、氨苄青霉素、磺胺、卡那霉素或氯霉素的抗性。有时在单一质粒中会存在多个抗性决定因子，从而赋予其对几种抗生素的多重耐药性（图 8.10）。携带这些质粒的细菌具有重要的医学意义，不仅因为它们具有多重耐药性，而且因为质粒易于转移至其他细菌中。

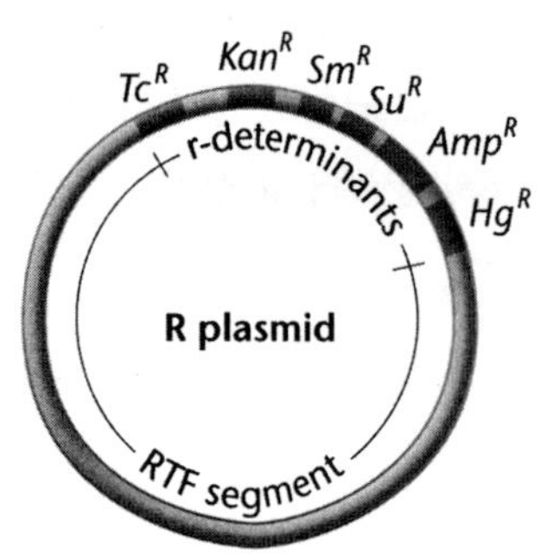

FIGURE 8.10 An R plasmid containing resistance transfer factors (RTFs) and multiple r-determinants (Tc, tetracycline; Kan, kanamycin; Sm, streptomycin; Su, sulfonamide; Amp, ampicillin; and Hg, mercury).

图 8.10 R 质粒，含抗药性转移因子（RTFs）和多个抗性决定因子（Tc，四环素；Kan，卡那霉素；Sm，链霉素；Su，磺胺；Amp，氨苄青霉素；Hg，汞）。

The first known case of such a plasmid occurred in Japan in the 1950s in the bacterium *Shigella*, which causes dysentery. In hospitals, bacteria were isolated that were resistant to as many as five of the above antibiotics. Obviously, this phenomenon represents a major health threat. Fortunately, a bacterial cell sometimes contains r-determinant plasmids but no RTF. Although such a cell is resistant, it cannot transfer the genetic information for resistance to recipient cells. The most commonly studied plasmids, however, contain the RTF as well as one or more r-determinants.

有关质粒抗性的首个已知病例是20世纪50年代发生在日本并引发痢疾的志贺氏杆菌。在医院里，分离得到的细菌对上述抗生素中多达五种都具有抗性。显然，这一现象表明存在重大的健康威胁。幸运的是，细菌细胞有时含抗性决定因子质粒，而不含RTF。尽管这种细胞具有抗性，但无法将抗性遗传信息传递给受体细胞。然而，最常用于研究的质粒通常含RTF以及一个或多个抗性决定因子。

The **Col plasmid**, ColE1 (derived from *E. coli*), is clearly distinct from the R plasmid. It encodes one or more proteins that are highly toxic to bacterial strains that do not harbor the same plasmid. These proteins, called **colicins**, can kill neighboring bacteria, and bacteria that carry the plasmid are said to be *colicinogenic*. Present in 10 to 20 copies per cell, a gene in the Col plasmid encodes an immunity protein that protects the host cell from the toxin. Unlike an R plasmid, the Col plasmid is not usually transmissible to other cells.

Col 质粒，即源自大肠杆菌的ColE1质粒，与R质粒存在明显不同。它编码一种或多种对含不同质粒的细菌菌株都具有高毒性的蛋白质，即**大肠菌素**，该蛋白质可以杀死相邻的细菌。携带Col质粒的细菌被称为大肠菌素产生菌。Col质粒上还编码一种保护宿主细胞免受毒素侵害的免疫蛋白，并在每个细胞中存在10—20个拷贝。与R质粒不同，Col质粒通常不能传递给其他细胞。

Interest in plasmids has increased dramatically because of their role in recombinant DNA research. Specific genes from any source can be inserted into a plasmid, which may then be inserted into a bacterial cell. As the altered cell replicates its DNA and undergoes division, the foreign gene is also replicated, thus cloning the foreign genes.

由于质粒在重组DNA研究中的用途广泛，因此人们对于质粒的研究兴趣日益增长。任何来源的特定基因均可插入质粒中，随后可继续将其引入细菌细胞中。当被改造的细胞复制DNA并进行细胞分裂时，外源基因同时也会被复制，从而使得外源基因也被克隆。

ESSENTIAL POINT

Plasmids, such as the F factor, are

基本要点

质粒，例如F因子，是位于细菌细胞

autonomously replicating DNA molecules found in the bacterial cytoplasm, sometimes containing unique genes conferring antibiotic resistance as well as the genes necessary for plasmid transfer during conjugation.

质中可进行自主复制的DNA分子，有时含一些独特基因以赋予细菌抗生素抗性或者是接合过程中质粒转移的必需基因。

8.4 Transformation Is Another Process Leading to Genetic Recombination in Bacteria

8.4 转化是细菌进行遗传重组的另一种方式

Transformation provides another mechanism for recombining genetic information in some bacteria. Small pieces of extracellular DNA are taken up by a living bacterium, potentially leading to a stable genetic change in the recipient cell. We discuss transformation in this chapter because in those bacterial species where it occurs, the process can be used to map bacterial genes, though in a more limited way than conjugation. Transformation has also played a central role in experiments proving that DNA is the genetic material.

转化为某些细菌的遗传信息重组提供了另一种机制。一小段细胞外DNA被活细菌摄入后，可能导致受体细胞发生稳定的遗传变化。我们在本章讨论转化是因为在发生转化的细菌物种中，该过程可用于定位细菌基因，尽管该方法与接合相比有局限性。此外，转化在证明DNA是遗传物质的实验中也发挥了核心作用。

The process of transformation consists of numerous steps divided into two categories: (1) entry of DNA into a recipient cell and (2) recombination of the donor DNA with its homologous region in the recipient chromosome. In a population of bacterial cells, only those in the particular physiological state of **competence** take up DNA. Entry is thought to occur at a limited number of receptor sites on the surface of the bacterial cell. Passage into the cell is an active process that requires energy and specific transport molecules. This model is supported by the fact that substances that inhibit

转化过程包括可分为两个阶段的许多步骤：（1）DNA进入受体细胞；（2）供体DNA与受体细胞染色体中的同源区发生重组。在细菌细胞群体中，只有那些处于特定生理状态（**感受态**）的细胞才能摄入DNA。DNA进入细胞仅发生在细菌细胞表面存在的数量有限的受体位点上。DNA进入细胞的动态过程需要能量和特定的转运分子，通过抑制受体细胞的能量产生或蛋白质合成可以抑制转化过程的实验事实为该模式提供了支持。

energy production or protein synthesis in the recipient cell also inhibit the transformation process.

During entry, one of the two strands of the double helix is digested by nucleases, leaving only a single strand to participate in transformation. The surviving strand of DNA then aligns with its complementary region of the bacterial chromosome. In a process involving several enzymes, the segment replaces its counterpart in the chromosome, which is excised and degraded. For recombination to be detected, the transforming DNA must be derived from a different strain of bacteria that bears some genetic variation, such as a mutation. Once it is integrated into the chromosome, the recombinant region contains one host strand (present originally) and one mutant strand. Because these strands are from different sources, this helical region is referred to as a **heteroduplex**. Following one round of DNA replication, one chromosome is restored to its original configuration, and the other contains the mutant gene. Following cell division, one untransformed cell (nonmutant) and one transformed cell (mutant) are produced.

在 DNA 进入细胞的过程中，DNA 双螺旋两条链中的一条被核酸酶消化降解，仅留下一条单链参与转化。然后，幸存的 DNA 单链与细菌染色体上与之互补的区域进行平行对齐。在多种酶的参与下，该片段取代其在染色体中的对应片段，而对应片段则被切除并降解。为了检测重组，转化 DNA 必须来自含某些遗传变异如突变的不同细菌菌株。一旦整合入染色体中，重组区域将包含一条宿主链（原本存在）和一条突变链。由于这些 DNA 链来源不同，因此该双螺旋区域被称为**异源双链**。经过一轮 DNA 复制后，一条染色体恢复其原始结构，另一条则包含突变基因。细胞分裂后，将会产生一个未转化细胞（非突变体）和一个转化细胞（突变体）。

Transformation and Linked Genes

In early transformation studies, the most effective exogenous DNA contained 10,000 to 20,000 nucleotide pairs, a length sufficient to encode several genes. Genes adjacent to or very close to one another on the bacterial chromosome can be carried on a single segment of this size. Consequently, a single transfer

转化与连锁基因

在早期转化研究中，最有效的外源 DNA 包含 10 000—20 000 个核苷酸对，其长度足以编码多个基因。在细菌染色体上彼此相邻或非常接近的基因可以同时存在于如此大小的单个 DNA 片段上。因此，单个转化事件可以同时导致多个基因的**共转化**。彼此距离接近、足以进行共转化的基

event can result in the **cotransformation** of several genes simultaneously. Genes that are close enough to each other to be cotransformed are *linked*. In contrast to *linkage groups* in eukaryotes, which consist of all genes on a single chromosome, note that here *linkage* refers to the proximity of genes that permits cotransformation (i.e., the genes are next to, or close to, one another).

因被称为连锁。与真核生物中由单一染色体上所有基因组成的连锁群不同，这里的连锁是指基因间彼此相邻或距离靠近，允许共转化的发生。

If two genes are not linked, simultaneous transformation occurs only as a result of two independent events involving two distinct segments of DNA. As in double crossovers in eukaryotes, the probability of two independent events occurring simultaneously is equal to the product of the individual probabilities. Thus, the frequency of two unlinked genes being transformed simultaneously is much lower than if they are linked.

如果两个基因不连锁，那么同时转化则仅发生在涉及两个不同 DNA 片段的独立事件同时进行时。与真核生物的双交换一样，两个独立事件同时发生的概率等于各单一事件发生概率的乘积。因此，两个不连锁基因同时被转化的频率要比两个连锁基因低得多。

ESSENTIAL POINT

Transformation in bacteria, which does not require cell-to-cell contact, involves exogenous DNA that enters a recipient bacterium and recombines with the host's chromosome. Linkage mapping of closely aligned genes is possible during the analysis of transformation.

基本要点

细菌转化不需要细胞 - 细胞间直接接触，仅涉及外源 DNA 进入受体细菌并与宿主染色体发生重组。在转化分析中，可以对距离相近的基因进行连锁作图。

8.5 Bacteriophages Are Bacterial Viruses

8.5 噬菌体是细菌病毒

Bacteriophages, or **phages** as they are commonly known, are viruses that have bacteria as their hosts. During their reproduction, phages can be involved in still another mode of bacterial

噬菌体，众所周知是以细菌为宿主的病毒。在繁殖过程中，噬菌体可以参与另一种细菌遗传重组模式——**转导**。要了解这一过程，我们必须了解噬菌体遗传学，

genetic recombination called **transduction**. To understand this process, we must consider the genetics of bacteriophages, which themselves undergo recombination.

即噬菌体自身的遗传重组。

A great deal of genetic research has been done using bacteriophages as a model system, making them a worthy subject of discussion. In this section, we will first examine the structure and life cycle of one type of bacteriophage. We then discuss how these phages are studied during their infection of bacteria. Finally, we contrast two possible modes of behavior once the initial phage infection occurs. This information is background for our discussion of transduction.

大量的遗传研究已经使用噬菌体作为模式系统，从而使得噬菌体成为一个值得被探讨的主题。在本节中，我们将首先详细地了解一类噬菌体的结构和生命周期，然后讨论如何在细菌感染过程中研究这些噬菌体。最后，我们将对比一旦噬菌体进行感染可能发生的两种行为模式。这些内容均是我们讨论转导的背景知识。

Phage T4: Structure and Life Cycle

Bacteriophage T4, which has *E. coli* as its host, is one of a group of related bacterial viruses referred to as T-even phages. It exhibits the intricate structure shown in Figure 8.11. Its genetic material, DNA, is contained within an icosahedral (referring to a polyhedron with 20 faces) protein coat, making up the head of the virus. The DNA is sufficient in quantity to encode more than 150 average-sized genes. The head is connected to a complex tail structure consisting of a collar, an outer contractile sheath that surrounds an inner spike-like tube, which

T4 噬菌体：结构与生命周期

T4 噬菌体是以大肠杆菌为宿主的噬菌体，并且也是 T 偶数系列噬菌体相关细菌病毒中的一种。它的复杂结构如图 8.11 所示。它的遗传物质——DNA 被包裹在呈二十面体的蛋白质外壳中，构成了病毒的头部。DNA 的量足以编码 150 多个平均大小的基因。头部与复杂的尾部结构相连接，尾部结构包括颈环和外部包裹可收缩尾鞘的尖刺状尾管等。尾管与基板连接，尾丝附着在基板上并向外伸出。基板结构非常复杂，由 15 种不同的蛋白质组成，大多数以多拷贝的形式存在。基板协调宿主细胞识别，

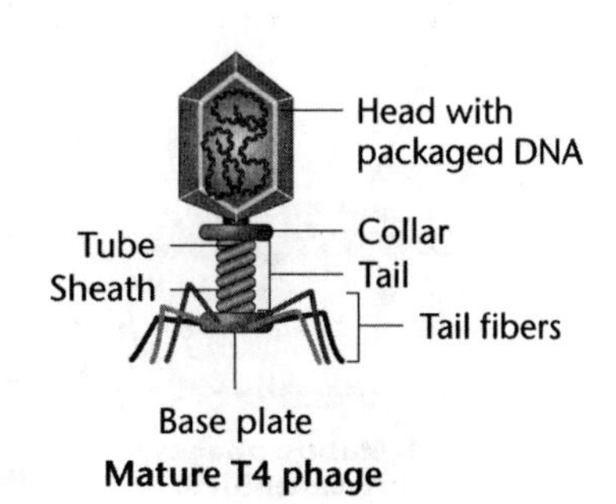

FIGURE 8.11 The structure of bacteriophage T4 includes an icosahedral head filled with DNA, a tail consisting of a collar, tube, sheath, base plate, and tail fibers. During assembly, the tail components are added to the head, and then tail fibers are added.

图 8.11 T4 噬菌体结构包括包裹 DNA 呈二十面体的头部及含颈环、尾管、尾鞘、基板和尾丝的尾部。在组装时，尾部组分与头部相连接，然后附着上尾丝。

sits atop a base plate from which tail fibers protrude. The base plate is an extremely complex structure, consisting of 15 different proteins, most present in multiple copies. The base plate coordinates the host cell recognition and is involved in providing the signal whereby the outer sheath contracts, propelling the inner tube across the cell membrane of the host cell.

并参与信号提供，使外部尾鞘收缩，推动尾管刺穿宿主细胞的细胞膜。

The life cycle of phage T4 (Figure 8.12) is initiated when the virus binds to the bacterial host cell. Then, during contraction of the outer sheath, the DNA in the head is extruded, and it moves across the cell membrane into the bacterial cytoplasm. Within minutes, all bacterial DNA, RNA, and protein synthesis in the host cell is inhibited, and synthesis of viral molecules begins. At the same time, degradation of the host DNA is initiated.

病毒与细菌宿主细胞相结合开启了 T4 噬菌体的生命周期（图 8.12）。然后，在外部尾鞘收缩时，头部的 DNA 会被挤出，并穿过细胞膜进入细菌细胞质中。在几分钟内，宿主细胞中所有细菌 DNA、RNA 和蛋白质的合成均被抑制，开始合成病毒分子，并同时启动宿主 DNA 的降解。

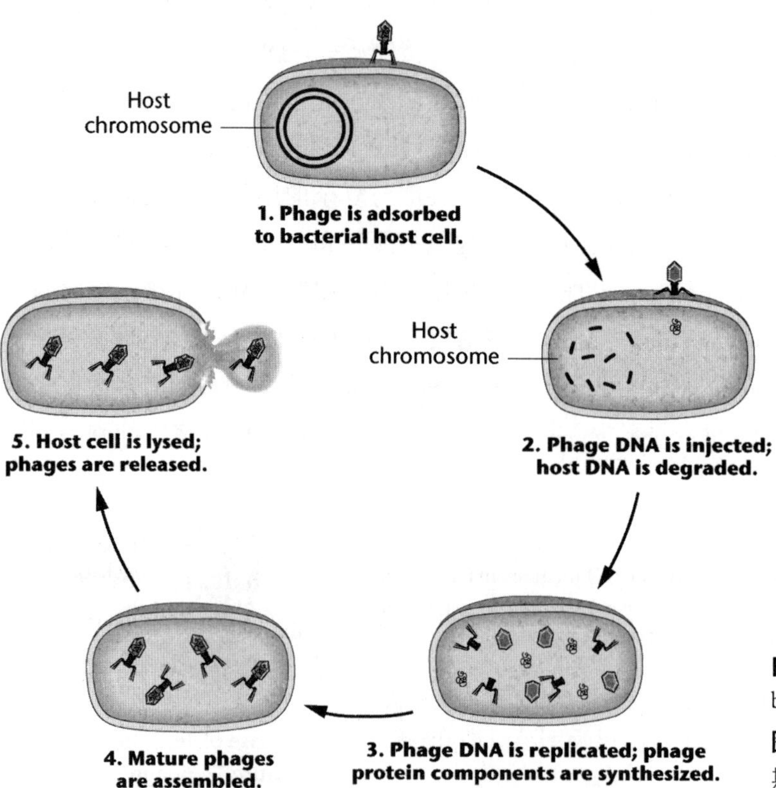

FIGURE 8.12 Life cycle of bacteriophage T4.

图 8.12 T4 噬菌体的生命周期。

A period of intensive viral gene activity characterizes infection. Initially, phage DNA replication occurs, leading to a pool of viral DNA molecules. Then, the components of the head, tail, and tail fibers are synthesized. The assembly of mature viruses is a complex process that has been well studied by William Wood, Robert Edgar, and others. Three sequential pathways occur: (1) DNA packaging as the viral heads are assembled, (2) tail assembly, and (3) tail fiber assembly. Once DNA is packaged into the head, it combines with the tail components, to which tail fibers are added. Total construction is a combination of self-assembly and enzyme-directed processes.

病毒基因活力加强标志着感染发生。最初，噬菌体DNA进行复制，产生大量的病毒DNA分子。然后，合成病毒的各个组成部分，包括头部、尾部和尾丝。成熟病毒的组装是一个复杂的过程，威廉·伍德、罗伯特·埃德加等人已经对此进行了深入研究。具体来讲，包括三个连续的过程：(1)通过DNA包装组装病毒头部；(2)组装尾部；(3)组装尾丝。一旦DNA被包装至病毒头部，即会与病毒尾部组分相结合，并添加尾丝。总组装过程是自组装和酶促组装过程的结合。

When approximately 200 new viruses have been constructed, the bacterial cell is ruptured by the action of the enzyme lysozyme (a phage gene product), and the mature phages are released from the host cell. The new phages infect other available bacterial cells, and the process repeats itself over and over again.

当组装形成大约200个新生病毒时，细菌细胞将在溶菌酶（一种噬菌体基因产物）的作用下发生破裂，成熟噬菌体将从宿主细胞中释放出来。新生噬菌体可以继续感染其他细菌细胞，这一过程周而复始地进行着。

The Plaque Assay

Bacteriophages and other viruses have played a critical role in our understanding of molecular genetics. During infection of bacteria, enormous quantities of bacteriophages can be obtained for investigation. Often, over 10^{10} viruses are produced per milliliter of culture medium. Many genetic studies rely on the ability to quantify the number of phages produced following infection under specific culture conditions. The **plaque assay** is a routinely used technique, which is invaluable in mutational and

噬斑测定

噬菌体和其他病毒在我们理解分子遗传学的过程中发挥着关键作用。在细菌感染过程中，可以获得大量噬菌体用于研究。通常，每毫升培养基可以产生数量在10^{10}个以上的病毒。许多遗传学研究依赖于对特定培养条件下感染产生的噬菌体进行定量的能力。**噬斑测定**是一种常规使用的技术，在噬菌体的突变研究和重组研究中十分有用。

recombinational studies of bacteriophages.

This assay is shown in Figure 8.13, where actual plaque morphology is also illustrated. A serial dilution of the original virally infected bacterial culture is performed first. Then, a 0.1-mL sample (an *aliquot*) from a dilution is added to melted nutrient agar (about 3 mL) into which a few drops of a healthy bacterial culture have been added. The solution is then poured evenly over a base of solid nutrient agar in a petri dish and allowed to solidify before incubation. A clear area called a **plaque** occurs wherever a single virus initially infected one bacterium in the culture (the lawn) that has grown up during incubation. The plaque represents clones of the single infecting bacteriophage, created as reproduction cycles are repeated. If the dilution factor is too low, the plaques are plentiful, and they will fuse, lysing the entire lawn—which has occurred in the 10^{-3} dilution of Figure 8.13. On the other hand, if the dilution factor is increased, plaques can be counted and the density of viruses in the initial culture can be estimated as

Initial phage density =
(plaque number/mL) × (dilution factor)

Using the results shown in Figure 8.13, 23 phage plaques are derived from the 0.1 mL aliquot of the 10^{-5} dilution. Therefore, we estimate that there are 230 phages/mL at this dilution (since the initial aliquot was 0.1 mL). The initial phage density in the undiluted sample, factoring in the 10^{-5} dilution, is then calculated as

图 8.13 展示了该检测方法，同时也展示了实际的噬斑形态。首先对原始病毒感染的细菌培养物进行系列稀释。然后，将稀释后的 0.1 mL 样品（一等分样品）添加到已加入几滴健康细菌培养物的半固体营养琼脂中（约 3 mL）。随后混匀，并将溶液均匀地倒入培养皿中的固体琼脂培养基上，待半固体营养琼脂凝固后进行培养。在培养过程中，每当一个病毒感染培养生长的菌苔中的细菌时，就会出现一个透明区域，我们称之为**噬斑**。噬斑代表单个感染噬菌体的克隆，由周而复始的病毒生命周期而产生。如果稀释倍数过低，则噬斑数量会很多，彼此间发生融合，裂解整个菌苔，正如图 8.13 中 10^{-3} 稀释度所看到的情形。此外，如果增加稀释倍数，则可以对噬斑计数，并估算初始培养物中的病毒密度：

初始病毒密度 =
（噬斑个数 /mL）×（稀释倍数）

利用图 8.13 所示结果，0.1 mL 10^{-5} 稀释液产生 23 个噬斑。因此，我们估算在该稀释度下有 230 个噬菌体 / mL（因为初始等分样品为 0.1 mL）。未稀释样品的初始噬菌体密度（以 10^{-5} 稀释度为例）计算为：

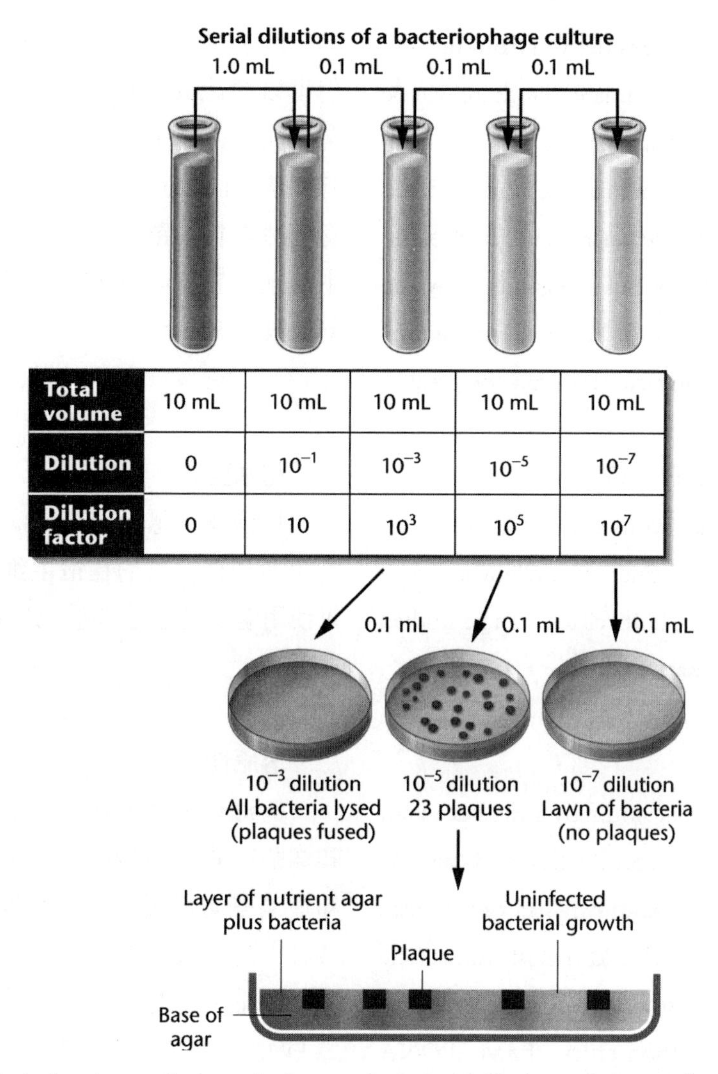

Total volume	10 mL	10 mL	10 mL	10 mL	10 mL
Dilution	0	10^{-1}	10^{-3}	10^{-5}	10^{-7}
Dilution factor	0	10	10^{3}	10^{5}	10^{7}

FIGURE 8.13 A plaque assay for bacteriophage analysis. Serial dilutions of a bacterial culture infected with bacteriophages are first made. Then three of the dilutions (10^{-3}, 10^{-5}, and 10^{-7}) are analyzed using the plaque assay technique. Each plaque represents the initial infection of one bacterial cell by one bacteriophage. In the 10^{-3} dilution, so many phages are present that all bacteria are lysed. In the 10^{-5} dilution, 23 plaques are produced. In the 10^{-7} dilution, the dilution factor is so great that no phages are present in the 0.1 mL sample, and thus no plaques form. From the 0.1 mL sample of the 10^{-5} dilution, the original bacteriophage density is calculated to be (230/ml) × (10^{5}) phages/mL (23×10^{6}, or 2.3×10^{7}).

图 8.13 用于噬菌体分析的噬斑测定。首先对原始病毒感染的细菌培养物进行系列稀释，然后使用噬斑测定技术分析三个稀释度（10^{-3}、10^{-5} 和 10^{-7}）。一个噬斑代表起始感染过程中一个噬菌体感染一个细菌细胞。在 10^{-3} 稀释度下，存在的噬菌体数量很多以至于所有细菌均被裂解。在 10^{-5} 稀释度下，产生了 23 个噬斑。在 10^{-7} 稀释度下，由于稀释倍数太大，在 0.1 mL 样品中没有噬菌体存在，所以也没有噬斑形成。从 0.1 mL 10^{-5} 稀释度样品所得结果，可以计算起始培养物中的病毒密度为 (230/mL)×(10^{5}) 噬菌体 /mL（23×10^{6}，或 2.3×10^{7}）。

Initial phage density = (230/mL) × (10^{5})

= 230 × 10^{5}/mL

初始病毒密度 = (230/mL) × (10^{5})

= 230 × 10^{5}/mL

Because this figure is derived from the 10^{-5} dilution, we can also estimate that there will be only 0.23 phage/0.1 mL in the 10^{-7} dilution. Thus, when 0.1 mL from this tube is assayed, it is predicted that no phage particles will be present. This possibility is borne out in Figure 8.13, which depicts an intact lawn of bacteria lacking any plaques. The dilution factor is simply too great.

由于该数据来自 10^{-5} 稀释液，因此我们可以预计 10^{-7} 稀释液将仅产生 0.23 个噬菌体 /0.1 mL。因此，当从该 10^{-7} 稀释液试管中取 0.1 mL 进行检测时，预计将不存在噬菌体颗粒。图 8.13 证实了这种可能，图中展示的完整菌苔上没有任何噬斑，即稀释倍数的确太大了。

ESSENTIAL POINT

Bacteriophages (viruses that infect bacteria) demonstrate a well-defined life cycle where they reproduce within the host cell and can be studied using the plaque assay.

基本要点

噬菌体（感染细菌的病毒）具有明确的生命周期，它们在宿主细胞内增殖，并可使用噬斑测定对其进行研究。

Lysogeny

Infection of a bacterium by a virus does not always result in viral reproduction and lysis. As early as the 1920s, it was known that some viruses can enter a bacterial cell and coexist with it. The precise molecular basis of this relationship is now well understood. Upon entry, the viral DNA is integrated into the bacterial chromosome instead of replicating in the bacterial cytoplasm, a step that characterizes the developmental stage referred to as **lysogeny**. Subsequently, each time the bacterial chromosome is replicated, the viral DNA is also replicated and passed to daughter bacterial cells following division. No new viruses are produced, and no lysis of the bacterial cell occurs. However, under certain stimuli, such as chemical or ultraviolet light treatment, the viral DNA loses its integrated status and initiates replication, phage reproduction, and lysis of the bacterium.

溶原性

病毒感染细菌并不总是导致病毒增殖和溶菌。早在 20 世纪 20 年代，人们就已经知道某些病毒可以进入细菌细胞内并与之共存。这种关系的精确分子基础现在已经了解得非常清楚。在进入细菌细胞后，病毒 DNA 会整合入细菌染色体中，而不在细菌细胞质中进行复制，这一步骤表征了**溶原性**的发展阶段。随后，每次细菌染色体进行复制时，病毒 DNA 也会随之复制，并在细胞分裂后传递给子代细菌细胞。没有新生病毒产生，也不发生溶菌。但是，在某些特定刺激下，例如化学试剂或紫外线处理，病毒 DNA 将失去其整合状态，开始复制，噬菌体增殖，细菌裂解。

Several terms are used to describe this relationship. The viral DNA that integrates into the bacterial chromosome is called a **prophage**. Viruses that either lyse the cell or behave as a prophage are **temperate phages**. Those that only lyse the cell are referred to as **virulent phages**. A bacterium harboring a prophage is said to be **lysogenic**; that is, it is capable of being lysed as a result of induced viral reproduction.

有几个专业术语用于描述这种关系。整合入细菌染色体中的病毒 DNA 被称为**原噬菌体**。既可裂解细胞，也可成为原噬菌体的病毒被称为**温和噬菌体**。只能裂解宿主细胞的噬菌体被称为**烈性噬菌体**。携带原噬菌体的细菌被称为**溶原菌**，就是说，诱导病毒增殖可以导致其被裂解。

ESSENTIAL POINT

Bacteriophages can be lytic, meaning they infect the host cell, reproduce, and then lyse it, or in contrast, they can lysogenize the host cell, where they infect it and integrate their DNA into the host chromosome, but do not reproduce.

基本要点

噬菌体可以裂解，这意味着它们可以感染宿主细胞、繁殖然后溶菌；与之相对的是，它们也可以溶原化宿主细胞，然后感染宿主并将其 DNA 整合到宿主染色体中，但不增殖。

8.6 Transduction Is Virus-Mediated Bacterial DNA Transfer

8.6 转导是由病毒介导的细菌 DNA 转移

In 1952, Norton Zinder and Joshua Lederberg were investigating possible recombination in the bacterium *Salmonella typhimurium*. Although they recovered prototrophs from mixed cultures of two different auxotrophic strains, investigation revealed that recombination was occurring in a manner different from that attributable to the presence of an F factor, as in *E. coli*. What they had discovered was a process of bacterial recombination mediated by bacteriophages and now called **transduction**.

1952 年，诺顿 · 津德和乔舒亚 · 莱德伯格研究了鼠伤寒沙门菌的遗传重组。尽管他们从两种不同营养缺陷型菌株的混合培养物中获得了原养型后代，但研究表明重组的发生方式与依赖 F 因子进行的大肠杆菌遗传重组不同。他们发现了由噬菌体介导的细菌遗传重组过程，现在被称为**转导**。

The Lederberg-Zinder Experiment

Lederberg and Zinder mixed the *Salmonella*

莱德伯格 - 津德实验

莱德伯格和津德将两株鼠伤寒沙门菌

auxotrophic strains LA-22 and LA-2 together, and when the mixture was plated on minimal medium, they recovered prototrophic cells. The LA-22 strain was unable to synthesize the amino acids phenylalanine and tryptophan (*phe*$^-$, *trp*$^-$), and LA-2 could not synthesize the amino acids methionine and histidine (*met*$^-$, *his*$^-$). Prototrophs (*phe*$^+$, *trp*$^+$, *met*$^+$, *his*$^+$) were recovered at a rate of about $1/10^5$ (10^{-5}) cells.

营养缺陷型菌株 LA-22 和 LA-2 进行混合培养，当将混合培养物涂布在基本培养基上进行培养时，他们得到了原养型细胞。LA-22 菌株无法合成苯丙氨酸和色氨酸（*phe*$^-$、*trp*$^-$），而 LA-2 无法合成甲硫氨酸和组氨酸（*met*$^-$、*his*$^-$）。获得原养型细胞（*phe*$^+$、*trp*$^+$，*met*$^+$、*his*$^+$）的概率约为 $1/10^5$（10^{-5}）。

Although these observations at first suggested that the recombination involved was the type observed earlier in conjugative strains of *E. coli*, experiments using the Davis U-tube soon showed otherwise (Figure 8.14). The two auxotrophic strains were separated by a

尽管最初的这些观察结果表明所涉及的遗传重组与较早在大肠杆菌菌株接合过程中观察到的是一类遗传重组类型，但很快戴维斯 U 形管实验就表明并非如此（图 8.14）。这两株营养缺陷型菌株被烧结玻璃滤板分隔开来，从而防止细胞接触，但允

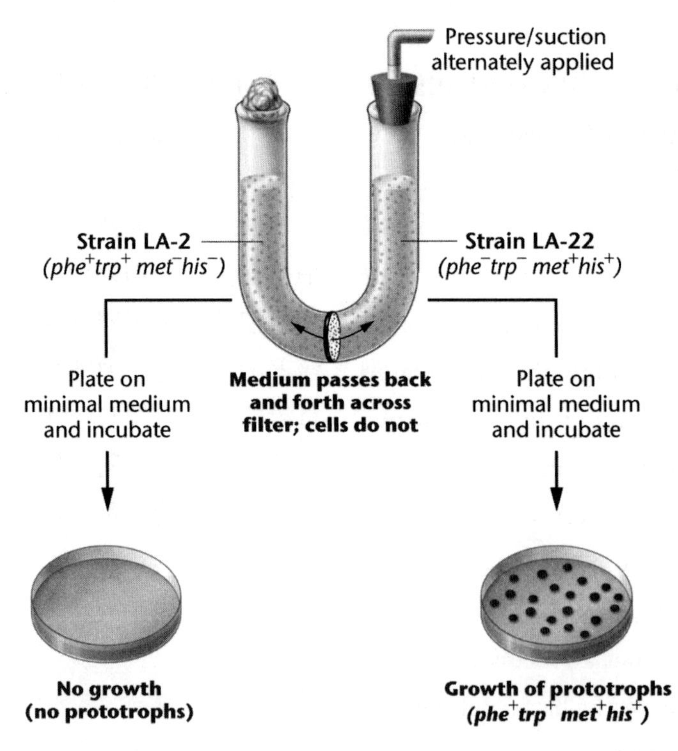

FIGURE 8.14 The Lederberg-Zinder experiment using *Salmonella*. After placing two auxotrophic strains on opposite sides of a Davis U-tube, Lederberg and Zinder recovered prototrophs from the side with the LA-22 strain, but not from the side containing the LA-2 strain.

图 8.14 鼠伤寒沙门菌的莱德伯格 - 津德实验。在戴维斯 U 形管的两侧分别培养两种缺陷型菌株，莱德伯格和津德在培养 LA-22 菌株的一侧获得了原养型细胞，而在培养 LA-2 菌株的一侧未能获得。

sintered glass filter, thus preventing cell contact but allowing growth to occur in a common medium. Surprisingly, when samples were removed from both sides of the filter and plated independently on minimal medium, prototrophs were recovered, but only from the side of the tube containing LA-22 bacteria. Recall that if conjugation were responsible, the conditions in the Davis U-tube would be expected to *prevent* recombinational together (see Figure 8.3).

许细胞在生长过程中共享同样的培养基。出乎意料的是，当从滤板两侧取样并分别接种在基本培养基上培养时，可以得到原养型细菌，并且仅能从含 LA-22 细菌的试管一侧获得。要知道，如果遗传重组的方式是接合，那么戴维斯 U 形管将阻止重组发生（见图 8.3）。

Since LA-2 cells appeared to be the source of the new genetic information (phe^+ and trp^+), how that information crossed the filter from the LA-2 cells to the LA-22 cells, allowing recombination to occur, was a mystery. The unknown source was designated simply as a filterable agent (FA).

由于 LA-2 细胞似乎是新遗传信息 (phe^+ 和 trp^+) 的来源，那么该信息如何从 LA-2 细胞一侧穿过滤板进入 LA-22 细胞一侧，从而导致重组发生呢？这是一个谜。这种未知的来源物被简单表示为可过滤物 (FA)。

Three observations were used to identify the FA:

三个观察结果被用于确定 FA 的本质：

1. The FA was produced by the LA-2 cells only when they were grown in association with LA-22 cells. If LA-2 cells were grown independently and that culture medium was then added to LA-22 cells, recombination did not occur. Therefore, LA-22 cells play some role in the production of FA by LA-2 cells and do so only when they share a common growth medium.

1. 只有当 LA-2 细胞与 LA-22 细胞一起生长时，才产生 FA。如果将 LA-2 细胞独立培养，然后将其培养基加入 LA-22 细胞中，则不发生遗传重组。因此，LA-22 细胞在 LA-2 细胞产生 FA 中起一定作用，而且仅当它们共享培养基时才可以。

2. The addition of DNase, which enzymatically digests DNA, did not render the FA ineffective. Therefore, the FA is not naked DNA, ruling out transformation.

2. 添加可以酶促消化 DNA 的 DNA 酶，不会使 FA 失效。因此，FA 不是裸露的 DNA，从而排除了转化的可能性。

3. The FA could not pass across the filter of the Davis U-tube when the pore size was reduced below the size of bacteriophages.

3. 当滤板的孔径减小至噬菌体尺寸大小以下时，FA 无法通过戴维斯 U 形管的滤板。

Aided by these observations and aware that temperate phages can lysogenize *Salmonella*, researchers proposed that the genetic recombination event is mediated by bacteriophage P22, present initially as a prophage in the chromosome of the LA-22 *Salmonella* cells. They hypothesized that P22 prophages rarely enter the vegetative or lytic phase, reproduce, and are released by the LA-22 cells. Such P22 phages, being much smaller than a bacterium, then cross the filter of the U-tube and subsequently infect and lyse some of the LA-2 cells. In the process of lysis of LA-2, these P22 phages occasionally package a region of the LA-2 chromosome in their heads. If this region contains the phe^+ and trp^+ genes and the phages subsequently pass back across the filter and infect LA-22 cells, these newly lysogenized cells will behave as prototrophs. This process of transduction, whereby bacterial recombination is mediated by bacteriophage P22, is diagrammed in Figure 8.15.

借助于这些观察结果，同时基于对温和噬菌体可以溶原化鼠伤寒沙门菌的了解，研究人员推测遗传重组事件由噬菌体 P22 介导发生。该噬菌体最初以原噬菌体形式存在于 LA-22 沙门氏菌染色体中。他们猜测极少的 P22 原噬菌体进入裂解周期，进行增殖并从 LA-22 细胞中释放出来。这样的 P22 噬菌体比细菌小得多，可以穿过 U 形管的滤板，随后感染并裂解一些 LA-2 细胞。在 LA-2 裂解过程中，这些 P22 噬菌体随机将 LA-2 染色体片段包装入病毒头部。如果该片段含 phe^+ 和 trp^+ 基因，则噬菌体随后通过滤板并感染 LA-22 细胞，这些新产生的溶原化细胞将表现为原养型。这一转导过程由噬菌体 P22 介导，使细菌发生遗传重组，如图 8.15 所示。

Transduction and Mapping

Like transformation, transduction was used in linkage and mapping studies of the bacterial chromosome. The fragment of bacterial DNA involved in a transduction event is large enough to include numerous genes. As a result, two genes that closely align (are linked) on the bacterial chromosome can be simultaneously transduced, a process called **cotransduction**. Two genes that are not close enough to one another along the chromosome to be included on a single DNA fragment require two independent events

转导与作图

和转化一样，转导也可用于细菌染色体的连锁和作图研究。如果涉及转导事件的细菌 DNA 片段足够大，包含多个基因，那么位于细菌染色体上距离较近（连锁）的两个基因可以同时被转导，这一过程称为**共转导**。如果沿着染色体彼此距离较远的两个基因不能被包含在单一 DNA 片段中，那么则需要两个独立事件才能被转导至同一细胞中。因为这种情况发生的可能性远比共转导低得多，所以由此可以确定连锁。

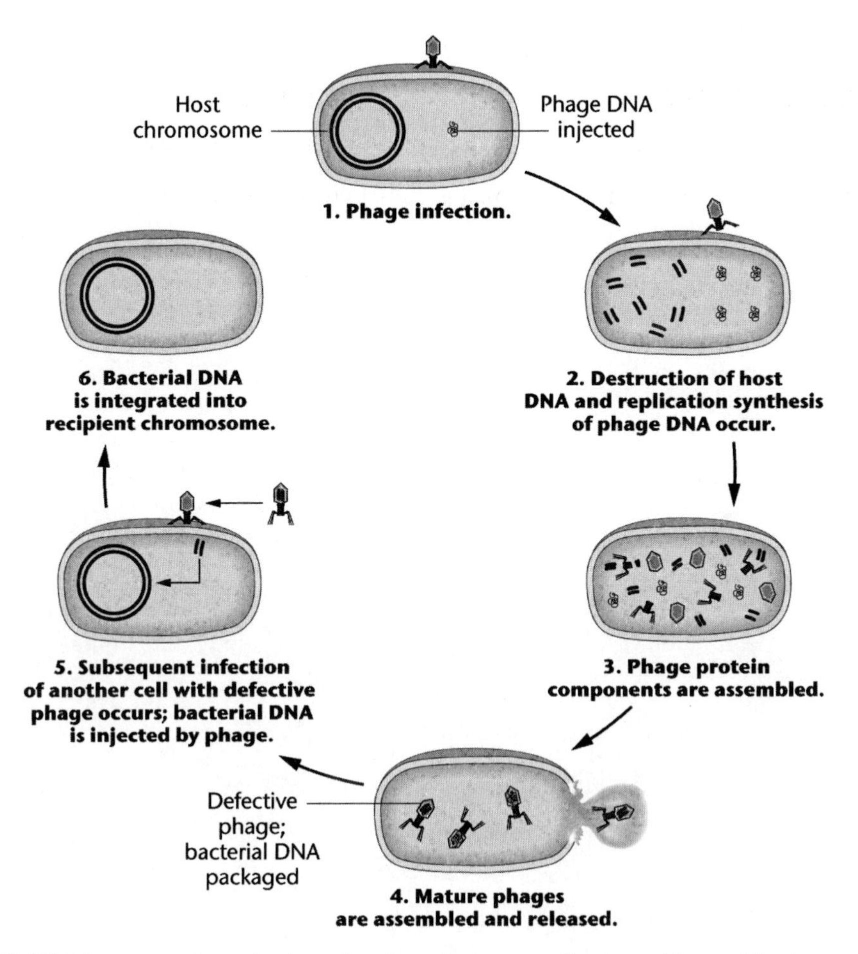

FIGURE 8.15 The process of transduction, where bacteriophages mediate bacterial recombination.

图 8.15 转导过程中噬菌体介导细菌重组。

to be transduced into a single cell. Since this occurs with a much lower probability than cotransduction, linkage can be determined.

By concentrating on two or three linked genes, transduction studies can also determine the precise order of these genes. The closer linked genes are to each other, the greater the frequency of cotransduction. Mapping studies involving three closely aligned genes can thus be executed, and the analysis of such an experiment is predicated on the same rationale underlying other mapping techniques.

通过同时研究两个或三个连锁基因，转导研究也可用于确定这些基因的精确顺序。彼此之间连锁越紧密的基因，共转导频率越高。因此可以进行涉及三个紧密连锁基因的作图研究，此类实验分析是基于与其他遗传作图技术相同的原理来进行的。

ESSENTIAL POINT

Transduction is virus-mediated bacterial DNA transfer and can be used to map phage genes.

基本要点

转导是病毒介导的细菌 DNA 转移，可用于定位噬菌体基因。

9 Epigenetics

9 表观遗传学

Epigenetics can be defined as the study of phenomena and mechanisms that cause chromosome-associated heritable changes to gene expression that are not dependent on changes in DNA sequence.

表观遗传学研究不依赖于 DNA 序列变化而引起的染色体相关的可遗传的基因表达变化现象及其机制。

Until recently, it was thought that most regulation of gene expression is coordinated by *cis*-regulatory elements along with DNA-binding proteins and transcription factors. However, these classical regulatory mechanisms cannot fully explain how some phenotypes arise. For example, monozygotic twins have identical genotypes but often have different phenotypes. In other instances, although one allele of each gene is inherited maternally, and one is inherited paternally, only the maternal or paternal allele is expressed, while the other remains transcriptionally silent. Investigations of such phenomena have led to the emerging field of epigenetics, which is providing us with a molecular basis for understanding how heritable genomic alterations other than those encoded in the DNA sequence can alter patterns of gene expression and influence phenotypic variation (Figure 9.1).

直至近来，人们曾一直认为大多数的基因表达调控是通过顺式调控元件、DNA 结合蛋白和转录因子共同协调实现的。然而，这些经典的调控机制无法完全解释某些表型的产生。例如，同卵双生双胞胎个体含有完全相同的基因型，但是他们通常拥有不同的表型。再如，尽管每种基因的一个等位基因拷贝遗传自母本，另一个等位基因拷贝遗传自父本，但是在个体中只有来自其中一个亲本的等位基因拷贝进行表达，而对应的另一个等位基因拷贝却保持沉默，不进行转录。对于这些现象的研究促成新兴领域——表观遗传学的诞生，从而为我们解析基因组中非 DNA 序列编码的可遗传改变如何影响基因表达模式和表型变异提供了分子基础（图 9.1）。

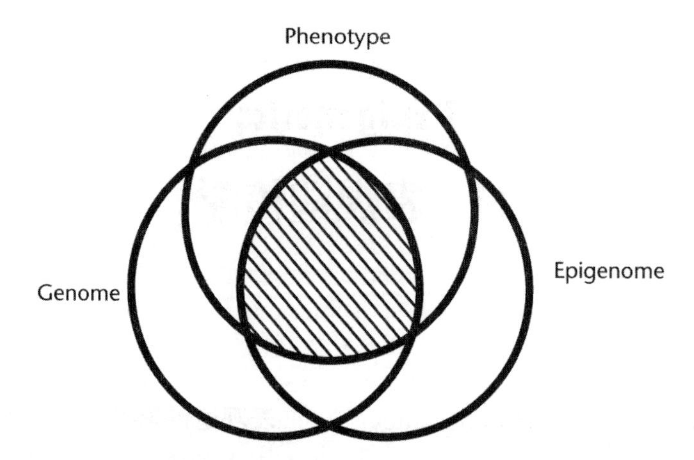

FIGURE 9.1 The phenotype of an organism is the product of interactions between the genome and the epigenome (hatched areas). The genome is constant from fertilization throughout life, but cells, tissues, and the organism develop different epigenomes because of epigenetic reprogramming of gene activity in response to environmental stimuli. These reprogramming events lead to phenotypic changes throughout the life cycle.

图 9.1 生物表型是基因组与表观基因组相互作用的结果（阴影区域）。基因组从受精开始终其一生保持恒定不变。但是为了响应环境刺激，细胞、组织和生物通过基因活性的表观重编程而形成不同的表观基因组。这些重编程事件在整个生命周期中产生表型变化。

The **epigenome** refers to the specific pattern of epigenetic modifications present in a cell at a given time. During its life span, an organism has one genome, which can be modified at different times to produce many different epigenomes.

表观基因组是指定时刻细胞内存在的特定表观遗传学修饰方式。在整个生命阶段中，生物体仅有一个基因组，但是该基因组在不同时间阶段经修饰后可以产生多种表观基因组。

Knowledge of the mechanisms of epigenetic modifications to the genome, how these modifications are maintained and transmitted, and their relationship to basic biological processes is important to enhance our understanding of reproduction and development, disease processes, and the evolution of adaptations to the environment, including behavior.

生物体基因组的表观遗传学修饰机制、这些表观遗传修饰如何保持和传递、它们与基本生物学过程的关系等知识对于加深我们理解生物体的繁殖与发育、疾病发生、生物体对环境的适应性进化以及生物体行为等均具有十分重要的意义。

Current research efforts are focused on several aspects of epigenetics: (1) how an epigenomic state arises in developing and differentiated cells and (2) how these epigenetic states are transmitted via mitosis and meiosis, making them heritable traits. The fruits of these

目前表观遗传学的研究主要集中在以下几个方面：（1）发育细胞或分化细胞中的表观遗传状态是如何形成的；（2）这些表观遗传状态如何通过有丝分裂或减数分裂进行传递，从而使其成为可遗传的性状。这些研究成果将构成本章的主要焦点。此

efforts will be a major focus of this chapter. In addition, because epigenetically controlled alterations to the genome are associated with common diseases such as cancer and diabetes, efforts are also directed toward developing drugs that can modify or reverse disease-associated epigenetic changes in cells.

外，由于表观遗传控制的基因组变化与癌症、糖尿病等一些常见疾病相关，因此人类还在积极研发药物用于修复或逆转细胞内疾病相关的表观遗传改变。

9.1 Molecular Alterations to the Genome Create an Epigenome

9.1 基因组的分子变化产生了表观基因组

Unlike the genome, which is identical in all cell types of an organism, the epigenome is cell-type specific and changes throughout the life cycle in response to environmental cues. Like the genome, the epigenome can be transmitted to daughter cells by mitosis and to future generations by meiosis. In this section, we will examine mechanisms that shape the epigenome.

There are three major epigenetic mechanisms: (1) reversible modification of DNA by the addition or removal of methyl groups; (2) chromatin remodeling by the addition or removal of chemical groups to histone proteins; and (3) regulation of gene expression by noncoding RNA molecules. We now will look at each of these molecular activities in turn.

生物体内所有类型的细胞，基因组都完全相同，而表观基因组则与之不同，它具有细胞类型特异性，并且在整个生命周期中都随着环境变化而发生改变。同时，表观基因组又与基因组类似，同样可以通过有丝分裂传递给子代细胞并通过减数分裂传递给下一世代个体。在这一部分，我们将对表观基因组的形成机制进行论述。

表观遗传学共有三个主要机制：（1）通过添加或移除甲基化基团对 DNA 进行可逆修饰；（2）通过对组蛋白进行化学基团的添加或移除来实现染色质重塑；（3）利用非编码 RNA 分子进行基因表达调控。下面我们将依次介绍这些分子活动。

DNA Methylation and the Methylome

The set of methylated nucleotides present in an organism's genome at a given time is known as the **methylome**. The methylome is cell- and tissue-specific, but is not fixed, and changes as cells are called upon to respond to changing conditions. In mammals, **DNA methylation**

DNA 甲基化与甲基化组

在指定时间内生物体基因组中成组存在的甲基化核苷酸被称为**甲基化组**。甲基化组具有细胞特异性和组织特异性，但是它并非固定不变，会随着细胞响应外界环境变化而发生改变。在哺乳动物中，**DNA 甲基化**发生在 DNA 复制后和细胞分化过程

takes place after DNA replication and during cell differentiation. This process involves the addition of a methyl group (—CH_3) to cytosine on the 5-carbon of the cytosine nitrogenous base (Figure 9.2), resulting in 5-methylcytosine (5mC), a reaction catalyzed by a family of enzymes called DNA methyltransferases (DNMTs). In humans, 5mC is present in about 1.5 percent of the genomic DNA.

中。这一过程涉及在胞嘧啶含氮碱基的 5 号碳原子上添加 1 个甲基基团 (—CH_3)（图 9.2）形成 5- 甲基胞嘧啶（5mC），该反应由 DNA 甲基转移酶（DNMTs）家族成员催化。在人类中，5mC 大约占整个基因组 DNA 的 1.5%。

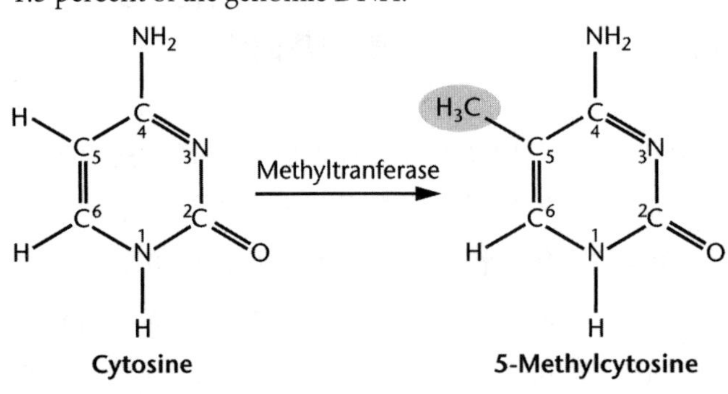

FIGURE 9.2 In DNA methylation, methyltransferase enzymes catalyze the transfer of a methyl group from a methyl donor to cytosine, producing 5-methylcytosine.

图 9.2 在 DNA 甲基化中，甲基转移酶催化甲基基团从甲基供体转移至胞嘧啶，生成 5- 甲基胞嘧啶。

Methylation takes place almost exclusively on cytosine bases located adjacent to a guanine base, a combination called a CpG dinucleotide:

$$5'—\overset{m}{C}pG—3'$$
$$3'—Gp\underset{m}{C}—5'$$

Many of these dinucleotide sites are clustered in regions called CpG islands, located in promoter and upstream sequences (Figure 9.3).

甲基化几乎毫无例外地发生在与鸟嘌呤相邻的胞嘧啶碱基上，这种组合也被称为 CpG 二核苷酸：

$$5'—\overset{m}{C}pG—3'$$
$$3'—Gp\underset{m}{C}—5'$$

许多这样的二核苷酸位点成簇存在于位于启动子和上游调控序列中的 CpG 岛区（图 9.3）。

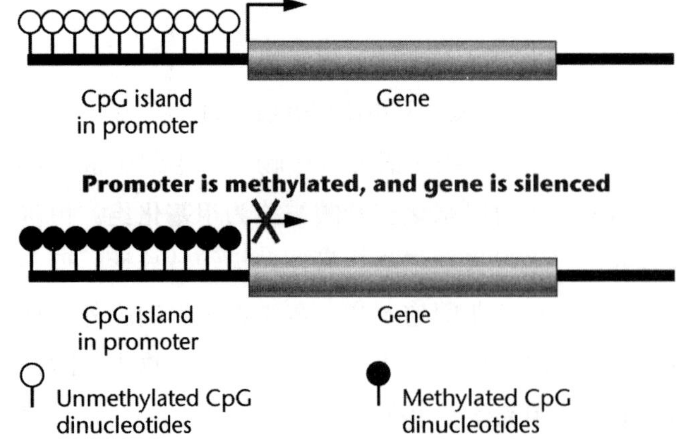

FIGURE 9.3 Methylation patterns of CpG dinucleotides in promoters control activity of adjacent genes. CpG islands outside and within genes also have characteristic methylation patterns, contributing to the overall level of genome methylation.

图 9.3 启动子中 CpG 二核苷酸的甲基化模式决定了相邻基因的活性。基因外和基因内的 CpG 岛也存在特征性的甲基化模式，从而形成基因组甲基化的整体水平。

CpG islands and promoters adjacent to essential genes (housekeeping genes) and cell-specific genes are unmethylated, making them available for transcription. Genes with adjacent methylated CpG islands and methylated CpG sequences within promoters are transcriptionally silenced. The added methyl groups occupy the major groove of DNA and silence genes by blocking the binding of transcription factors and other proteins necessary to form transcription complexes.

与必需基因（持家基因）和细胞特异性基因相邻的 CpG 岛和启动子是非甲基化的，因此可以被转录。与被甲基化的 CpG 岛或含甲基化 CpG 序列的启动子相邻的基因则发生转录沉默。添加的甲基基团会占据 DNA 的大沟并通过阻止转录因子和其他转录复合体形成所需的蛋白质与基因的结合来使基因沉默。

However, the bulk of methylated CpG dinucleotides are not adjacent to genes; instead they are in the repetitive DNA sequences of heterochromatic regions of the genome, including the centromere. Methylation of these sequences contributes to silencing of transcription and replication of transposable elements such as LINE and SINE sequences which constitute a major portion of the human genome. Heterochromatic methylation is important in maintaining chromosome stability by preventing translocations and related abnormalities.

然而，大多数甲基化 CpG 二核苷酸并不与基因相邻；而是存在于基因组中含着丝粒在内的异染色质区的 DNA 重复序列中。这些序列的甲基化有助于转录沉默和人类基因组中大量存在的如 LINE 序列与 SINE 序列等转座元件的复制。异染色质甲基化对阻止转座及相关染色体畸变的发生，从而维护染色体的稳定性具有十分重要的意义。

In mammals and other vertebrates, methylation of cytosine to form 5mC is the most common epigenetic modification of DNA. However, in the genomes of some eukaryotes, including the algae *Chlamydomonas reinhardtii* and the nematode *Caenorhabditis elegans*, 5mC is absent or, as in *Drosophila*, may be present at almost undetectable levels. Recent work has shown that although the methylomes of these and some other eukaryotes may not contain 5mC, they do contain adenine that

在哺乳动物及其他脊椎动物中，胞嘧啶通过甲基化形成 5mC 是 DNA 最为常见的表观遗传修饰方式。然而，一些真核生物基因组中并不存在 5mC，如绿藻（*Chlamydomonas reinhardtii*）和线虫（*Caenorhabditis elegans*）；或者也许存在但水平很低，几乎不能被检测出，如果蝇（*Drosophila*）。近期的研究工作已经表明尽管这些物种和其他真核生物的表观基因组中不含 5mC，但是它们的确含有 6 号氮原子上被甲基化的腺嘌呤（6mA），这种化

has been methylated at its N6 position (6mA), a modification that may have epigenetic functions. At this early stage, further research is needed in these species to fully explore the details of how 6mA controls gene expression. In addition, the extent to which 6mA is present in the methylomes of other organisms including mammals, and has epigenetic functions, remains to be determined.

学修饰可能同样具有表观遗传功能。在当前起步阶段，还需进一步深入探索这些物种中 6mA 控制基因表达的更多细节。此外，在其他生物包括哺乳动物的表观基因组中，6mA 存在的程度及其表观遗传功能依然尚待确定。

Histone Modification and Chromatin Remodeling

组蛋白修饰与染色质重塑

Interaction of DNA with proteins that facilitate transcription is controlled by two processes: (1) chromatin remodeling, which involves the action of ATP-powered protein complexes that move, remove, or alter nucleosomes, and (2) histone modifications, which are covalent posttranslational modifications of amino acids near the N-terminal ends of histone proteins. Together, these two processes activate or repress transcription and act as one of the primary methods of gene regulation.

促进转录的 DNA- 蛋白质相互作用受到两个过程的调控：（1）染色质重塑，包含由 ATP 驱动的蛋白质复合体的移动、移除或改变核小体的行为；（2）组蛋白修饰，即在组蛋白 N 末端附近的氨基酸上进行共价翻译后修饰。这两个过程共同激活或阻抑转录，成为基因调控的基本方法之一。

Normally, DNA is wound tightly around nucleosomes to form chromatin, which is further coiled and packaged to form chromosomes. In this state, regulatory regions of DNA and the genes themselves are unable to interact with proteins that facilitate transcription.

通常，DNA 紧密缠绕核小体形成染色质，并进一步卷曲、包装形成染色体。此时，DNA 的调控区和基因本身均无法与促进转录发生的蛋白质相互作用。

The N-terminal region of each histone extends beyond the nucleosome, forming a tail. Amino acids in these tails can be covalently modified in several ways, and a number of different proteins are involved in the process.

每个组蛋白的 N 末端区域延伸至核小体之外，形成尾巴。这些尾巴上的氨基酸可通过多种方式进行共价修饰，并且有大量不同的蛋白质参与该修饰过程。这些蛋白质有的能在组蛋白上添加化学基团（“书

These include proteins that add chemical groups to histones (“writers”), proteins that interpret those modifications (“readers”), and proteins that remove those chemical groups (“erasers”).

写者”），有的可以解读修饰信息（“阅读者”），还有的能移除化学基团（“擦除者”）。

Over 20 different chemical modifications can be made to histones, but the major changes include the addition of acetyl, methyl, and phosphate groups. Such additions alter the structure of chromatin, making genes on nucleosomes with modified histones accessible or inaccessible for transcription. Histone acetylation, for example, relaxes the grip of histones on DNA and makes genes available for transcription [Figure 9.4(a)]. Furthermore, acetylation is reversible. Removing (erasing)

目前人们已经发现超过 20 种的发生在组蛋白上的化学修饰方式，其中主要的化学修饰方式包括乙酰化、甲基化和磷酸化。这些化学基团的添加会改变染色质结构，从而使含组蛋白修饰的核小体上的基因更易于被转录或更难于被转录。例如，组蛋白乙酰化可以释放组蛋白对 DNA 的束缚，从而使基因能够进行转录 [图 9.4(a)]。而且，乙酰化是可逆的。移除乙酰基团有助于改变染色质，使其由“开放”构象变为“关闭”状态，因此通过使其无法进行转录而让基

“Open” configuration.
DNA is unmethylated, and histones are acetylated.
Genes can be transcribed.

(a)

“Closed” configuration.
DNA is methylated at CpG islands (black circles), and histones are deacetylated.
Genes cannot be transcribed.

(b)

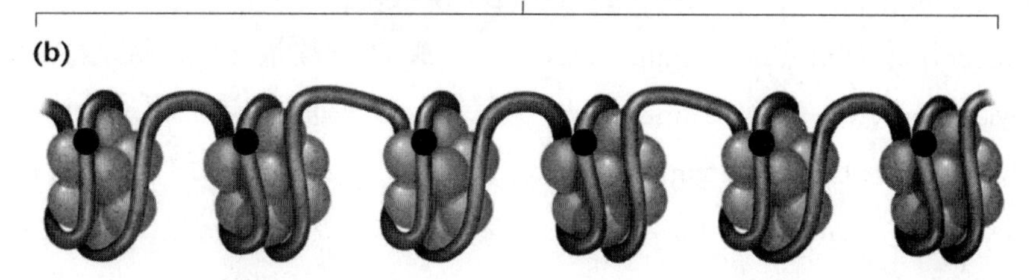

FIGURE 9.4 Epigenetic modifications to the genome alter the spacing of nucleosomes. (a) In the open configuration, chromatin remodeling shifts nucleosome positions, CpGs are unmethylated, and the genes on the DNA are available for transcription. (b) In the closed configuration, DNA is tightly wound onto the nucleosomes, CpGs are methylated, chemical groups have been removed from histones, and genes on the DNA are unavailable for transcription.

图 9.4 基因组的表观遗传修饰改变核小体间距离。(a) 在开放构象中，染色质重塑改变了核小体位置，CpG 未被甲基化，DNA 上的基因能够转录。(b) 在关闭构象中，DNA 紧密缠绕在核小体上，CpG 发生甲基化，化学基团从组蛋白上被移除，DNA 上的基因无法转录。

acetyl groups contributes to changing chromatin from an “open” configuration to a “closed” state, thereby silencing genes by making them unavailable for transcription [Figure 9.4(b)].

因沉默 [图 9.4(b)]。

Histone modifications occur at specific amino acids in the N-terminal tail of histones 2A, 2B, H3, and H4. Many combinations of histone modifications are possible within and between histone molecules, and the sum of their complex patterns and interactions is called the **histone code**. The basic idea behind a histone code is that reversible enzymatic modification of histone amino acids (by writers and erasers) recruits nucleoplasmic proteins (readers) that either further modify chromatin structure or regulate transcription.

组蛋白修饰发生在组蛋白 2A、2B、H3 和 H4 的 N 末端尾巴的特定氨基酸上。在组蛋白分子内部或组蛋白分子之间可能存在多种组蛋白修饰的不同组合方式，这些复杂模式和相互作用的总和被称为**组蛋白密码**。组蛋白密码的核心思想是组蛋白氨基酸的可逆酶学修饰（由“书写者”或“擦除者”参与），并招募可进一步修饰染色质结构或调控基因转录的核质蛋白（“阅读者”）。

The code is represented in a shorthand as follows:

■ Name of the histone (e.g., H3)

■ Single-letter abbreviation for the amino acid (e.g., K for lysine)

■ Position of the amino acid in the protein (e.g., 27)

■ Type of modification (ac = acetyl, me = methyl, p = phosphate, etc.)

■ Number of modifications (amino acids can be methylated one, two, or three times)

Thus, H3K27me3 represents a trimethylated lysine at position 27 from the N-terminus of histone H3.

组蛋白密码利用以下速记方法进行表示：

■ 组蛋白名称（如：H3）

■ 使用单字符缩写表示氨基酸（如：用 K 表示赖氨酸）

■ 氨基酸在蛋白质中的位置（如：27）

■ 修饰类型(ac = 乙酰化; me = 甲基化; p = 磷酸化；等)

■ 修饰数量（如：氨基酸可以被单甲基化、二甲基化或者三甲基化）

因此，H3K27me3 代表组蛋白 H3 的 N 末端第 27 位赖氨酸发生三甲基化。

Short and Long Noncoding RNAs

In addition to messenger RNA (mRNA), genome transcription produces several classes of **noncoding RNAs (ncRNAs)**, which are

短非编码 RNA 和长非编码 RNA

除信使 RNA（mRNA）之外，基因组转录还会产生几类**非编码 RNA (ncRNAs)**，它们由 DNA 转录而来，但不能被翻译成蛋

transcribed from DNA but not translated into proteins. The ncRNAs related to epigenetic regulation include two groups: (1) short ncRNAs (less than 31 nucleotides) and (2) long ncRNAs (greater than 200 nucleotides). Both types of ncRNAs have several roles, including the formation of heterochromatin, histone modification, site-specific DNA methylation, and gene silencing. They are also important in epigenetic regulatory networks.

白质。与表观遗传学调控相关的 ncRNAs 有两种：（1）短 ncRNAs（31 个核苷酸以内）；（2）长 ncRNAs（200 个核苷酸以上）。这两类 ncRNAs 均扮演着多种角色，包括异染色质形成、组蛋白修饰、位点特异性 DNA 甲基化和基因沉默。此外，它们在表观遗传调控网络中也十分重要。

There are three classes of short ncRNAs: miRNAs (microRNAs), siRNAs (short interfering RNAs), and piRNAs (piwi-interacting RNAs). miRNAs and siRNAs are transcribed as precursor molecules about 70 to 100 nucleotides long that contain a double-stranded stem-loop and single-stranded regions. After several processing steps that shorten the RNAs to lengths of 20 to 25 ribonucleotides, these RNAs act as repressors of gene expression. The origin of piRNAs is unclear, but they interact with proteins to form RNA-protein complexes that participate in epigenetic gene silencing in germ cells.

短 ncRNAs 共有三种：miRNAs（微 RNA）、siRNAs(干扰短 RNA)和 piRNAs(piwi 相互作用 RNAs）。miRNAs 和 siRNAs 被转录成长度为 70—100 个核苷酸的前体分子，含一个双链茎环结构和多个单链区。后经多个加工步骤，其 RNA 分子长度被缩短至 20—25 个核苷酸，这些 RNA 分子可作为基因表达的抑制因子。piRNAs 的来源尚不清楚，但是它们通过与蛋白质相互作用形成 RNA- 蛋白质复合物，参与生殖细胞中表观遗传基因沉默过程。

Long noncoding RNAs (lncRNAs) share many properties with mRNAs; they often have 5’-caps and 3’ poly-A tails and are spliced. What distinguishes lncRNAs from coding (mRNA) transcripts is the lack of an extended open reading frame that codes for the insertion of amino acids into a polypeptide.

长非编码 RNAs（lncRNAs）与 mRNAs 具有很多相似之处，比如：它们通常都含有 5’ 端 - 帽子和 3’ 端 - 多聚腺嘌呤尾巴，而且能够被剪切。lncRNAs 与编码（mRNA）转录本的区别在于其缺少一段延伸的用以编码可翻译成多肽链氨基酸序列的开放阅读框。

RNA genomic sequencing has identified more than 15,000 lncRNA genes in the human genome. lncRNAs are found in the nucleus and the cytoplasm, and through a variety of

RNA 基因组测序在人类基因组中已鉴定出 15 000 个以上的 lncRNA 基因。lncRNAs 存在于细胞核与细胞质中，并通过多种机制参与基因表达的转录调控和转录

mechanisms, are involved in both transcriptional and post-transcriptional regulation of gene expression. As epigenetic initiators, lncRNAs bind to chromatin-modifying enzymes and direct their activity to specific regions of the genome. At these sites, the lncRNAs direct chromatin modification, altering the pattern of gene expression.

后调控。作为表观遗传的启动者，lncRNAs 与染色质修饰酶相结合，并引导它们作用于基因组中的特定区域。在这些区域，lncRNAs 指导染色质修饰，并改变基因的表达模式。

In summary, epigenetic modifications alter chromatin structure by several mechanisms: DNA methylation, reversible covalent modification of histones, and action of short and long RNAs, all without changing the sequence of genomic DNA. This suite of epigenetic changes creates an epigenome that, in turn, can regulate normal development and generate changes in gene expression as a response to environmental signals.

总之，表观遗传修饰通过多种机制改变染色质结构：DNA 甲基化、组蛋白可逆共价修饰、短非编码 RNA 和长非编码 RNA 的作用，所有这些均不会改变基因组 DNA 的序列信息。这一系列表观遗传改变产生了表观基因组，由此调控个体正常发育，并为响应环境因子产生基因表达变化。

9.2 Epigenetics and Monoallelic Gene Expression

9.2 表观遗传学与单等位基因表达

Mammals inherit a maternal and a paternal copy of each gene, and aside from genes on the inactivated X chromosome in females, both copies of these genes are usually expressed at equal levels in the offspring. However, in some cases only one allele is transcribed, while the other allele is transcriptionally silent. This phenomenon is called **monoallelic expression** (**MAE**).

哺乳动物的每种基因分别从母本和父本获得一个基因拷贝，除雌性个体内位于失活 X 染色体上的基因外，这些基因的两个拷贝通常在后代中表达相同水平。然而，在某些情况下，只有一个等位基因拷贝被转录，而另一个等位基因拷贝则转录沉默，这种现象被称为**单等位基因表达**（MAE）。

There are three major classes of MAE. In one class, genes are expressed in a *parent-of-origin pattern*; that is, certain genes show expression of only the maternal allele or the

单等位基因表达可分为三大类。在第一类中，基因以亲本模式进行表达，也就是说，某些基因仅表达母本来源的等位基因或父本来源的等位基因，这种现象被称

paternal allele, a phenomenon called **genomic imprinting**. The remaining two classes involve *random* monoallelic expression. First is the random inactivation of one X chromosome in the cells of mammalian females, which compensates for their increased dosage of X-linked genes (recall that mammalian males have only one X chromosome). Second, a randomly generated pattern of allele inactivation, independent of parental origin, is observed in a significant number of autosomal genes. We will look at each of these three classes in turn.

为**基因组印记**。其余两类则涉及随机单等位基因表达：其一，雌性哺乳动物细胞中的一条X染色体发生随机失活，从而补偿它们多出的X连锁基因的基因剂量（雄性哺乳动物细胞中仅含一条X染色体）；其二，在大量常染色体基因中观察到的随机产生的等位基因失活模式，这与基因的亲本来源无关。下面我们将依次讨论这三类单等位基因表达模式。

Parent-of-Origin Monoallelic Expression: Imprinting

Parentally imprinted genes are marked in male and female germ-line cells during gamete formation; the fertilized egg thus has different marks on the copies of certain genes that came from the mother or the father. How is this marking accomplished?

To begin with, the DNA carried by sperm and eggs are highly methylated. However, shortly after fertilization, most of the methylation marks are erased. This modification of the DNA resets embryonic cells to a pluripotent state, allowing them to undergo new epigenetic modifications to form the more than 200 cell types found in the adult body. About the same time the embryo is implanting in the wall of the uterus, cells take on tissue-specific epigenetic identities, and methylation patterns and histone modifications change rapidly to reflect those seen in differentiated cells.

Some genomic regions, however, escape

亲本来源单等位基因表达：印记

亲本印记基因在配子形成阶段的雄性和雌性生殖细胞中被标记；因此，受精卵中某些基因拷贝携带来自母本或父本的不同标记。那么如何完成这些标记过程呢？

最初，精子和卵子携带的DNA被高度甲基化。然而，受精后不久，大部分甲基化标记被擦除。DNA的这种修饰方式将胚胎细胞重置为多潜能状态，允许它们进行新的表观遗传修饰，从而形成成年个体中所发现的200多种细胞类型。大约在胚胎植入子宫壁的同时，细胞呈现出组织特异性的表观遗传特征，并且甲基化模式和组蛋白修饰迅速发生变化，正如在分化细胞中所见到的情形那样。

然而，某些基因组区域在这些多次整

these rounds of global demethylation and remethylation. The genes contained in these regions remain imprinted with the methylation marks of the maternal and/or paternal chromosomes. These original parental patterns of methylation produce allele-specific imprinting. Imprinted alleles remain transcriptionally silent during embryogenesis and later stages of development. For example, if the allele inherited from the father is imprinted, it is silenced, and only the allele from the mother is expressed.

体去甲基化和再甲基化过程中发生逃逸。这些区域所含的基因仍保留来自其母本和/或父本染色体上的甲基化标记。这些原始亲本模式的甲基化会产生等位基因特异性印记，被打上印记的等位基因在胚胎形成和随后的发育阶段中将保持转录沉默。例如，如果遗传自父本的等位基因被打上印记，转录被沉默，那么仅有来自母本的等位基因会表达。

In humans, imprinted genes are usually found in clusters on the same chromosome and can occupy more than 1000 kb of DNA. Because these genes are located near each other at a limited number of sites in the genome, mutation in one imprinted gene can often affect the function of adjacent or coordinately controlled imprinted genes, thereby amplifying the mutation's phenotypic impact. These mutations in imprinted genes can arise through changes in the DNA sequence or by resultant dysfunctional epigenetic changes, called **epimutations**, both of which can cause heritable changes in gene activity.

在人类中，印记基因通常成簇存在于同一条染色体上，并且占据 1000 kb 以上的 DNA 片段。由于这些基因在相对有限的基因组区域中彼此靠近，因此一个印记基因发生突变通常也会影响其相邻的或协调控制的其他印记基因的功能，从而放大了该突变的表型影响。这些印记基因突变可通过 DNA 序列变化产生，或由导致功能失调的表观遗传变化（即**表观突变**）产生，二者均能引起基因活性的可遗传变化。

Occasionally, the imprinting process goes awry and is dysfunctional. In such cases, the imprinting defects can cause human disorders such as Beckwith-Wiedemann syndrome, Prader-Willi syndrome, Angelman syndrome, and several other diseases (Table 9.1). However, given the number of imprinting-susceptible candidate genes and the possibility that additional imprinted genes remain to be discovered, the overall number of imprinting-

有时，印记过程会发生错误，产生功能失调。在这种情况下，印记缺陷将引发人类疾病，例如贝 - 维综合征、普拉德 - 威利综合征、快乐木偶综合征和其他疾病（表 9.1）。但是，考虑到印记敏感的候选基因数量以及尚存在有待发现的其他印记基因的可能性，印记相关遗传疾病种类的总数可能更多。

related genetic disorders may be much higher.

TABLE 9.1 Some Imprinting Disorders in Humans

表 9.1 一些人类印记缺陷

Disorder	Locus
Albright hereditary osteodystrophy	20q13
Angelman syndrome	15q11–q15
Beckwith-Wiedemann syndrome	11p15
Prader-Willi syndrome	15q11–q15
Silver-Russell syndrome	Chromosome 7
Uniparental disomy 14	Chromosome 14

In humans, most known imprinted genes encode growth factors or other growth-regulating genes. An autosomal dominant disorder associated with imprinting, Beckwith – Wiedemann syndrome (BWS) occurs in about 1 in 13,700 births and offers insight into how disruptions of epigenetic imprinting can lead to an abnormal phenotype. BWS is a prenatal overgrowth disorder typified by abdominal wall defects, enlarged organs, high birth weight, and a predisposition to cancer. The genes associated with BWS are in a cluster of epigenetically imprinted genes on the short arm of chromosome 11 (Figure 9.5). BWS is not caused by mutation in the DNA sequence of the gene, nor is it associated with any chromosomal aberrations. Instead, BWS is a disorder of imprinting and is caused by abnormal patterns of DNA methylation resulting in altered patterns of gene expression, and not by mutations that change the nucleotide sequence of the genes involved.

在人类中，大多数已知的印记基因编码生长因子或其他生长调节基因。贝 - 维综合征（BWS）是一种印记相关的常染色体显性遗传疾病，大约每 13 700 个新生儿中会有 1 例患病，它为表观遗传印记失败如何导致表型异常的研究提供了视角。BWS 是一种产前过度生长疾病，具有腹壁缺损、器官增大、新生儿超重和易患癌症等特征。BWS 相关基因位于人类 11 号染色体短臂的表观遗传印记基因簇中（图 9.5）。BWS 并非由基因 DNA 序列发生突变所致，也与任何染色体畸变无关。事实上，BWS 是一种基因印记紊乱，由 DNA 甲基化异常最终导致基因表达模式改变，而并非由相关基因的核苷酸序列突变所致。

All the genes in this cluster are known to regulate growth during prenatal development. Two closely linked genes in this cluster are

研究发现该基因簇中所有基因均参与调节胚胎发育过程中的生长。其中，两个紧密连锁的基因分别是胰岛素样生长因

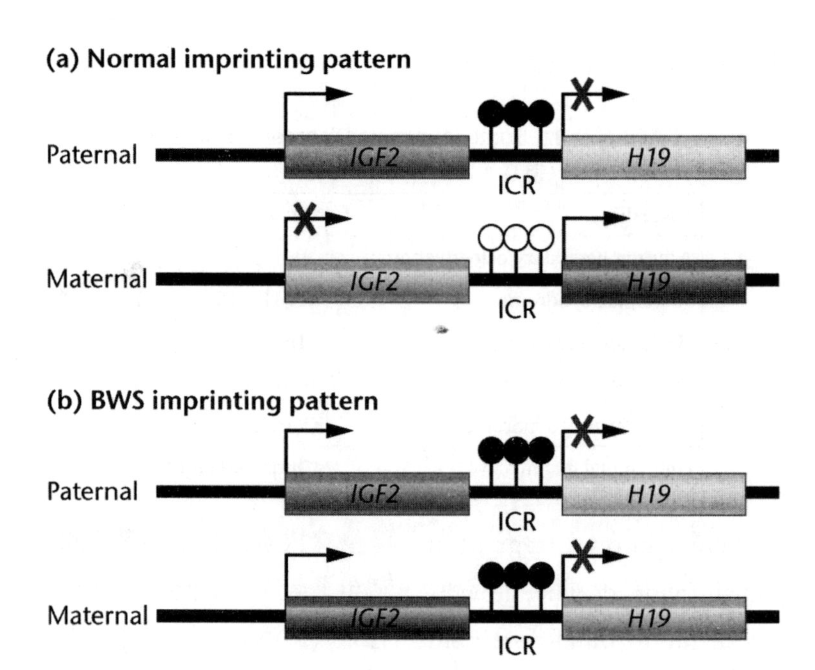

FIGURE 9.5 The imprinted region of human chromosome 11. (a) In normal imprinting, the ICR on the paternal chromosome is methylated (filled circles); the *IGF2* allele is active and the *H19* allele is silent. The ICR on the maternal chromosome is not methylated (open circles), and the *IGF2* allele is silent while the *H19* allele is active. (b) In one form of BWS, both the maternal and paternal ICRs are methylated (filled circles), both *IGF2* alleles are active, and both *H19* alleles are silent. The result is dysregulation of cell growth, resulting in the overgrowth of structures that are characteristic of BWS.

图 9.5 人类 11 号染色体的印记区域。(a) 在正常印记过程中，父本染色体上的 ICR 被甲基化（实心圆），*IGF2* 等位基因有活性，*H19* 等位基因发生沉默。母本染色体上的 ICR 不发生甲基化（空心圆），*IGF2* 等位基因发生沉默，*H19* 等位基因有活性。(b) 在一种形式的 BWS 中，母本和父本染色体上的 ICR 均被甲基化（实心圆），*IGF2* 等位基因均有活性，而 *H19* 等位基因均发生沉默。结果导致细胞生长调控失调，造成 BWS 的典型特征，即结构过度生长。

insulin-like growth factor 2 (*IGF2*), whose encoded protein plays an important role in growth and development, and *H19*, which is transcribed into an ncRNA. These two genes are separated by an imprinting control region (ICR), which controls the expression of both genes. Normally, the ICR on the paternal copy of chromosome 11 is methylated, allowing expression of the paternal *IGF2* allele but maintaining the paternal *H19* allele in a silenced state [Figure 9.5(a)]. Reciprocally, on the maternal copy of chromosome 11, the ICR is unmethylated allowing for the expression

子 -2（*IGF2*）和 *H19* 基因，*IGF2* 基因编码的蛋白质在生长发育中发挥着重要作用；*H19* 基因则转录成 ncRNA。这两个基因被一个同时调控二者表达的印记控制区（ICR）相隔开。通常，父本 11 号染色体上的 ICR 区会被甲基化，使得父本 *IGF2* 等位基因表达，而父本 *H19* 等位基因则保持沉默状态［图 9.5（a）］。与此相反的是，母本 11 号染色体上的 ICR 不发生甲基化，使得母本 *H19* 等位基因得以表达，而母本 *IGF2* 等位基因则保持沉默状态。在 BWS 的一种情形中，ICR 的两个亲本拷贝均被甲基化［图 9.5（b）］，则母本和父本的 *IGF2* 等位基因均

of the maternal *H19* allele, while the maternal *IGF2* allele is maintained in a silenced state. In one form of BWS, both copies of the ICR are methylated [Figure 9.5(b)] and both the maternal and paternal *IGF2* alleles are transcribed, resulting in the overgrowth of tissues that are characteristic of this disease. The transcription of both *IGF2* alleles is accompanied by silencing of both copies of the *H19* allele, further compounding the overgrowth of tissues.

被转录，从而导致该疾病所具有的组织过度生长特征。两个 *IGF2* 等位基因的转录伴随着两个 *H19* 等位基因的沉默，进一步加剧了组织的过度生长。

The known number of imprinted genes represents less than 1 percent of the mammalian genome, but they play major roles in regulating growth and development during prenatal stages. Because they act so early in life, any external or internal factors that disturb the epigenetic patterns of imprinting or the expression of these imprinted genes can have serious phenotypic consequences.

已知印记基因的数量仅占哺乳动物基因组的不足 1%，但是它们在产前阶段调节胚胎生长发育中起着重要作用。由于它们在生命的发育早期发挥作用，因此任何干扰印记表观遗传模式或印记基因表达的外部因素或内部因素均会产生严重的表型后果。

Random Monoallelic Expression: Inactivation of the X Chromosome

随机单等位基因表达：X 染色体失活

The random inactivation of an X chromosome in cells of female mammals was the first example of epigenetic allele-specific regulation to be identified. At an early stage of development, about half of embryonic cells randomly inactivate the maternal X chromosome and the other half inactivate the paternal X chromosome, effectively silencing almost all the 900 or so genes on whichever homolog is inactivated. Once inactivated, the same X chromosome remains silenced in all cells descended from this progenitor cell.

雌性哺乳动物细胞中的一条 X 染色体发生随机失活是被证实的表观遗传等位基因特异性调控的首个例子。在发育的早期阶段，大约一半的胚胎细胞随机失活母本 X 染色体，而另一半则失活父本 X 染色体，从而有效沉默被失活的 X 染色体上几乎所有的约 900 种基因。一旦发生失活，同一亲本来源的 X 染色体将在该细胞的所有后代细胞中均保持沉默。

How does X inactivation occur? Several

X 失活是如何发生的呢？几种 lncRNAs

lncRNAs play a key role in this process. Two of the major contributors are Xist (X inactive specific transcript), and Tsix (Xist spelled backward), which are sense and antisense transcripts of the same gene (transcribed in opposite directions). The Xist lncRNA is expressed on the inactivated X chromosome and coats the entire chromosome, converting it into a Barr body (see Chapter 5), which is a highly condensed and genetically silent chromatin structure. The lncRNA Tsix is expressed on the active X chromosome and represses expression of the Xist lncRNA, thus preventing the active X chromosome from being silenced.

在该过程中起到了关键作用。其中两个主要贡献者是 Xist（X 染色体失活特异转录因子）和 Tsix（Xist 逆向拼写），它们分别是同一基因的有义链和反义链的转录物（转录方向相反）。Xist lncRNA 在失活 X 染色体上进行表达并包裹整条染色体，将其转变为巴氏小体（请参阅第 5 章）。巴氏小体是一种高度聚集压缩且遗传沉默的染色质结构。Tsix lncRNA 分子则在有活性的 X 染色体上表达，并抑制 Xist lncRNA 的表达，从而防止有活性的 X 染色体发生沉默。

Random Monoallelic Expression of Autosomal Genes

常染色体基因的随机单等位基因表达

Genome-wide analysis of allele-specific expression in mice and humans led to the surprising discovery that monoallelic expression (MAE) is a widespread event, involving 10 to 20 percent of autosomal genes in a range of different cell types.

小鼠和人类的等位基因特异性表达全基因组分析使人们惊奇地发现单等位基因表达（MAE）是广泛存在的事件，涉及不同细胞类型中 10%—20% 的常染色体基因。

Unlike imprinted genes, which are in clusters, autosomal MAE genes are scattered throughout the genome. Because autosomal MAE is a random process, four states of expression for a gene are possible in cells of a given tissue: (1) expression of both alleles (biallelic expression), (2) expression of only the maternal allele, (3) expression of only the paternal allele, or (4) expression of neither allele. These different patterns of expression, all present in the same tissue, can have an impact on the phenotype and may offer a molecular

与基因簇中的印记基因不同，常染色体 MAE 基因分散存在于整个基因组中。由于常染色体 MAE 是一种随机发生的过程，因此在指定组织的细胞中，一种基因有四种可能的表达状态：（1）两个等位基因均表达（双等位基因表达）；（2）仅母本等位基因表达；（3）仅父本等位基因表达；（4）两个等位基因均不表达。这些不同的基因表达模式同时存在于同一组织中，将对表型产生影响，并为某些遗传疾病中观察到的性状外显不全现象提供分子解释。（有关外显不全的讨论请参阅第 4 章。）

explanation for the incomplete penetrance of traits observed in some genetic disorders. (See Chapter 4 for a discussion of incomplete penetrance.)

By analyzing several epigenetic marks present in a wide range of cell types, researchers established that two modifications of histone 3, H3K27me3 and H3K36me3, explain most of the difference between cells with monoallelic expression of a given gene and cells with biallelic expression of the same gene. In MAE cells, the H3K27me3 marker, associated with gene silencing, is linked to the inactive allele, while the H3K36me3 marker, associated with transcription, is linked to the active allele. This chromatin signature is a powerful and reliable predictor of MAE activity in many cell types and offers a way of exploring the relationship between epigenetics and disease.

通过分析大量不同细胞类型中存在的不同表观遗传标记，研究人员确定了组蛋白 3 的两种修饰方式，即 H3K27me3 和 H3K36me3，这可以解释指定基因单等位基因表达细胞和同一基因双等位基因表达细胞的大部分差异。在 MAE 细胞中，基因沉默相关的 H3K27me3 标记与失活的等位基因相关联，而转录相关的 H3K36me3 标记则与有活性的等位基因相关联。这种染色质特征标记是许多细胞类型中预测 MAE 活性的有力且可靠的工具，并为探索表观遗传学与疾病间的相互关系提供了途径。

Assisted Reproductive Technologies (ART) and Imprinting Defects

In the United States, **assisted reproductive technologies** (**ART**), including *in vitro* fertilization (IVF), are now used in over 1 percent of all births. Over the past decade, several studies have suggested that children born using ART have an increased risk for imprinting errors (epimutations) caused by the manipulation of gametes or embryos.

For example, the use of ART results in a four- to nine- fold increased risk of Beckwith-Wiedemann syndrome (BWS); in addition, there are increased risks for Prader-Willi syndrome (PWS) and Angelman syndrome (AS). Studies of

辅助生殖技术（ART）和印记缺陷

目前在美国，超过 1%的新生儿采用了包括体外受精（IVF）在内的**辅助生殖技术**（**ART**）。在过去十年中，多项研究已经表明，借助 ART 出生的儿童因配子或胚胎操作引发印记错误（表观突变）的风险会增加。

例如，使用 ART 导致贝 - 维综合征（BWS）的患病风险增加 4—9 倍；此外，普拉德 - 威利综合征（PWS）和快乐木偶综合征（AS）的患病风险也有所增加。对通过体外受精受孕且患有 BWS 或 AS 的儿童

children with BWS or AS conceived by IVF have shown that they have reduced levels or complete loss of maternal-specific methylation at known imprinting sites in the genome, confirming the role of epigenetics in these cases. ART procedures are done at times when the oocyte and the early embryo are undergoing epigenetic reprogramming. It appears that disturbances in epigenetic programming at sensitive times during development may be responsible for the increased risk of these disorders.

开展的研究表明，他们基因组中已知印记位点的母本特异性甲基化水平降低甚至完全消失，从而证实了表观遗传学在这些病例中的作用。ART 相关操作正处于卵母细胞和早期胚胎进行表观遗传重新调整的过程中。这样看来，在发育的敏感阶段出现的表观遗传运行紊乱可能是造成这些疾病风险增加的原因。

Although imprinting errors are uncommon in the general population (BWS occurs in only about 1 in 13,700 births), epimutations may be a significant risk factor for those conceived by ART.

尽管印记错误在总人群中并不常见（BWS 的发病率仅为 1/13 700），但表观突变可能是 ART 受孕个体最为显著的风险因素。

9.3 Epigenetics and Cancer

9.3 表观遗传学与癌症

Originally it was thought that cancer is clonal in origin and begins in a single cell that has accumulated a suite of mutations that allow it to escape control of the cell cycle. Subsequent mutations allow cells of the tumor to become metastatic, spreading the cancer to other locations in the body where new malignant tumors appear. However, converging lines of evidence are now clarifying the importance of epigenetic changes in the initiation and maintenance of malignancy. These findings are helping researchers understand the properties of cancer cells that are difficult to explain by the action of mutant alleles alone. Evidence for the role of epigenetic changes in cancer has established epigenomic changes as a major

人们最初认为癌症起源于无性繁殖系，它始于已积累系列基因突变并成功逃脱细胞周期控制的单个细胞。随后的突变使癌细胞具有转移性，并将癌症扩散至体内其他位置，引发新的癌症。然而，目前越来越多的证据表明了表观遗传变化在癌症发生和持续发展中的重要性。这些发现正帮助研究人员了解癌细胞所具有的仅利用等位基因突变作用难以解释的特性。有关癌症中表观遗传变化功能的相关证据已证实表观基因组变化是癌细胞形成和扩散的主要途径。

pathway for the formation and spread of malignant cells.

DNA Methylation and Cancer

As far back as the 1980s, researchers observed that cancer cells had much lower levels of methylation than normal cells derived from the same tissue. Subsequent research by many investigators showed that complex changes in DNA methylation patterns are associated with cancer. These studies showed that genomic hypomethylation is a property of all cancers examined to date.

DNA hypomethylation reverses the silencing of genes, leading to unrestricted transcription of many gene sets— including those associated with the development of cancer. It also relaxes control over imprinted genes, causing cells to acquire new growth properties. Hypomethylation of repetitive DNA sequences in heterochromatic regions increases chromosome rearrangements and changes in chromosome number, both of which are characteristic of cancer cells.

Even though cancer cells are characterized by global hypomethylation, selected regions of their genome are hypermethylated when compared to normal cells. Selective hypermethylation of promoter-associated CpG islands silences certain genes, including tumor-suppressor genes, often in a tumor-specific fashion (Table 9.2). Analysis of these patterns provides a way to identify tumor types and subtypes and predict the sites to which the tumor may metastasize.

DNA 甲基化与癌症

早在 20 世纪 80 年代，研究人员就发现了癌细胞的甲基化水平远低于源自同一组织的正常细胞。随后开展的许多研究表明 DNA 甲基化模式的复杂变化与癌症有关。这些研究显示，基因组低甲基化是迄今研究的所有癌症的一个共同特征。

DNA 低甲基化将基因沉默逆转，导致许多成组基因不受限制地进行转录，包括那些癌症发生相关的基因。它还会放松对印记基因的管控，导致细胞获得新的生长特性。异染色质区内重复 DNA 序列的低甲基化将增加染色体重排和染色体数目变化，二者均为癌细胞所具有的特征。

尽管癌细胞的特征是整体低甲基化，但是其基因组中选定区域与正常细胞相比却是超甲基化的。启动子相关的 CpG 岛选择性地超甲基化通常以癌症特异性方式沉默包括肿瘤抑制基因在内的某些基因（表 9.2）。分析这些模式为鉴定癌症类型和亚型并预测癌症可能的转移部位提供了途径。

TABLE 9.2 Some Human Cancer-Related Genes Inactivated by Hypermethylation
表 9.2 一些人类癌症相关基因通过超甲基化发生失活

Gene	Locus	Function	Related Cancers
BRCA1	17q21	DNA repair	Breast, ovarian
APC	5q21	Nucleocytoplasmic signaling	Colorectal, duodenal
MLH1	3p21	DNA repair	Colon, stomach
RB1	13q14	Cell-cycle control point	Retinoblastoma, osteosarcoma
AR	Xq11–12	Nuclear receptor for androgen; transcriptional activator	Prostate
ESR1	6q25	Nuclear receptor for estrogen; transcriptional activator	Breast, colorectal

For example, the promoter region of the breast cancer gene *BRCA1* is hypomethylated in normal cells but is hypermethylated and inactivated in many cases of breast and ovarian cancer. In another example, silencing of the DNA repair gene *MLH1* by hypermethylation is a key step in the development of some forms of colon cancer.

例如，乳腺癌基因 *BRCA1* 的启动子区在正常细胞中是低甲基化的，但在许多乳腺癌和卵巢癌病例中却被超甲基化和失活。在另一个例子中，通过超甲基化使 DNA 修复基因 *MLH1* 发生沉默是某些类型结肠癌发展的关键步骤。

In fact, cancer is now viewed as a disease that usually results from the accumulation of both genetic and epigenetic changes (Figure 9.6). For example, in a bladder cancer cell line, one allele of a tumor-suppressor gene, *CDKN2A*, is mutated, and the other, normal, allele is silenced by hypermethylation. Because both alleles are inactivated (although by different mechanisms), cells can escape control of the cell cycle and divide continuously. Even more striking, in ovarian cancer, mutations in nine specific genes are predominant, but promoter hypermethylation is observed in 168 genes. These genes are epigenetically silenced, and their reduced expression is linked to the development and maintenance of this cancer.

实际上，目前癌症已被认为是一种通常由基因突变和表观遗传变化共同参与并累积所致的疾病（图 9.6）。例如，在一株膀胱癌细胞系中，肿瘤抑制基因 *CDKN2A* 的一个等位基因发生突变，而另一个正常的等位基因则因超甲基化而发生基因沉默。由于两个等位基因均失活（尽管基因失活机制不同），所以细胞能够逃脱细胞周期调控并持续进行细胞分裂。更令人惊讶的是，在卵巢癌中，有 9 种特定基因突变显著存在，但是还发现有 168 种基因存在启动子超甲基化现象。这些基因通过表观遗传学方式被沉默，并且这些基因表达水平降低与该癌症的发生和发展紧密相关。

The broad pattern of hypermethylation seen in cancer cells and the many functions of

癌细胞中观察到的广泛超甲基化模式及相关基因的许多功能表明这一现象可能

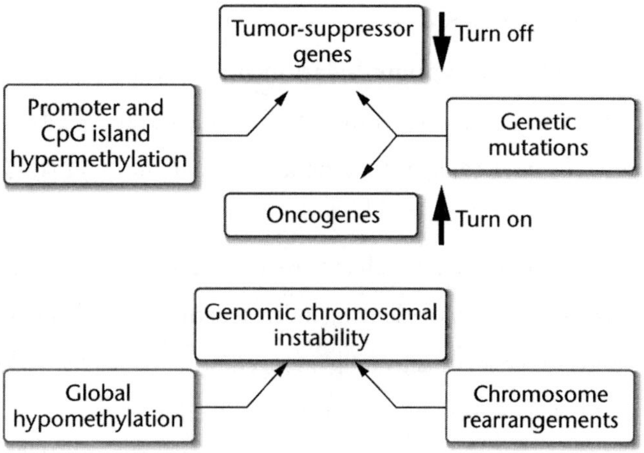

FIGURE 9.6 The development and maintenance of malignant growth in cancer involves the interaction of gene mutations, hypomethylation, hypermethylation, overexpression of oncogenes, and the silencing of tumor-suppressor genes.

图 9.6 癌症恶性生长的发展和持续包括基因突变、低甲基化、超甲基化、原癌基因过表达和抑癌基因沉默间的相互作用。

the affected genes suggest that this phenomenon may result from a widespread deregulation of the methylation process rather than a targeted event.

源自甲基化过程全面调控失常，而并非具有针对性的特定调控事件。

At the present time, many of the mechanisms that cause epigenetic changes in cancer cells are not well understood, partly because the changes take place very early in the conversion of a normal cell to a cancerous one, and partly because by the time the cancer is detected, alterations in methylation patterns have already occurred. The DNA repair gene *MLH1*, for example, plays an important role in genome stability, and silencing this gene by hypermethylation (as described earlier) causes instability in repetitive microsatellite sequences, which, in turn, is an important step in the development of colon cancer and several other cancers. In some individuals with colon cancer, the *MLH1* promoter in normal cells of the colon is already silenced by hypermethylation, indicating that this epigenetic event occurs very early in tumor formation before the development of downstream genetic mutations.

目前，引发癌细胞表观遗传变化的许多机制依然尚未得到很好的了解，部分原因是这些变化发生于正常细胞向癌细胞转变的极早期阶段，还有部分原因则是当癌症被检测到时，甲基化模式变化已然发生。例如，DNA 修复基因 *MLH1* 在维护基因组稳定中起着重要作用，而超甲基化使该基因沉默（如前所述），进而导致重复微卫星序列不稳定，而这正是结肠癌和其他几种癌症发展的重要环节。在一些结肠癌患者体内，结肠正常细胞的 *MLH1* 基因启动子已被超甲基化沉默，表明这一表观遗传学事件发生于下游基因突变发生之前，癌症形成的极早期阶段。

In summary, several lines of evidence support the role of epigenetic alterations in

总之，多组证据支持表观遗传变化在癌症发生中的作用：

cancer:

1. Global hypomethylation may cause genomic instability and the large-scale chromosomal changes that are a characteristic feature of cancer.

1. 广泛存在的低甲基化可能导致基因组不稳定和大规模染色体畸变，这些都是癌症的典型特征。

2. Epigenetic mechanisms can replace mutations as a way of silencing individual tumor-suppressor genes or activating oncogenes.

2. 表观遗传机制可以取代基因突变成为沉默单个肿瘤抑制基因或激活原癌基因的一种途径。

3. Epigenetic modifications can silence multiple genes, making them more effective in transforming normal cells into malignant cells than sequential mutations of single genes.

3. 表观遗传修饰可以沉默多个基因，从而比单个基因依次突变更加有效地将正常细胞转化为癌细胞。

Chromatin Remodeling and Histone Modification in Cancer

癌症中的染色质重塑与组蛋白修饰

In addition to abnormal regulation of methylation, many cancers also have altered patterns of chromatin remodeling. One form of remodeling is controlled by the reversible covalent modification of histone proteins in nucleosome cores. Recall that this process involves three classes of enzymes: *writers* that add chemical groups to histones; *erasers* that remove these groups; and *readers* that recognize and read the epigenetic marks. Abnormal regulation of each of these enzyme classes results in disrupted histone profiles and is associated with a variety of cancer subtypes.

除甲基化调控异常外，许多癌症也会改变染色质重塑模式。重塑的一种方式受核小体核心中组蛋白的可逆共价修饰过程调控。首先我们回顾一下该过程中所涉及的三种酶类：将化学基团添加到组蛋白上的“书写者”；移除这些化学基团的“擦除者”；识别并解读这些表观遗传标记的“阅读者”。这些酶类中的任何一种发生调控异常都会导致组蛋白信息遭受破坏，并与多种癌症亚型相关联。

Epigenetic Cancer Therapy

表观遗传癌症治疗

The fact that unlike genetic alterations, which are almost impossible to reverse, epigenetic changes are potentially reversible has inspired researchers to look for new classes of drugs to treat cancer. The focus of epigenetic

与基因改变几乎不可逆转的情形不同，表观遗传变化具有潜在可逆性，这一事实鼓舞着研究人员寻找治疗癌症的新型药物。在第一代药物开发中，表观遗传疗法着眼于重新激活被甲基化或组蛋白修饰沉默的

therapy in the development of first-generation drugs has been the reactivation of genes silenced by methylation or histone modification, essentially reprogramming the pattern of gene expression in cancer cells.

基因，其本质在于对癌细胞中的基因表达模式重新进行设定。

The U.S. Food and Drug Administration has approved several epigenetic drugs, and another 18 or more drugs are in clinical trials. One approved drug, Vidaza (azacitidine), is used in the treatment of myelodysplastic syndrome, a precursor to leukemia, and acute myeloid leukemia. This drug is an analog of cytidine and is incorporated into DNA during replication during the S phase of the cell cycle. Methylation enzymes (methyltransferases) bind irreversibly to this analog, preventing methylation of DNA at many other sites, effectively reducing the amount of methylation in cancer cells.

美国食品与药品监督管理局（FDA）已经审批通过多种表观遗传药物，另有18种或更多的药物正在临床试验。获批药物维达扎（阿扎胞苷）被用于治疗骨髓增生异常综合征，这种疾病是白血病和急性髓系白血病的前期病症。该药物属于胞苷类似物，在细胞周期的S期随着DNA复制被整合入DNA分子中。甲基化酶类（甲基转移酶）可以与该类似物发生不可逆结合，阻止DNA的许多其他位点发生甲基化，从而有效减少癌细胞内的甲基化数量。

Other drugs that inhibit histone deacetylases (HDACs) have been approved by the FDA for use in epigenetic therapy. Laboratory experiments with cancer cell lines indicate that inhibiting HDAC activity results in the re-expression of tumor-suppressor genes. HDAC inhibitors like Zolinza (vorinostat) are used to treat certain forms of lymphoma.

FDA批准的具有抑制组蛋白脱乙酰酶（HDACs）作用的其他药物也被用于表观遗传治疗。利用癌细胞系进行的相关实验室研究表明，抑制HDAC活性可使肿瘤抑制基因重新启动表达。HDAC抑制剂如佐林扎（伏立诺他）已被用于治疗某些类型的淋巴瘤。

The development of epigenetic drugs for cancer therapy is still in its infancy. The approved epigenetic drugs are only moderately effective on their own and are best used in combination with other anticancer drugs. To develop more effective drugs, several important questions remain to be answered: What causes cancer cells to respond to certain epigenetic drugs? Which combinations of chromatin remodeling drugs,

用于癌症治疗的表观遗传药物开发仍处于起步阶段。获批的表观遗传药物本身仅在一定程度上有效，而且最好与其他抗癌药物联合使用。为了开发更有效的药物，尚有一些重要问题亟待解答：癌细胞对某些表观遗传药物产生响应的原因是什么？对于特定癌症，染色质重塑药物、组蛋白修饰药物和常规抗癌药物如何组合才能达到最佳治疗效果？哪些表观遗传标记可以

histone modification drugs, and conventional anticancer drugs are most effective on specific cancers? Which epigenetic markers will be effective in predicting sensitivity or resistance to newly developed drugs? Further research into the mechanisms and locations of epigenetic genome modification in cancer cells will allow the design of more potent drugs to target epigenetic events as a form of cancer therapy.

有效预测新开发药物的敏感性或耐药性？癌细胞中表观遗传基因组修饰机制和修饰位点的深入研究将有助于设计更加有效的靶向表观遗传过程的药物，并成为癌症治疗的一种途径。

9.4 Epigenetic Traits Are Heritable

9.4 表观遗传性状具有可遗传性

Environmental agents including nutrition, exposure to chemicals, medical or recreational drugs, as well as social interactions and exercise can alter gene expression by affecting the epigenome. In humans it is difficult to determine the relative contributions of the environment as factors in altering the epigenome, but there is indirect evidence that changes in nutrition and exposure to agents that affect a developing fetus can have detrimental effects during adulthood.

营养、化学试剂暴露、医用或娱乐性毒品、社会互动和运动等环境因素均可通过影响表观基因组来改变基因的表达。在人类中，很难确定环境因子在改变表观基因组过程中的相对贡献程度，但是间接证据表明，营养变化和接触影响胎儿发育的药剂对于个体的成年阶段将造成有害影响。

During World War II, a famine in the western part of the Netherlands lasted from November 1944 to May 1945. During this time, daily food intake for adults was limited to 400 to 800 calories, well below the normal levels of 1800 to 2000 calories. Studies were conducted for decades afterward on the health of adult children of women who were pregnant or became pregnant during the famine. Overall, the findings show that the severity of health effects was correlated with prenatal time of exposure to famine conditions. Adults who were exposed early in prenatal development (an F_1 generation)

在第二次世界大战期间，发生在荷兰西部的饥荒从 1944 年 11 月一直持续至 1945 年 5 月。在此期间，成年人的每日食物摄入量仅为 400—800 cal（1 cal ≈ 4.1859 J），大大低于正常水平的 1800—2000 cal。此后数十年间，人们对于饥荒期间处于孕期或成功受孕的成年女性的子女成年后的健康状况进行了研究。总体而言，结果表明对健康影响的严重程度与处于饥荒期间的产前阶段有关。胚胎发生早期处于饥荒期的成年个体（F_1 代）相对于发育后期处于饥荒期的成年个体患某些疾病（包括肥胖症、心脏病和乳腺癌）的概率更高，死亡率也

had higher rates of several disorders— including obesity, heart disease, and breast cancer— and higher mortality rates than adults exposed later in development. In addition, as adults, there was increased risk for schizophrenia and other neuropsychiatric disorders for those with early exposure, perhaps related to nutritional deficiencies during development of the brain and nervous system. Some effects persisted in the F_2 generation, where adults had abnormal patterns of growth and increased rates of obesity. Other studies in China and Africa on the adult children of women who were pregnant or became pregnant during times of famine confirm the deleterious impact of poor maternal nutrition during pregnancy on offspring in subsequent generations.

更高。此外，胚胎发生早期处于饥荒期的成年个体患精神分裂症和其他神经精神疾病的风险会增加，这可能与大脑和神经系统发育过程中的营养缺乏有关。有些影响在 F_2 代个体中仍然持续存在，个体成年后生长模式异常，并且肥胖率增加。在中国和非洲发生的饥荒时期怀孕或受孕女性的成年子女相关研究同样证实，怀孕期间母体营养不良对后代个体将产生有害影响。

More direct evidence for the role of environmental factors in modifying the epigenome comes from studies in experimental animals. One dramatic example of how epigenome modifications affect the phenotype comes from the study of coat color in mice, where color is controlled by the dominant allele *Agouti* (A). In homozygous AA mice, the allele is active only during a specific time during hair development, resulting in a yellow band on an otherwise black hair shaft, producing the agouti phenotype. A nonlethal mutant allele (A^{vy}) causes yellow pigment formation along the entire hair shaft, resulting in yellow fur color. This allele is the result of the insertion of a transposable element near the transcription start site of the *Agouti* gene. A promoter element within the transposon is responsible for this change in gene

实验动物研究为环境因子在修饰表观基因组中的作用提供了更多的直接证据。表观基因组修饰影响表型的一个生动例子来自小鼠皮毛颜色的研究，小鼠皮毛颜色性状由显性等位基因 *Agouti*（A）控制。在纯合 AA 小鼠中，该等位基因仅在毛发发育的特定时期有活性，从而在黑色毛干上产生黄色条带，最终形成灰色表型。非致死突变基因（A^{vy}）会使小鼠沿着整个毛干形成黄色色素，从而形成黄色皮毛。该等位基因是在 *Agouti* 基因的转录起始位点附近插入一个转座元件所导致的，转座子内的启动子元件使基因表达发生了这般变化。

expression.

Researchers found that the degree of methylation in the transposon's promoter is related to the amount of yellow pigment deposited in the hair shaft and that the amount of methylation varies from individual to individual. The result is variation in coat color phenotypes even in genetically identical mice. In these mice, coat colors range from yellow (unmethylated promoter) to pseudoagouti (highly methylated promoter). In addition to a gradation in coat color, there is also a gradation in body weight. Yellow mice are more obese than the brown pseudoagouti mice and are more likely to be diabetic. Alleles such as A^{vy} that show variable expression from individual to individual in genetically identical strains caused by different patterns of epigenetic modifications to the alleles are called *metastable epialleles*. "Metastable" refers to the variable nature of the epigenetic modifications, and "epiallele" refers to the heritability of the epigenetic status of the altered gene. In other words, the epigenetic modifications to the A^{vy} allele can be passed on to offspring; this is a clear example of transgenerational inheritance.

研究人员发现转座子的启动子内甲基化程度与沉积在毛干中的黄色色素的量有关，并且甲基化程度在个体与个体之间存在差异，因此即使基因型相同的小鼠，它们的皮毛颜色也会有所不同。在这些小鼠中，皮毛颜色会从黄色（启动子未甲基化）变化到伪黑色（启动子高度甲基化）。除了皮毛颜色变化具有层次外，小鼠体重同样也存在梯度变化，黄色小鼠比伪黑色小鼠更肥胖，也更易患糖尿病。像A^{vy}等位基因这样，由于表观遗传修饰模式不同而使得基因型完全相同的个体之间基因表达多样化的等位基因被称为亚稳表观等位基因。“亚稳”是指表观遗传修饰具有多变特性，而“表观等位基因”则指出被改变基因的表观遗传状态具有可遗传性。换句话说，A^{vy}等位基因的表观遗传修饰可以传递给后代，并且这也是跨代遗传的一个很好的例子。

To evaluate the role of environmental factors in modifying the epigenome, the diet of pregnant A^{vy} mice was supplemented with methylation precursors, including folic acid, vitamin B_{12}, and choline. In the offspring, variation in coat color was reduced and shifted toward the pseudoagouti (highly methylated) phenotype. The shift in coat color was accompanied by increased methylation of the

为了评估环境因素在修饰表观基因组中的作用，研究人员在孕期A^{vy}小鼠的饮食中添加了甲基化前体，包括叶酸、维生素B_{12}和胆碱。在后代中，人们观察到皮毛颜色的变异程度降低，并向伪黑色（高度甲基化）表型转变，即皮毛颜色的变化伴随着转座子的启动子甲基化程度的增加。这些发现可以应用于人类表观遗传疾病的治疗，例如，一种类型的结直肠癌患病风险

transposon's promoter. These findings have applications to epigenetic diseases in humans. For example, the risk of one form of colorectal cancer is linked directly to increased methylation of the DNA repair gene *MLH1*.

即与 DNA 修复基因 *MLH1* 的甲基化水平升高直接相关。

Stress-Induced Behavior Is Heritable

A growing body of evidence shows that epigenetic changes, including alterations in DNA methylation and histone modification, have important effects on behavioral phenotypes.

One of the most significant findings in the epigenetics of behavior is that stress-induced epigenetic changes that occur prenatally or early in life can influence behavior (and physical health) later in adult life and can potentially be transmitted to future generations. A classic study showed that newborn rats raised with low levels of maternal nurturing (low-MN) did not adapt well to stress and anxiety inducing situations in adulthood. In rats and humans, the hypothalamic region of the brain mediates stress reactions by controlling levels of glucocorticoid hormones via the action of cell-surface glucocorticoid receptors (GRs).

In rats exposed to high levels of maternal nurturing care early in life (high-MN), GR expression is increased, and adults are stress adaptive. However, low-MN rats had reduced levels of GR transcription and were less able to adapt to stress. Differences in GR expression were associated with differences in histone acetylation and DNA methylation levels in the GR gene promoter. Low-MN rats had significantly higher levels of promoter methylation than high-MN

压力应激行为具有可遗传性

越来越多的证据表明，表观遗传变化，包括 DNA 甲基化和组蛋白修饰的变化，对行为表型有重要影响。

行为表观遗传学中最重要的发现之一是孕期或生命早期压力诱发的表观遗传变化将影响成年后的行为（和身体健康），并可能传递给后代。一项经典研究表明，新生大鼠的母亲如果养育质量低（low-MN），新生大鼠在成年后将不能很好地适应易引发压力和焦虑的环境。在大鼠和人类中，大脑的下丘脑区通过细胞表面糖皮质激素受体（GRs）来控制糖皮质激素水平，从而调节应激反应。

在生命早期，如果新生大鼠得到高质量的母亲护理（high-MN），GR 表达增加，则成年后将具有较强的压力适应性。然而，low-MN 大鼠的 GR 转录水平降低，适应压力的能力较弱。GR 表达差异与该基因启动子中组蛋白乙酰化和 DNA 甲基化水平的差异有关。low-MN 大鼠中该启动子甲基化水平明显高于 high-MN 大鼠（图 9.7）。

rats (Figure 9.7).

Subsequent research showed that differences in DNA methylation are present in hundreds of genes across the genome, all of which show differential expression in low-MN and high-MN adults. Later studies showed that these behavioral phenotypes can be transmitted across generations. Female rats raised by more nurturing mothers are more attentive to their own newborns, whereas those raised by less nurturing mothers are much less attentive and less nurturing to their offspring.

随后的研究表明，整个基因组中数百种基因都存在 DNA 甲基化差异，并且所有这些基因在 low-MN 成年鼠和 high-MN 成年鼠中都会出现差异表达。后续研究显示这些行为表型可以跨代传递，受到高质量母亲养育的雌鼠对自己的新生儿更加用心，而受到母亲抚养较少的雌鼠则对子女养育更少，也更不专心。

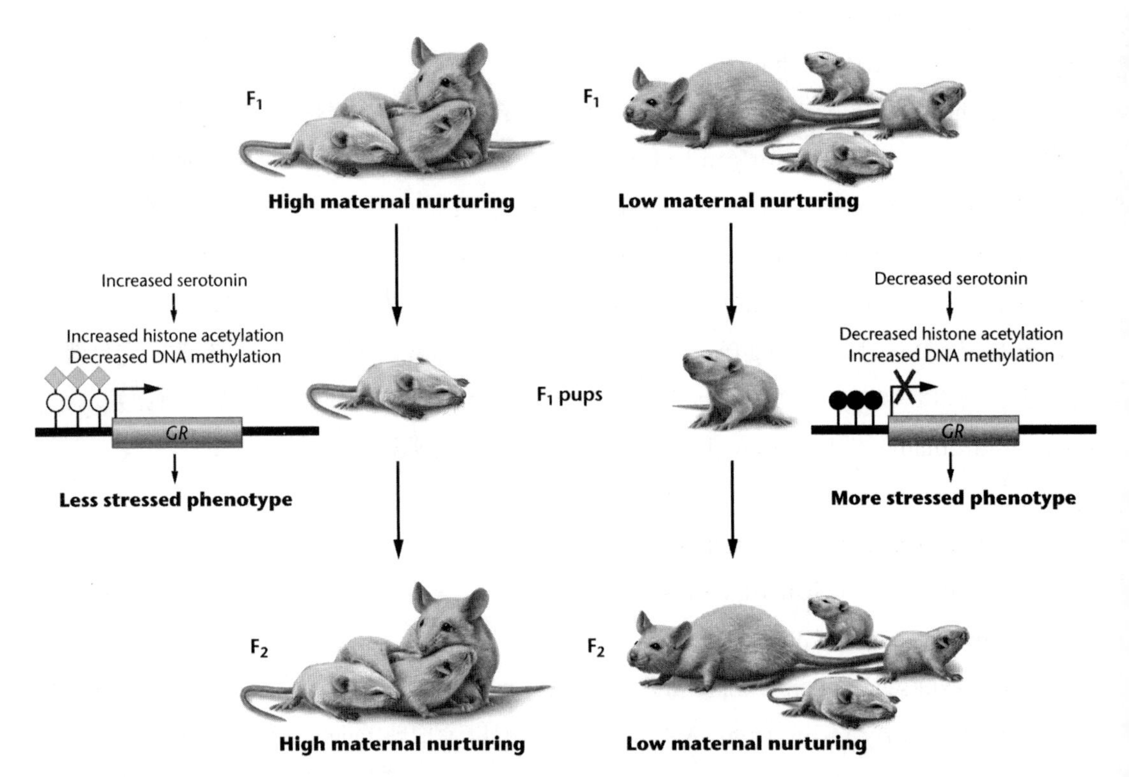

FIGURE 9.7 Style of maternal care is transmitted across rat generations through epigenetic events that take place early in postnatal life. High maternal nurturing induces high levels of serotonin in the brain, leading to DNA hypomethylation, histone acetylation, and increased expression of GR. In adulthood, high levels of GR expression increase adaptation to stress and, in females, passes on the high-MN phenotype. Rat pups experiencing low levels of maternal nurturing had higher levels of promoter methylation and reduced levels of GR expression. In adulthood, this led to poor stress adaptation and, in females, perpetuation of low levels of nurturing her pups.

图 9.7 母亲护理模式通过产后早期阶段的表观遗传世代相传。高质量的母亲护理会诱发大脑高水平的 5- 羟色胺，从而导致 DNA 低甲基化、组蛋白乙酰化、GR 表达增加。在成年阶段，GR 高水平表达可以提升个体面对压力的适应性，雌性个体将遗传该高质量母亲护理（high-MN）表型。如果幼年小鼠遭遇低质量的母亲护理，则启动子甲基化水平提高，GR 表达水平降低。在成年阶段，这将导致较差的压力适应性，雌性个体将延续低质量护理子女的能力。

9.5 Epigenome Projects and Databases

As the role of the epigenome in disease has become increasingly clear, researchers across the globe have formed multidisciplinary projects to map all the epigenetic changes that occur in the normal genome and established databases to study the role of the epigenome in specific diseases. We will discuss some of these projects and their goals and will summarize the major findings of a few of the large-scale projects.

The NIH Roadmap Epigenomics Project was established to elucidate the role of epigenetic mechanisms in human biology and disease. The project has two main goals: (1) provide a set of at least 1000 reference epigenomes in a range of cell types from healthy and diseased individuals, and (2) delineate the epigenetic differences in conditions such as Alzheimer disease, autism, and schizophrenia.

In 2015, the project published an analysis of the first 111 reference genomes collected, representing the most comprehensive map of the human epigenome to date. One of the important results of this study establishes that genetic variants associated with several complex human disorders such as Alzheimer disease, cancer, and autoimmune disorders are enriched in tissue-specific epigenomic marks, identifying relevant cell types associated with these and other disorders.

The Human Epigenome Atlas, which is part of the Roadmap Project, collects and

9.5 表观基因组计划与数据库

随着表观基因组在疾病中扮演的角色日益明确，全球研究人员已建立多学科合作项目来绘制正常基因组中的所有表观遗传变化，并建立数据库以研究表观基因组在特定疾病中的作用。我们将讨论其中的一些项目及其目标，并对其中一些大型项目的主要发现进行汇总。

“美国国立卫生研究院路线图表观基因组计划”旨在阐明表观遗传机制在人类生物学和人类疾病中的作用。该项目有两个主要目标：（1）提供来自健康个体和患病个体不同类型细胞的至少1000个参考表观基因组；（2）绘制诸如阿尔茨海默病、孤独症和精神分裂症等疾病的表观遗传差异。

2015年，该项目发表了首批收集的111个参考基因组的分析，代表了迄今为止人类表观基因组最全面的图谱。这项研究最重要的成果之一是证实了与多种复杂人类疾病（例如阿尔茨海默病、癌症和自身免疫性疾病）相关的遗传变体富含组织特异性表观基因组标记，可用于鉴定这些疾病和其他疾病相关的细胞类型。

作为路线图项目的一部分，人类表观基因组图集对特定基因座、不同细胞类型、

catalogs detailed information about epigenomic modifications at specific loci, in different cell types, different physiological states, and different genotypes. These data allow researchers to perform comparative analysis of epigenomic data across genomic regions or entire genomes.

不同生理状态和不同基因型的表观基因组修饰的详细信息进行收集和分类。这些数据使研究人员能够对基因组内不同区域或整个基因组的表观基因组数据开展比较分析。

The International Human Epigenome Consortium (IHEC) is a global program established to coordinate the collection of epigenome maps for 1000 human cell populations. Several projects are contributing to the program, each specializing in different cell types and/or approaches. The U.S. Reference Epigenome Mapping Centers are using stem cells and tissue samples from healthy donors, and the Germany-based DEEP Project is collecting 70 reference epigenomes of human and mouse tissues associated with metabolic and inflammatory diseases. The European BLUEPRINT Project is collecting epigenomic profiles from several different types of blood cells related to specific diseases.

国际人类表观基因组联盟（IHEC）是一项全球计划，旨在协调收集 1000 个人类细胞群体的表观基因组图。一些项目正致力于该计划的实施，并且每个项目专注于不同细胞类型和 / 或不同技术。美国参考表观基因组图谱中心正在使用健康供体的干细胞和组织样本，而德国 DEEP 项目正从事代谢疾病和炎性疾病相关的来自人和小鼠组织样本的 70 个参考表观基因组的收集工作。欧洲 BLUEPRINT 项目正从特定疾病相关的几种不同类型血细胞中收集表观基因组图谱信息。

To complement the efforts of IHEC in mapping the epigenomes of primary cell lines collected directly from tissues, the Encyclopedia of DNA Elements (ENCODE) project is focused on collecting epigenome maps for cell lines grown under laboratory conditions. To compare the epigenomes of normal cells with cancer cells, the International Cancer Genome Consortium (ICGC) is mapping the epigenomes and the transcriptome profiles of 50 different cancer types.

为了与 IHEC 直接从组织中收集原代细胞系绘制表观基因组的工作相互补，DNA 元件百科全书（ENCODE）项目则着重于收集实验室培养细胞系的表观基因组图谱。为了比较正常细胞与癌细胞的表观基因组，国际癌症基因组联盟（ICGC）正在绘制 50 种不同类型癌症的表观基因组图谱和转录组图谱。

Although these projects are still in progress, the information already available strongly

尽管这些项目仍在进行中，但是已获得的信息强烈地表明我们正处于一个遗传

suggests that we are on the threshold of a new era in genetics, one in which we can study the development of disease at the genomic level and understand the impact of epigenetic factors on gene expression.

学新时代的门槛处，我们将能在基因组水平研究疾病发生并了解表观遗传因子对基因表达的影响。

Visit the Study Area in Mastering Genetics (http://www.masteringgenetics.com) for a list of further readings on this topic, including journal references and selected Web sites.

访　问 Mastering Genetics (http://www.masteringgenetics. com) 中的学习模块以获得与本专题相关，包括期刊参考文献和推荐网站在内的更多拓展阅读材料清单。

10 Genetic Testing

10 遗传检测

Genetic testing, including genomic analysis by DNA sequencing, is transforming medical diagnostics. Technologies for genetic testing have had major impacts on the diagnosis of disease and are revolutionizing medical treatments based on the development of specific and effective pharmaceuticals.

包括利用 DNA 测序进行基因组分析在内的遗传检测正在改变医用诊断。遗传检测技术已对疾病诊断产生巨大影响，并且正在基于特异性有效药物的开发为医学治疗带来革新。

Because of the Human Genome Project and related advances in genomics, researchers have been making rapid progress in identifying genes involved in both single-gene diseases and complex genetic traits. As a result, **genetic testing**—the ability to analyze DNA, and increasingly RNA, for the purposes of identifying specific genes or sequences associated with different genetic conditions—has advanced very rapidly.

由于人类基因组计划以及基因组学的相关进展，研究人员已在单基因疾病和复杂遗传性状的基因鉴定方面取得了快速进展。因此，**遗传检测**——为了鉴定不同遗传疾病相关的特定基因或基因序列而开展的分析DNA以及越来越多RNA的能力——的发展十分迅猛。

In this chapter we provide an overview of applications that are effective for the genetic testing of children and adults and examine historical and modern methods. We consider the impact of different genetic technologies on the diagnosis of human diseases and disease treatment. Finally, we consider some of the social, ethical, and legal implications of genetic testing.

在本章中，我们将简要介绍对儿童和成人的遗传检测十分有效的一些应用，并详细介绍一些传统和现代方法。我们将思考不同遗传技术对人类疾病诊断和疾病治疗的影响。最后，我们还将思考遗传检测带来的一些社会、伦理和法律影响。

10.1 Testing for Prognostic or Diagnostic Purposes

Genetic testing was one of the first successful applications of recombinant DNA technology, and currently more than 900 tests are in use that target a specific gene or sequence. Increasingly, scientists and physicians can directly examine an individual's DNA for mutations associated with disease, including through DNA sequencing, as we will discuss in Section 10.5. These tests usually detect gene alterations associated with single-gene disorders. But, only about 3900 genes have been linked to such disorders. Examples include sickle-cell anemia, cystic fibrosis, Huntington disease, hemophilia, and muscular dystrophy. Other tests have been developed for disorders that may involve multiple genes such as certain types of cancers.

Gene tests are used for prenatal, childhood, and adult prognosis and diagnosis of genetic diseases; to identify carriers; and to identify genetic diseases in embryos created by *in vitro* fertilization, among other applications. For genetic testing of adults, DNA from white blood cells is commonly used. Alternatively, many genetic tests can be carried out on cheek cells, collected by swabbing the inside of the mouth, or on hair cells. Some genetic testing can be carried out on gametes.

What does it mean when a genetic test is performed for *prognostic* purposes, and how does this differ from a *diagnostic* test? A prognostic test predicts a person's likelihood

10.1 预后检测和诊断检测

遗传检测是重组 DNA 技术最早的成功应用之一，目前已有 900 多种针对特定基因或基因序列的测试项目正在使用中。正如我们在 10.5 节中将要讨论的，科学家和医师能够越来越多地通过 DNA 测序在内的技术直接检查个体 DNA 中与疾病相关的突变。这些技术通常检测单基因疾病相关的基因突变。但是，目前仅有大约 3900 种基因与此类疾病相关联，这样的例子包括镰状细胞贫血、囊性纤维化、亨廷顿病、血友病和肌营养不良。目前也已研发出一些可用于多基因相关疾病（例如某些类型的癌症）的检测技术。

遗传检测可用于产前、儿童和成人遗传疾病的预后和诊断；鉴定遗传疾病基因携带者；鉴定体外受精胚胎或其他应用中的遗传疾病。在成人的遗传检测中，通常使用来自白细胞的 DNA。另外，许多遗传检测也可利用擦拭口腔内侧收集的口腔上皮细胞或毛囊细胞。还有一些遗传检测则利用配子细胞。

在遗传检测中，什么是预后检测？它与诊断检测有何不同？预后检测可预测个体罹患特定遗传疾病的可能性；而针对某种遗传疾病的诊断检测则用于鉴定引发该

of developing a particular genetic disorder. A diagnostic test for a genetic condition identifies a particular mutation or genetic change that causes the disease or condition. Sometimes a diagnostic test identifies a gene or mutation associated with a condition, but the test will not be able to determine whether the gene or mutation is the cause of the disorder or is a genetic variation that results from the condition.

疾病或遗传症状的特定突变或遗传改变。有时诊断检测可以鉴定疾病相关的基因或突变，但该检测无法确定该基因或突变是该疾病的发病原因还是由该疾病导致的遗传变异结果。

10.2 Prenatal Genetic Testing to Screen for Conditions

10.2 用于遗传筛查的产前遗传检测

Although genetic testing of adults is increasing, over the past two decades more genetic testing has been used to detect genetic conditions in babies than in adults. In newborns, a simple prick of a baby's heel produces a few drops of blood that are used to check the newborn for many genetic disorders. In the United States, all states now require genetic testing, often called *newborn screening*, for certain medical conditions (the number of diseases screened for is set by the individual state). There are currently about 60 conditions that can be detected, although many of these tests detect proteins or other metabolites and are not DNA-or RNA-based genetic tests.

尽管成人遗传检测项目种类不断增加，但在过去 20 年中，用于婴儿遗传筛查的遗传检测项目要比成人的多。在新生儿中，在婴儿脚后跟简单地用取血针一扎就能获得少量血样，用于新生儿的许多遗传疾病筛查。在美国，所有州现在均要求对某些疾病进行遗传检测，这通常被称为新生儿筛查（筛查疾病的种类由各州自行确定）。目前可以对大约 60 种遗传疾病进行遗传检测，尽管其中许多检测方法是检测蛋白质或其他代谢产物，而不是基于 DNA 或 RNA 的遗传检测方法。

Prenatal genetic tests, performed before a baby is born, are used for certain disorders in which waiting until birth is not desirable. For prenatal testing, fetal cells are obtained by **amniocentesis** or **chorionic villus sampling (CVS)**. Figure 10.1 shows the procedure for amniocentesis, in which a small volume of the

在婴儿出生前进行的**产前遗传检测**用于检测某些疾病，以便在分娩之前即可及时发现。对于产前检查，胎儿细胞可通过**羊膜腔穿刺术**或**绒毛膜绒毛吸取术**（CVS）获得。图 10.1 显示了羊膜腔穿刺术的步骤，在此过程中包裹在胎儿周围的少量羊水被取出，羊水中所含的胎儿细胞可用于核型分

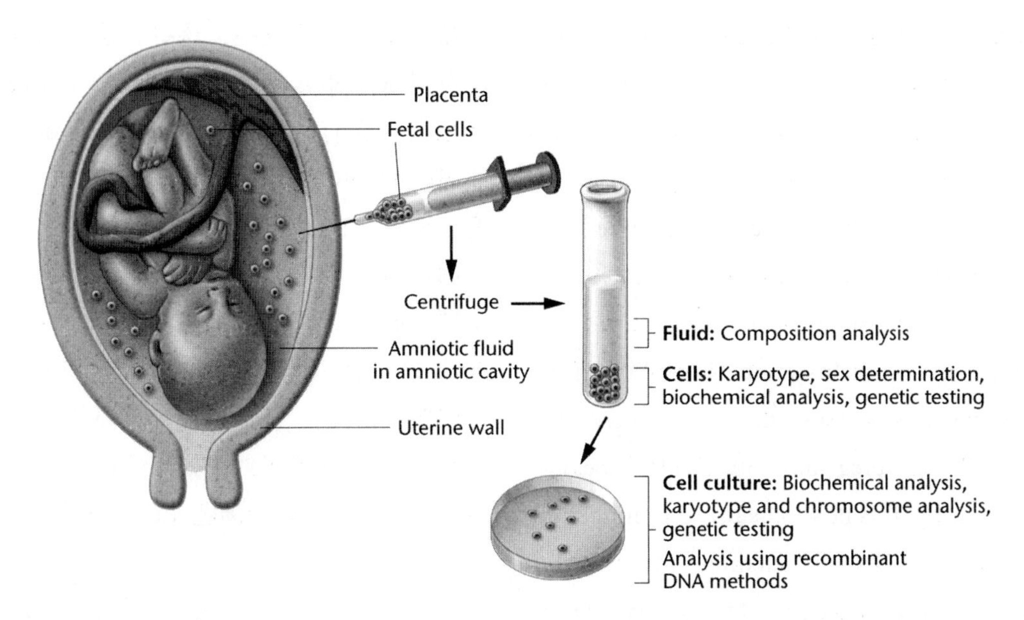

FIGURE 10.1 For amniocentesis, the position of the fetus is first determined by ultrasound, and then a needle is inserted through the abdominal and uterine walls to recover amniotic fluid containing fetal cells for genetic or biochemical analysis.

图 10.1 在羊膜腔穿刺术中，首先通过超声检查确定胎儿位置，然后使用针穿透腹壁和子宫壁获得含胎儿细胞的羊水，用于遗传分析或生化分析。

amniotic fluid surrounding the fetus is removed. Amniotic fluid contains fetal cells that can be used for karyotyping, genetic testing, and other procedures. For chorionic villus sampling, cells from the fetal portion of the placental wall (the chorionic villi) are sampled through a vacuum tube, and analyses can be carried out on this tissue. Captured fetal cells can then be subjected to genetic analysis by techniques that involve PCR (such as allele-specific oligonucleotide testing, described in Section 10.3) or DNA sequencing (described in Section 10.5). In Section 10.3 we discuss a fetal screening approach called preimplantation genetic diagnosis.

析、遗传检测和其他测试。在绒毛膜绒毛吸取术中，胎盘壁（绒毛膜绒毛）上属于胎儿部分的细胞通过真空管被取出，该组织可用于各种分析。获取的胎儿细胞随后利用 PCR（如 10.3 节中介绍的等位基因特异的寡核苷酸检测）或 DNA 测序（如 10.5 节中所述）等技术进行遗传分析。在 10.3 节中，我们将讨论一种胎儿筛查方法——植入前遗传学诊断。

Noninvasive Procedures for Genetic Testing of Fetal DNA

Noninvasive procedures are also being

胎儿 DNA 的非侵入性遗传检测

用于胎儿 DNA 产前遗传检测的非侵入

developed for prenatal genetic testing of fetal DNA. These procedures reduce the risk to the fetus. Circulating in each person's bloodstream is DNA that is released from the person's dead and dying cells. This so-called cell-free DNA (cfDNA) is cut up into small fragments by enzymes in the blood. The blood of a pregnant woman also contains snippets of cfDNA from the fetus. It is estimated that ~3 to 6 percent of the DNA in a pregnant mother's blood belongs to her baby. It is now possible to analyze these traces of fetal DNA to determine if the baby has certain types of genetic conditions such as Down syndrome. Such tests require about a tablespoon of blood.

性技术也正在研发中，这些方法可以降低检测过程给胎儿带来的风险。个体血液循环系统内含有从个体死亡和垂死细胞中释放出来的DNA，血液中所含的酶将这些游离DNA（cfDNA）酶切成DNA小片段。孕妇血液中同样含有来自胎儿的游离DNA片段。据估计，孕妇血液中有3%—6%的DNA来自胎儿。现在可以分析这些痕量存在的胎儿DNA，从而确定胎儿是否患有某些类型的遗传疾病，例如唐氏综合征。类似的检测大约需要一汤匙（15 mL）的血液。

DNA in the blood is sequenced to analyze **haplotypes**— contiguous segments of DNA that do not undergo recombination during gamete formation—that distinguish which cfDNA segments are maternal and which are from the fetus (see Figure 10.2). If a fetal haplotype contained a specific mutation, this would also be revealed by sequence analysis. Nearly complete fetal genome sequences have been assembled from maternal blood. These are developed by sequencing cfDNA fragments from maternal blood and comparing those fragments to sequenced genomes from the mother and father. Bioinformatics software is then used to organize the genetic sequences from the fetus in an effort to assemble the fetal genome. Currently, this technology does not capture the entire fetal genome; it results in an assembled genome sequence with segments missing. It has been shown, however, that **whole genome sequencing**

血液中的DNA经测序后可用于分析**单体型**，即配子形成过程中未发生重组的DNA连续片段，从而区分哪些cfDNA片段来自母体，哪些来自胎儿（参见图10.2）。如果胎儿单体型中含特定突变，通过DNA序列分析即可发现。母体血液中获得的cfDNA片段经组装后几乎可得到完整的胎儿基因组序列。这一过程通过对母体血液中cfDNA片段进行测序，然后将这些片段序列与母亲和父亲的基因组序列进行比较，最后使用生物信息学软件对来自胎儿的基因序列进行拼接，从而组装获得胎儿基因组。目前，该技术并不能捕获整个胎儿基因组，因此组装得到的基因组序列存在DNA片段缺失。但是实践证明，母体血浆cfDNA的**全基因组测序**（WGS）可用于胎儿整个外显子组的精确测序。

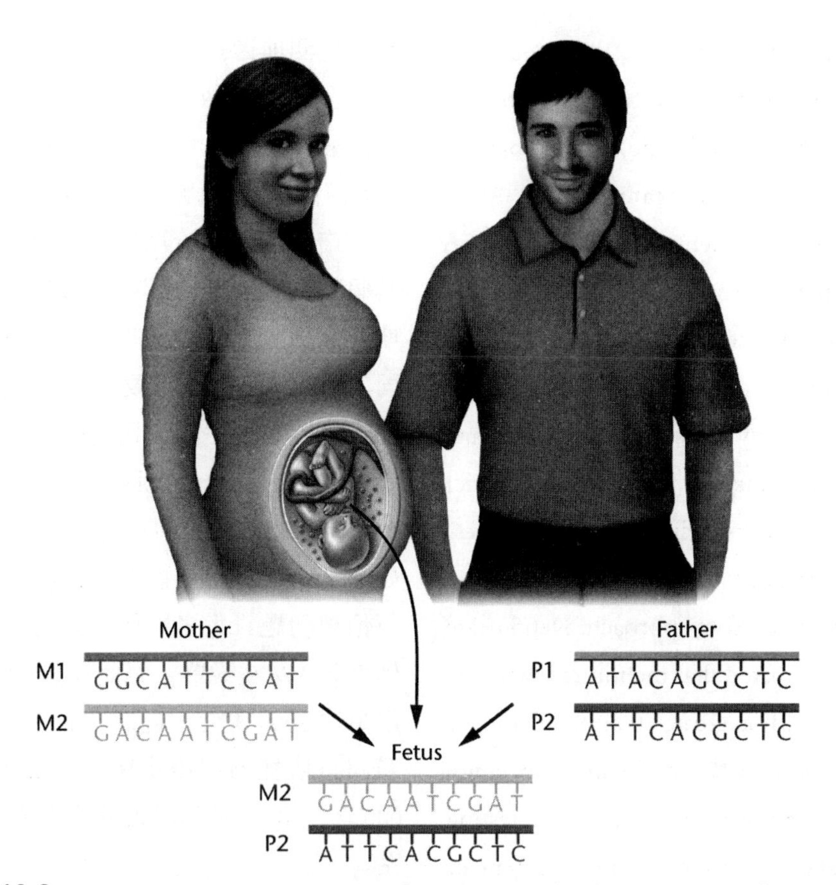

FIGURE 10.2 Deducing fetal genome sequences from maternal blood. For any given chromosome, a fetus inherits one copy of a haplotype from the mother (maternal copies, M1 or M2) and another from the father (paternal copies, P1 or P2). For simplicity, a single-stranded sequence of DNA from each haplotype is shown. These haplotype sequences can be detected by WGS. Here the fetus inherited haplotypes M2 and P2 from the mother and father, respectively. DNA from the blood of a pregnant woman would contain paternal haplotypes inherited by the fetus (P2), maternal haplotypes that are not passed to the fetus (M1), and maternal haplotypes that are inherited by the fetus (M2). The maternal haplotype inherited by the fetus (M2) would be present in excess amounts relative to the maternal haplotype that is not inherited (M1).

图 10.2 从母体血液获取胎儿基因组序列。对于任何一条指定染色体，胎儿从母亲遗传获得一个单体型拷贝（母本拷贝，M1 或 M2），从父亲遗传获得另一个单体型拷贝（父本拷贝，P1 或 P2）。为了简便起见，图中仅显示每个单体型 DNA 的单链序列，这些单体型序列可通过 WGS 进行检测。这里，胎儿分别从母亲和父亲遗传获得 M2 和 P2 单体型。孕妇血液 DNA 中可能含胎儿遗传得到的父本单体型（P2），母本单体型 M1 未被遗传至胎儿，母本单体型 M2 遗传至胎儿。遗传至胎儿的母本单体型 M2 在含量上要比未被遗传的母本单体型 M1 相对为多。

(**WGS**) of maternal plasma cfDNA can be used to accurately sequence the entire exome of a fetus.

Tests for fetal genetic analysis based on maternal blood samples started to arrive on the market in 2011. Sequenom of San Diego, California, was one of the first companies to

基于母体血液样本进行胎儿遗传分析的检测项目于 2011 年开始投放市场。位于加利福尼亚州圣地亚哥的西格诺公司即是最早开展此类检测项目的公司之一，其开展的 MaterniT® 21 PLUS 是一种唐氏综合征

launch such a test—MaterniT®21 PLUS, a Down syndrome test that can also be used to test for trisomy 13 (Patau syndrome) and trisomy 18 (Edwards syndrome). MaterniT® 21 PLUS analyzes 36-bp fragments of DNA to identify chromosome 21 from the fetus. Sequenom claims that this test is highly accurate with a false positive rate of just 0.2 percent. The test can be done as early as week 10 (about the same time CVS sampling can be performed, which is about 4 to 6 weeks earlier than amniocentesis can be performed). Several companies have followed the Sequenom approach. Nationwide, it has been estimated that the future market for these tests could be greater than $1 billion. As discussed in Section 10.7, there are many ethical issues associated with prenatal genetic testing. Most insurance companies are not yet paying for WGS of maternal blood, which can cost as much as $2000 for a single test. Recently, California agreed to subsidize noninvasive prenatal testing for women through the state's genetic diseases program, which screens ~400,000 women each year.

筛查检测项目，也可用于筛查13三体（帕托综合征）和18三体（爱德华综合征）。MaterniT® 21 PLUS分析长度为36 bp的DNA片段，从而鉴定胎儿的21号染色体。西格诺公司称该检测技术准确率非常高，其假阳性率仅为0.2%。该检测最早可在孕期第10周进行（大约与CVS取样的时间相同，比羊膜腔穿刺术要早4—6周）。多家公司也紧随西格诺公司的脚步开展了此类检测项目。在全美范围内，这些检测服务的未来市场据估计可能超过10亿美元。正如10.7节所述，产前遗传检测存在许多道德伦理问题。大多数保险公司尚不会为母体血样WGS买单，而此项检测的单次费用高达2000美元。近期，加利福尼亚州已经同意依托州遗传疾病项目每年为约400 000名女性开展非侵入性产前筛查进行补贴。

Originally these noninvasive tests were offered to women older than 35 years of age, or if they were identified as at risk based on family history of birth-related complications. Now these tests are being marketed to women with low-risk pregnancies as well, and their value, given the cost, has been questioned. Recent figures indicate that sales for these tests exceeded $600 million annually.

最初，这些无创检查面向35岁以上的女性或者依据出生相关并发症家族史被确定为有风险的女性。现在，这些检查项目同时面向低风险孕妇群体，只是鉴于其检测成本，它们的价值受到了质疑。最新数据显示，这些检测项目的年销售额已经超过6亿美元。

In Section 10.7 we will discuss preconception testing and recent patents for

在10.7节中，我们将讨论拟生育测试和旨在预测后代遗传潜力（命运测试）的

computing technologies designed to predict the genetic potential of offspring (destiny tests).

计算技术相关的最新专利。

10.3 Genetic Testing Using Allele-Specific Oligonucleotides

10.3 利用等位基因特异的寡核苷酸进行遗传检测

When genetic testing of adults was initiated, one of the first methods was **restriction fragment length polymorphism (RFLP) analysis**. For example, historically, RFLP analysis was the primary method to detect **sickle-cell anemia**. As discussed in previous chapters, this disease is an autosomal recessive condition common in people with family origins in areas of West Africa, the Mediterranean basin, parts of the Middle East, and India. It is caused by a single amino acid substitution in the β -globin protein, as a consequence of a single-nucleotide substitution in the corresponding gene.

最初在开启成人遗传检测时，所采用的一种技术是**限制性片段长度多态性（RFLP）分析**。例如，历史上，RFLP 分析是检测**镰状细胞贫血**的主要方法。正如前面的章节所述，该病是一种常染色体隐性遗传病，常见于有西非、地中海盆地、中东部分地区和印度家族起源的人群。它是由 β - 珠蛋白中单个氨基酸发生取代所致，而这又是相应基因中单核苷酸替代所导致的结果。

The single-nucleotide substitution also eliminates a restriction site in the β -globin gene for the restriction enzymes *Mst* II and *Cvn* I. As a result, the mutation alters the pattern of restriction fragments seen on Southern blots. These differences in restriction sites could be used to diagnose sickle-cell anemia prenatally and to establish the parental genotypes and the genotypes of other family members who may be heterozygous carriers of this condition.

单核苷酸替代也使得 β - 珠蛋白基因中原有的限制性内切酶 *Mst* II 和 *Cvn* I 的酶切位点消失。因此，该突变改变了 DNA 印迹所显示的限制性酶切所得片段的模式。这些限制性内切酶酶切位点的差异可用于产前诊断镰状细胞贫血，确定父母基因型以及其他家庭成员中可能为杂合型突变基因携带者的基因型。

But only about 5 to 10 percent of all point mutations can be detected by RFLP analysis because most mutations occur in regions of the genome that do not contain restriction enzyme sites. However, since the Human Genome Project (HGP) was completed and

但是所有点突变中仅有 5%—10% 能通过 RFLP 分析检测到，因为基因组中大多数突变发生区域并不含有限制性内切酶酶切位点。然而，由于人类基因组计划（HGP）已经完成，更多疾病相关突变已被人所知，因此遗传学家可以采用 PCR 技术和合成寡

many more disease-associated mutations became known, geneticists employed PCR and synthetic oligonucleotides to detect these mutations, including the use of synthetic DNA probes known as **allele-specific oligonucleotides (ASOs)**.

核苷酸序列来检测这些突变，包括使用合成 DNA 探针，即**等位基因特异的寡核苷酸（ASOs）**。

This rapid, inexpensive, and accurate technique is used to diagnose a wide range of genetic disorders caused by point mutations. In contrast to RFLP analysis, which is limited to cases for which a mutation changes a restriction site, ASOs detect single-nucleotide changes called **single-nucleotide polymorphisms (SNPs)**.

这种快速、价廉和准确的技术可用于诊断由点突变引起的多种遗传疾病。与仅限于点突变引发限制性内切酶酶切位点改变的 RFLP 分析不同，ASOs 可检测单核苷酸变化，即**单核苷酸多态性（SNPs）**。

An ASO is a short, single-stranded fragment of DNA designed to hybridize to a complementary specific allele in the genome. Under proper conditions, an ASO will hybridize only with its complementary DNA sequence and not with other sequences, even those that vary by as little as a single nucleotide.

ASO 是一段短的单链 DNA 片段，可与基因组中与之互补的特异性等位基因进行杂交。在合适条件下，ASO 仅与其互补的 DNA 序列相杂交，而不与其他序列进行杂交，即使是那些仅相差一个核苷酸的序列。

Genetic testing using ASOs and PCR analysis is available to screen for many disorders in adults and newborns and for prenatal screening. In the case of sickle-cell anemia screening, DNA is extracted (either from a maternal blood sample or from fetal cells obtained by amniocentesis), and a region of the β-globin gene is amplified by PCR. A small amount of the amplified DNA is spotted onto strips of a DNA-binding membrane, and each strip is hybridized to an ASO synthesized to resemble the relevant sequence from either a normal or mutant β-globin gene. The ASO is tagged with a molecule that is either radioactive

使用 ASO 和 PCR 分析进行的遗传检测可用于成人和新生儿的许多疾病筛查以及产前筛查。在镰状细胞贫血的筛查中，首先要从母体血液样本或通过羊膜腔穿刺术获得的胎儿细胞中提取 DNA，然后利用 PCR 技术扩增 β-珠蛋白基因片段。从扩增获得的 DNA 中取少量样品点样至条状 DNA 结合膜上，然后将每个条状膜与 β-珠蛋白的正常基因或突变基因相关序列一致的合成 ASO 进行杂交。由于 ASO 被含放射性或荧光的分子标记了，所以与膜上 DNA 能够杂交的 ASO 可以被可视化。图 10.3 展示了该方法的原理。这种快速、价廉和准确的技术可用于诊断由点突变引起的多种遗传疾

or fluorescent, to allow for visualization of the ASO hybridized to DNA on the membrane. Figure 10.3 illustrates the principle behind this approach. This rapid, inexpensive, and accurate technique is used to diagnose a wide range of genetic disorders caused by point mutations.

病。

Although ASO testing is highly effective, SNPs can affect ASO probe binding leading to false positive or false negative results that may not reflect a genetic disorder, particularly if precise hybridization conditions are not

尽管ASO检测技术十分有效，但是SNPs会影响ASO探针结合，导致产生假阳性或假阴性结果，最终无法反映遗传疾病，尤其是在未使用精确杂交条件的情况下。有时，需要对已扩增基因片段进行DNA序

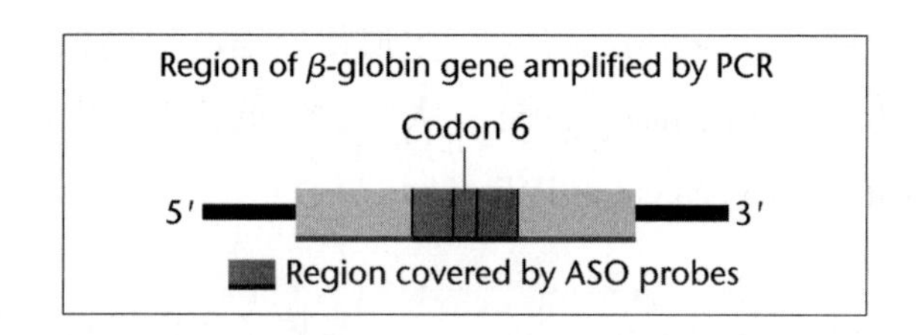

DNA is spotted onto binding membranes and hybridized with ASO probe

(a) Genotypes *AA* *AS* *SS*

Normal (β^A) ASO: 5′ – CTCCTG**A**GGAGAAGTCTGC – 3′

(b) Genotypes *AA* *AS* *SS*

Mutant (β^S) ASO: 5′ – CTCCTG**T**GGAGAAGTCTGC – 3′

FIGURE 10.3 Allele-specific oligonucleotide (ASO) testing for sickle-cell anemia. (a) Results observed if the three possible β-globin genotypes are hybridized to an ASO for the normal β-globin allele: *AA*-homozygous individuals have normal hemoglobin that has two copies of the normal β-globin gene and will show heavy hybridization; *AS*-heterozygous individuals carry one normal β-globin allele and one mutant allele and will show weaker hybridization; *SS*-homozygous sickle-cell individuals carry no normal copy of the β-globin gene and will show no hybridization to the ASO probe for the normal β-globin allele. (b) Results observed if DNA for the three genotypes are hybridized to the probe for the sickle-cell β-globin allele: no hybridization by the *AA* genotype, weak hybridization by the heterozygote (*AS*), and strong hybridization by the homozygous sickle-cell genotype (*SS*).

图 10.3 镰状细胞贫血的等位基因特异寡核苷酸检测（ASO）。（a）三种可能的β-珠蛋白基因型与正常β-珠蛋白基因型ASO进行杂交可以观察到的结果：*AA*，含正常珠蛋白的纯合个体，即含两个拷贝的正常β-珠蛋白等位基因，因此呈现深印迹杂交结果；*AS*，杂合个体，含一个拷贝的正常β-珠蛋白等位基因和一个拷贝的突变β-珠蛋白等位基因，呈现印迹较弱的杂交结果；*SS*，纯合镰状细胞贫血个体，不含正常β-珠蛋白等位基因，不呈现与正常β-珠蛋白等位基因ASO探针的杂交。（b）三种基因型DNA与镰状细胞β-珠蛋白等位基因探针进行杂交可以观察到的结果：*AA*不发生杂交；杂合子（*AS*）弱杂交；纯合镰状细胞贫血基因型（*SS*）强杂交。

used. Sometimes DNA sequencing is carried out on amplified gene segments to confirm identification of a mutation.

列测定来进一步确定被鉴定的突变。

Preimplantation Genetic Diagnosis

Because ASO testing makes use of PCR, only small amounts of DNA are required for analysis. As a result, ASO testing is ideal for **preimplantation genetic diagnosis (PGD)**, also called preimplantation genetic testing or screening. PGD involves the genetic analysis of cells from embryos created by IVF (Figure 10.4). PGD has been used for over 25 years, typically when there is concern about a particular genetic defect. In the United States, ~ 25 percent of IVF attempts use PGD.

胚胎植入前遗传学诊断

由于 ASO 检测使用 PCR 技术，仅需少量 DNA 样本即可进行分析，因此，ASO 检测非常适于**胚胎植入前遗传学诊断（PGD）**，也被称为胚胎植入前遗传学检测或胚胎植入前遗传学筛查。PGD 涉及对体外受精产生的胚胎细胞进行遗传分析（图 10.4）。PGD 的应用已超过 25 年，通常应用于对特定遗传缺陷心存疑虑的情况下。在美国，大约 25% 的体外受精会使用 PGD。

When sperm and eggs are mixed to create zygotes, the early-stage embryos are grown in culture. A single cell can be removed from an early-stage embryo using a vacuum pipette to gently aspirate one cell away from the embryo (Figure 10.4, top). This could possibly kill the embryo, but if it is done correctly, the embryo will often continue to divide normally. DNA from the single cell is then typically analyzed by fluorescence in situ hybridization for chromosome analysis or by ASO testing (Figure 10.4, bottom). The genotypes for each early-stage embryo can be tested to decide which embryos will be implanted into the uterus.

当精子和卵子混合形成合子时，早期阶段的胚胎会在培养基中生长。研究人员使用真空吸管从早期胚胎中轻轻吸取一个细胞，从而分离得到单细胞（图 10.4，顶部）。这个过程可能杀死胚胎，但是如果操作正确，胚胎通常将继续进行正常的细胞分裂。然后，分离得到的单细胞所含 DNA 会通过荧光原位杂交进行染色体分析或使用 ASO 检测技术（图 10.4，底部）进行分析。每个早期阶段胚胎个体的基因型均可被检测，从而确定选择哪些胚胎进行子宫植入。

PGD can also involve removing 5 to 7 cells from the trophectoderm, the outermost layer of cells in a blastocyst-stage embryo which forms the placenta, as an attempt to determine if the embryo has a normal complement of

PGD 还可以通过从最终形成胎盘的囊胚期胚胎的最外层细胞，即滋养外胚层中移取 5—7 个细胞，以此为样本来确定胚胎是否含有正常的染色体组成。但是这种分析并不总能反映胚胎的遗传学特征，因为

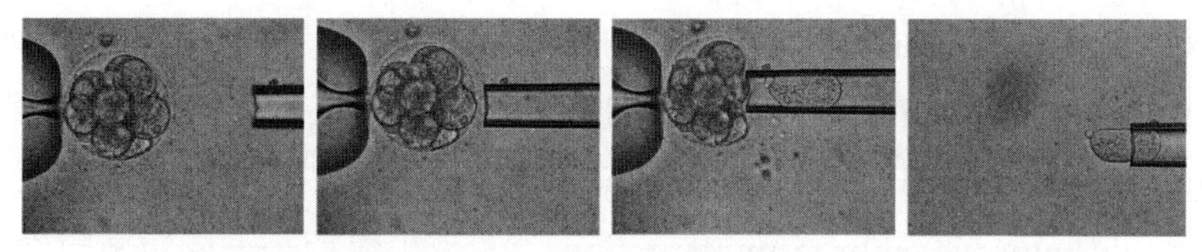

At the 8- to 16-cell stage, one cell from an embryo is gently removed with a suction pipette. The remaining cells continue to grow in culture.

↓

DNA from an isolated cell is amplified by PCR with primers for the β-globin gene. Small volumes of denatured PCR products are spotted onto two separate DNA-binding membranes.

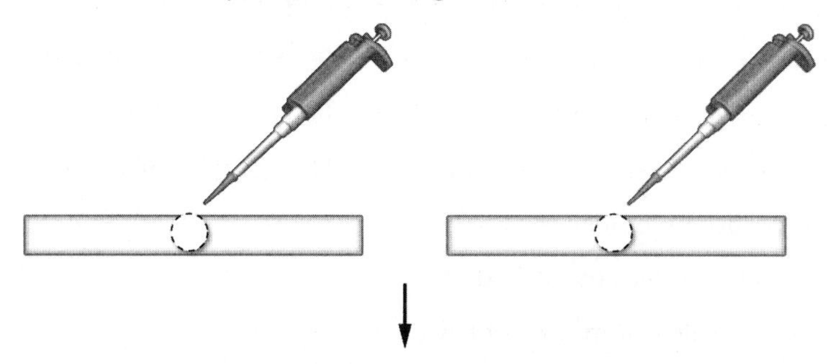

↓

One membrane is hybridized to a probe for the normal β-globin allele (β^A), and the other membrane is hybridized to a probe for the mutant β-globin allele (β^S).

Membrane hybridized to a probe for the normal β-globin allele (β^A)

Membrane hybridized to a probe for the mutant β-globin allele (β^S)

In this example, hybridization of the PCR products to the probes for both the β^A and β^S alleles reveals that the cell analyzed by PGD has a carrier genotype ($\beta^A\beta^S$) for sickle-cell anemia.

FIGURE 10.4 A single cell from an early-stage human embryo created by *in vitro* fertilization can be removed and subjected to preimplantation genetic diagnosis (PGD) by ASO testing. DNA from the cell is isolated, amplified by PCR with primers specific for the gene of interest, and then subjected to ASO analysis. In this example, a region of the gene was amplified and analyzed by ASO testing to determine the sickle-cell genotype for this cell.

图 10.4 从体外受精产生的早期人类胚胎中取出一个单细胞采用 ASO 检测进行胚胎植入前遗传学诊断(PGD)。从细胞中提取 DNA，使用目的基因的特异性引物进行 PCR 扩增，然后进行 ASO 检测。在本例中，基因的一个区域得到扩增，并用 ASO 检测确定细胞的镰状细胞基因型。

chromosomes. But such analysis does not always reflect the genetics of the embryo, which develops from a cluster of cells within the blastocyst called the inner cell mass.

胚胎由囊胚内细胞团的细胞簇发育而来。

Any alleles that can be detected by ASO testing can be used during PGD. Sickle-cell

任何可利用 ASO 进行检测的等位基因均可在 PGD 中得到应用。镰状细胞贫血、

anemia, cystic fibrosis, and dwarfism are often tested for by PGD, but alleles for many other conditions are also often analyzed. In theory, PGD should improve embryo implantation success rates and reduce miscarriages for couples—and success rates have improved, particularly for older women undergoing IVF. But this turns out not to be true for all couples because PGD cannot be used for identifying epigenetic changes that affect fertility. As we learn more about epigenetic influences on fertilization, it is expected that PGD will be expanded to incorporate epigenetic analysis in the future. Another limitation of PGD is that the genetics of a single cell may not provide a complete snapshot of the genetic health of an embryo.

囊性纤维化和侏儒症常使用 PGD 进行检测，许多其他疾病相关的等位基因也常被分析。理论上，PGD 应该能提高胚胎着床成功率并减少流产——成功率的确有所提高，尤其对于年龄较大的进行体外受精的女性来说。但事实表明并非所有夫妇都是如此，因为 PGD 不能用于鉴定影响生育力的表观遗传变化。随着我们对表观遗传对于受精的影响了解得越来越多，可以预测将来的 PGD 还将扩展至联合表观遗传学分析。PGD 的另一个局限性在于单细胞遗传学可能无法提供胚胎基因健康的全貌。

Also, as you will learn in Section 10.5, it is now possible to carry out WGS on individual cells. This method is being applied for PGD of cells from an embryo created by IVF.

而且，正如你将在 10.5 节中所了解到的，现在单细胞 WGS 已成为可能。该方法也正应用于体外受精胚胎细胞的 PGD 中。

10.4 Microarrays for Genetic Testing

10.4 用于遗传诊断的微阵列

ASO analysis is an effective method of screening for one, or a small number, of mutations within a gene. However, there is a significant demand for genetic tests that detect complex mutation patterns or previously unknown mutations in a single gene associated with genetic diseases and cancers. For example, the gene that is responsible for cystic fibrosis (the *CFTR* gene) contains 27 exons and encompasses 250 kb of genomic DNA. Of the 1000 known mutations of the *CFTR* gene, about half are

ASO 分析是筛选基因内一个或少量突变的有效方法。然而，检测与遗传性疾病和癌症相关的复杂突变模式或目前尚未知的单基因突变仍然存在巨大的市场需求。例如，囊性纤维化相关基因（*CFTR* 基因）含 27 个外显子，占据基因组 DNA 长达 250 kb。在 1000 个已知的 *CFTR* 基因突变中，约一半是点突变、插入突变和缺失突变，它们广泛地分布于整个基因中。类似地，肿瘤抑制因子 *TP53* 基因中已知存在 500 种以上的不同突变，这些突变中的任何一种

point mutations, insertions, and deletions—and they are widely distributed throughout the gene. Similarly, over 500 different mutations are known to occur within the tumor suppressor *TP53* gene, and any of these mutations may be associated with, or predispose a patient to, a variety of cancers. In order to screen for mutations in these genes, comprehensive, high-throughput methods are required.

都可能与多种癌症有关，或易于诱发多种癌症。为了筛选这些基因突变，研究人员需要综合的、高通量的方法。

One high-throughput screening technique is based on the use of **DNA microarrays** (also called DNA or gene chips). The numbers and types of DNA sequences on a microarray are dictated by the type of analysis that is required. For example, each spot or field (sometimes also called a feature) on a microarray contain a DNA sequence derived from each might member of a gene family, sequence variants from one or several genes of interest, or a sequence derived from each gene in an organism's genome.

一种高通量的筛选技术即基于 **DNA 微阵列**（也称为 DNA 芯片或基因芯片）的使用。微阵列上 DNA 序列的数量和类型取决于所需分析的种类。例如，微阵列上的每个点或域（有时也称为特征）可能包含源自基因家族每个成员衍生而来的一段 DNA 序列，一个或多个目标基因的序列变体，或从生物体基因组中每个基因衍生而来的序列。

In the recent past, DNA microarrays have been used for a wide range of applications, including the detection of mutations in genomic DNA and the detection of gene-expression patterns in diseased tissues. However, in the near future, whole genome sequencing, **exome sequencing**, and **RNA sequencing** are expected to replace most applications involving microarrays and render this technology obsolete.

最近，DNA 微阵列已得到广泛应用，包括检测基因组 DNA 突变和检测患病组织的基因表达模式。然而，在不久的将来，全基因组测序、**外显子组测序**和 **RNA 测序**有望取代微阵列相关的大多数应用，进而导致该技术被淘汰。

But because of the impact microarrays have had on genetic testing, it is still valuable to discuss applications of this method. What makes DNA microarrays so useful is the immense amount of information that can be simultaneously generated from a single array. DNA microarrays

但是由于微阵列对遗传检测已经产生影响，因此讨论这种方法的应用仍然具有价值。DNA 微阵列应用价值如此之大的原因在于单个阵列可以同时生成海量信息。邮票大小（刚好超过 1 cm^2）的 DNA 微阵列能够包含多达 500 000 个不同的域，每个

the size of postage stamps (just over 1 cm square) can contain up to 500,000 different fields, each representing a different DNA sequence. Human genome microarrays containing probes for most human genes are available, including many disease-related genes, such as the *TP53* gene, which is mutated in a majority of human cancers, and the *BRCA1* gene, which, when mutated, predisposes women to breast cancer and men to breast and prostate cancer.

域代表一段不同的 DNA 序列。目前可提供的人类基因组微阵列含大多数人类基因探针，其中包括许多疾病相关基因，如在大多数人类癌症中发生突变的 *TP53* 基因，以及突变后使女性易患乳腺癌、男性易患乳腺癌和前列腺癌的 *BRCA1* 基因。

In addition to testing for mutations in single genes, DNA microarrays can include probes that detect SNPs. SNPs crop up in an estimated 15 million positions in the genome where these single-based changes reveal differences from one person to the next. SNP sequences as probes on a DNA microarray allow scientists to simultaneously screen thousands of mutations that might be involved in single-gene diseases as well as those involved in disorders exhibiting multifactorial inheritance. This technique, known as **genome scanning**, makes it possible to analyze a person's DNA for dozens or even hundreds of disease alleles, including those that might predispose the person to heart attacks, asthma, diabetes, Alzheimer disease, and other genetically defined disease subtypes. Genome scans are now occasionally used when physicians encounter patients with chronic illnesses where the underlying cause cannot be diagnosed.

除了检测单基因突变外，DNA 微阵列还包括可检测 SNPs 的探针。基因组中估计存在 1500 万个 SNPs 位点，这些单碱基变化揭示了人与人之间的差异。在 DNA 微阵列上，科学家以 SNP 序列作为探针可以同时筛选单基因疾病或多因素遗传疾病相关的数以千计的突变。这项技术被称为**基因组扫描**，这使得分析个体 DNA 中数十种甚至上百种疾病的等位基因成为可能，包括可能诱发个体患心脏病、哮喘、糖尿病、阿尔茨海默病和其他遗传相关疾病亚型的等位基因。如今，当医生遇到无法诊断根本病因的慢性病患者时，有时会采用基因组扫描。

In contrast to genome scanning microarrays that detect mutations in DNA, **gene-expression microarrays** detect gene-expression patterns for specific genes. This can be an effective approach for diagnosing genetic

与检测 DNA 突变的基因组扫描微阵列不同，**基因表达微阵列**用以检测特定基因的基因表达模式。这对遗传疾病诊断来讲可能是一种行之有效的途径，因为人体组织从健康状态发展至患病状态几乎总是伴随

diseases because the progression of a tissue from a healthy to a diseased state is almost always accompanied by changes in mRNA expression of hundreds to thousands of genes. Gene-expression microarrays may contain probes for only a few specific genes thought to be expressed differently in different cell types or may contain probes representing each gene in the genome. Although microarray techniques provide novel information about gene expression, it should be emphasized that DNA microarrays do not directly provide us with information about protein levels in a cell or tissue. We often infer what predicted protein levels may be based on mRNA expression patterns, but this may not always be accurate.

着成百上千种基因mRNA表达水平的变化。基因表达微阵列所含探针可能仅针对一些被认为在不同类型细胞中表达水平有差异的特定基因，也可能代表基因组中的每种基因。尽管微阵列技术提供了基因表达方面的全新信息，但应该强调的是，DNA微阵列并不能直接提供细胞或组织中蛋白质水平的相关信息。我们通常基于mRNA表达模式来推断待测蛋白质的表达水平，但这可能并不总是准确。

In one type of gene-expression micro-array analysis, mRNA is isolated from two different cell or tissue types—for example, normal cells and cancer cells arising from the same cell type. The mRNA samples contain transcripts from each gene that is expressed in that cell type. Some genes are expressed at higher levels than others. The expression level of each mRNA can be used to develop a *gene-expression profile* that is characteristic of the cell type. To do this, isolated mRNA molecules are converted into cDNA molecules, using reverse transcriptase. The cDNAs from the normal cells are tagged with fluorescent dye-labeled nucleotides (for example, green), and the cDNAs from the cancer cells are tagged with a different fluorescent dye-labeled nucleotide (for example, red).

在基因表达微阵列分析的一种类型中，研究人员分别从两种不同类型的细胞或组织中分离提取mRNA——例如源自同一类型细胞的正常细胞和癌细胞。这些mRNA样本包含了该类型细胞中表达的每种基因的转录物。一些基因的表达水平较高，而另一些基因的表达水平较低。每种mRNA的表达水平均可用于产生该类型细胞特有的基因表达图谱。为此，分离得到的mRNA分子被逆转录酶逆转录为cDNA分子。来自正常细胞的cDNAs用荧光染料标记的核苷酸（例如绿色）进行标记，而来自癌细胞的cDNAs则用不同荧光染料标记的核苷酸（例如红色）进行标记。

The labeled cDNAs are mixed together and applied to a DNA microarray. The cDNA

被标记的cDNAs混合后与DNA微阵列相作用。cDNA分子与微阵列上的互补单

molecules bind to complementary single-stranded probes on the microarray but not to other probes. Keep in mind that each field or feature does not consist of just one probe molecule but rather contains thousands of copies of the probe. After washing off the nonbinding cDNAs, scientists scan the microarray with a laser, and a computer captures the fluorescent image pattern for analysis. The pattern of hybridization appears as a series of colored dots, with each dot corresponding to one field of the microarray.

链探针相结合，但不与其他探针结合。需要提醒大家的是，每一域或每一特征并不是只含一个探针分子，而是同时存在数以千计的探针分子拷贝。将未结合的 cDNAs 漂洗后，科学家用激光扫描微阵列，同时用计算机捕获荧光图像模式并加以分析。杂交模式显示为一系列有颜色的点，每个点均对应微阵列的一个域。

This color pattern representation of results is often referred to as a *heat map*, because the color (or intensity of brightness) of a particular spot provides a sensitive measure of the relative levels of each cDNA in the mixture. If an mRNA is present only in normal cells, the probe representing the gene encoding that mRNA will appear as a green dot because only "green" cDNAs have hybridized to it. Similarly, if an mRNA is present only in the cancer cells, the microarray probe for that gene will appear as a red dot. If both samples contain the same cDNA, in the same relative amounts, both cDNAs will hybridize to the same field, which will appear yellow. Intermediate colors indicate that the cDNAs are present at different levels in the two samples.

这种以色彩模式呈现的结果通常被称为热图，因为特定斑点的颜色（或亮度的强度）提供了混合物中每种 cDNA 分子相对含量的敏感度。如果一种 mRNA 分子仅存在于正常细胞中，那么代表编码该 mRNA 的基因探针将显示为绿色斑点，因为仅有标记为“绿色”的 cDNAs 与微阵列进行了杂交。同理，如果某种 mRNA 分子仅存在于癌细胞中，那么微阵列中该基因的探针将显示为红色斑点。如果两个样本中含有同种 cDNA 分子且含量相当，则来源于两个样本的 cDNAs 均会与微阵列中的同一域进行杂交，从而呈现黄色。其他过渡颜色则表明两个样本中 cDNAs 的含量不同。

Gene-expression microarray analysis has revealed that certain cancers have distinct patterns of gene expression and that these patterns correlate with factors such as the cancer's stage, clinical course, or response to treatment. For example, scientists examined gene

基因表达微阵列分析已表明某些癌症具有不同的基因表达模式，并且这些模式与癌症的发展阶段、临床病程或对治疗的响应等因素相关。例如，科学家同时检测正常白细胞的基因表达和一种白细胞癌——弥漫大 B 细胞淋巴瘤（DLBCL）的白细胞

expression in both normal white blood cells and in cells from a white blood cell cancer known as *diffuse large B-cell lymphoma* (*DLBCL*). About 40 percent of patients with DLBCL respond well to chemotherapy and have long survival times. The other 60 percent respond poorly to therapy and have short survival.

基因表达。DLBCL 患者中大约 40％对化疗响应良好，并获得长的生存期；而另外 60％的患者则对治疗响应较差，生存期短。

The investigators assayed the expression profiles of 18,000 genes and discovered that there were two types of DLBCL, with almost inverse patterns of gene expression. One type of DLBCL, called GC B-like, had an expression pattern dramatically different from that of a second type, called activated B-like. Patients with the activated B-like pattern of gene expression had much lower survival rates than patients with the GC B-like pattern. The researchers concluded that DLBCL is actually two different diseases with different outcomes.

研究人员分析了 18 000 种基因的表达谱，发现存在两种类型的 DLBCL，并且它们的基因表达模式几乎相反。一种类型的 DLBCL（被称为 GC B-like）具有与第二种类型（被称为激活型 B-like）截然不同的基因表达模式。具有激活型 B-like 基因表达模式的患者，他们的生存概率远低于具有 GC B-like 模式的患者。因此研究人员指出，DLBCL 实际上是两种具有不同结果的不同疾病。

Once this type of profiling analysis is introduced into routine clinical use, it may be possible to adjust therapies for each group of cancer patients and to identify new specific treatments based on gene-expression profiles. Similar gene-expression profiles have been generated for many other cancers, including breast, prostate, ovarian, and colon cancer, providing tremendous insight into both substantial and subtle variations in genetic diseases.

一旦将这类表达谱分析引入常规临床应用，那么有可能会针对每类癌症患者群体都进行治疗方法的调整，并且可以基于基因表达谱来确定新的特异性治疗方法。研究人员已为许多其他癌症，包括乳腺癌、前列腺癌、卵巢癌和结肠癌，绘制了类似的基因表达谱，从而为遗传疾病的实质性变异和细微变异提供了非常深刻的见解。

Several companies are now promoting **nutrigenomics** services in which they claim to use gene-expression analysis to identify allele polymorphisms and gene-expression patterns for genes involved in nutrient metabolism. For

日前一些公司正在推广**营养基因组学**服务，他们声称可以使用基因表达分析来识别与营养代谢有关的等位基因多态性和基因表达模式。例如，脂代谢相关的载脂蛋白 A 基因（*APOA1*）、叶酸代谢相关的

example, polymorphisms in genes such as that for apolipoprotein A (*APOA1*), involved in lipid metabolism, and for methylenetetrahydrofolate reductase (*MTHFR*), involved in metabolism of folic acid, have been implicated in cardiovascular disease. Nutrigenomics companies claim that analysis of a patient's DNA for genes such as these enables them to judge whether allele variations or gene-expression profiles warrant dietary changes to potentially improve that person's health and reduce the risk of diet-related diseases.

亚甲基四氢叶酸还原酶基因（*MTHFR*）等在心血管疾病中存在基因多态性现象。营养基因组学公司指出，对患者DNA中此类基因进行分析使他们能够判断等位基因变异或基因表达谱是否可以通过改变饮食来改善个体健康状况并降低饮食相关疾病的患病风险。

10.5 Genetic Analysis of Individual Genomes by DNA Sequencing

10.5 运用DNA测序进行个体基因组遗传分析

Because of the relatively low cost of quickly and accurately sequencing individual genomes—what we call **personal genomics**—the ways that scientists and physicians evaluate a person's genetic information is rapidly changing. WGS is being utilized in medical clinics at an accelerating rate. Many major hospitals around the world are setting up clinical sequencing facilities for use in screening for the causes of rare diseases.

由于快速准确的个体基因组测序，即**个人基因组学**的成本相对较低，因此科学家和医生评估个体遗传信息的方式正在迅速发生变化。WGS正以不断加快的速度应用于医疗诊所中。世界各地许多大型医院正在建立临床测序设施，用以筛查罕见疾病的致病原因。

Recently, WGS has provided new insights into the genetics of anorexia, Alzheimer disease, and autism, among other disorders. Already there have been some very exciting success stories whereby WGS of individual genomes has led to improved treatment of diseases in children and adults. For example, native Newfoundlanders have one of the highest incidences in the world of *arrhythmogenic right ventricular dysplasia/cardiomyopathy* (*ARVD/C*), a rare condition

近期，WGS已为厌食症、阿尔茨海默病、孤独症等疾病的遗传学机制提供了全新的见解。凭借个体基因组的WGS改善儿童和成人疾病治疗方案已经产生了一些非常令人兴奋的成功案例。例如，纽芬兰原住民是世界上心律失常性右心室发育不全/心肌病（ARVD/C）发病率最高的人群之一，患有这种罕见病的个体通常在没有任何发病症状的情况下，由于心脏内不规则的电脉冲而突然死亡。

in which affected individuals often have no symptoms but then die suddenly from irregular electrical impulses within the heart.

Through individual genome sequencing, a mutation in the *AVRD5* gene has been identified as the cause of such cases of premature death. Of those with this mutation, approximately 50 percent of males and 5 percent of females die by age 40, and 80 percent of males and 20 percent of females die by age 50. Individuals carrying this mutation are now being implanted with internal cardiac defibrillators that can restart their hearts if electrical impulses stop or become irregular.

通过个体基因组测序，已经确定*AVRD5*基因的一个突变是此类疾病患者早逝的原因。在含该基因突变的患者中，大约50%的男性和5%的女性在40岁前死亡，80%的男性和20%的女性在50岁前死亡。现在携带该基因突变的个体可以通过植入内部心脏除颤器在电脉冲停止或变得不规则时重新启动心脏跳动。

Diseases that are caused by multiple genes are much harder to diagnose and treat based on sequencing data. For example, WGS of individuals affected by **autism spectrum disorder** (**ASD**) has revealed the involvement of more than 100 different genes. The genetics of ASD is particularly complex because of the broad range of phenotypes associated with this disorder. While WGS of individuals affected by ASD has revealed inherited mutations, it has also identified sporadic *de novo* mutations. In the future, sequence-based knowledge of these mutations may help physicians develop patient-specific treatment strategies.

由多基因引发的疾病更难得到诊断，也更难依据测序数据进行治疗。例如，一些**孤独症谱系障碍**（ASD）患病个体的WGS揭示该病症与100种以上的不同基因相关。由于该疾病相关的表型类型十分广泛，因此ASD的遗传机制尤其复杂。虽然ASD患病个体的WGS揭示了一些可遗传突变，但是同时也鉴定发现了一些零星存在的新生突变。未来这些突变相关的基于序列的科学知识可能会帮助医生开发患者特异性治疗策略。

Recall the concept of **whole exome sequencing** (**WES**). This alternative to WGS has also produced promising results in clinical settings. For example, from the time he was born, Nicholas Volmer had to live with unimaginable discomfort from an undiagnosed condition that was causing intestinal fistulas (holes from his gut to outside of his body) that were leaking

下面回想一下**全外显子组测序**(WES)的相关概念。该技术作为WGS的替代技术在临床情况中已产生了可喜的结果。例如，尼古拉斯·沃尔默自出生以来即患有一种无法确诊的导致肠瘘（肠道与体表形成不正常的通道）的疾病，其体液和粪便持续不断地从体内渗漏出来，生活极其不便，并且需要进行持续的手术治疗。在3岁之

body fluids and feces and requiring constant surgery. By 3 years of age, Nicholas had been to the operating room more than 100 times. A team at the Medical College of Wisconsin decided to have Nicholas's exome sequenced. Applying bioinformatics to compare his sequence to that of the general population, they identified a mutation in a gene on the X chromosome called *X-linked inhibitor of apoptosis* (*XIAP*). *XIAP* is known to be linked to another condition that can often be corrected by a bone marrow transplant. In 2010 a bone marrow transplant saved Nicholas's life and largely restored his health. Shortly thereafter the popular press described Nicholas as the first child saved by DNA sequencing.

前，尼古拉斯已经经历了 100 次以上的手术治疗。威斯康星医学院的一个研究小组决定对尼古拉斯的外显子组进行测序。他们应用生物信息学将其外显子序列与普通人群的序列进行比对，发现 X 染色体上 X 连锁凋亡抑制因子（*XIAP*）基因中存在一个突变。*XIAP* 同时也与另一种通常可以通过骨髓移植进行治疗的疾病相关。2010 年，骨髓移植手术挽救了尼古拉斯的生命，并在很大程度上恢复了他的健康。此后不久，尼古拉斯被媒体称为是通过 DNA 测序被拯救的第一个孩子。

Recently WGS and WES have been used to identify mutations of the *NGLY1* gene (which encodes a protein processing enzyme) that are associated with a very rare condition in children sharing certain development delays, liver disease, and a phenotype notable because of the inability to produce tears and thus the inability to cry. This list of confirmed patients with this mutation is less than two dozen, but this is the first time any definitive diagnosis has been available for these children despite a multitude of different clinical analyses and evaluations from doctors around the country.

近期，WGS 和 WES 已被用于鉴定 *NGLY1* 基因（编码一种蛋白质加工酶）的突变位点，这些突变与一种非常罕见的儿童疾病有关。患有该疾病的儿童均存在一定的发育迟缓并患有肝病，同时具有一种显著表型，即无法产生眼泪，进而无法哭泣。携带这种基因突变的确诊患者不足 20 人，但这是这些儿童首次获得明确诊断，尽管全美各地医生所给出的临床分析和评估存在众多不同。

It is worth noting that sequencing can also be applied in genetic analysis to identify human pathogens. While this is not a form of human genetic testing, it does represent an example of a genetic testing application.

值得注意的是测序也可用于遗传分析，从而鉴定人类病原体。尽管这并不是一种人类遗传检测，但确实也代表了遗传检测应用的一类实例。

Genetic Analysis of Single Cells by DNA and RNA Sequencing

We now have the ability to sequence the genome from a single cell! **Single-cell sequencing** (**SCS**) typically involves isolating genomic DNA from a single cell that is then subjected to *whole genome amplification* (*WGA*) by PCR to produce sufficient DNA to be sequenced. Amplification of the genome to produce enough DNA for sequencing without introducing errors remains a major challenge that researchers are working on so that SCS can become a more reliable and accurate technique for genetic testing.

Genomic sequencing from single cells is valuable for analyzing both *somatic cell mutations* (for example, mutations that arise in somatic cells such as in a skin cancer, which are not heritable) as well as *germ-line mutations* (heritable mutations that are transmitted to offspring via gametes). Sequencing genomes from individual egg or sperm cells, especially for couples undergoing *in vitro* fertilization, can identify carrier conditions or specific germ-line mutations that could result in a disorder in the offspring.

SCS allows scientists to explore genetic variations from cell to cell. These studies are revealing that different mutant genes can vary greatly between individual cells. In particular, cancer cells from a tumor often show genetic diversity, a fact that is increasingly being appreciated by researchers and clinicians. Understanding variations in genetic diversity and gene expression by individual cells within

利用 DNA 和 RNA 测序来进行单细胞遗传分析

现在，我们已经具备对单个细胞基因组进行测序的能力了！**单细胞测序**（SCS）通常包括从单个细胞中分离基因组 DNA，然后通过 PCR 对其进行全基因组扩增（WGA）来获得充足的 DNA 用于测序。在基因组扩增产生可用于测序的足量 DNA 的过程中如何不引入错误依然是研究人员正在努力将单细胞测序发展成为更可靠、更准确的遗传检测技术所面临的主要挑战。

单细胞基因组测序对分析体细胞突变（不可遗传的突变，如皮肤癌中体细胞发生的突变）和生殖细胞突变（可通过配子传递给后代的可遗传突变）均具有重要的价值。对单个卵子细胞或精子细胞进行基因组测序，尤其是对进行体外受精的夫妇的生殖细胞进行测序，可以确定可能导致后代遗传疾病的基因携带状况或特定的生殖细胞突变。

SCS 使科学家能够探索细胞与细胞之间的遗传变异。这些研究正不断揭示单细胞间不同突变基因的差异可以很大。尤其是癌细胞，通常表现出遗传多样性，这一事实越来越受到研究人员和临床医生的重视。理解癌症中单细胞的遗传多样性和基因表达差异有助于选择更好的、特异性更高的治疗方案。

a tumor could lead to better and more specific treatment options.

The contribution of individual cells to the phenotype of a tissue or organ affected by a genetic disorder is increasingly of interest. **RNA sequencing** (**RNA-seq**) is becoming a powerful tool for transcriptome-wide analysis of genes expressed by cells within a population, thus allowing researchers to differentiate genetic variations between cells.

单细胞对患有遗传疾病的组织或器官表型的促成作用日益受到人们的关注。**RNA 测序**（**RNA-seq**）正成为群体中在转录组范围内分析细胞基因表达的强大工具，从而使研究人员能够区分细胞之间的遗传变异。

Until recently, researchers or clinicians had to analyze the genome and transcriptome from cells independently. But now, it is possible to isolate DNA and RNA from the same cells, sequence the DNA, and, thanks to **single-cell RNA sequencing** (**scRNA-seq**), sequence the RNA. This enables a comparison of the genes present in a cell and the relative levels of expression for each transcript encoded by the genome.

直到不久之前，研究人员或临床医生还不得不将细胞的基因组和转录组分别进行独立分析。然而现在，研究人员可以从同一细胞中分离获得 DNA 和 RNA，对 DNA 进行测序，并且利用**单细胞 RNA 测序**（**scRNA-seq**）同时对 RNA 进行测序。这使得能够对细胞内存在的基因、基因组编码的每种转录物的相对表达水平进行比较。

Many disease treatments are designed to target cells, such as those in a tumor, as if all cells are homogeneous in genotype and phenotype. In fact, often such cells are quite heterogeneous genetically. scRNA-seq is now being applied to reveal the heterogeneity of cell types in tumors and other conditions, to then help plan better treatment approaches based on genetics of the cell types and their relative abundance.

许多疾病治疗是针对靶细胞，例如癌细胞进行设计的，似乎所有细胞的基因型和表型均为同质的、无差别的。实际上，这些细胞在遗传上通常非常不同。scRNA-seq 现在正被用于揭示癌症和其他疾病中细胞类型的异质性，从而根据细胞类型的遗传学特征及其相对丰度来协助设计更好的治疗方案。

Sequencing DNA and RNA from the same cell type typically requires the use of PCR to amplify genomic DNA (to sequence DNA) or mRNA, which is reverse transcribed into cDNA and then subsequently incorporated into a library and sequenced. scRNA-seq also then provides a

对同一类型的细胞进行 DNA 和 RNA 测序通常需要使用 PCR 来扩增基因组 DNA，再对 DNA 进行测序；或者扩增 mRNA，然后将其逆转录为 cDNA，随后建立 cDNA 文库并进行测序。scRNA-seq 还能提供定量转录组分析，从而确定单个细胞

quantitative transcriptome analysis in which the relative levels of RNA expressed in a cell can be determined. scRNA-seq provides quantitative data about RNA expression, similar to gene-expression microarray analysis. But scRNA-seq reveals all transcripts expressed in a cell, whereas the transcripts analyzed by microarrays are limited by the probes present on the array. These are among the reasons why scRNA-seq is likely to eventually replace microarrays for transcriptome analysis.

中 RNA 表达的相对水平。scRNA-seq 提供有关 RNA 表达的定量数据，类似于基因表达微阵列分析。但是 scRNA-seq 展示了细胞中所有表达的转录物，而利用微阵列分析的转录物则受限于微阵列上存在的探针。这些也是 scRNA-seq 最终可能取代微阵列进行转录组分析的原因之一。

Innate lymphoid cells (ILCs) are a relatively recently identified group of immune cells that reside in bone marrow and other tissues of the body. ILCs can differentiate into a variety of different immune cell types. They resemble T lymphocytes (T cells), although they lack antigen-specific immune response capability, and play important roles in immunity and the regulation of inflammation. Abnormalities in ILCs are involved in conditions such as autoimmune diseases, allergic responses, and asthma. As a result, ILCs have emerged as important cellular targets for medical interventions designed to manipulate the immune system— approaches often called *immunotherapy*.

先天淋巴样细胞（ILCs）是较近期才被鉴定的位于骨髓和其他组织中的免疫细胞群，可以分化为多种类型的免疫细胞。尽管它们缺乏抗原特异性的免疫应答能力，但是它们类似于 T 淋巴细胞（T 细胞），在免疫和炎症调节中扮演着重要角色。自身免疫病、过敏反应和哮喘等疾病均涉及 ILCs 异常。因此，ILCs 已经成为重要的细胞靶标，旨在通过设计医学干预措施来控制免疫系统——这些方法通常被称为免疫疗法。

scRNA-seq of mouse bone marrow progenitor cells enabled identification of different subsets of ILCs by their RNA-expression patterns. In this experiment, levels of RNA expression are color-coded in "heat map" fashion similar to the way gene-expression results are displayed for microarray analysis.

小鼠骨髓祖细胞的 scRNA-seq 能够通过它们的 RNA 表达模式鉴定 ILCs 的不同亚类。在该实验中，RNA 表达水平以“热图”方式进行色彩编码，类似于微阵列分析中基因表达结果的展示方式。

Computational analysis applies algorithms

计算分析应用可以生成聚类的算法。

that result in *clustering*—the grouping of cells based on similar patterns of gene-expression data. Such analysis reveals similarities in gene expression but significant differences in the transcriptomes of ILCs that, by phenotype, might appear to be the same. This delineation of ILCs based on transcriptome analysis reveals important genetic differences and offers the potential to develop new therapeutic approaches to manipulate these cells and maximize their immune responses.

聚类即根据相似的基因表达数据模式对细胞进行分组。此类分析揭示了表型上可能相同的ILCs基因虽表达相似，但转录组却存在显著差异。这些基于转录组分析的ILCs研究揭示了重要的遗传差异，并为开发新型治疗手段以控制这些细胞，使其免疫应答能力达到最大化提供了可能。

Screening the Genome for Genes or Mutations You Want

在基因组中筛选所需基因或突变

While we have thus far focused on genetic testing and identifying genes involved in disease, genomic analysis is also revealing genome diversity that confers beneficial attributes or phenotypes to humans. This can have a role in medical diagnosis because it allows scientists and physicians to understand why genetic differences account for resistance to certain diseases in some individuals compared to others.

至此，我们一直专注于遗传检测和鉴定疾病相关基因，但是基因组分析也在不断发现赋予人类有益特征或表型的基因组多样性。这在医用诊断中同样十分重要，因为它让科学家和医生理解了为什么基因差异会使得一些个体相比于另一些个体对某些疾病更具抵抗力。

For instance, researchers at the Broad Institute in Boston were studying elderly, overweight individuals who by all conventional medical diagnostic approaches should have shown symptoms of diabetes, yet they were not diabetic. Instead of seeking diabetes-causing mutations, the Broad group employed genomic analysis to search for mutations associated with protection from diabetes. Their efforts were rewarded when they determined that individuals with loss-of-function mutations in the *SLC30A8* gene (*solute carrier family 30, member 8* gene,

例如，波士顿博德研究所的研究人员正在研究老年超重个体，这些个体如果依据所有传统医用诊断方法则应表现出糖尿病症状，但是他们并未患糖尿病。博德研究小组并未去寻找糖尿病致病基因突变，而是采用基因组分析来寻找糖尿病预防相关的突变。他们的努力得到了回报。*SLC30A8*基因是溶质载体家族30成员8基因，编码参与胰岛素分泌的锌转运蛋白。他们发现该基因功能缺失突变的个体患糖尿病的可能性会降低65%，即使他们具有类似肥胖等高风险相关因素。

which encodes a zinc transport protein involved in insulin secretion) are 65 percent less likely to develop diabetes even when they have highly associated risk factors such as obesity.

In this spirit, increasingly geneticists are analyzing "natural or healthy knockouts" — the fortunate few individuals who may lack a specific gene or have a mutation in a disease-causing gene that provides a health benefit, such as protecting against development of a particular disease. Identifying such genetic variations may make it easier to help combat infections and disease.

本着这种精神，越来越多的遗传学家开始分析“天然的或健康的基因敲除”——幸运的极少数个体可能缺失某特定基因或某致病基因含有突变，从而为个体提供健康益处，比如预防某特定疾病的发生。鉴定这些遗传变异将更易于帮助抵御感染和疾病。

For example, through mutation many viruses that infect humans can evade drugs used to combat them. But these same viruses are defenseless against a rare mutation in the human gene *ISG15*. Individuals with mutations in *ISG15* fight off many if not most viruses. (Estimates suggest that less than 1 person in 10 million has this mutation.) *ISG15* mutations knock out a function that helps to dampen inflammation, so individuals with this mutation have a heighted inflammatory system, which helps fight off viruses. It is thought that this elevated response prevents viruses from replicating to levels that typically cause illness.

例如，许多能够感染人类的病毒可以通过基因突变来逃避用于抵抗它们的药物。但是，同样的病毒对一种罕见的人类*ISG15*基因突变则毫无防御能力。携带*ISG15*基因突变的个体可以抵抗即使不是大多数也是许多种类的病毒。（据估计，每1000万人中只有不到1人携带此突变。）*ISG15*突变导致抑制炎症功能的缺失，因此携带此突变的个体拥有增强的炎症系统，从而有助于抵抗病毒。这种炎症增强反应可以阻止病毒通过复制达到通常能引发疾病的水平。

Based on this knowledge, researchers are trying to develop drugs that might mimic the effects of the *ISG15* mutation as a future treatment strategy. As you will learn in Chapter 11—Gene Therapy, this strategy has been used to mimic naturally occurring mutations in the *CCR5* receptor gene that provides a rare subset of individuals (a percentage of northern Europeans) with complete

基于这些知识，研究人员正在尝试研发可以模拟*ISG15*突变效应的药物以作为未来的治疗策略。正如你将在第11章——基因治疗中了解的内容，该策略已被用于模拟*CCR5*受体基因天然发生的突变，该突变为人类罕见群体（占北欧人的1%）提供了对HIV感染的完全免疫力。

immunity to HIV infection.

A project called the Exome Aggregation Consortium (ExAC) is cataloging genetic variation from exome sequences of more than 60,000 individuals from diverse ancestries with the purpose of identifying naturally occurring gene knockouts. People with such knockouts or those who carry disease-causing genes but don't develop a particular illness are of significant interest to geneticists. Clearly this is an area of research that will continue to advance rapidly.

外显子组整合数据库（ExAC）项目正在对来自不同祖先的超过 60 000 人的外显子序列的遗传变异进行分类，目的在于鉴定天然存在的基因敲除。具有这些基因敲除的个体或携带致病基因但未患某特定疾病的个体对于遗传学家来说均非常重要。显然，这是一个将继续快速发展的研究领域。

10.6 Genome-Wide Association Studies Identify Genome Variations That Contribute to Disease

10.6 全基因组关联分析鉴定导致疾病的基因组变异

Many of the genetic testing approaches we have discussed so far have focused on analyzing genes in individuals or relatively small numbers of people. Microarray-based genomic analysis and WGS have led geneticists to employ a powerful strategy called **genome-wide association studies (GWAS)** in their quest to analyze populations of people for disease genes.

到目前为止，我们讨论的许多遗传检测方法均集中于分析个体或相对数量较少人群的基因。基于微阵列的基因组分析和 WGS 已经引领遗传学家采用**全基因组关联分析**（GWAS）这一强大策略，旨在分析人类群体中的疾病基因。

GWAS of relatively large populations of people for diagnostic or prognostic purposes often enables scientists to identify multiple genes that may influence disease risk. During the past decade there has been a dramatic expansion in the number of GWAS being reported. For example, GWAS for autism, obesity, diabetes, macular degeneration, myocardial infarction, arthritis, hypertension, several cancers, bipolar disease, autoimmune diseases, Crohn disease, schizophrenia (a recent publication noted

用于诊断或预后目的而进行的相对较大人类群体的 GWAS 通常使科学家能够鉴定可能影响疾病风险的多个基因。在过去 10 年中，被报道的 GWAS 数量飞速增加。例如孤独症、肥胖症、糖尿病、黄斑变性、心肌梗死、关节炎、高血压、多种癌症、双相情感障碍、自身免疫病、克罗恩病、精神分裂症（最新文献指出 100 多个遗传基因位点参与促进该疾病风险）、肌萎缩侧索硬化和多发性硬化的 GWAS 均属于科学文献和大众媒体中广泛宣传和报道的众多

more than 100 genetic loci contributing to disease risk), amyotrophic lateral sclerosis, and multiple sclerosis are among the many GWAS that have been widely publicized in the scientific literature and popular press. Behavioral traits such as intelligence have also been analyzed by GWAS. For example, recently, a high-profile, controversial genome-wide association study reported genetic markers influencing cognitive ability and attempted to relate these markers to differences in educational attainment between people. Other studies have identified more than 50 genes that may influence intelligence.

GWAS 中的一部分。GWAS 也被用于分析智力等行为特征。例如，最近一项备受瞩目并引发争议的全基因组关联分析报道了影响认知能力的遗传标记，并试图将这些标记和人与人之间的受教育程度差异相关联。其他研究已经鉴定出 50 多种可能影响智力的基因。

In a genome-wide association study, the genomes from several hundred, or several thousand (if available), unrelated individuals with a particular disease are analyzed, and the results are compared with genomes of individuals without the disease. The goal is to identify genetic variations that may confer risk of developing the disease. Many GWAS involve large-scale use of SNP microarrays that can probe on the order of 500,000 SNPs to evaluate results from different individuals. Other approaches of GWAS use WGS to look for specific gene differences, evaluate CNVs, or search for changes in the epigenome, such as methylation patterns. By determining which SNPs, CNVs, or epigenome changes are present in individuals with the disease, scientists can calculate the disease risk associated with each variation. Analysis of GWAS results requires statistical analysis to predict the relative potential impact (association or risk) of a particular genetic variation on the development of a disease phenotype.

在全基因组关联分析中，来自成百或上千（如果可以获得）患有特定疾病但相互之间毫无关联的个体基因组将被分析，并且这些结果将与未患该疾病的个体基因组进行比较。目的在于鉴定可能赋予该疾病发病风险的遗传变异。许多 GWAS 涉及大规模使用 SNP 微阵列，依次探测 500 000 个 SNPs 位点，从而评估来自不同个体的检测结果。GWAS 的其他方法则是使用 WGS 寻找特定基因差异，评估 CNVs，或者搜寻表观基因组变化，如甲基化模式。通过确定患病个体存在的 SNPs、CNVs 或表观基因组变化，科学家可以计算与每种变异相关的疾病发病风险。GWAS 结果需要利用统计分析来预测一种特定遗传变异对于疾病表型发生的相对潜在影响 (关联或风险)。

The scatterplot representation, called a Manhattan plot, is a typical representation of one way that results from GWAS are commonly reported. It is used to display data with a large number of data points. Particular positions in the genome are plotted on the *x*-axis; in this case loci on each chromosome are plotted in a different color. The results of a genotypic association test are plotted on the *y*-axis. There are several ways that associations can be calculated. A negative log of *p* values for loci is determined to be significantly associated with a particular condition. The top line establishes a threshold value for significance. Marker sequences with significance levels exceeding the threshold *p* value of 10^{-5}, corresponding to 5.0 on the *y*-axis, are likely disease-related sequences.

散点图（称为曼哈顿图）是一种常见的报道 GWAS 结果的典型表示方法。它使用大量数据点来展示数据。基因组中的特定位点被绘制在 *x* 轴上；在这种情况下，每条染色体的基因座用不同颜色进行标识。基因型关联检测结果则绘制在 *y* 轴上。多种方法可用于计算关联性。基因座 *p* 值的负对数确定与某特定性状显著相关。顶线确定了显著性的阈值。显著性水平超过阈值，即 *p* 值为 10^{-5}（*y* 轴上对应于 5.0）的标记序列即可能为疾病相关序列。

One prominent study that brought the potential of GWAS to light involved research on 4587 patients in Iceland and the United States with a history of myocardial infarction (MI), and 12,767 control patients. This work was done with microarrays containing 305,953 SNPs. Among the most notable results, the study revealed variations in two tumor-suppressor genes (*CDKN2A* and *CDKN2B*) on chromosome 9. Twenty-one percent of individuals with a history of MI were homozygous for deleterious mutations in both genes, and these individuals showed a 1.64 odds ratio of MI compared with noncarriers, including individuals homozygous for wild-type alleles. These variations were correlated with those in people of European descent, but interestingly, these same mutant alleles are not prominent in African-Americans.

一项揭示 GWAS 潜力的著名研究是涉及具有心肌梗死（MI）患病史的 4587 例来自冰岛与美国的患者和 12 767 例对照患者的研究。该项工作的完成借助了包含 305 953 个 SNPs 的微阵列。在最显著的结果中，该研究发现了两个位于人类 9 号染色体上的肿瘤抑制因子基因（*CDKN2A* 和 *CDKN2B*）的变异。在 21%具有 MI 患病史的个体中，这两个基因的有害突变均为纯合的，并且这些个体与未携带有害基因突变的个体（包括野生型等位基因纯合个体）相比，比值为 1.64。这些变异与个体拥有欧洲血统相关，但有趣的是，同种突变型等位基因在非裔美国人中并不显著。这是否意味着这些遗传标记在后者所在种群中并不是心肌梗死的风险因子呢？

Does this mean that these genetic markers are not MI risk factors among the latter ethnicity?

Examples such as this raise questions and ethical concerns about patients' emotional responses to knowing about genetic risk data. For example, GWAS often reveal dozens of DNA variations, but many variations have only a modest effect on risk. How does one explain to a person that he or she has a gene variation that changes a risk difference for a particular disease from 12 to 16 percent over an individual's lifetime? What does this information mean? Similarly, if the sum total of GWAS for a particular condition reveals about 50 percent of the risk alleles, what are the other missing elements of heritability that may contribute to developing a complex disease? In some cases, risk data revealed by GWAS may help patients and physicians develop diet and exercise plans designed to minimize the potential for developing a particular disease.

诸如此类的例子引发了人们对患者获悉遗传风险数据时情绪反应的相关疑问和伦理方面的担忧。例如，GWAS 通常发现数十种 DNA 变异，但是许多变异对风险的影响不大。如何向一个个体解释他或她所拥有的一种基因变异在整个生命历程中使其患某种特定疾病的风险从 12%变为 16%？这样的信息意味着什么？同样，如果特定疾病的 GWAS 整体数据揭示了大约 50%的风险等位基因，那么其他可能导致引发复杂疾病的可遗传但尚未知的元件又有哪些？在某些案例中，GWAS 所揭示的风险数据可能会帮助患者和医生制订饮食和锻炼计划，从而降低罹患特定疾病的潜在风险。

GWAS are showing us that, unlike single-gene disorders, complex genetic disease conditions involve a multitude of genetic factors contributing to the total risk for developing a condition. We need such information to make meaningful progress in disease diagnosis and treatment, which is ultimately a major purpose of what GWAS are all about.

GWAS 向我们展示了复杂遗传疾病与单基因疾病不同，它们包括众多导致患病总风险的遗传因素。我们需要这些信息使疾病诊断和治疗取得有意义的进展，这也正是 GWAS 最终想要达到的主要目的。

10.7 Genetic Testing and Ethical, Social, and Legal Questions

10.7 遗传检测与伦理、社会和法律问题

Applications of genetic testing raise important ethical, social, and legal issues that

遗传检测应用引发了许多重要的伦理、社会和法律问题，这些问题必须予以确定、

must be identified, debated, and resolved. Here we present a brief overview of some current ethical debates concerning genetic testing.

讨论和解决。在这里，我们简要概述一些当前遗传检测相关的伦理之争。

Genetic Testing and Ethical Dilemmas

When the Human Genome Project was first discussed, scientists and the general public voiced concerns about how genome information would be used and how the interests of both individuals and society can be protected. To address these concerns, the **Ethical, Legal, and Social Implications (ELSI) Research Program** was established by the National Human Genome Research Institute [a division of the National Institutes of Health (NIH)]. The ELSI Program focuses on four areas: (1) privacy and fairness in the use and interpretation of genetic information, (2) the transfer of genetic knowledge from the research laboratory to clinical practice, (3) ways to ensure that participants in genetic research know and understand the potential risks and benefits of their participation and give informed consent, and (4) enhancement of public and professional education.

遗传检测与伦理困境

当首次讨论人类基因组计划时，科学家和公众即对如何使用基因组信息以及如何保护个人及社会利益有所担忧。为了解决这些问题，美国国立卫生研究院（NIH）的分支部门美国国立人类基因组研究所建立了**伦理、法律和社会影响（ELSI）研究计划**。ELSI 计划关注四个领域：（1）使用和诠释遗传信息的隐私权和公平性；（2）将遗传学知识从科研实验室向临床实践的转化；（3）确保遗传研究参与者了解并理解他们的参与可能存在的潜在风险和收益，并获得他们的知情同意；（4）加强公共教育和专业教育。

The majority of the most widely applied genetic tests that have been used to date have provided patients and physicians with information that improves quality of life. One example involves prenatal testing for phenylketonuria (PKU) and implementing dietary restrictions to diminish the effects of the disease. But many of the potential benefits and consequences of genetic testing are not always clear. For example,

迄今为止，大多数使用最为广泛的遗传检测已为患者和医生提供了可提升生命质量的有效信息。一个实例就是针对苯丙酮尿症（PKU）进行产前检测，并实施饮食控制以减少疾病造成的影响。但是，遗传检测的诸多潜在益处和后果依然并不十分清楚。例如：

- We have the technologies to test for

- 有些遗传病，虽然我们有技术能

genetic diseases for which there are no effective treatments. *Should we test people for these disorders?*

■ With current genetic tests, a negative result does not necessarily rule out future development of a disease, nor does a positive result always mean that an individual will get the disease. *How can we effectively communicate the results of testing and the actual risks to those being tested?*

■ *What information should people have before deciding to have a genome scan or a genetic test for a single disorder or to have their whole genome sequenced?*

■ Sequencing fetal genomes from the maternal bloodstream has revealed examples of mutations in the fetal genome (for example, a gene involved in Parkinson disease). *How might parents and physicians use this information?*

■ Because sharing patient data through electronic medical records is a significant concern, *what issues of consent need to be considered?*

■ *How can we protect the information revealed by genetic tests?*

■ *How can we define and prevent genetic discrimination?*

Let's consider a specific example. In 2011, a case in Boston revealed the dangers of misleading results based on genetic testing. A prenatal ultrasound of a pregnant woman revealed a potentially debilitating problem (Noonan syndrome) involving the spinal cord of the woman's developing fetus. Physicians ordered a DNA test, which came back positive for a

够检测，但是目前尚无有效的治疗方法。我们应该为人们提供这些疾病的检测服务吗？

■ 目前的遗传检测，阴性结果并不一定排除未来患病的风险，阳性结果也并不总意味着个体一定会患上该种疾病。我们如何有效地与被测试者进行检测结果和实际患病风险相关信息的交流？

■ 在决定进行基因组扫描或针对某疾病进行遗传检测或进行全基因组测序前，人们应该了解哪些信息？

■ 从母体血液中测序胎儿基因组可以发现胎儿基因组信息中的突变（例如，帕金森病相关基因）。父母和医生该如何使用这些信息？

■ 通过电子病历共享患者数据受到广泛关注，因此需要考虑的患者知情同意相关问题应该有哪些？

■ 我们如何保护遗传检测所得的信息？

■ 我们如何定义和阻止遗传歧视？

让我们来看一个真实事例。2011 年波士顿发生的一个案例揭示了基于基因测试产生误导性结果的危害。一位孕妇的产前超声检查显示该女性腹中胎儿的脊髓存在潜在的功能发育不佳（努南综合征）。医生安排了 DNA 检测，检测结果表明数据库中努南综合征相关基因中的一种基因变体检测结果呈阳性。因此这对夫妇选择终止

gene variant in a database that listed the gene as implicated in Noonan syndrome. The parents chose to terminate the pregnancy. Months later it was learned that the locus linked to Noonan was not involved in the disease, yet there was no effective way to inform the research and commercial genetic testing community.

妊娠。数月后获悉该努南综合征相关基因座与该疾病的发生无关，到目前为止尚无有效途径与研究团体和商业化遗传检测界进行沟通。

To minimize these kinds of problems in the future, the NIH National Center for Biotechnology Information (NCBI) has developed a database called ClinVar (see www.ncbi.nlm. nih.gov/clinvar/), which integrates data from clinical genetic testing labs and research literature to provide an updated resource for researchers and physicians.

为了减少将来此类问题的发生，美国国立卫生研究院国家生物技术信息中心（NCBI）开发了ClinVar数据库（请参见www.ncbi.nlm.nih.gov/clinvar/），该数据库整合了来自临床遗传检测实验室的数据和科研文献，从而为科研人员和医生提供最新资源。

Disclosure of *incidental results* is another ethically challenging issue. When someone has his or her genome sequenced or has a test done involving a particular locus thought to be involved in a disease condition, the analysis sometimes reveals other mutations that could be of significance to the patient. Researchers and clinicians are divided on whether such information should be disclosed to the patient or whether patients should be asked for consent to receive all results from such tests. For example, a recent study considered 26 pregnant women who underwent prenatal genetic testing and learned they had genes associated with certain cancers and cognitive disorders as well as sex-chromosome abnormalities associated with reduced fertility. Again, this raises ethical issues about what type of consent women should consider when having these tests. Should these results be disclosed to these women? What do you think?

公开伴随检测获得的偶然结果则是另一个挑战伦理的问题。当某人进行基因组测序或针对某疾病相关特定基因座进行检测时，结果分析有时可能会揭示对患者有显著意义的其他突变。是否应向患者披露此类信息，或者是否应在接受所有检测结果上都征得患者同意，研究人员和临床医生在这些问题上存在分歧。例如，最近一项研究对26名孕妇进行了产前遗传检测，发现她们含某些癌症相关基因、认知障碍相关基因以及育性下降相关的性染色体畸变。这一事件再次引发道德伦理问题，即这些检测应该考虑允许哪些志愿者参加。这些检测结果是否应该向这些女性志愿者透露？你是如何认为的？

Earlier in this chapter we discussed preimplantation genetic diagnosis (PGD), which provides couples with the ability to screen embryos created by *in vitro* fertilization for genetic diseases. As we learn more about genes involved in human traits, will other, non-disease-related genes be screened for by PGD? Will couples be able to select embryos with certain genes encoding desirable traits for height, weight, intellect, or other physical or mental characteristics? What do you think of using genetic testing to purposely select for an embryo with a genetic disorder? There have been several well-publicized cases of couples seeking to use prenatal diagnosis or PGD to select for embryos with dwarfism and deafness.

在本章的前面，我们讨论了胚胎植入前遗传学诊断（PGD），它为夫妇提供了对体外受精胚胎进行遗传病筛查的能力。随着我们对人类性状相关基因的了解越来越多，PGD 是否会被用于筛选其他与疾病无关的基因呢？这些夫妇能否选择由某些特定基因决定的、包含他们想要的理想性状的胚胎，如身高、体重、智力或其他生理或心理特征？你如何看待使用遗传检测有目的地对胚胎进行特定遗传疾病筛查？在几例广为人知的案例中，人们使用产前诊断或 PGD 来选择患侏儒症和耳聋的胚胎。

As identification of genetic traits becomes more routine in clinical settings, physicians will need to ensure genetic privacy for their patients. There are significant concerns about how genetic information could be used in negative ways by employers, insurance companies, governmental agencies, or the general public. Genetic privacy and prevention of genetic discrimination will be increasingly important in the coming years. In 2008, the **Genetic Information Nondiscrimination Act (GINA)** was signed into law in the United States. This legislation is designed to prohibit the improper use of genetic information in health insurance and employment, but not life insurance.

随着遗传性状的鉴定在临床机构中越来越常规，医生需要确保患者的遗传隐私。雇主、保险公司、政府机构或普通民众如何采用消极的方式使用这些遗传信息已经引起人们的极大关注。未来，遗传隐私和防止遗传歧视将变得越来越重要。2008 年，**反基因歧视法**（GINA）在美国签署并立法。该法旨在禁止基因信息在健康保险和雇佣中被不当使用，但并不包括人寿保险。

Direct-to-Consumer Genetic Testing and Regulating the Genetic Test Providers

面向消费者的遗传检测和遗传检测提供方的管理

The past decade has seen dramatic

在过去 10 年中，**面向消费者**（DTC）

developments in **direct-to-consumer (DTC) genetic tests**. A simple Web search will reveal many companies offering DTC genetic tests. As of 2018, there were over 2000 diseases for which such tests are now available (in 1993 there were about 100 such tests). Most DTC tests require that a person mail a saliva sample, hair sample, or cheek cell swab to the company. For a range of pricing options, DTC testing companies largely use SNP-based tests such as ASO tests to screen for different mutations. For example, in 2007, Myriad Genetics, Inc., began a major DTC marketing campaign of its tests for *BRCA1* and *BRCA2*. Mutations in these genes increase the risk of developing breast and ovarian cancer. DTC testing companies report absolute risk, the probability that an individual will develop a disease, but how such risks are calculated is highly variable and subject to certain assumptions.

的遗传检测已经取得了巨大发展。一个简单的网络搜索即可发现许多可提供DTC遗传检测服务的公司。截至2018年，此类服务已可提供2000多种疾病检测（1993年约有100种此类检测）。大多数DTC检测要求被检人员将唾液样本、头发样本或口腔表皮细胞拭子邮寄至检测公司。DTC检测公司大多使用基于SNP的检测，例如ASO测试来筛选不同的基因突变，并提供一系列定价选项。例如，在2007年，万基遗传科技公司开启了一项针对*BRCA1*和*BRCA2*检测的大型DTC营销活动。这些基因突变将增加个体患乳腺癌和卵巢癌的风险。DTC检测公司对绝对风险，即个体罹患该种疾病的可能性出具报告，但是此类风险的计算方法高度可变并受到某些假设的影响。

Such tests are controversial for many reasons. For example, the test is purchased online by individual consumers and requires no involvement of a physician or other health-care professionals such as a nurse to administer the test or a genetic counselor to interpret the results. There are significant questions about the quality, effectiveness, and accuracy of such products because currently the DTC testing industry is largely self-regulated. The FDA does not regulate DTC genetic tests. There is at present no comprehensive way for patients to make comparisons and evaluations about the range of tests available and their relative quality.

由于诸多原因，这类检测备受争议。例如，此类检测服务由个人消费者通过网络在线购买，没有医生或其他卫生保健专业人员如护士参与实施检测，也没有遗传咨询顾问解读检测结果。当前的DTC检测行业在很大程度上是自主管理的，所以此类产品的质量、有效性和准确性均存在很大问题。FDA并不参与规范DTC遗传检测服务。目前，患者尚无综合全面的方法对此类检测的应用范围、相对质量进行比较和评估。

Most companies make it clear that they

大多数公司明确表示，他们并不是对

are not trying to diagnose or prevent disease, nor are they offering health advice, so what is the purpose of the information that test results provide to the consumer? Web sites and online programs from DTC testing companies provide information on what advice a person should pursue if positive results are obtained. But is this enough? If results are not understood, might negative tests not provide a false sense of security? Just because a woman is negative for *BRCA1* and *BRCA2* mutations does not mean that one cannot develop breast or ovarian cancer. An example of a personal decision that actress Angelina Jolie made based on the results of a genetic test is a question deserved to be discussed.

疾病进行诊断或预防，他们也不会提供健康方面的建议，那么相关检测结果提供给消费者的目的是什么呢？ DTC 检测公司的网站和在线程序为检测结果呈阳性的个体提供可以遵循的建议等相关信息。但是这些是否足够呢？如果检测结果不能被解读，那么检测结果呈阴性是否会提供错误的安全感呢？仅仅因为 *BRCA1* 和 *BRCA2* 基因突变检测呈阴性是不足以说明该女性将不会罹患乳腺癌或卵巢癌的。有关女演员安吉丽娜 · 朱莉根据遗传检测结果做出个人决定的案例是一个值得被讨论的问题。

Whether the FDA will oversee DTC genetic tests in the future is unclear. However, at the time of publication of this edition, the FDA has not revealed any definitive plans to regulate or oversee DTC genetic tests. But because some DTC genetic testing companies, such as 23andMe, offer health-related analyses or health reports, they do fall under FDA regulations. The FDA continues to issue warnings to DTC testing companies to provide what the FDA considers to be appropriate health-related interpretations of genetic tests. For example, in 2017 for the first time the FDA approved a DTC saliva test from 23andMe that can test for genetic mutations associated with ten conditions including Parkinson disease and Alzheimer disease. There are varying opinions on the regulatory issue. Some believe that the FDA has no business regulating DTC tests and that consumers should

FDA 将来是否参与监督 DTC 遗传检测尚不清楚。但是，截至本书此版本出版时，FDA 尚未透露任何规范或监督 DTC 遗传检测的明确计划。但是由于某些 DTC 遗传检测公司（例如 23andMe 公司），提供健康相关分析或健康报告，因此它们的确接受 FDA 监管。FDA 持续向 DTC 检测公司发出警告，要求其提供 FDA 认为合理的、健康相关的遗传检测解读。例如，2017 年，FDA 首次批准 23andMe 公司的 DTC 唾液检测项目，该项目可以测试包括帕金森病、阿尔茨海默病在内的 10 种疾病的相关基因突变。关于监管问题，也有人提出不同意见。一些人认为 FDA 无权规范 DTC 检测业务，消费者应该根据个人需求或兴趣自由购买检测服务产品。另一些人则坚持认为，为了保护消费者的利益，FDA 必须监管 DTC 检测业务。

be free to purchase products based on their personal needs or interests. Others insist that the FDA must regulate DTCs in the interest of protecting consumers.

Genetic Testing and Patents

Intellectual property (IP) rights are being debated as an aspect of the ethical implications of genetic engineering, genomics, and biotechnology in many ways. For example, patents on IP for isolated genes, recombinant cell types, and GMOs can be potentially lucrative for the patent-holders but may also pose ethical and scientific problems. Why is protecting IP important for companies? Consider this issue. If a company is willing to spend millions or billions of dollars and several years doing research and development (R&D) to produce a valuable product, then shouldn't it be afforded a period of time to protect its discovery so that it can recover R&D costs and made a profit on its product?

Genes in their natural state as products of nature cannot be patented. Consider the possibilities for a human gene that has been cloned and then patented by the scientists who did the cloning. The person or company holding the patent could require that anyone attempting to do research with the patented gene pay a licensing fee for its use. Should a diagnostic test or therapy result from the research, more fees and royalties may be demanded, and as a result the costs of a genetic test may be too high for many patients to afford. But limiting or preventing the holding of patents for genes or genetic tools could reduce the incentive for

遗传检测与专利

作为基因工程、基因组学和生物技术的伦理涵义的一个方面，人们正以多种方式对**知识产权**（IP）开展讨论。例如，分离获得的基因、重组细胞类型和GMOs的知识产权专利对于专利持有人可能具有潜在利润，但是也可能带来道德伦理和科学问题。为什么保护知识产权对于公司十分重要呢？让我们探讨一下这个问题。如果一家公司愿意投入数百万或数十亿美元并历时数年时间进行研发（R&D），生产一款有价值的产品，那么难道不应该为公司提供一段时间的保护从而使公司能够收回研发成本并从研发产品中获利吗？

自然存在状态下的基因作为大自然的产物无法申请专利。试想一种人类基因被成功克隆，然后被完成基因克隆实验的科学家申请了专利。持有专利的个人或公司可能会要求任何尝试研究该专利基因的个人支付使用该专利的许可费用。如果诊断检测或治疗方案源自该项研究，更多的费用和专利使用费将被要求支付，最终遗传检测可能会对许多患者而言费用太高且无力承担。但是，限制或阻止基因或遗传工具的专利持有会降低对基因和遗传工具等产品投入研究的动力，尤其是对那些需要从研究中获利的公司来讲。

pursuing the research that produces such genes and tools, especially for companies that need to profit from their research.

Patenting of genetic tests is under increased scrutiny in part because of concerns that a patented test can create monopolies in which patients cannot get a second opinion if only one company holds the rights to conduct a particular genetic test. Recent analysis has estimated that as many as 64 percent of patented tests for disease genes make it very difficult or impossible for other groups to propose a different way to test for the same disease.

遗传检测的专利授予受到越来越严格的审查，部分原因是担心获得专利的检测服务形成垄断。如果仅有一家公司拥有从事某项特定遗传检测的权利，那么患者将无法获得第二种专业意见和建议。最新分析预计多达64%的疾病基因专利检测使得其他团体很难甚至不可能提出其他不同方法用于检测同一疾病。

In 2010 a landmark case brought by the American Civil Liberties Union against Myriad Genetics contended that Myriad could not patent the *BRCA1* and *BRCA2* gene sequences used to diagnose breast cancer. Myriad's BRACAnalysis® product has been used to screen over a million women for *BRCA1* and *BRCA2* during its period of patent exclusivity. A U.S. District Court judge ruled Myriad's patents invalid on the basis that DNA in an isolated form is not fundamentally different from how it exists in the body. Myriad was essentially accused of having a monopoly on its tests, which have existed for a little over a decade based on its exclusive licenses in the United States.

2010年，美国公民自由联盟抗议万基遗传科技公司，称万基公司不能将用于诊断乳腺癌的*BRCA1*和*BRCA2*基因序列申请专利，该事件具有里程碑式意义。万基公司的产品BRACAnalysis®在专利独占期内已被数以百万计的女性用于*BRCA1*和*BRCA2*基因筛查。美国地方法院法官裁定万基公司的专利无效，其依据是分离获得的DNA与体内存在的DNA在本质上并无区别。事实上，万基公司由于在检测项目上存在垄断而被指控，这种情况基于其在美国的独家许可已有十余年的历史。

This case went to the Supreme Court in 2013, which rendered a 9—0 ruling against Myriad, stripping it of five of its patent claims for the *BRCA1* and *BRCA2* genes, largely based on the view that natural genes are a product of nature and just because they are isolated does not mean they can be patented.

此案于2013年提交至美国联邦最高法院，法院最终对万基公司做出9—0一致裁决，剥夺其对*BRCA1*和*BRCA2*基因的5项专利声明，主要基于天然基因是自然产物，不能仅因为获得分离即意味着可以使用它们申请专利。同时法院还裁定，实验室获得的cDNA序列可以继续被专利保护，

The Court ruled that cDNA sequences produced in a lab can continue to be patentable. Myriad still holds about 500 valid claims related to *BRCA* gene testing.

因此万基公司仍然拥有约500项与*BRCA*基因检测相关的有效专利。

Whole Genome Sequence Analysis Presents Many Ethical Questions

全基因组测序分析存在很多伦理问题

In the next decade and beyond, it is expected that WGS analysis of adults and babies will become increasingly common in clinical settings. The Newborn Sequencing in Genomic Medicine and Public Health (NSIGHT) research program initiated by the NIH is under way to sequence the genomes of more than 1500 babies. Both infants with illnesses and babies who are healthy will be part of this study. This initiative will allow scientists to carry out comparative genomic analyses of specific sequences to help identify genes involved in disease conditions.

在接下来的10年及未来，预计成人和婴儿的WGS分析将在临床机构中越来越普遍。由NIH启动的基因组医学和公共卫生中的新生儿测序（NSIGHT）研究计划正在实施中，该项目拟对1500多名婴儿进行基因组测序。患病婴儿和健康婴儿均为本研究的组成部分。该项目将使科学家针对特定序列开展比较基因组分析，从而有助于鉴定疾病相关基因。

Screening of newborns is important to help prevent or minimize the impacts of certain disorders. Each year routine blood tests from a heel prick of newborn babies reveal rare genetic conditions in several thousand infants in the United States alone. A small number of states allow parents to opt out of newborn testing. In the future, should DNA sequencing at the time of birth be universally required? Do we really know enough about which human genes are involved in disease to help prevent disease in children? Estimates suggest that sequencing can identify approximately 15 to 50 percent of children with diseases that currently cannot be diagnosed by other methods. What is the value of having sequencing data for healthy children?

新生儿筛查对于有效预防或最大限度降低某些疾病的影响非常重要。每年通过新生儿足跟取血进行的常规血液检查仅在美国即可发现数千名患有罕见遗传疾病的婴儿。美国的少数州允许父母选择不接受新生儿检测。将来是否应该普遍要求新生儿在出生时即进行DNA测序？我们是否真的足够了解人类疾病相关基因，从而有助于预防儿童疾病？预计结果表明，测序可以鉴定15%—50%目前尚无法通过其他方法进行诊断的儿童疾病。那么对于健康儿童来说，拥有测序数据有何价值呢？

Personal genomics adds another layer of complexity to this discussion. When people donate their DNA for WGS projects, should they have access to the raw data from their sequence analysis? Currently, such volunteers are refused access to their genetic data.

个人基因组学为这一讨论又增加了一层复杂性。当人们将其DNA捐赠给WGS项目时，他们是否可以得到序列分析的原始数据？目前，这些志愿者被拒绝访问本人的基因数据。

As exciting as this period of human genetics and medicine is becoming, most of the WGS studies of individuals are happening in a largely unregulated environment. This raises significant ethical concerns especially with respect to DNA collection, the variability and quality control of DNA handling protocols, sequence analysis, storage, and confidentiality of genetic information.

随着当前人类遗传学和医学的发展日益令人兴奋，大多数的个体WGS研究在很大程度上是在不受监管的环境中进行的。这引发了显著的伦理担忧，特别是在DNA收集，DNA操作手册的变化和质量控制，遗传信息的序列分析、存储和保密等方面。

Preconception Testing, Destiny Predictions, and Baby-Predicting Patents

拟生育检测、命运预测和婴儿预测相关专利

Companies are now promoting the ability to do *preconception* testing and thus make "destiny predictions" about the potential phenotypes of hypothetical offspring based on computational methods for analyzing sequence data of parental DNA samples. The company 23andMe has been awarded a U.S. patent for a computational method called the *Family Traits Inheritance Calculator* to use parental DNA samples to predict a baby's traits, including eye color and the risk of certain diseases. This patent includes applications of technologies to screen sperm and ova for IVF.

目前，许多公司正在提升拟生育检测能力，并通过基于计算方法分析亲本DNA样本的序列数据对未来后代的可能表型进行"命运预测"。23andMe公司的计算方法"家族特征遗传计算器"被授予美国专利，该方法使用父母的DNA样本来预测婴儿的特征，包括眼睛颜色和某些疾病的患病风险。该专利还包括为体外受精筛查精子和卵子的相关技术应用。

More than 5 million babies have been born by IVF. Woman are having children at a later age, couples involved in IVF may be same sex or transgender, and the use of sperm and egg

借助体外受精出生的婴儿已经超过500万名。女性生育年龄较大、选择体外受精的夫妇可能是同性伴侣或变性人、使用精子供体和卵子供体以及代孕母亲等情形都

donors, as well as surrogates, is also increasing. Currently, gender selection of embryos generated by IVF is very common. But could preconception testing lead to the selection of “designer babies” as people look to customize their offspring? Fear of *eugenics* surrounds these conversations, particularly as genetic analysis starts moving away from disease conditions to nonmedical traits such as hair color, eye color, other physical traits, and potentially behavioral traits. The patent has been awarded for a process that will compare the genotypic data of an egg provider and a sperm provider to suggest gamete donors that might result in a baby or hypothetical offspring with particular phenotypes of interest to a prospective parent. What do you think about this?

在不断增加。当前，对体外受精产生的胚胎进行性别选择的现象非常普遍。但是拟生育检测是否可能导致人们为了选择所期望的“设计婴儿”而定制他们的后代呢？优生学的恐惧围绕着这些话题，尤其当遗传分析开始从疾病相关性状向一些非医学特征转移时，例如头发颜色、眼睛颜色、其他生理特征以及潜在的行为特征。比较卵子供体和精子供体的基因型数据，从而向准父母暗示这些配子供体可能产生的婴儿或假想后代所具有的特定表型，这一过程已被授予专利。你如何看待这一事件？

A company called GenePeeks claims to have a patent-pending technology for reducing the risk of inherited disorders by “digitally weaving” together the DNA of prospective parents. GenePeeks plans to sequence the DNA of sperm donors and women who want to get pregnant to inform women about donors who are most genetically compatible for the traits they seek in offspring. Their proprietary computing technology is intended to use sequence data to examine “virtual” eggs and sperm from donor-client pairings to estimate the likelihood of about 10,000 specific diseases in hypothetical offspring from prospective parents. Will technologies such as this become widespread and attract consumer demand in the future? What do you think? Would you want this analysis done before deciding whether to have a child with a particular person?

一家名为 GenePeeks 的公司声称拥有一项正在申请专利的技术，该技术通过对准父母的 DNA 信息进行“数字编织”来降低遗传性疾病的发病风险。GenePeeks 计划对精子供体和备孕女性的 DNA 进行测序，然后告知备孕女性这些精子供体中哪些在遗传上与预期获得的后代性状最为匹配。他们专有的计算技术旨在使用序列数据考察供体 - 客户配对中的“虚拟”卵子和供体精子，从而预测准父母的假想后代大约 10 000 种特定疾病的发病可能性。类似技术将来会普及并吸引消费者的需求吗？你是如何认为的？当你决定与某人一起孕育孩子前，是否想要进行此类分析呢？

11 Gene Therapy

11　基因治疗

The treatment of a human genetic disease by gene therapy is the ultimate application of genetic technology.

利用基因治疗对人类遗传疾病进行治疗是基因工程的终极应用。

Although drug treatments can be effective in controlling symptoms of genetic disorders, the ideal outcome of medical treatment is to cure a disease. This is the goal of **gene therapy**—the delivery of therapeutic genes into a patient's cells to correct genetic disease conditions caused by a faulty gene or genes. The earliest attempts at gene therapy focused on the delivery of normal, *therapeutic copies* of a gene to be expressed in such a way as to override or negate the effects of the disease gene and thus minimize or eliminate symptoms of the genetic disease. But in recent years newer methods for removing, correcting, inhibiting, or silencing defective genes have shown promise. However, no approach to gene therapy has generated more excitement for its potential than genome-editing applications involving CRISPR-Cas. We will consider several recent examples of genome editing to target specific genes.

尽管药物治疗能够有效控制遗传疾病的症状，但是医学治疗的理想结果还是治愈疾病。这即是**基因治疗**的目标——将具有治疗作用的基因传送至患者细胞中，纠正由一个或多个错误基因引发的遗传疾病。基因治疗的最早尝试集中于传送正常的、具治疗作用的基因拷贝，并使这些基因表达超越或抵消致病基因的效应，从而减少甚至消除遗传疾病的症状。但是近些年来，一些新的用于移除、纠正、抑制或沉默缺陷基因的方法呈现出巨大的应用前景。然而，没有一种基因治疗方法比 CRISPR-Cas 进行基因组编辑更具有应用潜力，也更令人兴奋。我们在这里介绍几个针对特定靶基因进行基因组编辑的最新实例。

Gene therapy is one of the goals of **translational medicine**— taking a scientific discovery, such as the identification of a disease-

基因治疗是**转化医学**的众多目标之一——以科学发现为例，如鉴定致病基因，将科学发现转化为有效的治疗手段，从而

causing gene, and translating the finding into an effective therapy, thus moving from the laboratory bench to a patient's bedside to treat a disease. In theory, the delivery of a therapeutic gene is rather simple, but in practice, gene therapy has been very difficult to execute. In spite of over 25 years of trials, this field has not lived up to its expectations. However, gene therapy is currently experiencing a fast-paced resurgence, with several high-profile new successes and potentially exciting new technologies sitting on the horizon. Gene therapy is an example of **precision medicine** (see Chapter 15—Genomics and Precision Medicine). It is hoped that gene therapy will soon become part of mainstream medicine. The treatment of a human genetic disease by gene therapy is the ultimate application of genetic technology. In this chapter we will explore how gene therapy is executed, and we will highlight selected examples of successes and failures as well as discuss new approaches to gene therapy. Finally, we will consider ethical issues regarding gene therapy.

将研究成果从实验室转移至患者病床旁以用于疾病治疗。理论上讲，传送治疗基因相当简单，但在实践中，基因治疗的实施却非常困难。尽管已经历了 25 年以上的努力，但该领域仍未达到原本的预期。然而，随着近期几项备受瞩目的成功案例和极具潜力且令人兴奋的新技术的产生，基因治疗正在迅速复苏。基因治疗属于**精准医疗**(请参阅第 15 章——基因组学和精准医疗)。人们期望基因治疗能快速发展为主流医学的一部分。通过基因治疗来治疗人类遗传疾病是基因工程的终极应用。在本章中，我们将探讨基因治疗如何实施，重点选取并介绍一些成功与失败的案例，同时讨论基因治疗的新方法。最后，我们还将探讨基因治疗相关的伦理问题。

11.1 What Genetic Conditions Are Candidates for Treatment by Gene Therapy?

11.1 哪些遗传疾病有望使用基因治疗?

Two essential criteria for gene therapy are that the gene or genes involved in causing a particular disease have been identified and that the gene can be cloned or synthesized in a laboratory. As a result of the Human Genome Project, the identification of human disease genes and their specific DNA sequences has greatly

基因治疗的两个基本条件是导致特定疾病的一种或多种基因已被鉴定，并且这些基因可以在实验室中进行克隆或合成。人类基因组计划的成果之一便是人类疾病基因及其特定 DNA 序列的鉴定大大增加了基因治疗试验的候选基因数量。几乎所有的早期基因治疗试验和大多数基因治疗方

increased the number of candidate genes for gene therapy trials. Almost all of the early gene therapy trials and most gene therapy approaches have focused on treating conditions caused by a single gene.

法都集中于治疗单基因引起的遗传疾病。

The cells affected by the genetic condition must be readily accessible for treatment by gene therapy. For example, blood disorders such as leukemia, hemophilia, and other conditions have been major targets of gene therapy because it is relatively routine to manipulate blood cells outside of the body and return them to the body in comparison to treating cells in the brain and spinal cord, skeletal or cardiac muscle, and organs with heterogeneous populations of cells such as the pancreas.

受遗传疾病影响的细胞必须易于接受基因治疗。例如，白血病、血友病和其他的血液类疾病已成为基因治疗的主要靶标，因为与治疗脑、脊髓、骨骼肌、心肌或含有细胞异质群体的器官如胰腺等相比，体外操作血细胞并将其重新输入体内的技术已相对常规和成熟。

In the past decade, every major category of genetic disease has been targeted by gene therapy (Figure 11.1). A majority of recently approved clinical trials are for cancer treatment. Gene therapy approaches are also currently being investigated for the treatment of hereditary blindness; hearing loss; neurodegenerative diseases including Alzheimer disease, Parkinson disease, and amyotrophic lateral sclerosis (ALS); cardiovascular disease; muscular dystrophy; hemophilia; and infectious diseases, such as HIV; among many other conditions, including depression and drug and alcohol addiction. Worldwide, over 2300 approved gene therapy clinical trials have occurred or recently been initiated.

在过去 10 年中，遗传疾病的每种主要类别均已开展基因治疗（图 11.1）。近期被批准的大多数临床试验均针对癌症治疗。目前同时在进行的基因疗法研究还包括治疗遗传性失明，听力损失，神经系统变性疾病 [包括阿尔茨海默病、帕金森病和肌萎缩侧索硬化（ALS）]，心血管疾病，肌营养不良，血友病，艾滋病等传染性疾病，抑郁症和毒品或酒精成瘾在内的许多其他疾病。在全球范围内，已经实施或近期启动的获批基因治疗临床试验已经超过 2300 项。

In the United States, proposed gene therapy clinical trials must first be approved by review boards at the institution where they will

在美国，提议的基因治疗临床试验必须首先获得实施该研究的所属机构评审委员会的批准，然后该基因治疗方案必须再

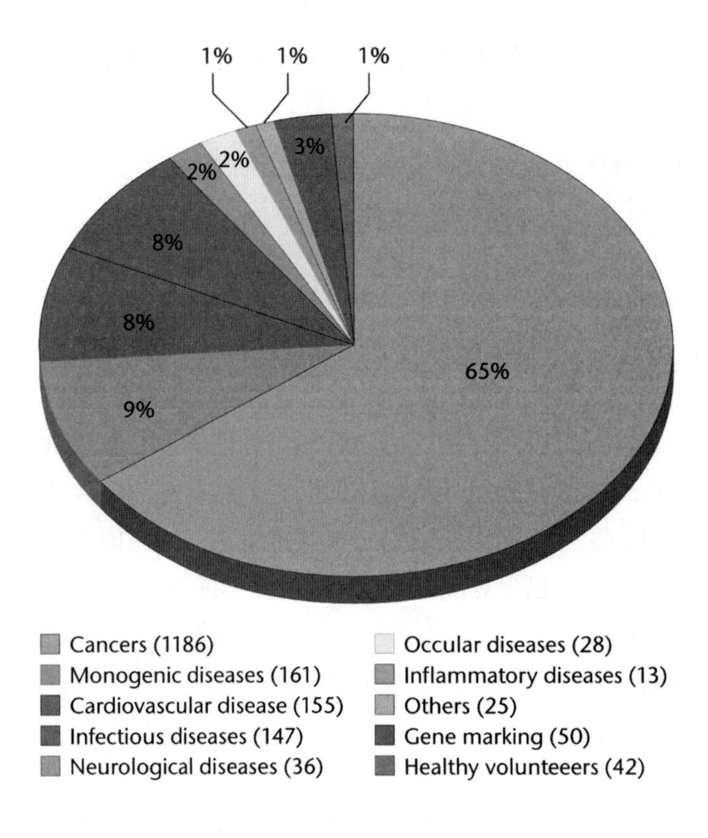

FIGURE 11.1 Graphic representation of different genetic conditions being treated by gene therapy clinical trials worldwide. Notice that cancers are the major target for treatment.

图 11.1 全球范围内基因治疗临床试验所涉及的不同遗传疾病的比例图。注意癌症是主要的治疗靶标。

be carried out, and then the protocols must be approved by FDA.

获得 FDA 的批准。

11.2 How Are Therapeutic Genes Delivered?

11.2 如何传送治疗基因?

In general, there are two broad approaches for delivering therapeutic genes to a patient being treated by gene therapy, *ex vivo gene therapy* and *in vivo gene therapy* (Figure 11.2). In *ex vivo* gene therapy, cells from a person with a particular genetic condition are removed, treated in a laboratory by adding either normal copies of a therapeutic gene or a DNA or RNA sequence that will inhibit expression of a defective gene, and then these cells are transplanted back into the person. Genetically altered cells treated in this manner can be transplanted back into the patient

通常有两种广泛使用的方法可将治疗基因传送至接受基因治疗的患者体内：离体基因疗法和体内基因疗法（图 11.2）。在离体基因疗法中，患特定遗传疾病的个体细胞从体内取出，在实验室中加入治疗基因的正常拷贝或能够抑制缺陷基因表达的 DNA 或 RNA 序列，然后再将这些细胞重新移植回患者体内。通过这种方式处理后的遗传改造细胞在重新移植回患者体内时，不必担心免疫系统排斥，因为这些细胞原本即来自患者本人。

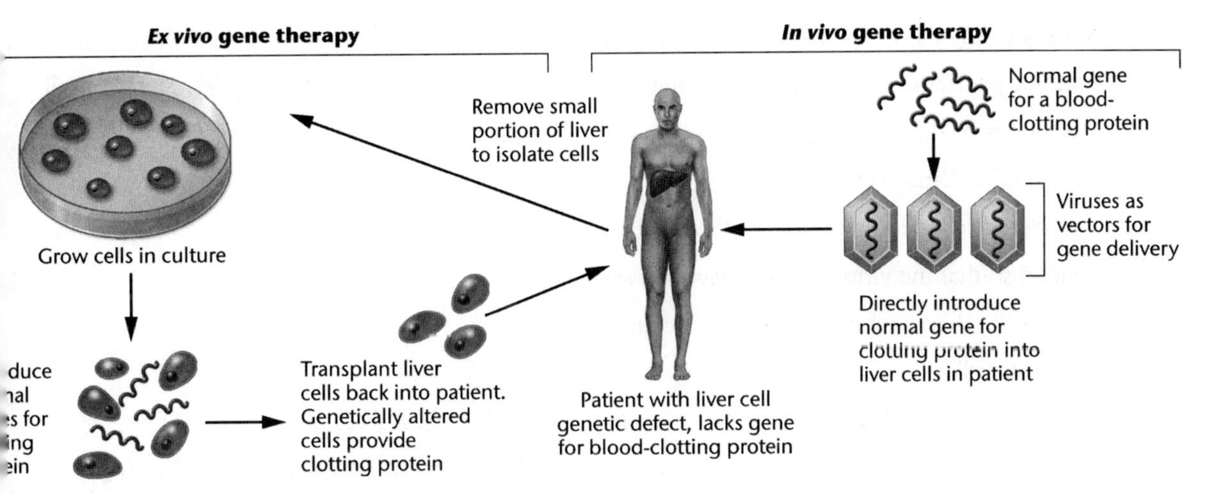

FIGURE 11.2 *Ex vivo* and *in vivo* gene therapy for a patient with a liver disorder. *Ex vivo* gene therapy involves isolating cells from the patient, introducing normal copies of a therapeutic gene (encoding a blood-clotting protein in this example) into these cells, and then returning cells to the body where they will produce the required clotting protein. *In vivo* approaches involve introducing DNA directly into cells while they are in the body.

图 11.2 采用离体基因疗法和体内基因疗法对肝病患者进行治疗。离体基因疗法涉及从患者体内分离细胞，将治疗基因（本例中是编码凝血因子的蛋白质）的正常拷贝引入患者细胞，然后将这些细胞重新移植回患者体内产生机体所需的凝血蛋白。体内基因疗法则将 DNA 直接导入机体细胞内。

without fear of immune system rejection because these cells were derived from the patient initially.

In vivo gene therapy does not involve removal of a person's cells. Instead, therapeutic DNA is introduced directly into affected cells of the body. One of the major challenges of *in vivo* gene therapy is restricting the delivery of therapeutic genes to only the intended tissues and not to all tissues throughout the body.

体内基因疗法则不涉及个体细胞的转移，而是将治疗 DNA 直接引入人体内受影响的细胞中。体内基因疗法的主要挑战之一是限制治疗基因仅传送至预期组织中，而不是全身所有组织。

Viral Vectors for Gene Therapy

For both *in vitro* and *ex vivo* approaches, the key to successful gene therapy is having a delivery system to transfer genes into a patient's cells. Because of the relatively large molecular size and electrically charged properties of DNA, most human cells do not take up DNA easily. Therefore, delivering therapeutic DNA molecules into human cells is challenging.

用于基因治疗的病毒载体

对于体外和离体方法，基因治疗成功的关键是拥有一套将基因转移至患者细胞内的传递系统。由于 DNA 的分子量较大并具有带电特性，大多数人类细胞并不容易接收外源 DNA。因此，将治疗 DNA 分子传递至人类细胞内极具挑战性。从早期基因治疗开始，基因工程病毒作为载体已成为将治疗基因传递至人类细胞的主要工具。

Since the early days of gene therapy, genetically engineered viruses as vectors have been the main tools for delivering therapeutic genes into human cells. Viral vectors for gene therapy are engineered to carry therapeutic DNA as their payload so that the virus infects target cells and delivers the therapeutic DNA without causing damage to cells.

用于基因治疗的病毒载体经工程改造后携带治疗 DNA 作为其有效载荷，从而随着病毒感染靶细胞传递治疗 DNA 且不会引起细胞损伤。

In a majority of gene therapy trials around the world, scientists have used genetically modified *retroviruses* as vectors. Retroviruses such as HIV contain an RNA genome that scientists use as a template for the synthesis of a complementary DNA molecule. **Retroviral vectors** are created by removing replication and disease-causing genes from the virus and replacing them with a cloned human gene. After the altered RNA has been packaged into the virus, the recombinant viral vector containing the therapeutic human gene is used to infect a patient's cells. Technically, virus particles are carrying RNA copies of the therapeutic gene. Once inside a cell, the virus cannot replicate itself, but the therapeutic RNA is reverse transcribed into DNA, which enters the nucleus of cells and *integrates* into the genome of the host cells' chromosome. If the inserted therapeutic gene is properly expressed, it produces a normal gene product that may be able to ameliorate the effects of the mutation carried by the affected individual.

在全球大多数基因治疗试验中，科学家使用基因改造的逆转录病毒作为载体。逆转录病毒（如人类免疫缺陷病毒）含一个 RNA 基因组，科学家以其为模板合成与之互补的 DNA 分子。**逆转录病毒载体**是通过移除逆转录病毒的复制元件和致病基因，并用克隆的人类基因进行替换而创建的。改造后的 RNA 被包装入病毒中，这些含治疗性人类基因的重组病毒载体可用于感染患者细胞。从技术上讲，病毒颗粒携带治疗基因的 RNA 拷贝。一旦进入细胞，病毒无法进行自我复制，但治疗 RNA 被逆转录为 DNA，然后进入细胞核内并整合入宿主细胞染色体基因组中。如果插入基因组的治疗基因能够正确表达，那么将合成正常的基因产物，该产物可能能够改善患病个体所携带的基因突变带来的影响。

One advantage of retroviral vectors is that they provide long-term expression of delivered genes because they integrate the therapeutic gene into the genome of the patient's cells. But a

逆转录病毒载体的优势之一是它们可以提供被转移基因的长效表达，因为治疗基因被整合到了患者细胞基因组中。但是逆转录病毒载体存在的一个主要问题是它

major problem with retroviral vectors is that they have produced random, unintended alterations in the genome, in some cases due to *insertional mutations*. Retroviral vectors generally integrate their genome into the host-cell genome at random sites. Thus, there is the potential for retroviral integration that randomly inactivates genes in the genome or gene-regulatory regions such as a promoter sequence.

们在一些情况下由于插入突变，会在基因组中产生随机的、意想不到的改变。通常逆转录病毒载体将自身基因组整合入宿主细胞基因组的位点是随机的。因此，逆转录病毒的整合可能会随机失活基因组内的基因或启动子序列等基因调控区功能。

In many early gene therapy trials, **adenovirus vectors** were the retrovirus vector of choice. An advantage of these vectors is that they are capable of carrying large therapeutic genes. But because many humans produce antibodies to adenovirus vectors, they can mount immune reactions that render the virus and its therapeutic gene ineffective or cause significant side effects to the patient. A related virus called **adeno-associated virus (AAV)** is now widely used as a gene therapy vector. In its native form, AAV infects about 80 to 90 percent of humans during childhood, causing symptoms associated with the common cold. Disabled forms of AAV are popular for gene therapy because the virus is nonpathogenic, so it usually does not elicit a major response from the immune system of treated patients. AAV also does not typically integrate into the host-cell genome, so there is little risk of the insertional mutations that have plagued retroviruses, although modified forms of AAV have been used to deliver genes to specific sites on individual chromosomes. Most forms of AAV deliver genes into the host-cell nucleus where it forms small hoops of DNA called *episomes* that are expressed under the control of

在许多早期基因治疗试验中，**腺病毒载体**被选作逆转录病毒载体。这些载体的优点在于它们能够携带较大的治疗基因。但是，由于许多患者会产生腺病毒载体抗体，因此他们会产生免疫反应，使腺病毒及其所携带的治疗基因无效，或对患者造成严重的副作用。现在，一种被称为**腺相关病毒**（**AAV**）的病毒被广泛用作基因治疗载体。天然形式的 AAV 在儿童时期会感染 80%—90% 的人类，引发的症状与普通感冒相似。失去功能的 AAV 在基因治疗中很受欢迎，因为该病毒是非病原性的，因此它通常不会诱发被治疗患者的免疫系统产生巨大反应。AAV 通常也不会整合入宿主细胞基因组中，尽管经改造的 AAV 已被用于将基因传送至患者染色体上的特定位点，但是 AAV 几乎没有困扰逆转录病毒的插入突变风险。大多数形式的 AAV 会将基因送至宿主细胞核内，形成名为“附加体”的环形 DNA 小分子，该小分子会在病毒基因组启动子序列的控制下进行表达。但是由于 AAV 传递的治疗 DNA 通常不整合入基因组，因此在宿主细胞分裂时这些基因不会随之复制。这对治疗大脑或视网膜中不进行分裂的特定细胞来讲是可以的，但是治疗快速分裂细胞则通常需要重复不断的

promoter sequences contained within the viral genome. But because therapeutic DNA delivered by AAV does not usually become incorporated into the genome, it is not replicated when host cells divide. This is fine for certain cells in the brain or the retina that do not divide, but treating rapidly dividing cells typically requires repeated, ongoing applications to be successful.

应用才能获得成功。

Work with **lentivirus vectors** is an active area of gene therapy research. Lentivirus is a retrovirus that can accept relatively large pieces of genetic material. Another positive feature of lentivirus is that it is capable of infecting nondividing cells, whereas other viral vectors often infect cells only when they are dividing. It is still not possible to control where lentivirus integration occurs in the host-cell genome, but the virus does not appear to gravitate toward gene-regulatory regions the way that other retroviruses do. Thus the likelihood of causing insertional mutations appears to be much lower than for other vectors.

慢病毒载体相关工作是基因治疗研究中的一个活跃领域。慢病毒是一种逆转录病毒，可以接纳相对较大片段的遗传物质。慢病毒的另一个优势特征是能够感染非分裂细胞，而其他病毒载体通常仅在细胞分裂时才能实现感染。当前尚无法控制慢病毒在宿主细胞基因组中的整合位点，但是该病毒似乎并不像其他逆转录病毒那样易于整合入基因调控区。因此，慢病毒引发插入突变的可能性似乎比其他载体低得多。

The human immunodeficiency virus (HIV) responsible for acquired immunodeficiency syndrome (AIDS) is a type of lentivirus. It may surprise you that HIV could be used as a vector for gene therapy. For any viral vector, scientists must be sure that the vector has been genetically engineered to render it inactive. Modified forms of HIV, strains lacking the genes necessary for reconstitution of fully functional viral particles, are being used for gene therapy trials. HIV has evolved to infect certain types of T lymphocytes (T cells) and macrophages, making it a good vector for delivering therapeutic genes into the

导致获得性免疫缺陷综合征（AIDS，即艾滋病）的人类免疫缺陷病毒（HIV）就是一种慢病毒。利用 HIV 病毒作为基因治疗载体可能会让你感到惊讶。对于任何一种病毒载体，科学家必须确保该载体已经过基因工程改造而使其失活。HIV 经改造后，缺乏重构完整功能病毒颗粒所需基因，该病毒株正被用于基因治疗试验。HIV 已进化为可以感染特定类型 T 淋巴细胞（T 细胞）和巨噬细胞的病毒，这使其成为将治疗基因传送至血液的优良载体。

bloodstream.

Nonviral Delivery Methods

Scientists continue to experiment with various *in vivo* and *ex vivo* strategies for trying to deliver so-called naked DNA into cells without the use of viral vectors. Nonviral methods include chemically assisted transfer of genes across cell membranes, nanoparticle delivery of therapeutic genes, and fusion of cells with artificial lipid vesicles called *liposomes*. Short-term expression of genes through "gene pills" is being explored, whereby a pill delivers therapeutic DNA to the intestines where the DNA is absorbed by cells that express the therapeutic protein and secrete it into the bloodstream.

非病毒传送方式

科学家继续尝试各种体内和离体策略将裸露的 DNA 传送至细胞中而不使用病毒载体。非病毒方法包括化学辅助基因跨细胞膜转移、纳米颗粒传送治疗基因，以及细胞与人工脂质囊泡——脂质体融合。通过“基因药丸”实现基因短期表达正在探索中，药丸将治疗 DNA 传送至肠道并被细胞吸收，再由细胞表达治疗性蛋白并分泌至血液中。

Stem Cells for Delivering Therapeutic Genes

Increasingly, viral and nonviral vectors are being used to deliver therapeutic genes into **stem cells**, usually *in vitro*, and then the stem cells are either introduced into the patient or differentiated *in vitro* into mature cell types before being transplanted into a patient. In particular, *hematopoietic stem cells* (*HSCs*), which are found in bone marrow and give rise to blood cells, are widely used for gene therapy in adults and children. There are many advantages to using HSCs: they are easily accessible and often taken from the marrow of the patient to be treated to avoid complications with tissue rejection by the immune system when they are reintroduced; they replicate quickly *in vitro*;

干细胞传送治疗基因

病毒和非病毒载体越来越多地用于体外传送治疗基因至**干细胞**内，然后将干细胞直接导入患者体内，或者在体外培养分化为成熟的细胞类型后再移植入患者体内。尤其是骨髓中发现的产生血细胞的造血干细胞（HSCs），已被广泛应用于成人和儿童的基因治疗。使用 HSCs 有许多优点：它们易于取用，通常从接受治疗的患者骨髓中取出以避免再次植入体内时免疫系统产生组织排斥，引发并发症；它们在体外增殖快速；它们具有相当长的寿命；它们既可以分化为红细胞，也可以分化为白细胞。此外，干细胞正被用于 CRISPR-Cas 和基因沉默技术，从而展现出它们在不同基因治疗方法中的应用价值。

they are fairly long lived; and they differentiate into both red blood cells and white blood cells (leukocytes). And as you will also learn, stem cells are being used in CRISPR-Cas and gene-silencing approaches, thus demonstrating their value for different gene therapy applications.

In late 2017 one of the most extraordinarily successful applications of stem cell therapy was reported. This approach combined gene therapy with stem cells to regenerate the epidermis of a 7-year-old boy with a genetic condition called junctional epidermolysis bullosa, who had lost nearly two-thirds of his skin due to bacterial infections. A team of stem cell scientists from the Center for Regenerative Medicine at the University of Modena and Reggio Emilia used a small biopsy of skin from the boy and introduced copies of the *LAMB3* gene via retroviral vectors and then grew cells in culture into sheets of skin. Through a series of three operations spanning 4 months, a surgical team attached the skin sheets to the boy's body. After 21 months, skin from these sheets had grown normally and replaced prior wounds, thus reconstructing ~80 percent of the boy's skin with healthy tissue.

在 2017 年年末，一项干细胞疗法最为成功的应用被报道了。这种方法结合基因疗法与干细胞技术再生了一位患有遗传疾病交界型大疱性表皮松解症的 7 岁男孩的表皮，这位患者由于细菌感染已经失去了将近 2/3 的皮肤。来自意大利摩德纳雷焦艾米利亚大学再生医学中心的干细胞科学家团队利用来自男孩身上的一小块活体皮肤组织，通过逆转录病毒载体导入 *LAMB3* 基因拷贝，然后在培养基中培养细胞获得成片的皮肤。经过历时 4 个月的 3 次系列手术，外科手术团队将这些皮肤片层移植到了男孩身上。21 个月后，这些被移植的成片皮肤已经能够正常生长并替换了先前的皮肤创伤，从而利用健康组织重建了男孩身上约 80%的皮肤。

11.3 The First Successful Gene Therapy Trial

11.3 首个基因治疗成功案例

In 1990 the FDA approved the first human gene therapy trial, which began with the treatment of a young girl named Ashanti DeSilva, who has a heritable disorder called **adenosine deaminase severe combined immunodeficiency** (**ADA-SCID**), a condition

1990 年 FDA 批准了首个人类基因治疗试验，接受治疗的是一位名为阿莎提 · 德席尔瓦的年轻女孩，她所患遗传疾病名为**腺苷脱氨酶重症联合免疫缺陷**（**ADA-SCID**），这种疾病的发病率约为每一百万个新生儿中有 1—9 个患病。患有 SCID 的个体免疫

affecting approximately 1 to 9 out of every 1 million live births. Individuals with SCID have no functional immune system and usually die from what would normally be minor infections. Ashanti has an autosomal form of SCID caused by a mutation in the gene encoding the enzyme *adenosine deaminase*. Her gene therapy began when clinicians isolated some of her white blood cells, called T cells (Figure 11.3). These cells, which are key components of the immune system, were mixed with a retroviral vector carrying an inserted copy of the normal *ADA* gene. The virus infected many of the T cells, and a normal copy of the *ADA* gene was inserted into the genome of some T cells.

系统功能缺陷，通常死于一些常见的轻微感染。阿莎提所患疾病属于常染色体形式的 SCID，致病原因是编码腺苷脱氨酶的基因发生了突变。在基因治疗过程中，首先需要分离获得她的一些白细胞，即T细胞(图 11.3)，这些细胞是免疫系统的关键组成部分。然后将这些细胞与携带正常 *ADA* 基因拷贝插入的逆转录病毒载体混合。这些病毒会转染其中多数 T 细胞，并将正常 *ADA* 基因拷贝插入一些 T 细胞的基因组中。

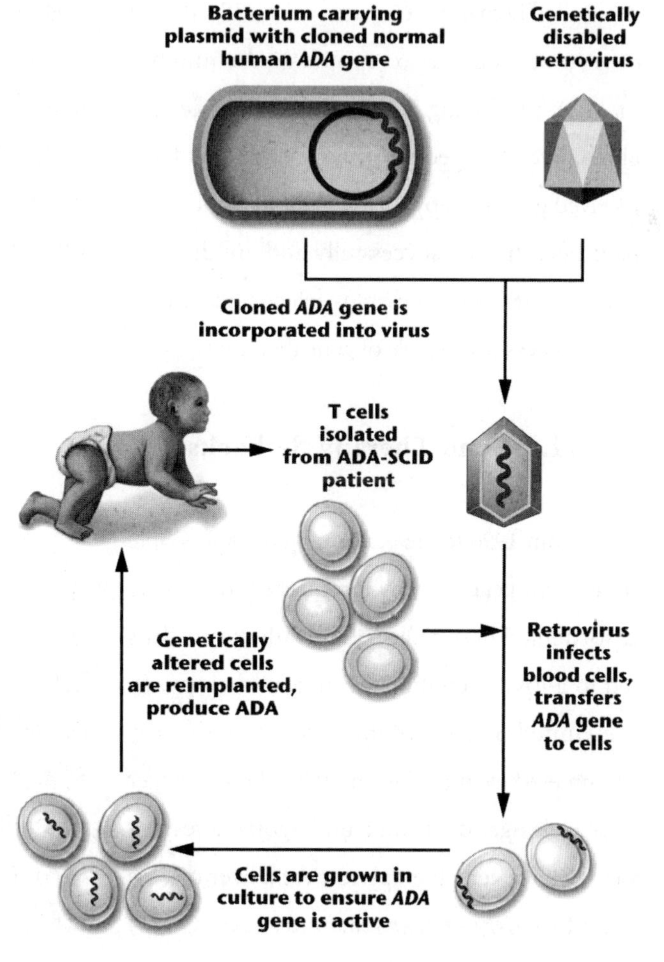

FIGURE 11.3 The first successful gene therapy trial. To treat ADA-SCID using gene therapy, a cloned human *ADA* gene was transferred into a viral vector, which was then used to infect white blood cells removed from the patient. The transferred *ADA* gene was incorporated into a chromosome, after which the cells were cultured to increase their numbers. Finally, the cells were inserted back into the patient, where they produce ADA, allowing the development of an immune response.

图 11.3 首个基因治疗成功案例。为了利用基因疗法治疗 ADA-SCID，经克隆的人类 *ADA* 基因被转移至病毒载体中，随后用该病毒感染从患者体内分离获得的白细胞。转移的 *ADA* 基因被整合入染色体中，然后对细胞进行培养以增加细胞数量。最后，这些细胞将被重新输入患者体内，它们将产生 ADA 并使免疫应答得以发展。

After being mixed with the vector, the T cells were grown in the laboratory and analyzed to make sure that the transferred *ADA* gene was expressed (Figure 11.3). Then a billion or so genetically altered T cells were injected into Ashanti's bloodstream. Repeated treatments were required to produce a sufficient number of functioning T cells. In addition, Ashanti also periodically received injections of purified ADA protein throughout this process, so the exact effects of gene therapy were difficult to discern. Ashanti continues to receive supplements of the ADA enzyme to allow her to lead a normal life.

与载体混合后，这些T细胞会在实验室中进行培养和分析，从而确保转移的*ADA*基因正常表达（图11.3）。然后将数十亿左右基因被改造的T细胞注入阿莎提的血液中，再经多次治疗产生足够数量的功能T细胞。此外，在此过程中，阿莎提还定期注射纯的ADA蛋白，因此基因治疗的确切效果其实难以评估。后续阿莎提还会继续接受ADA酶补充剂以保证她的正常生活。

Subsequent gene therapy treatments for SCID have focused on using bone marrow stem cells called hematopoietic stem cells (HSCs), and *in vitro* approaches to repopulate the number of ADA-producing T cells. In the past decade alone, over 100 people (mostly children) have received gene therapy for ADA-SCID, and most have been treated successfully and are disease-free. ADA-SCID treatment is still considered the most successful example of gene therapy.

随后SCID的基因治疗方法集中于使用骨髓干细胞，即造血干细胞（HSCs），用体外方法产生大量能够合成ADA的T细胞。仅在过去10年中，就有超过100人（主要是儿童）接受了ADA-SCID基因治疗，并且大多数个体治疗成功并获痊愈。ADA-SCID治疗迄今仍被认为是基因治疗最为成功的例子。

11.4 Gene Therapy Setbacks

11.4 基因治疗的逆境

From 1990 to 1999, more than 4000 people underwent gene therapy for a variety of genetic disorders. These trials often failed and thus led to a loss of confidence in gene therapy. In the United States, optimism for gene therapy plummeted even further in 1999 when teenager Jesse Gelsinger died while undergoing a test for the safety of gene therapy to treat a liver disease called *ornithine transcarbamylase* (OTC)

从1990年到1999年，已有超过4000人由于各种遗传疾病而接受了基因治疗。这些试验通常是失败的，从而导致人们对基因治疗丧失信心。在美国，人们对于基因治疗抱有的乐观态度在1999年进一步受到重创，当时一位名为杰西·基尔辛格的青少年在接受肝脏疾病鸟氨酸氨甲酰基转移酶（OTC）缺乏症的基因治疗安全性测试时去世。大量携带*OTC*基因的腺病毒载

deficiency. Large numbers of adenovirus vectors bearing the *OTC* gene were injected into his hepatic artery. The vectors were expected to target his liver, enter liver cells, and trigger the production of OTC protein. In turn, it was hoped that the OTC protein might correct his genetic defect and cure him of his liver disease.

体被注入患者的肝动脉中，这些载体预期靶向肝脏，进入肝脏细胞，并触发OTC蛋白表达。因此人们期望OTC蛋白可以纠正他的遗传缺陷并治愈他的肝病。

Researchers had previously treated 17 people with the therapeutic virus, and early results from these patients were promising. But the 18th patient, Jesse Gelsinger, within hours of his first treatment, developed a massive immune reaction. He developed a high fever, his lungs filled with fluid, multiple organs shut down, and he died four days later of acute respiratory failure. Jesse's severe response to the adenovirus may have resulted from how his body reacted to a previous exposure to the virus used as the vector for this protocol.

在此之前，研究人员已经使用这种治疗性病毒对17人进行了治疗，这些患者的早期结果令人鼓舞。但是第18位患者杰西·基尔辛格在首次接受治疗的几个小时内即产生了严重的免疫反应：高烧、肺部充满积液、多个器官功能衰竭，并在4天后由于急性呼吸衰竭而死亡。杰西对腺病毒的强烈反应可能是由于他先前被治疗过程中用作载体的病毒感染过。

In the aftermath of the tragedy, several government and scientific inquiries were conducted. Investigators learned that in the clinical trial scientists had not reported other adverse reactions to gene therapy, and that some of the scientists were affiliated with private companies that could benefit financially from the trials. It was also determined that serious side effects seen in animal studies were not explained to patients during informed-consent discussions. The FDA subsequently scrutinized gene therapy trials across the country, halted a number of them, and shut down several gene therapy programs. Other groups voluntarily suspended their gene therapy studies. Tighter restrictions on clinical trial protocols were imposed to correct

在悲剧发生后，进行了多次政府和科学调查。调查人员了解到，在临床试验中，科学家并未报告基因治疗的其他不良反应，并且一些科学家隶属于从试验中可获取经济利益的私人公司。调查人员还查明在知情同意的讨论过程中，动物研究中呈现的严重副作用并未向患者进行说明解释。随后FDA在全国范围内对基因治疗试验进行了详细审核，叫停了其中一些试验，并关闭了一些基因治疗项目。其他一些试验小组自愿暂停了他们的基因治疗研究。更为严格的限制被施加于临床试验方案中以纠正基尔辛格病例中出现的一些程序问题。杰西的去世给苦苦挣扎的基因治疗领域带来了重创——直至第二次悲剧袭来时，基因治疗仍未从此次打击中恢复过来。

some of the procedural problems that emerged from the Gelsinger case. Jesse's death had dealt a severe blow to the struggling field of gene therapy—a blow from which it was still reeling when a second tragedy hit.

The outlook for gene therapy brightened momentarily in 2000, when a group of French researchers reported what was hailed as the first large-scale success in gene therapy. Children with a fatal X-linked form of ADA-SCID (X-SCID, also known as "bubble boy" disease) developed functional immune systems after being treated with a retroviral vector carrying a normal gene. But elation over this study soon turned to despair, when it became clear that 5 of the 20 patients in two different trials developed leukemia as a direct result of their therapy. One of these patients died as a result of the treatment, while the other four went into remission from the leukemia. In two of the children examined, their cancer cells contained the retroviral vector, inserted near a gene called *LMO2*. This *insertional mutation* activated the *LMO2* gene, causing uncontrolled white blood cell proliferation and development of leukemia. The FDA immediately halted 27 similar gene therapy clinical trials, and once again gene therapy underwent a profound reassessment.

基因治疗的前景在 2000 年出现了短暂辉煌，法国研究人员报道了被誉为"基因治疗的首次大规模成功"案例。患有致命 X 连锁 ADA-SCID 的儿童（X-SCID，也被称为"泡泡男孩"疾病）在接受携带正常基因的逆转录病毒载体治疗后，功能性免疫系统得到重建。但是，乐极生悲的是，在两项不同的临床试验中，20 名患者中有 5 名接受治疗后直接引发白血病，其中 1 名患者在治疗过程中死亡，另外 4 名患者的白血病症状有所缓解。在接受检查的两名儿童中，其癌细胞含逆转录病毒载体并且插入位点位于 *LMO2* 基因附近。这种插入突变激活了 *LMO2* 基因，导致白细胞增殖不受控制，进而引发白血病。FDA 立即终止了 27 项类似的基因治疗临床试验，并再次对基因治疗展开了深入的评估。

On a positive note, long-term survival data from trials in the UK to treat X-SCID and ADA-SCID using HSCs from the patients' bone marrow for gene therapy have shown that 14 of 16 children have had their immune system restored at least nine years after the treatment. These children formerly had life expectancies

积极的一面是，在英国，使用患者自身骨髓 HSCs 对 X-SCID 和 ADA-SCID 进行基因治疗试验得到的一些患者长期生存数据显示，16 名儿童中有 14 名的免疫系统在治疗后得到恢复并持续了至少 9 年，而这些孩子原先的预期寿命不足 20 岁。仅过去 5 年，在意大利、英国和美国的 3 项进展良好的

of less than 20 years. In the past five years alone, more than 40 patients have been treated for ADA-SCID in three well-developed programs in Italy, the United Kingdom, and the United States. All individuals treated have survived, and 75 percent of those treated are disease-free.

项目中，已有超过 40 名患者接受了 ADA-SCID 基因治疗。所有接受治疗的个体均存活，并且其中 75% 的患者已痊愈。

Problems with Gene Therapy Vectors

基因治疗载体的相关问题

Most of the problems associated with gene therapy, including the Jesse Gelsinger case and the French X-SCID trial, have been traced to the viral vectors used to transfer therapeutic genes into cells. These vectors have been shown to have several serious drawbacks.

与基因治疗有关的大多数问题，包括杰西·基尔辛格病例和法国 X-SCID 临床试验，均可追溯至将治疗基因转入细胞中的病毒载体。这些载体已经显示出存在多种严重缺陷。

■ First, integration of retroviral genomes, including the human therapeutic gene into the host cell's genome, occurs only if the host cells are replicating their DNA. But only a small number of cells in any tissue are dividing and replicating their DNA.

■ 第一，逆转录病毒基因组，包括人类治疗基因，只有在宿主细胞复制自身 DNA 的过程中才会被整合入宿主细胞基因组中。但是任意组织中仅有少数细胞正处于分裂期并正在复制自身 DNA。

■ Second, the injection of massive quantities of most viral vectors, but particularly adenovirus vectors, is capable of causing an adverse immune response in the patient, as happened in Jesse Gelsinger's case.

■ 第二，大多数病毒载体尤其腺病毒载体大剂量注入人体内时会导致患者体内发生不良免疫反应，正如杰西·基尔辛格病例中发生的情形。

■ Third, insertion of viral genomes into host chromosomes can activate or mutate an essential gene, as in the case of the French patients. Viral integrase, the enzyme that allows for viral genome integration into the host genome, interacts with chromatin- associated proteins, often steering integration toward transcriptionally active genes.

■ 第三，病毒基因组插入宿主染色体时可以激活或突变必需基因，正如法国患者病例。催化病毒基因组整合入宿主基因组的病毒整合酶通过与染色质相关蛋白质相互作用，通常会引导整合发生在转录活跃的基因处。

■ Fourth, AAV vectors cannot carry DNA sequences larger than about 5 kb, and

■ 第四，AAV 载体无法携带 5 kb 以上的 DNA 序列，逆转录病毒不能携带 10 kb

retroviruses cannot carry DNA sequences much larger than 10 kb. Many human genes exceed the 5- to 10-kb size range.

以上的 DNA 序列。而许多人类基因大小则超出 5—10 kb 的范围。

■ Finally, there is a possibility that a fully infectious virus could be created if the inactivated vector were to recombine with another unaltered viral genome already present in the host cell.

■ 第五，如果灭活载体与宿主细胞中已存在的未被改造的病毒基因组发生重组，则可能产生具有完全传染性的病毒。

To overcome these problems, new viral vectors and strategies for transferring genes into cells are being developed in an attempt to improve the action and safety of vectors. No new technology has had a greater impact on gene therapy than gene targeting, especially by CRISPR-Cas. Fortunately, gene therapy has experienced a resurgence in part because of several promising new trials and successful treatments.

为了克服这些问题，研究人员正在开发新型病毒载体和将基因导入细胞的新策略，从而改良载体作用及其安全性。没有任何一项新技术对基因治疗的影响比基因靶向更大，尤其是 CRISPR-Cas。幸运的是，由于一些具有前瞻性的新尝试和成功的治疗案例，基因治疗在一定程度上再次兴起。

11.5 Recent Successful Trials by Conventional Gene Therapy Approaches

11.5 传统基因治疗技术的近期成功尝试

Treating Retinal Blindness

In recent years, patients being treated for blindness have greatly benefited from gene therapy approaches. Congenital retinal blinding conditions affect about 1 in 2000 people worldwide, many of which are the result of a wide range of genetic defects. Over 165 different genes have been implicated in various forms of retinal blindness. Currently there are over two dozen active gene therapy trials for at least 10 different retinal diseases.

治疗视网膜失明

近些年来，因失明接受治疗的患者在基因治疗技术中受益匪浅。先天性视网膜失明在全世界范围内的患病概率约为 1/2000，其中许多患者是由于广泛存在的遗传缺陷而导致的。已有超过 165 种不同基因显示与各种形式的视网膜失明有关。目前正在积极开展的 20 余项基因治疗试验即针对其中至少 10 种视网膜疾病。

Successful gene therapy has been achieved in subsets of patients with **Leber congenital**

针对**莱伯氏先天性黑朦（LCA）**患者亚群进行的基因治疗已获成功，该疾病属

amaurosis (**LCA**), a degenerative disease of the retina that affects 1 in 50,000 to 1 in 100,000 infants each year and causes severe blindness. Gene therapy treatments for LCA were originally pioneered in dogs. Based on the success of these treatments, the protocols were adapted and applied to human gene therapy trials.

于视网膜退行性疾病，每年新生儿患病率为 1/100 000—1/50 000，并且会导致严重失明。LCA 的基因治疗最初以狗为试验对象进行了开创性试验。基于这些治疗的成功，研究人员对治疗方案进行了修正，并应用于人类的基因治疗试验。

LCA is caused by alterations to photoreceptor cells (rods and cones), light-sensitive cells in the retina, due to 18 or more genes. One gene in particular, *RPE65*, has been the gene therapy target of choice. The protein product of the *RPE65* gene metabolizes retinol, which is a form of vitamin A that allows the rod and cone cells of the retina to detect light and transmit electrical signals to the brain. In one of the earliest trials, young adult patients with defects in the *RPE65* gene were given injections of the normal gene incorporated into an AAV vector. Several months after a single treatment, many adult patients, while still legally blind, could detect light, and some of them could read lines of an eye chart. This treatment approach for LCA was based on injecting AAV-carrying *RPE65* at the back of the eye directly under the retina. The therapeutic gene enters about 15 to 20 percent of cells in the retinal pigment epithelium, the layer of cells just beneath the visual cells of the retina.

LCA 是由视网膜中感光细胞（视杆细胞与视锥细胞）和光敏细胞的改变而导致的，涉及的基因有 18 种或更多。特别是 *RPE65* 基因，已成为基因治疗选择的靶标。*RPE65* 基因的蛋白质产物可以代谢视黄醇，视黄醇是维生素 A 的一种，可使视网膜的视杆细胞和视锥细胞感应光信号并将电信号传输至大脑。在一项早期试验中，整合正常 *RPE65* 基因的 AAV 载体被注入有该基因缺陷的年轻成年患者体内。单次治疗数月后，许多成年患者虽仍是法定意义上的盲人，但对光有了感应，其中一些个体甚至可以看到视力表中的线条。LCA 的这种治疗方法基于在眼球背面视网膜正下层注射携带 *RPE65* 基因的 AAV。治疗基因进入 15%—20% 的视网膜色素上皮细胞中，该层细胞恰好位于视网膜视细胞下方。

Adults treated by this approach have shown substantial improvements in a variety of visual functions tests, but the greatest improvement has been demonstrated in children, all of whom have gained sufficient vision to allow them to be ambulatory. Researchers think the success

接受这种方法治疗的成年人在各种视觉功能测试中均呈现出实质性改善，但是最大的改善还是体现在接受治疗的儿童中，所有儿童均已获得足够满足其行走所需的视力。研究人员认为儿童治疗获得成功是因为年轻患者失去的感光细胞尚不及年长

in children has occurred because younger patients have not lost as many photoreceptor cells as older patients. In January 2018, the FDA approved a *RPE65* gene therapy approach by Spark Therapeutics as the first treatment to target a genetic disease caused by mutations in a single gene.

患者那么多。2018年1月，FDA批准了星火治疗公司的 *RPE65* 基因治疗方法，这成为靶向单基因突变遗传病的首个治疗方法。

Because of the small size of the eye and the relatively small number of cells that need to be treated, the prospects for gene therapy to become routine treatment for eye disorders appears to be very good. Retinal cells are also very long lived; thus, AAV delivery approaches can be successful for long periods of time even if the gene does not integrate.

由于眼睛体积小，需要接受治疗的细胞数量相对较少，因此基因治疗成为眼部疾病常规治疗的前景非常好。而且视网膜细胞的寿命很长，因此即使基因没有整合，AAV传递法在较长时间内也有希望获得成功。

Successful Treatment of Hemophilia B

A very encouraging gene therapy trial in England successfully treated a small group of adults with hemophilia B, a blood disorder caused by a deficiency in the coagulation protein human factor IX. This, and other similar trials, are based largely on approaches derived from successful gene therapy to treat hemophilia B in dogs. Currently, most hemophilia B patients are treated several times each week with infusions of concentrated doses of the factor IX protein. In the gene therapy trial, six adult patients received, *in vivo*, a single dose of an adenovirus vector (AAV8) carrying normal copies of the human factor IX gene introduced into liver cells. Of six patients treated, four were able to stop factor IX infusion treatments after the gene therapy trial. Several other trials of this AAV treatment approach are under way, and expectations are

血友病B的成功治疗

英格兰的一项非常令人鼓舞的基因治疗试验成功治疗了一小组血友病B成年患者。血友病B是一种由于缺乏人类凝血因子IX而引起的血液疾病。该试验和其他类似试验主要都是基于对患血友病B的狗进行的成功基因治疗方法。目前，大多数血友病B患者每周都要接受多次浓缩IX因子蛋白的输液治疗。在基因治疗试验中，6名成年患者接受了单剂量携带可导入肝细胞的正常人源IX因子基因拷贝的腺病毒载体（AAV8）。接受治疗的6名患者中，4名在基因治疗试验后能够停止IX因子输液治疗。这种AAV治疗方法的其他几项试验也正在进行中，人们对于使用基因治疗治愈血友病B逐步变为现实的期望很高。

high that a gene therapy cure for hemophilia B is close to becoming a routine reality.

HIV as a Vector Shows Promise in Recent Trials

Researchers at the University of Paris and Harvard Medical School reported that two years after gene therapy treatment for **β-thalassemia**, a blood disorder involving the β-globin gene that reduces the production of hemoglobin, a young man no longer needed transfusions and appeared to be healthy. A modified, disabled HIV was used to carry a copy of the normal β-globin gene. Although this trial resulted in activation of the growth factor gene called *HMGA2*, reminiscent of what occurred in the French X-SCID trials, activation of the transcription factor did not result in an overproduction of hematopoietic cells or create a condition of preleukemia.

In 2013, researchers at the San Raffaele Telethon Institute for Gene Therapy in Milan, Italy, first reported two studies using lentivirus vectors derived from HIV in combination with HSCs to successfully treat children with either **metachromatic leukodystrophy** (**MLD**) or **Wiskott-Aldrich syndrome** (**WAS**). MLD is a neurodegenerative disorder affecting storage of enzymes in lysosomes and is caused by mutation in the arylsulfatase A (*ARSA*) gene that results in an accumulation of fats called sulfatides. These are toxic to neurons, causing progressive loss of the myelin sheath (demyelination) surrounding neurons in the brain, leading to a loss of cognitive functions and motor skills. There is no cure for

近期试验中 HIV 作为载体已呈现出应用前景

β - 地中海贫血是一种涉及 β - 珠蛋白基因的血液疾病，可导致血红蛋白的产量降低。巴黎大学和哈佛医学院的研究人员报道称，对一位年轻患者开展该疾病基因治疗两年后，他不再需要输血，而且表现健康。一种经改造后不致病的 HIV 病毒被用于携带正常 β - 珠蛋白基因拷贝。尽管该试验导致生长因子基因 *HMGA2* 被激活，这让人联想起法国 X-SCID 试验，但转录因子激活并未导致造血细胞的过度增殖或引发白血病前期。

2013 年，意大利米兰圣拉斐尔 Telethon 基因治疗研究所的研究人员首次报道了两项利用源自 HIV 的慢病毒载体与 HSCs 联合成功治疗**异染性脑白质营养不良**（**MLD**）或**威斯科特 - 奥尔德里奇综合征**（**WAS**）儿童患者的研究。MLD 是一种神经系统变性疾病，影响溶酶体中酶的存储，由芳基硫酸酯酶 A（*ARSA*）基因突变引起，该突变会导致硫苷脂脂肪积累。硫苷脂对神经元具有毒性，会导致大脑神经元周围髓鞘逐渐丧失（脱髓鞘作用），引发认知功能和运动技能丧失。MLD 无法治愈，MLD 患儿在出生时表现健康，但最终会发展出 MLD 症状。

MLD. Children with MLD appear healthy at birth but eventually develop MLD symptoms.

Researchers used an *ex vivo* approach with a lentivirus vector to introduce a functional *ARSA* gene into bone marrow-derived HSCs from each patient and then infused treated HSCs back into patients. Four years after the start of a trial involving 10 patients with MLD, data from 6 patients analyzed 18 to 24 months after gene therapy indicated that the trials are safe and effective. Treatment halted disease progression as determined by magnetic resonance images of the brain and through tests of cognitive and motor skills. Patients with MLD in the first group treated have already lived past the expected lifetime normally associated with this disease. Additional patients are now being treated. This approach took over 15 years of research and a team of over 70 people, including researchers and clinicians, which is indicative of the teamwork approach typical of gene therapy trials.

研究人员使用慢病毒载体采用离体方法将 *ARSA* 功能基因引入每位患者的骨髓 HSCs 中，然后将治疗后的 HSCs 输回患者体内。一项含 10 名 MLD 患者的试验自开始 4 年后，来自 6 名患者基因治疗后 18—24 个月的数据分析表明，该试验安全有效。大脑磁共振成像以及认知和运动能力测试表明治疗阻止了疾病进展。第一组接受治疗的 MLD 患者寿命已经超过该疾病患者通常预期的寿命。目前其他患者也正在陆续接受治疗中。该研究历时 15 年，研究团队超过 70 人，包括研究人员和临床医生，这也正是基因治疗试验中典型的团队合作特点。

The trial was technically complicated because it required that HSCs travel through the bloodstream and release the ARSA protein that is taken up into neurons. A major challenge was to create enough engineered cells to produce a sufficient quantity of therapeutic ARSA protein to counteract the neurodegenerative process.

该试验技术复杂，因为它要求 HSCs 随血液进行移动并释放神经元可吸收的 ARSA 蛋白。主要挑战在于产生充足的工程细胞，以产生足够数量的治疗性 ARSA 蛋白来抵抗神经系统变性过程。

Similar results were reported for treating patients with WAS, an X-linked condition resulting in defective platelets that make patients more vulnerable to infections, frequent bleeding, autoimmune diseases, and cancer. Genome sequencing of MLD and WAS patients treated in these trials showed no evidence of genome

据报道 WAS 患者治疗也获得了类似结果。WAS 是一种 X 连锁疾病，会导致血小板缺陷，从而使患者更易被感染、频繁出血、患自身免疫病和癌症。在这些试验中接受治疗的 MLD 患者和 WAS 患者的基因组测序结果未显示原癌基因附近存在基因组整合。类似地，患者无 HSC 过量产生的现象，

integration near oncogenes. Similarly, patients showed no evidence of HSC overproduction, suggesting that this lentivirus delivery protocol produced a safe and stable delivery of the therapeutic genes.

表明这种慢病毒传送是一种安全稳定的治疗基因传送方案。

11.6 Genome-Editing Approaches to Gene Therapy

11.6 基因治疗中的基因编辑法

The gene therapy approaches and examples we have highlighted thus far have focused on the addition of a therapeutic gene that functions along with the defective gene. However, rapid progress is being made with **genome editing**—the removal, correction, and/or replacement of a mutated gene. Genome editing by CRISPR-Cas, in particular, has shown great potential and provided renewed optimism for scientists and physicians involved in gene therapy as well as patients.

自此，我们重点介绍的基因治疗方法和实例都集中在引入治疗基因，使其与缺陷基因共同行使功能。但是，实现突变基因移除、校正和 / 或替换的**基因组编辑**正迅速发展。尤其，CRISPR-Cas 基因组编辑已显示出巨大潜力，使参与基因治疗的科学家、医生以及患者重拾信心。

DNA-Editing Nucleases

For nearly 20 years, scientists have been working on modifications of restriction enzymes and other nucleases to engineer proteins capable of editing the genome with precision, including the ability to edit one or a few bases or to replace specific genes. The concept is to combine a nuclease with a sequence-specific DNA binding domain that can be precisely targeted for digestion. In 1996 researchers fused DNA-binding proteins with a zinc-finger motif and DNA cutting domain from the restriction enzyme *Fok*I to create enzymes called **zinc-finger nucleases** (**ZFNs**). The zinc-finger

DNA 编辑核酸酶

近 20 年来，科学家一直致力于改良限制性内切酶和其他核酸酶，以改造蛋白质，使其能够精确编辑基因组，包括编辑一个或多个碱基，或替换特定基因的能力。这项工作基于的理念是将核酸酶与其精确靶向消化的序列特异性 DNA 结合域相结合。1996 年，研究人员将含锌指模体结构的 DNA 结合蛋白与限制性内切酶 *Fok*I 的 DNA 切割域相融合，得到了**锌指核酸酶**（**ZFNs**）。锌指模体存在于许多转录因子中，含有与锌原子相结合的由两个半胱氨酸、两个组氨酸构成的簇，可以与特定 DNA 序列相作用。通过将锌指模体结构与多肽 DNA 切割

motif is found in many transcription factors and consists of a cluster of two cysteine and two histidine residues that bind zinc atoms and interact with specific DNA sequences. By coupling zinc-finger motifs to DNA cutting portions of a polypeptide, ZFNs provide a mechanism for modifying sequences in the genome in a sequence-specific targeted way.

域相偶联，ZFNs 提供了一种通过靶向特异性序列来改造基因组序列的机制。

The DNA-binding domain of the ZFN can be engineered to attach to any sequence in the genome. The zinc fingers bind with a spacing of 5 to 7 nucleotides, and the nuclease domain of the ZFN cleaves between the binding sites.

ZFN 的 DNA 结合结构域经工程改造后可与基因组中任意序列相结合。锌指与 5—7 个核苷酸距离的序列相结合，且 ZFN 的核酸酶结构域在结合位点间进行剪切。

Another category of DNA-editing nucleases called **transcription activator-like effector nucleases (TALENs)** was created by adding a DNA-binding motif identified in transcription factors from plant pathogenic bacteria known as transcription activator-like effectors (TALEs) to nucleases to create TALENs. TALENs also cleave as dimers. The DNA-binding domain is a tandem array of amino acid repeats, with each TALEN repeat binding to a specific single base pair. The nuclease domain then cuts the sequence between the dimers, a stretch that spans about 13 bp.

另一类 DNA 编辑核酸酶是**类转录激活因子效应物核酸酶**（TALENs），该酶由植物病原菌转录因子即类转录激活因子效应物（TALEs）中已被鉴定的 DNA 结合模体与核酸酶融合而成。TALENs 也可以二聚体形式实施剪切功能。DNA 结合结构域是由氨基酸重复序列组成的串联排列，每一个 TALEN 重复与特定的单个碱基对相结合，然后核酸酶结构域切割位于二聚体中间大约 13 bp 核苷酸序列的伸展区域。

ZFNs and TALENs have shown promise in animal models and cultured cells for gene replacement approaches that involve removing a defective gene from the genome. ZFNs, TALENs, and CRISPR-Cas, as we will discuss shortly, all create double-stranded breaks in the DNA and then are mended by either nonhomologous end joining or homologous recombination. These enzymes can create site-

ZFNs 和 TALENs 在动物模型和培养细胞中均显示出从基因组中移除缺陷基因，实施基因替换的应用前景。ZFNs、TALENs 和随后我们将讨论的 CRISPR-Cas 都可在 DNA 中产生双链断裂，然后通过非同源末端连接或同源重组进行修补。这些酶可以在基因组中产生位点特异性双链切割。当与特定的整合酶协作时，ZFNs 和 TALENs 可通过切除缺陷基因序列，使用重组将同

specific double-stranded cleavage in the genome. When coupled with certain integrases, ZFNs and TALENs may lead to genome editing by cutting out defective sequences and using recombination to introduce homologous sequences into the genome that replace defective sequences. Although this technology has not yet advanced sufficiently for reliable use in humans, there have been several promising trials.

源基因序列引入基因组中替代缺陷基因序列以实现基因组编辑。尽管该项技术尚未发展到足以在人类中可靠使用的程度，但已经开展了一些很有应用前景的试验。

For example, ZFNs are actively being used in clinical trials for treating patients with HIV. Scientists are exploring ways to deliver immune system-stimulating genes that could make individuals resistant to HIV infection or cripple the virus in HIV-positive persons. In 2007, Timothy Brown, a 40-year-old HIV-positive American, had a relapse of acute myeloid leukemia and received a stem cell transplant. Because he was HIV-positive, Brown's physician selected a donor with a mutation in both copies of the *CCR5* gene, which encodes an HIV co-receptor carried on the surface of T cells to which HIV must bind to enter T cells (specifically CD4 + cells). People with naturally occurring mutations in both copies of the *CCR5* gene are resistant to most forms of HIV. Brown relapsed again and received another stem cell transplant from the *CCR5*-mutant donor. Eventually, the cancer was contained, and by 2010, levels of HIV in his body were still undetectable even though he was no longer receiving immune-suppressive treatment. Brown is generally considered to be the first person to have been cured of an HIV infection.

例如，ZFNs 正被积极应用于治疗 HIV 患者的临床试验中。科学家正在探索传送免疫系统激活基因的途径，从而使个体对 HIV 感染具有抵抗能力，或者削弱 HIV 阳性患者的病毒毒性。2007 年，HIV 检测呈阳性、现年 40 岁的美国人蒂莫西 · 布朗的急性髓系白血病复发，并接受了干细胞移植。因为布朗的 HIV 检测呈阳性，所以他的医生选择了 *CCR5* 基因的两个拷贝均含突变的供体。该基因编码位于 T 细胞表面的 HIV 辅助受体，即 HIV 必须与之结合才能进入 T 细胞（尤其是 CD4 + 细胞）。*CCR5* 基因两个拷贝均含突变的个体对大多数形式的 HIV 病毒都具有天然抵抗力。布朗病情再次复发，并再次接受了 *CCR5* 突变供体的干细胞移植。最终，他的癌症得以被控制，并且直至 2010 年，尽管不再接受免疫抑制治疗，但布朗体内的 HIV 病毒仍然处于无法被检测到的水平。布朗通常被认为是首位被治愈的 HIV 感染病人。

This example encouraged researchers to

这一病例激励了研究人员推动基因治

press forward with a gene therapy approach to modify the *CCR5* gene of HIV patients. In the first genome-editing trial to treat people with HIV, T cells were removed from HIV-positive men, and ZFNs were used to disrupt the *CCR5* gene *ex vivo*. The modified cells were then reintroduced into patients. In five of six patients treated, immune-cell counts rose substantially and viral loads also decreased following the therapy. To date, more than 90 people have been treated by this approach. What percentage of immune cells would have to be treated this way to significantly inhibit spread of the virus is still not known.

疗以修饰 HIV 患者的 *CCR5* 基因。在首个治疗 HIV 感染者的基因组编辑试验中，人们从 HIV 检测呈阳性的男性个体体内取出 T 细胞，并使用 ZFNs 在体外破坏 *CCR5* 基因，然后将被修饰细胞重新引入患者体内。在接受治疗的 6 名患者中，5 名患者的免疫细胞计数大幅上升，并且治疗后的病毒载量也下降了。迄今为止，已有 90 余人接受了这种治疗。目前尚不清楚必须采用这种方法处理多少百分比的免疫细胞才能显著抑制病毒传播。

Recently, researchers working with human cells used TALENs to remove defective copies of the *COL7A1* gene, which causes recessive dystrophic epidermolysis bullosa (RDEB), an incurable and often fatal disease that presents as excessive blistering of the skin, pain, and severely debilitating skin damage. Researchers at the University of Minnesota used a TALEN to cut DNA near a mutation in the *COL7A1* gene in fibroblast cells taken from a patient with RDEB and supplied these cells with a functional copy of the *COL7A1* gene. These cells were then converted into a type of stem cell called *induced pluripotent stem cells* (*iPSCs*), which were then differentiated into skin cells that expressed the correct protein. This is a promising result, and researchers now plan to transplant these skin cells into patients in an attempt to cure them of RDEB. Another group has recently taken a similar approach using TALENs to repair cultured cells in order to correct the mutation

最近，研究人员尝试利用 TALENs 清除人体细胞中 *COL7A1* 基因的缺陷型拷贝，该缺陷基因会导致隐性营养不良型大疱性表皮松解症（RDEB），这是一种无法治愈的致命疾病，表现为皮肤过度起泡、疼痛和被严重破坏而造成损伤。明尼苏达大学的研究人员利用 TALEN 将 RDEB 患者成纤维细胞中位于 *COL7A1* 基因突变附近的 DNA 双链切开，为这些细胞提供有功能的 *COL7A1* 基因拷贝，并将这些细胞转化为一种类型的干细胞，即诱导多能干细胞（iPSCs），这些干细胞随后会分化为表达正确蛋白质的皮肤细胞。该结果令人鼓舞，研究人员正在计划将这些皮肤细胞移植到患者体内，以试图治愈 RDEB。另一个研究小组近期也采取了类似方法，使用 TALENs 修复培养细胞，以纠正迪谢内肌营养不良（DMD）的相关基因突变。研究人员对于将该方法很快应用于疾病治疗表示乐观。

in Duchenne muscular dystrophy (DMD). Researchers are optimistic that this approach can soon be adapted to treat patients.

CRISPR-Cas Method Revolutionizes Genome-Editing Applications and Renews Optimism in Gene Therapy

No method has created more excitement than the genome-editing technique known as **CRISPR-Cas** (clustered regularly interspaced short palindromic repeats-CRISPR-associated proteins). Identified in bacterial cells, the CRISPR-Cas system functions to provide bacteria and archaea immunity against invading bacteriophages and foreign plasmids. However, CRISPR-Cas has been adapted as a tool for genome editing in eukaryotic cells, which has revolutionized gene therapy strategies.

CRISPR-Cas is based on delivering a single-stranded guide RNA sequence (sgRNA) and a Cas endonuclease. The sgRNA is complementary to the target gene sequence in the genome and directs the endonucleas to cut that sequence. One commonly used nuclease is called Cas9 (Figure 11.4). At the same time as Cas9 and the sgRNA are delivered, a DNA donor template coding for a replacement sequence is also delivered. The sgRNA-Cas9 complex binds to the target DNA sequence through complementary base pairing, and Cas9 generates a double-stranded break in the DNA. In order for Cas9 to cut a target sequence, there must be a specific three-nucleotide sequence adjacent to the complementary sequence called a protospacer adjacent motif (PAM). As cells

CRISPR-Cas 法革新了基因组编辑的应用并重新给基因治疗带来了曙光

没有一种方法比基因组编辑技术 **CRISPR-Cas**（规律成簇间隔短回文重复 -CRISPR 相关蛋白）更令人激动。在细菌细胞中获得鉴定的 CRISPR-Cas 系统可为细菌和古细菌提供免疫力，以抵御外来入侵的噬菌体和外源质粒。然而，CRISPR-Cas 已被优化为可用于真核细胞基因组编辑的工具，这彻底革新了基因治疗策略。

CRISPR-Cas 基于传送单链指导 RNA 序列（sgRNA）和 Cas 核酸内切酶。sgRNA 与基因组中靶基因序列互补，并指导核酸内切酶对该序列进行切割。一种常用的核酸内切酶是 Cas9（图 11.4）。当 Cas9 和 sgRNA 被传送的同时，编码替换序列的 DNA 供体模板也会被传送。sgRNA-Cas9 复合物通过互补碱基配对与靶 DNA 序列相结合，并且 Cas9 会通过切割 DNA 产生双链断裂。为了让 Cas9 剪切靶序列，互补序列相邻处必须存在特定的三核苷酸序列，即前间隔序列邻近基序（PAM）。当细胞修复由 Cas9 引起的 DNA 损伤时，修复酶将供体模板 DNA 整合入基因组以替换靶 DNA 序列。

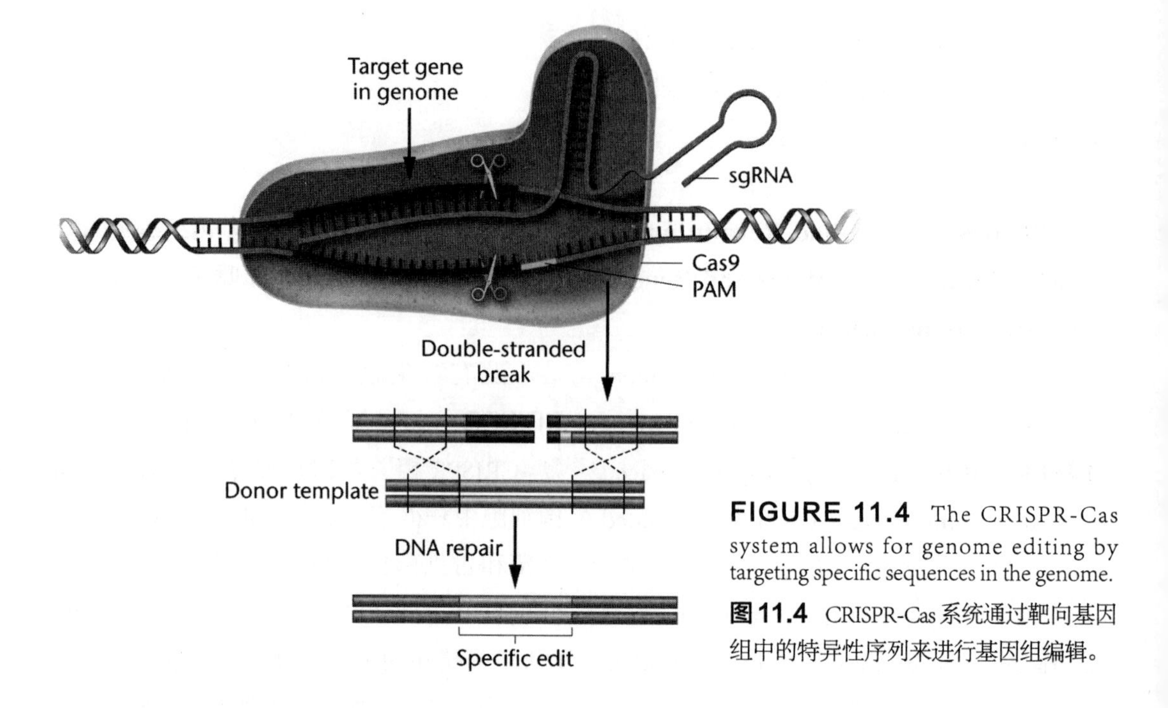

FIGURE 11.4 The CRISPR-Cas system allows for genome editing by targeting specific sequences in the genome.

图 11.4 CRISPR-Cas 系统通过靶向基因组中的特异性序列来进行基因组编辑。

repair the DNA damage caused by Cas9, repair enzymes incorporate donor template DNA into the genome, thus replacing the target DNA sequence.

While ZFNs and TALENs are sufficient for genome editing, CRISPR-Cas is more efficient and easier to design. Although ZFN and TALEN target specificity is provided by protein-DNA interactions, which can be difficult to engineer, CRISPR-Cas specificity is simply determined by complementary base pairing of the sgRNA and the target sequence. Within months of the technique being widely available, researchers around the world used CRISPR-Cas to target specific genes in human cells, mice, rats, bacteria, fruit flies, yeast, zebrafish, and dozens of other organisms. In one of the first reported applications of CRISPR-Cas for gene therapy, a team from the Massachusetts Institute of Technology (MIT) cured mice of a rare liver

虽然 ZFNs 和 TALENs 足以实施基因组编辑，但 CRISPR-Cas 则更加有效且更易于设计。尽管 ZFN 和 TALEN 的靶标特异性由蛋白质 -DNA 相互作用所决定，但是对于工程设计改造来讲则具有难度，而 CRISPR-Cas 的特异性仅由 sgRNA 与靶标序列的互补碱基配对确定。在该技术被广泛使用的几个月内，世界各地的研究人员使用 CRISPR-Cas 靶向位于人类细胞、小鼠、大鼠、细菌、果蝇、酵母、斑马鱼和许多其他生物中的特定基因。CRISPR-Cas 应用于基因治疗的首批报道之一是麻省理工学院（MIT）的一个研究团队通过基因组编辑治愈了小鼠的一种罕见肝脏疾病——I 型酪氨酸血症。酪氨酸血症在人类中的发病率约为 1/100 000，编码延胡索酰乙酰乙酸水

disorder, type I tyrosinemia, through genome editing. In tyrosinemia, a condition affecting about 1 in 100,000 people, mutation of the *FUH* gene encoding the enzyme fumarylacetoacetase prevents breakdown of the amino acid tyrosine. After an *in vivo* approach with a one-time treatment, roughly 1 in 250 liver cells accepted the CRISPR-Cas – delivered replacement of the mutant gene with a normal copy of the gene. But about 1 month later these cells proliferated and replaced diseased cells, taking over about one-third of the liver, which was sufficient to allow mice to metabolize tyrosine and show no effects of disease. Mice were subsequently taken off a low-protein diet and a drug normally used to disrupt tyrosine production.

解酶的 *FUH* 基因发生突变阻止了酪氨酸分解。经过体内方法进行一次治疗后，大约 1/250 的肝细胞中的突变基因被 CRISPR-Cas 传送的正常基因拷贝所替代。但大约 1 个月后，这些细胞通过增殖会陆续替换患病细胞，最后占据肝脏的 1/3 左右，从而足以使小鼠可以代谢酪氨酸，不再表现出疾病症状。随后小鼠不再采用低蛋白饮食，也不再使用通常用于阻断酪氨酸合成的药物。

In Chapter 15—Genomics and Precision Medicine—we will discuss a promising way to treat cancer called **immunetherapy**. This approach involves the genetic engineering of **T cells** (cells of the immune system), which can recognize, bind to, and destroy tumor cells. However, cancer cells have ways of evading T cells or of preventing their activation. The principle behind engineered T-cell therapies is to create T cells that are better equipped to find and target tumor cells for destruction. Two strategies for immunotherapy are to create recombinant **T-cell receptors** (**TCRs**) that specifically recognize antigens on or within cancer cells or **chimeric antigen receptor** (**CAR**)-**T cells** engineered to express receptors that can directly recognize antigens on the surface of the tumor cell without requiring T-cell activation by antigen-presenting cells. Immunotherapy

在第 15 章——基因组学和精准医疗中，我们将讨论一种具有应用前景的癌症治疗方法——**免疫疗法**。这种方法涉及**T细胞**(免疫系统细胞) 的基因工程，T 细胞可以识别、结合并破坏癌细胞。但是，癌细胞具有逃避 T 细胞或阻止其活性的方法。工程 T 细胞疗法的原理是创造更有能力找到并靶向癌细胞实施破坏的 T 细胞。免疫疗法有两种策略: 一是创造重组**T细胞受体** (**TCRs**)，特异性识别癌细胞表面或内部抗原；二是设计**嵌合抗原受体** (**CAR**) -**T 细胞**，经改造后的表达抗体可以直接识别癌细胞表面抗原，而不需要抗原呈递细胞来激活 T 细胞。免疫疗法在治疗某些类型白血病的过程中已显示出巨大前景，其应用是 CRISPR-Cas 通过简化基因组编辑可能会有效提高疗效的研究领域之一。

has shown great promise for the treatment of certain forms of leukemia, and its applications are one area where the ease of genome editing by CRISPR-Cas may improve its efficacy.

In 2017, the FDA approved the first ever immunotherapy— Novartis' CTL019 CAR-T treatment for children and young adults with B-cell acute lymphoblastic leukemia (ALL). ALL is the most common childhood cancer in the United States, and patients who relapse or fail to respond to chemotherapy have a low survival rate. But over 80% of 63 patients treated with CTL109 went into remission almost immediately after treatment began, and most remained cancer-free 6 months after treatment. CTL019 (brand name, Kymriah ™) is approved for ALL patients up to 25 years of age, and as of May 2018, is also approved for another cancer called diffuse large B-cell lymphoma. Kymriah ™ also made headlines for its sticker price of approximately $475,000 for a single treatment and estimates that the total cost of care with this drug could exceed $1.5 million.

2017 年，FDA 批准了首个免疫疗法——诺华 CTL019 CAR-T 治疗患有 B 细胞急性淋巴细胞白血病（ALL）的儿童和青少年。ALL 是美国最为常见的儿童期癌症，复发患者或化疗无效的患者生存率较低。但在接受 CTL109 治疗的 63 位患者中，80%以上的患者几乎在治疗初期病情就立即得到了缓解，并且大多数患者在治疗后 6 个月内癌症没有复发。CTL019（商品名为 Kymriah ™）已获批用于 25 岁以下的患者，并且在 2018 年 5 月，该药物也被批准用于另一种癌症——弥漫大 B 细胞淋巴瘤。Kymriah ™的单次治疗价格约为 475 000 美元，并且估计使用这种药物进行治疗的总费用可能超过 150 万美元，而这一标价也使得 Kymriah ™成为头条新闻。

More recently, a team from the University of Pennsylvania; the University of California, San Francisco; and the MD Anderson Cancer Center at the University of Texas plans to treat cancer patient T cells with a different immunotherapy strategy involving CRISPR-Cas. The T cells will have their *TCR* gene inactivated, and a recombinant *TCR* gene will be introduced to help these cells target destruction of tumor cells and to avoid nontumor cells. The trial will also use CRISPR-Cas to edit a gene called *program cell death 1* (*PD-1*), which expresses a protein on

最近，来自宾夕法尼亚大学、加利福尼亚大学旧金山分校、得克萨斯大学 MD 安德森癌症中心的一个研究团队计划采用涉及 CRISPR-Cas 的一种不同的免疫疗法来治疗癌症患者的 T 细胞。T 细胞的 *TCR* 基因将被失活，并且引入重组 *TCR* 基因以帮助这些细胞靶向癌细胞进行破坏并避免非癌细胞。该试验还使用 CRISPR-Cas 编辑程序性细胞死亡 1（*PD-1*）基因，该基因表达的蛋白质位于 T 细胞表面并常被癌细胞中和，从而降低免疫应答、抵御 T 细胞对癌细胞的破坏。改造后含 *PD-1* 突变基因的 T

the surface of T cells that can often be neutralized by cancer cells to minimize immune responses and ward off T-cell destruction of a tumor. The edited T cells with the mutant *PD-1* gene are expected to be able to recognize and attack lung tumor cells.

细胞预计能够识别和攻击肺癌细胞。

In 2016, a team from China reported the first CRISPR-Cas trial of human patients, designed to treat an aggressive form of cancer called metastatic non - small-cell lung cancer. An *ex vivo* approach was used to edit *PD-1* in T cells, which were then reintroduced into patients with the hope that these cells would target tumors in the lung for destruction without being disabled by the cancer cells. The main purpose of this initial trial is to determine whether this approach is safe. The success of this trial, which has enrolled at least 86 patients, had not been reported at the time this edition of *Essentials of Genetics* was published. Several more genome-editing trials are scheduled to begin at different centers around the world targeting kidney, bladder, and prostate cancer, among others.

2016 年，中国的一个研究团队报道了首例人类患者的 CRISPR-Cas 试验，该试验旨在治疗侵袭性癌症——转移性非小细胞肺癌。人们采用离体方法对 T 细胞的 *PD-1* 基因进行编辑，然后将其重新引入患者体内，希望这些细胞靶向、摧毁肺癌，同时不被癌细胞破坏。该初步试验的主要目的是确定方法是否安全。至少 86 名患者参与了该项试验，截至本版《基础遗传学》出版时，该试验是否获得成功尚未见报道。更多靶向肾癌、膀胱癌和前列腺癌等的基因编辑试验正计划在全球不同的研究中心启动。

Additional headline-grabbing examples of successful CRISPR-Cas applications in mice and humans that highlight the potential of this approach for gene therapy include:

其他能够突显 CRISPR-Cas 的基因治疗潜力，并在小鼠和人类中获得成功应用，且登上新闻头条的例子还包括：

■ AAV delivery of CRISPR-Cas9 to remove a defective exon from the *Dmd* gene in a mouse model (*mdx* mice) of DMD significantly restored muscle function in treated mice. It has been estimated that this approach has the potential to cure approximately 80 percent of human cases of DMD.

■ 在小鼠 DMD 模型（*mdx* 小鼠）中，通过 AAV 传送 CRISPR-Cas9 以移除 *Dmd* 基因中含缺陷的外显子，接受治疗的小鼠肌肉功能获得了显著恢复。据估计，该方法有可能治愈约 80%的人类 DMD 病患。

■ After successful trials in mice, *in vitro*

■ 在小鼠中获得试验成功后，体外修

repair of the human β-globin (*HBB*) gene in HSCs was used to treat sickle-cell disease and β-thalassemia in humans. It is expected that these preclinical studies will soon lead to the delivery of edited HSCs in humans.

复 HSCs 中的 β-珠蛋白（*HBB*）基因被用于治疗人类镰状细胞贫血和 β-地中海贫血。预计这些临床前研究将很快促进人类编辑 HSCs 的体内传送。

■ CRISPR-Cas9 targeting and replacement of the defective clotting factor IX gene in liver cells was used to cure mice of hemophilia B.

■ CRISPR-Cas9 靶向并置换肝细胞中缺陷型凝血因子 IX 基因已被用于治疗小鼠血友病 B。

■ HIV-infected patients are being recruited for a clinical study in China where CRISPR-Cas9 editing will be used to disable the *CCR5* gene (which we discussed as a ZFN treatment) to block HIV infection in modified cells.

■ 中国一项临床研究正在招募 HIV 感染患者参与，CRISPR-Cas9 编辑将用于破坏 *CCR5* 基因（该基因我们在 ZFN 治疗方法中讨论过），从而在被改造的细胞中阻止 HIV 感染。

■ A CRISPR-Cas approach to edit the *CEP290* gene to treat LCA type 10 (recall that this a form of blindness) is being developed in a joint effort by the pharmaceutical company Allergan and the genome-editing company Editas Medicine.

■ Allergan 制药公司和 Editas Medicine 基因组编辑公司正联合开发一种 CRISPR-Cas 方法编辑 *CEP290* 基因以治疗 LCA10，这是一种失明疾病。

■ CRISPR-Cas editing of mutations involved in genetic forms of hearing loss in mice have shown early potential.

■ CRISPR-Cas 编辑小鼠遗传性听力损失相关突变已显示出早期应用潜力。

Finally, the use of CRISPR-Cas to edit the human germ line (sperm and egg cells) and human embryos has been one of the most controversial potential applications of this technology, although similar concerns existed when ZFNs and TALENs were first being used. In 2015, a team of Chinese scientists reported using CRISPR-Cas9 to edit the *HBB* gene in 86 human embryos. The embryos were generated by *in vitro* fertilization but were donated for research because they were triploid and thus would only survive a few days. Two days after

最后，CRISPR-Cas 用于编辑人类生殖细胞系（精子和卵细胞）和人类胚胎一直是该技术最具争议的潜在应用之一，尽管 ZFNs 和 TALENs 首次应用时也存在类似担忧。2015 年，一个来自中国的科学家研究小组报道了使用 CRISPR-Cas9 编辑了 86 个人类胚胎的 *HBB* 基因。这些胚胎来自体外受精，但由于是三倍体，仅能存活数天，因此被捐赠用于研究。在 CRISPR-Cas 处理后两天，有 71 个胚胎存活，但其中只有 4 个携带了预期的 *HBB* 基因改变。出乎意料的是，由于该基因编辑处理，许多其他胚

CRISPR-Cas treatment, 71 embryos had survived, but only 4 of them carried the intended change to the *HBB* gene. Unexpectedly, many other embryos had acquired mutations in genes other than *HBB* as a result of the treatment. From this the Chinese research team concluded that genome-editing technology is not sufficiently developed for use in embryos.

胎获得了除 *HBB* 以外的其他基因突变。据此，中国研究小组认为基因组编辑技术的发展尚未成熟至可应用于胚胎的程度。

In 2016, the second published report of genome editing of human (triploid) embryos, also from a team in China, described the use of CRISPR-Cas9 to introduce a mutation in the *CCR5* gene to confer resistance to HIV infection. In this study, 4 of 26 embryos were successfully edited, but others contained undesirable mutations as a result of the treatment. While this demonstrated some proof of concept for creating HIV-resistant embryos, like the 2015 work, it also clearly showed that genome editing of embryos is neither precise nor safe at this point. However, a 2017 study on viable diploid embryos, created by *in vitro* fertilization for the purpose of embryo-editing research, reported a high degree of success in correcting a disease-causing mutation. Other researchers have disputed the claims of this study, and it has yet to be replicated.

在 2016 年，第二例基因组编辑人类三倍体胚胎的工作被报道，同样来自中国的研究团队。CRISPR-Cas9 将突变引入 *CCR5* 基因以赋予其抗 HIV 感染的能力。在这项研究中，26 个胚胎中的 4 个被成功编辑，但其他胚胎则由于该处理而携带上非预期的突变。尽管这项研究工作与 2015 年的工作一样为创建抗 HIV 胚胎的设想提供了一些证据，但也清楚地表明，此时此刻胚胎基因组编辑既不精确也不安全。然而，2017 年的一项胚胎编辑研究使用了来自体外受精、可存活的二倍体胚胎，研究报告指出纠正致病突变的成功率很高。其他研究人员对这项研究结果提出异议，并且该研究成果尚未能被重复。

These and other studies have stimulated significant ethical discussions about genetic engineering of embryos (which we discuss further in the final section of this chapter).

这些和其他一些研究均已引发胚胎遗传工程相关的重大伦理讨论（我们将在本章最后一节中进一步讨论）。

Also, as is the case with other gene therapy approaches, there is concern about CRISPR-Cas creating mutations at nontarget locations in the genome and also unwanted deletions and rearrangements in a target gene. But so far,

此外，与其他基因治疗方法一样，人们同样担心 CRISPR-Cas 在基因组中的非靶标位点产生突变，以及在靶标基因中产生有害的基因缺失和基因重排。但是到目前为止，CRISPR-Cas 显然是已开发的基因组

CRISPR-Cas is clearly the most promising tool for genome editing and gene therapy that has ever been developed, and in a short time, the pace of progress with this technique in animals and humans has been remarkable. Stay tuned!

编辑和基因治疗工具中最具应用前景的一种，并且在短期内，该技术在动物和人类中的进展已经非常引人注目。请大家拭目以待！

RNA-Based Therapeutics

Over the past decade, RNA-based therapeutics such as antisense RNAs, RNA interference, and microRNAs for gene therapy have received a great deal of attention. This is partly because these methods can be designed to be highly specific for a target RNA of interest to block or upregulate gene expression and are versatile because they can also be used to alter mRNA splicing, to target noncoding RNAs, and to express an exogenous RNA among other examples.

Attempts have been made to use **antisense oligonucleotides** to inhibit translation of mRNAs from defective genes (see Figure 11.5), thus blocking or "silencing" gene expression, but this approach to gene therapy has generally not yet proven to be reliable. The emergence of **RNA interference** (**RNAi**) as a powerful gene-silencing tool *in vitro* for research has reinvigorated interest in gene therapy approaches by gene silencing. RNAi is a form of gene-expression regulation. In animals, short, double-stranded RNA molecules are delivered into cells where the enzyme Dicer chops them into 21- to 25-nt-long pieces called **small interfering RNAs** (**siRNAs**). siRNAs then join with an enzyme complex called the **RNA-indued silencing complex** (**RISC**), which shuttles the siRNAs

基于 RNA 的治疗

在过去十几年中，基于 RNA 的疗法如反义 RNAs、RNA 干扰和微 RNA 用于基因治疗已经受到广泛关注。部分原因是这些方法可以针对关注的靶标 RNA 进行高度特异性设计，以阻断或上调基因表达。此外，由于它们还可以用于改变 mRNA 剪接、靶向非编码 RNAs 以及表达外源 RNA，所以用途十分广泛。

人们已经尝试使用**反义寡核苷酸**来抑制缺陷基因 mRNAs 的翻译（见图 11.5），从而阻断或“沉默”基因表达，但总的说来这种方法用于基因治疗的可靠性尚未得到证明。**RNA 干扰**（**RNAi**）是一种功能强大的体外基因沉默工具，它的出现重新激发了人们通过基因沉默开展基因治疗的兴趣。RNAi 是基因表达调控的一种形式。在动物中，短的双链 RNA 分子被传送至细胞内，并被 Dicer 酶切成 21—25 个核苷酸长度的小片段，即**干扰小 RNAs**（**siRNAs**）。然后，siRNAs 与 **RNA 诱导沉默复合物**（**RISC**）相结合，该酶复合物将 siRNAs 运送至其靶标 mRNA 处，二者通过互补碱基配对相结合。RISC 复合物可以阻止与 siRNA 结合的 mRNAs 翻译成蛋白质，或者导致与 siRNA 结合的 mRNAs 降解，从而使其无法翻译成

to their target mRNA, where they bind by complementary base pairing. The RISC complex can block siRNA-bound mRNAs from being translated into protein or can lead to degradation of siRNA-bound mRNAs so that they cannot be translated into protein (Figure 11.5).

蛋白质（图 11.5）。

A main challenge to RNAi-based therapeutics so far has been *in vivo* delivery of double-stranded RNA or siRNA. RNAs degrade quickly in the body. It is also hard to get RNA to cross lipid bilayers to penetrate cells in the target tissue. And how does one deliver RNA-based therapies to cancer cells but not to noncancerous, healthy cells? Two common delivery approaches are to inject the siRNA directly or to deliver them via a DNA plasmid vector that is taken in by cells and transcribed to make double-stranded RNA which Dicer can cleave into

基于 RNAi 的治疗目前面临的主要挑战是体内传送双链 RNA 或 siRNA。RNAs 分子在体内会迅速降解，也很难使 RNA 分子穿过脂双层结构进入目标组织细胞中。基于 RNA 的疗法如何只针对癌细胞而不影响非癌性的健康细胞呢？两种常见的传送方法是直接注射 siRNA 或者通过可被细胞吸收的 DNA 质粒载体进行传送，然后被转录成可被 Dicer 酶剪切形成 siRNAs 的双链 RNA。慢病毒、脂质体以及 siRNAs 与胆固醇和脂肪酸相结合是传送 siRNAs 的其他方法（图 11.5）。当使用脂质体或与脂类相结

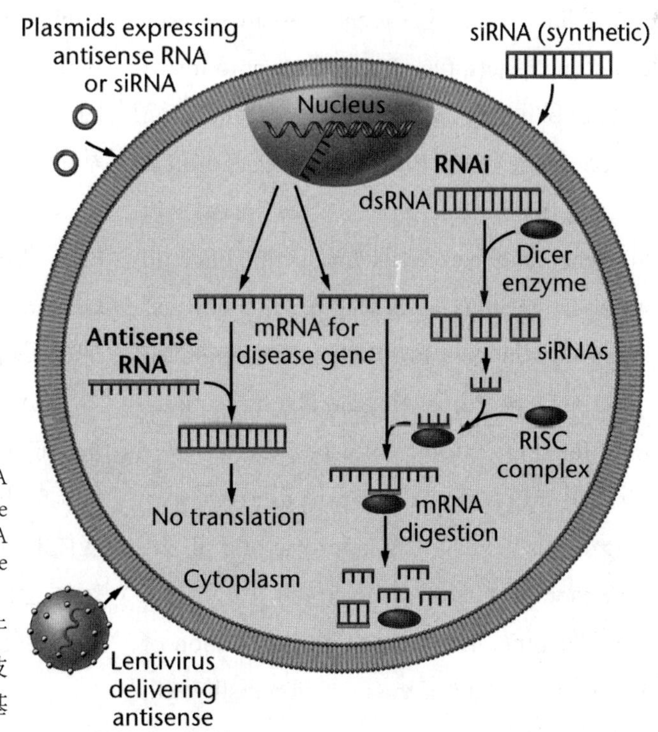

FIGURE 11.5 Antisense RNA and RNA interference (RNAi) approaches to silence genes for gene therapy. Antisense RNA technology and RNAi are two ways to silence gene expression and turn off disease genes.

图 11.5 基因治疗中反义 RNA 和 RNA 干扰（RNAi）可以沉默基因。反义 RNA 技术和 RNAi 是沉默基因表达和关闭疾病基因的两种途径。

siRNAs. Lentivirus, liposome, and attachment of siRNAs to cholesterol and fatty acids are other approaches being used to deliver siRNAs (Figure 11.5). When delivered in liposomes or attached to lipids, siRNAs are taken into the cell by endocytosis, but because of their charge, another challenge is getting therapeutic RNA out of the endosome and into the cytoplasm.

合进行传送时，siRNAs 通过胞吞作用进入细胞，但是由于它们有电荷，另一个挑战则是如何将治疗 RNA 从内吞体中移出，并使其进入细胞质中。

The same approaches used to deliver antisense RNAs and siRNAs can also be used to deliver vectors encoding **microRNAs** (**miRNAs**) or miRNAs themselves. In recent years a tremendous body of research literature has developed on the roles of miRNAs in silencing gene expression naturally in cells. The application of miRNAs for gene therapy is only in initial stages of development.

用于传送反义 RNAs 和 siRNAs 的方法同样适用于传送编码**微 RNAs（miRNAs）**的载体或 miRNAs 自身。近年来，miRNAs 在细胞中天然沉默基因表达的作用已被大量研究文献所报道。miRNAs 在基因治疗中的应用目前尚处于发展的初期阶段。

More than a dozen clinical trials involving RNAi are under way in the United States. Several RNAi clinical trials to treat blindness are showing promising results. One RNAi strategy to treat a form of blindness called macular degeneration targets a gene called *VEGF*. The VEGF protein promotes blood vessel growth. Overexpression of this gene, causing excessive production of blood vessels in the retina, leads to impaired vision and eventually blindness. In 2018, the FDA approved the drug Onpattro, by Alnylam Pharmaceuticals, Inc., for the treatment of a peripheral nerve disease in adults called polyneuropathy. Onpattro is the first FDA approval of a siRNA gene therapy treatment. It targets mRNA for the *TTR* gene, interfering with production of an abnormal form of the protein transthyretin (TTR) that contributes to polyneuropathy.

在美国正在进行的涉及 RNAi 的临床试验多达十几项。一些治疗失明的 RNAi 临床试验显示的结果令人期待。一种治疗黄斑变性失明的 RNAi 策略是靶向 *VEGF* 基因，其蛋白质产物可以促进血管生长。该基因过度表达会引起视网膜中血管过度生成，造成视力受损，最终失明。2018 年，FDA 批准了艾拉伦制药公司出品的药物 Onpattro，该药用于治疗成年人的一种周围神经疾病——多发性神经病。Onpattro 是首个 FDA 批准的 siRNA 基因疗法，它以 *TTR* 基因的 mRNA 为靶标，干扰导致多发性神经病的甲状腺素视黄质运载蛋白（TTR）突变体的合成。其他可采用 RNAi 治疗的候选疾病还包括一些癌症、糖尿病、肝病、多发性硬化和关节炎。

Other disease candidates for treatment by RNAi include several different cancers, diabetes, liver diseases, multiple sclerosis, and arthritis.

But many are predicting that RNA-based therapies will become much more successful and more widely adopted within the next decade. Antisense RNA is the oldest RNA-based therapeutic approach, but it initially did not live up to expectations. In the mid-2000s, many companies dropped this approach to gene therapy because of problems associated with delivering antisense RNA oligonucleotides across cell membranes and keeping them from being degraded while in the circulatory system. However, recent advances in RNA oligonucleotide chemistry have helped overcome these hurdles, and hundreds of clinical trials involving antisense RNAs are in the planning stages.

人们预测在未来10年内，基于RNA的疗法将更成功，并将得到更广泛的应用。反义RNA是最早的基于RNA的治疗方法，但最初并未达到预期。在21世纪第一个10年的中期，由于反义RNA寡核苷酸跨细胞膜传送及在循环系统中避免降解等相关问题，许多公司在基因治疗中放弃了这种方法。然而，RNA寡核苷酸化学的最新进展已帮助克服了这些障碍，并且数百项涉及反义RNAs的临床试验正处于计划阶段。

By late 2016, within a six-month span, antisense oligonucleotide trials were approved by the FDA for familial cholesterolemia, Duchenne muscular dystrophy, and spinal muscular atrophy (SMA). For example, the antisense oligonucleotide called Spinraza® (nusinersen), produced by Ionis Pharmaceuticals of Carlsbad, California, was approved as the first treatment for SMA. Affecting 1 in 10,000 to 12,000 children born, this disease is characterized by the loss of motor neurons in the spinal cord resulting in progressive muscle weakness and is a leading genetic cause of death for infants. In 2018, SMA was added to the Recommended Uniform Screening Panel for genetic testing which we discussed in Chapter 10—Genetic Testing.

截至2016年年底，在6个月的时间跨度内，FDA批准了家族性胆固醇血症、迪谢内肌营养不良和脊髓性肌萎缩（SMA）的反义寡核苷酸试验。例如，由位于加利福尼亚州卡尔斯巴德的伊奥尼斯制药公司生产的Spinraza®（诺西那生）反义寡核苷酸获批成为SMA的首个治疗药物。该疾病的新生儿发病率为1/12 000—1/10 000，其特征是脊髓运动神经元缺失导致进行性肌无力，并且是婴儿死亡的主要遗传原因。在2018年，SMA被加入遗传检测推荐的统一筛选方案中，我们在第10章——遗传检测中对此已进行了探讨。

Patients with SMA have a mutation in the *SMN1* gene that prevents production of the functional SMN protein required for normal motor neuron development. Spinraza is delivered into the cerebrospinal fluid. It is an 18-nt antisense oligonucleotide that targets *SMN2*, a homolog of *SMN1*. *SMN2* is normally mis-spliced at exon 7 to produce a truncated, largely nonfunctional SMN protein.

SMA 患者的 *SMN1* 基因突变阻止了正常运动神经元发育所需的功能 SMN 蛋白的产生。药物 Spinraza 被输送至脑脊液中，该药物是一段长 18 个核苷酸的反义寡核苷酸，靶向 *SMN1* 的同源基因 *SMN2*。*SMN2* 通常在第 7 个外显子处被错误剪切，从而产生截短的、基本无功能的 SMN 蛋白。

Spinraza binds to pre-mRNA of the *SMN2* gene, altering splicing to include exon 7 in the mature transcript leading to translation of a functional copy of the SMN protein. The drug showed such promise in two trials that the trials were halted early and considered successful because it was deemed unethical to continue to deny the drug to SMN-affected children in placebo groups. This antisense approach to alter mRNA splicing for gene therapy has also generated excitement because it has potential for treating Huntington disease, ALS, and other neurological conditions.

Spinraza 与 *SMN2* 基因的 mRNA 前体相结合，改变剪切方式，在成熟转录物中包含第 7 个外显子，使 SMN 蛋白的功能性拷贝得以被翻译。该药物在两项试验中均显示出较好的试验结果，所以试验在早期即被终止，因为继续在安慰剂组的 SMN 患儿中不使用该药物被认为是不道德的。这种使用反义法改变 mRNA 剪切方式的基因疗法同样使研究人员兴奋，因为它具有治疗亨廷顿病、ALS 和其他神经系统疾病的应用潜力。

11.7 Future Challenges and Ethical Issues

11.7　未来的挑战及伦理问题

Despite the progress that we have noted thus far, many questions remain to be answered before widespread application of gene therapy for the treatment of genetic disorders becomes routine:

尽管到目前为止我们论述了诸多进展，但是在广泛应用基因疗法治疗遗传疾病成为常规之前，仍有许多问题亟待回答：

■ What is the proper route for gene delivery in different kinds of disorders? For example, what is the best way to treat brain or muscle tissues?

■ 在不同种类的疾病中基因传递的合适途径是什么？例如，治疗脑组织或肌肉组织的最佳方法是什么？

■ What percentage of cells in an organ

■ 器官或组织中需要多少百分比的

or a tissue need to express a therapeutic gene to alleviate the effects of a genetic disorder?

细胞表达治疗基因才能缓解遗传疾病的影响？

■ What amount of a therapeutic gene product must be produced to provide lasting improvement of the condition, and how can sufficient production be ensured? Currently, many approaches provide only short-lived delivery of the therapeutic gene and its protein.

■ 为了持续改善病情，必须合成多少治疗基因产物，并且如何确保有充足的产量？当前，许多方法仅能提供治疗基因及其蛋白质产物的短期传送。

■ Will it be possible to use gene therapy to treat diseases that involve multiple genes?

■ 是否有可能使用基因疗法治疗涉及多基因的疾病？

■ Can expression or the timing of expression of therapeutic genes be controlled in a patient so that genes can be turned on or off at a particular time or as necessary?

■ 是否可以控制患者体内治疗基因的表达或表达时间，以便使基因表达可以在特定时间或根据需要进行开启或关闭？

■ Will genome-editing approaches become more widely used for gene therapy trials?

■ 基因组编辑方法是否将被更广泛地应用于基因治疗试验？

■ Will genome editing emerge as the safest and most reliable method of gene therapy, rendering other approaches obsolete, or will a combination of approaches (vector and nonvector delivery, RNA-based therapeutics, and genome editing) be necessary depending on the genetic condition being treated?

■ 基因组编辑是否会淘汰其他方法，成为最安全、最可靠的基因治疗方法，还是说会根据待治疗遗传疾病的情况，综合使用多种方法（如载体传送和非载体传送、基于 RNA 的疗法及基因组编辑）？

For many people, the question remains whether gene therapy can ever recover from past setbacks and fulfill its promise as a cure for genetic diseases. Clinical trials for any new therapy are potentially dangerous, and often, animal studies will not accurately reflect the reaction of individual humans to the methodology leading to the delivery of new genes. However, as the history of similar struggles encountered with life-saving developments such as the use of antibiotics and organ transplants show, there will be setbacks and even tragedies, but step by small step, we will move

对于许多人来说，问题仍然是基因疗法是否能走出过去的挫折，实现其有望治愈遗传疾病的愿望。任何新型疗法的临床试验都存在潜在危险，并且动物研究无法准确体现人类个体对新基因传送方法的反应。但是，正如在挽救生命的历史进程中（例如抗生素使用和器官移植）所遇到的类似挣扎，虽然有挫折甚至悲剧发生，但是一小步一小步地坚持，我们终将朝着未来为严重遗传病提供可靠、安全的治疗技术而迈进。

toward a technology that could—someday—provide reliable and safe treatment for severe genetic diseases.

Ethical Concerns Surrounding Gene Therapy

Gene therapy raises several ethical concerns, and many forms of gene therapy are sources of intense debate. At present, in the United States, all gene therapy trials are restricted to using somatic cells as targets for gene transfer. This form of gene therapy is called **somatic gene therapy**; only one individual is affected, and the therapy is done with the permission and informed consent of the patient or family.

Two other forms of gene therapy have not been approved, primarily because of the unresolved ethical issues surrounding them. The first is called **germ-line therapy**, whereby germ cells (the cells that give rise to the gametes— i.e., sperm and eggs) or mature gametes are used as targets for gene transfer or genome editing. In this approach, the transferred or edited gene is incorporated into all the future cells of the body, including the germ cells. As a result, individuals in future generations will also be affected, without their consent. Recently, ethical discussions about germ-line therapy have been accelerated by several reports of CRISPR-Cas being used for genome editing of human embryos *in vitro* (none were transferred to a uterus to develop). A report from the U.S. National Academy of Sciences and the National Academy of Medicine is recommending that germ-line therapy trials only be considered for serious conditions for which

基因治疗相关的伦理问题

基因治疗引发了一些伦理学担忧，并且许多形式的基因治疗成为激烈争论的缘由。目前在美国，所有基因治疗试验仅限于使用体细胞作为基因转移的靶标。这种形式的基因治疗被称为**体细胞基因治疗**，仅涉及一个个体，并且治疗需在患者或家属的许可和知情同意下进行。

基因治疗的另外两种形式目前尚未获批，主要是因为其相关的伦理问题尚未得到解决。第一种被称为**生殖细胞治疗**，即生殖细胞（产生配子的细胞，例如精子和卵细胞）或成熟配子被用作基因转移或基因组编辑的靶标。在这种方法中，被转移或被编辑的基因会整合入人体的所有未来细胞中，包括生殖细胞。因此，后代所有个体在未经本人同意的情况下也将受到影响。最近，CRISPR -Cas 被用于体外人类胚胎（并未将其转移至子宫内进行发育）基因组编辑的几篇报道进一步加速了有关生殖细胞治疗的伦理学讨论。美国国家科学院和美国国家医学院的报告建议，仅在没有合理可替代治疗方案选择，同时存在风险 - 收益选择和广泛监督的条件下，才能考虑对严重疾病进行生殖细胞治疗试验。该程序是否符合伦理道德？我们是否有权为后代做出这一决定？迄今为止，在美国，人们的担忧已经远超其潜在效益，因此国

there is no reasonable alternative treatment option, and where both the risk-benefit options and broad oversight are available. Is this kind of procedure ethical? Do we have the right to make this decision for future generations? Thus far, the concerns have outweighed the potential benefit in the United States where Congress has barred the FDA from approving clinical trials of germ-line gene therapy. Across the globe, the legality of editing human embryos varies widely. Some countries have banned the editing of human embryos, some have set some restrictions, and others have no restrictions at all. Also, there is gene doping, which is an example of **enhancement gene therapy**, whereby people may be "enhanced" for some desired trait.

会已经禁止 FDA 批准生殖细胞基因治疗的临床试验。在全球范围内，人类胚胎编辑的合法性差异很大，一些国家已禁止编辑人类胚胎，另一些国家则设置了一些限制，还有一些国家则完全没有限制。此外，还有基因兴奋剂，这是**增强型基因治疗**的例子，人们可以通过这种方式"强化"某些所需性状。

Gene therapy is currently a fairly expensive treatment. For rare conditions, the fewer the people treated, the more expensive the treatment will be. But what is the right price for a cure? It remains to be seen how health-care insurance providers will view gene therapy. But if gene therapy treatments provide a health-care option that drastically improves the quality of life for patients for whom there are few other options, it is likely that insurance companies will reimburse patients for treatment costs.

基因治疗目前是一种相当昂贵的治疗方法。对于罕见疾病，接受治疗人员越少，治疗费用则越高。但是治疗的合理价格应该是多少？这取决于卫生保健保险提供方如何评估基因治疗。但是，如果基因治疗提供卫生保健选择，那么对于几乎没有其他医疗方式可选的患者，他们的生活质量将得到极大改善，保险公司可能将向患者补偿治疗费用。

Finally, *whom* to treat by gene therapy is yet another ethically provocative consideration. In the Jesse Gelsinger case mentioned earlier, the symptoms of his OTC deficiency were minimized by a low-protein diet and drug treatments. Whether it was necessary to treat Jesse by gene therapy is a question that has been widely debated.

最后，通过基因治疗对谁进行治疗则是另一个在伦理上引发争议的考虑因素。在前面提到的杰西·基尔辛格病例中，低蛋白饮食和药物治疗可使其 OTC 缺乏症的症状减轻。是否需要通过基因疗法来治疗杰西是一个已被广泛争论的问题。

Jesse Gelsinger volunteered for the study

杰西·基尔辛格是自愿参加该项研究

to test the safety of the treatment for those with more severe disease. If a benefit was shown, it would have relieved him of an intense treatment regimen. Whether he should have been selected for the safety study is, of course, a matter to be debated. His tragic death due to unforeseen complications could not have been predicted at the time.

的，为那些患有更严重疾病的患者测试治疗安全性。如果测试结果良好，他将不再使用之前的强化治疗方案。当然，他是否应该被选择参与安全性研究也存在争议。他的悲剧性死亡源于当时尚无法预期的意外并发症。

12 Advances in Neurogenetics: The Study of Huntington Disease

12 神经遗传学进展：亨廷顿病的研究

"Driving with my father through a wooded road leading from Easthampton to Amagansett, we suddenly came upon two women, mother and daughter, both bowing, twisting, grimacing. I stared in wonderment, almost in fear. What could it mean?"

As the result of groundbreaking advances in molecular genetics and genomics made since the 1970s, new fields in genetics and related disciplines have emerged. One new field is **neurogenetics**—the study of the genetic basis of normal and abnormal functioning of the nervous system, with emphasis on brain functions. Research in this field includes the genes associated with neurodegenerative disorders, with the ultimate goal of developing effective therapies to combat these devastating conditions. Of the many such diseases, including Alzheimer disease, Parkinson disease, and amyotrophic lateral sclerosis (ALS), **Huntington disease (HD)** stands out as a model for the genetic investigation of neurodegenerative disorders. Not only is it monogenic and 100 percent penetrant, but nearly all analytical approaches in molecular genetics have been successfully applied to the study of HD, validating its significance as a

"当我和父亲驾车从伊斯特汉普顿至阿默甘西特的途中经过一条树木繁茂的公路时偶遇两位女性，她们是母女俩，她俩又是鞠躬、又是旋转、又是扮鬼脸，我惊讶地看着她们，心中充满了恐惧。这是什么情况？"

自20世纪70年代以来，分子遗传学和基因组学取得了诸多突破性进展，一些遗传学及其相关学科的崭新研究领域也随之出现。**神经遗传学**即是新生领域之一——研究神经系统正常功能和异常功能的遗传基础，其研究重点是脑功能。该领域的研究包括神经系统变性疾病相关基因的研究，最终目标是开发对抗这些破坏性疾病的有效疗法。此类疾病包括阿尔茨海默病、帕金森病和肌萎缩侧索硬化（ALS）等，**亨廷顿病**（**HD**）则是神经系统变性疾病遗传研究的模式疾病。HD不仅由单基因控制，外显率为100%，而且几乎所有分子遗传学分析方法均已成功地应用于HD的研究中，从而证实了其作为这些神经系统变性疾病的模型的重要性和显著意义。

model for these diseases.

HD is an autosomal dominant disorder characterized by adult onset of defined and progressive behavioral changes, including uncontrolled movements (chorea), cognitive decline, and psychiatric disturbances, with death occurring within 10 to 15 years after symptoms appear. HD was one of the first examples of complete dominance in human inheritance, with no differences in phenotypes between homozygotes and heterozygotes. In the vast majority of cases, symptoms do not develop until about age 45. Overall, HD currently affects about 25,000 to 30,000 people in North America.

HD 是一种常染色体显性遗传疾病，其特征是成年后发作，伴有明确的进行性行为变化，包括运动不受控制（舞蹈症）、认知能力下降和精神障碍，个体通常在症状出现后 10—15 年内死亡。HD 是人类遗传中最早被确定的完全显性遗传疾病中的一种，即纯合个体和杂合个体的表型没有差异。在绝大多数情况下，个体直至 45 岁左右才会出现相应症状。总体而言，目前在北美地区，HD 患者为 25 000—30 000 人。

The disease is named after George Huntington, a nineteenth-century physician. He was not the first to describe the disorder, but his account was so comprehensive and detailed that the disease eventually took on his name. Further, his observation of transgenerational cases in several families precisely matched an autosomal dominant pattern of inheritance. Shortly after the rediscovery of Mendel's work in the early twentieth century, pedigree analysis confirmed that HD is inherited as an autosomal dominant disorder.

该疾病以 19 世纪乔治 · 亨廷顿医生的名字命名。他并不是第一位记录该疾病的人，但是他的描述十分全面和详尽，所以该疾病最终以他的名字命名。此外，他对多个家庭跨世代病例的考察结果与常染色体显性遗传模式精确匹配。20 世纪初孟德尔的杰出工作重见天日后不久，系谱分析即证实 HD 是一种常染色体显性遗传疾病。

We will begin our consideration of Huntington disease by discussing the successful efforts to map, isolate, and clone the HD gene. We will then turn our attention to what we know about the molecular and cellular mechanisms associated with the disorder, particularly those discovered during the study of transgenic model systems. Finally, we will consider how this information is being used to develop a range of therapies.

我们将从 HD 基因的成功定位、分离和克隆开始对该疾病进行介绍。然后，我们将注意力转移至与该疾病相关的分子和细胞生物学机制上，尤其是转基因模型系统研究中的发现。最后，我们将讨论如何使用这些信息开发系列疾病疗法。

12.1 The Search for the Huntington Gene

Mapping the gene for Huntington disease was one of the first attempts to employ a method from a landmark 1980 paper by Botstein, White, and Davis in which the authors proposed that DNA sequence variations in humans could be detected as differences in the length of DNA fragments produced by cutting DNA with restriction enzymes. These differences, known as restriction fragment length polymorphisms (RFLPs), could be visualized using Southern blots (see Chapter 10—Genetic Testing for a discussion of RFLPs). The authors estimated that a collection of about 150 RFLPs distributed across the genome could be used with pedigrees to detect linkage anywhere in the genome between an RFLP marker and a disease gene of interest. In practical terms, this meant that it would be possible to map a disease gene with no information about the gene, its gene product, or its function—an approach referred to as reverse genetics.

12.1 寻找亨廷顿病基因

定位 HD 基因是博特斯坦、怀特和戴维斯于 1980 年发表的具有里程碑意义的论文中所涉及方法的首次应用尝试之一。在该论文中，作者提出利用限制性内切酶将 DNA 切割成长短不一的 DNA 片段，这些片段的长度差异可用于检测人类 DNA 的序列变异，即限制性片段长度多态性（RFLPs），且这些变异可使用 DNA 印迹法实现可视化（RFLPs 相关讨论请参阅第 10 章——遗传检测）。作者预计，收集分布于整个基因组中的约 150 个 RFLPs 位点，并与系谱分析一起即可检测存在于基因组中任何位置的 RFLP 标记与目标疾病基因间的连锁关系。在应用层面上，这意味着在不了解任何基因、基因产物及其功能相关信息的情况下即可对疾病基因进行定位——这种方法被称为反向遗传学。

Finding Linkage between Huntington Disease and an RFLP Marker

In the early 1980s, Huntington disease research was largely driven by the Hereditary Disease Foundation, established by the family of Leonore Wexler, who, along with her three brothers, died of Huntington disease. One daughter, Nancy, after learning about the proposal to map disease genes using DNA markers, used her awareness of a large population affected with Huntington disease in Venezuela to organize trips

寻找亨廷顿病与 RFLP 标记间的连锁

在 20 世纪 80 年代初期，HD 研究主要由莉奥诺 · 韦克斯勒家族建立的遗传疾病基金会推动。莉奥诺 · 韦克斯勒和她的三位兄弟均死于 HD。她的女儿南希在得知使用 DNA 标记定位疾病基因的研究计划后，利用其对委内瑞拉大量 HD 患病人群的了解，多次前往收集系谱信息并获得了可用于 DNA 连锁研究的血样。

to collect pedigree information and to obtain blood samples for DNA linkage studies.

About the same time Nancy Wexler began working on the Venezuelan pedigree, James Gusella began collecting RFLP markers to map the gene for Huntington disease. One of the RFLP markers developed in Gusella's lab, called G8, identified four possible patterns of DNA fragments when human DNA was cut with the restriction enzyme *Hin*dIII. These patterns, called haplotypes, were named A, B, C, and D. Using this marker and DNA from a large Venezuelan HD pedigree provided by Nancy Wexler, Gusella's team concluded there was linkage between the gene for Huntington disease and haplotype C (Figure 12.1). For confirmation, they sent their results to an

大约在南希 · 韦克斯勒开始研究委内瑞拉家族系谱的同时，詹姆斯 · 古塞拉开始收集 RFLP 标记以定位 HD 基因。G8 是古塞拉实验室开发的 RFLP 标记之一，可用于鉴定人类 DNA 经限制性内切酶 *Hin*dIII 剪切后形成的四种 DNA 片段的可能模式。这些模式称为单倍型，被命名为 A、B、C 和 D。通过使用该 DNA 标记和由南希 · 韦克斯勒提供的来自委内瑞拉的亨廷顿遗传家族 DNA 样本，古塞拉研究小组得出结论：HD 基因与单倍型 C 存在连锁关系（图 12.1）。为了证实这一点，他们将结果发送给连锁分析专家，后者证实 HD 和单倍型 C 之间连锁的证据确凿，古塞拉及其同事的确已经发现 HD 基因相关连锁。

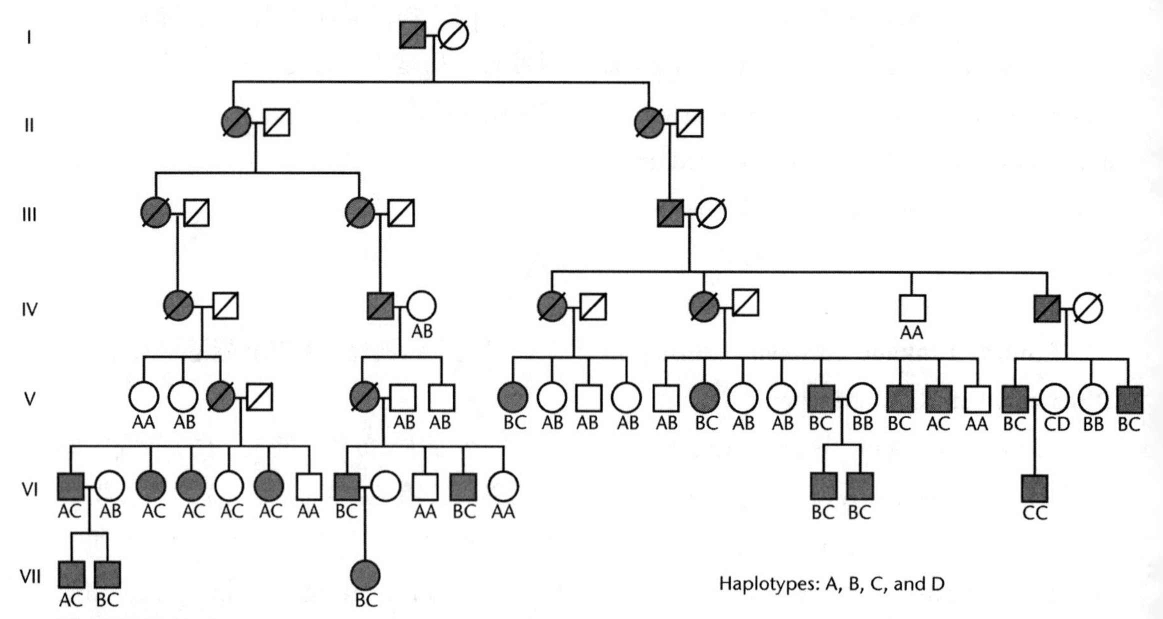

FIGURE 12.1 A part of the Venezuelan pedigree used in the search for linkage between RFLP markers and Huntington disease. Filled symbols indicate affected individuals. Deceased individuals are marked by diagonal slashes. In this pedigree, haplotype C of the G8 marker is coinherited with HD in all cases, indicating that the RFLP marker and the mutant HD allele are on the same chromosome.

图 12.1 用于研究 RFLP 标记与 HD 间连锁关系的部分委内瑞拉家族系谱。被填充的符号表示患病个体，已去世个体则通过在符号上画对角线来表示。在该系谱中，G8 标记的单倍型 C 在所有遗传事件中都与 HD 共遗传，表明该 RFLP 标记与 HD 突变基因位于同一条染色体上。

expert in linkage analysis, who verified that the evidence for linkage between HD and haplotype C was overwhelming and that Gusella and his colleagues had discovered linkage to HD.

Once the G8 probe was linked to the HD gene, the next task was to determine which human chromosome carried the G8 marker and the gene for Huntington disease.

一旦发现 G8 探针与 HD 基因相连锁，下一个任务就是确定 G8 标记和 HD 基因位于人类的哪条染色体上。

Assigning the HD Gene to Chromosome 4

A collection of somatic cell hybrids can be used to map DNA markers or genes to specific chromosomes. In this case, a panel of 18 mouse-human somatic cell hybrid cell lines, each of which contained a unique combination of human chromosomes, was used for mapping the G8 probe. On Southern blots from these hybrid cells digested with *Hin*dIII, G8 fragments were seen in all cells carrying human chromosome 4 and never seen when chromosome 4 was absent.

These results established that the G8 marker and the gene for Huntington disease were both on chromosome 4. This was the first time that an RFLP marker was used to map an autosomal disease gene to a specific chromosome. This discovery launched a whole new branch of genetics, called *positional cloning* (sometimes called reverse genetics).

将亨廷顿病基因定位于人类 4 号染色体上

体细胞杂交细胞系列可用于将 DNA 标记或基因定位到特定染色体上。在本研究中，一组鼠 - 人体细胞杂交细胞系共有 18 株用于定位 G8 探针，其中每株细胞随机含人类的若干染色体。再使用 *Hin*dIII 消化这些杂交细胞，通过 DNA 印迹法分析，结果显示在所有携带人类 4 号染色体的细胞中均可检测到 G8 片段，而在不含 4 号染色体的细胞中则均未检测出 G8 片段。

这些结果确定了 G8 标记和 HD 基因均位于第 4 号染色体上，这是首次使用 RFLP 标记将常染色体疾病基因定位到特定染色体上。这一发现开启了一个全新的遗传学分支，即定位克隆（有时也称为反向遗传学）。

The Identification and Cloning of the Huntington Gene

To identify and clone the gene for Huntington disease, researchers formed The Huntington’s Disease Collaborative Research

鉴定和克隆亨廷顿病基因

为了鉴定和克隆 HD 基因，研究人员成立了亨廷顿病合作研究小组（HDCRG），该研究小组由来自两大洲的 58 位科学家组

Group (HDCRG), consisting of 58 scientists on two continents. In spite of this massive effort, it took 10 years to identify the gene. First, the region most likely to contain the G8 marker (now renamed as *D4S10*) and the gene was narrowed to a small region near the tip of the short arm of chromosome 4. Expressed genes in this region were isolated and named "interesting transcripts (ITs)." One of the genes identified in this screening, called *IT15*, encoded a previously unknown protein containing the sequence CAG repeated a number of times. Populations unaffected with HD were found to carry many alleles of this gene, with the number of CAG repeats ranging from 11 to 34 copies. However, in individuals with HD, CAG repeats were significantly longer, ranging from 42 to 66 copies. None of the other genes identified in this region of chromosome 4 had any differences between affected and unaffected individuals that would implicate them as the HD gene. Because variations in the size of trinucleotide repeats had previously been identified as the causes of myotonic dystrophy and fragile-X syndrome, researchers proposed that variation in the number of CAG repeats was the cause of HD and that *IT15* encoded the HD gene.

成。尽管付出了巨大努力，但是鉴定该基因仍花费了10年时间。首先，将最有可能包含G8标记（现已更名为*D4S10*）和HD基因的区域缩小至邻近4号染色体短臂末端的一个狭小区域。该区域中的表达基因均被分离获得，并命名为"有趣的转录物（ITs）"。筛选中鉴定得到的系列基因中有一个名为"*IT15*"的基因含有多次重复的CAG序列，该基因编码一种未知的蛋白质。人们发现未患HD的人群携带该基因的许多等位基因，其中CAG序列重复次数介于11—34个拷贝之间。然而，在HD患者个体中，CAG重复序列明显更长，介于42—66个拷贝之间。4号染色体该区域中被鉴定的其他基因在HD患者和非患者个体之间没有任何可以体现其可能是HD基因的差异。由于先前已确定三核苷酸重复序列的长度变异是强直性肌营养不良和脆性X染色体综合征的致病原因，因此研究人员指出，CAG序列重复数量的变化是HD的致病原因，并且*IT15*编码HD基因。

A paper authored by all 58 members of the HDCRG ended the decade-long search for the gene, now called *HTT*, and its encoded protein, which they named huntingtin. Subsequent analysis of CAG repeat lengths in populations of unaffected and affected individuals clarified the relationship between repeat length and the onset of HD.

由HDCRG所有58位成员撰写的论文结束了长达10年的寻找HD基因（现称为*HTT*）及其编码蛋白质（命名为Huntingtin）的研究历程。随后对健康人群和患病人群中CAG重复序列长度的分析阐明了序列重复长度与HD发病之间的关系。

12.2 The *HTT* Gene and Its Protein Product

Huntington disease is caused by the expansion of a CAG repeat and is one of 14 known trinucleotide repeat disorders. Nine of these, including HD, are caused by the expansion of CAG repeats, each of which codes for the insertion of the amino acid glutamine in the protein product. Thus, these genetic conditions are known as polyglutamine or polyQ disorders (Q is the one-letter abbreviation for glutamine). In addition to carrying mutant alleles with expansion of CAG repeats, other polyQ disorders have many symptoms in common with HD, including adult onset, behavioral changes, neurodegeneration, and premature death.

The *HTT* gene encodes a large protein that is 348 to 350 kDa in size. In normal alleles, a region near the 5'-end of the gene contains 6 to 35 CAG repeats encoding a stretch of glutamines in the protein product. Disease-causing mutant alleles contain an expanded number of CAG repeats (>36) that increase the number of glutamine residues in the mutant protein. The normal HTT protein is expressed in most, if not all, cells of the body and is associated with many different cellular compartments and organelles, including the plasma membrane, nucleus, cytoskeleton, cytoplasm, endoplasmic reticulum, Golgi complexes, and mitochondria. In brain cells of the striatum and caudate nucleus, HTT is present at synapses. The HTT protein domains involved in protein-protein

12.2 *HTT* 基因与其蛋白质产物

HD 由 CAG 重复序列的扩增引起，也是 14 种已知三核苷酸重复异常所致疾病之一。这些疾病中的 9 种（包括 HD）由 CAG 重复序列扩增引起，每个 CAG 编码的谷氨酰胺都将插入蛋白质产物中。因此，这些遗传疾病被称为聚谷氨酰胺或 polyQ 疾病（Q 是谷氨酰胺英文单字符缩写）。除了携带含 CAG 重复序列扩增的突变型等位基因外，其他的 polyQ 疾病还具有许多与 HD 同样的症状，包括成年后发作、行为改变、神经退行性病变和早亡。

HTT 基因编码大小为（348—350）$\times 10^3$ 的大蛋白质。在正常等位基因中，基因 5' 端附近区含 6—35 个 CAG 重复序列，编码蛋白质产物中的一段谷氨酰胺序列。致病的突变型等位基因含重复次数扩增的 CAG 重复序列（>36），从而增加了突变蛋白中谷氨酰胺残基的数量。正常 HTT 蛋白在体内即使不是在全部也是在大多数细胞中表达，并与许多不同的细胞区域和细胞器相关，包括质膜、细胞核、细胞骨架、细胞质、内质网、高尔基复合体和线粒体。在纹状体和尾状核的脑细胞中，HTT 则存在于突触中。参与蛋白质 - 蛋白质相互作用的 HTT 蛋白结构域类似于调节从细胞核到细胞质进行分子转运的蛋白质。HTT 还具有在突触连接处促进神经冲动传递的作用。总之，正常 HTT 蛋白具有多种功能，在细

interactions are similar to those in proteins that regulate the transport of molecules from nucleus to cytoplasm. HTT also has a role in facilitating nerve impulse transmission at synaptic junctions. In sum, normal HTT is multifunctional, is present in many cellular locations, and has a number of specific roles in cellular processes.

胞中的定位多元化，并且在细胞过程中扮演着多种特定角色。

The HD mutation is a gain-of-function mutation. The extended polyQ region of the mutant HTT protein (called mHTT) causes misfolding and the formation of protein aggregates held together by hydrogen bonds. PolyQ regions of these aggregates bind to and inactivate regulatory molecules, disrupting a number of cellular functions, causing neurodegeneration in the striatum and caudate nucleus. In addition, toxic peptide fragments generated by breakdown of aggregates are transported into the nucleus where they accumulate and disrupt transcription and nucleocytoplasmic transport. The net result of these cellular changes is a gradually increasing degradation of cellular function that culminates in degeneration and death of nerve cells in specific brain regions.

HD 突变是一种功能获得性突变，HTT 突变蛋白（mHTT）的 polyQ 区延长导致蛋白质错误折叠并形成通过氢键聚集在一起的蛋白质聚集体。这些聚集体的 PolyQ 区与调节分子相结合并使其失活，从而破坏了许多细胞功能，导致纹状体和尾状核神经退行性病变。此外，由聚集体分解产生的有毒肽片段被转运至细胞核中，并在其中进行积累，破坏转录和核质转运。这些细胞过程变化的最终结果是细胞的功能衰退逐渐增加，最终导致特定大脑区域中的神经细胞变性和死亡。

12.3 Molecular and Cellular Alterations in Huntington Disease

12.3 亨廷顿病的分子变化和细胞变化

Although HD is caused by the mutation of a single gene, which was isolated and characterized over 25 years ago, the mechanisms by which mHTT protein cause HD are still largely unknown. In spite of decades of research employing a wide

尽管 HD 由单基因突变引起，且该基因已于 25 年前获得分离和鉴定，但是 mHTT 蛋白引发 HD 的机制目前在很大程度上尚不清楚。尽管几十年来研究人员采用了大量技术和模型系统对此展开研究，但是正常 HTT 蛋白和 mHTT 蛋白所具有的

number of techniques and model systems, the full range of functions carried out by the normal HTT protein and those of the mHTT protein have not been completely elucidated. The straightforward view is that the mutant allele encodes a toxic protein that causes cell death initially in the striatal region of the brain. Increasing loss of cells in the striatum and other regions results in progressive and degenerative changes in muscle coordination and behavior. Death usually occurs 10 to 15 years after symptoms appear.

全部功能尚未完全得到阐明。直接的观点是，突变等位基因编码一种毒性蛋白，该蛋白质最先导致大脑纹状体区域细胞的死亡。随着纹状体和其他区域中细胞损失的增加，肌肉协调和行为发生进行性和退行性变化。死亡通常发生在症状出现后 10—15 年。

Unraveling the functions of the normal and mutant versions of HTT is proving to be extremely difficult because HTT interacts with more than 180 different proteins. Protein-protein network maps indicate that network proteins are involved in many cellular processes including transcription, protein folding and degradation, synaptic transmission, and mitochondrial function. In the following four sections, we will review the malfunction of each of the processes in the presence of mHTT.

由于 HTT 与 180 多种蛋白质存在相互作用，因此要弄清正常 HTT 蛋白和突变 HTT 蛋白的功能十分困难。蛋白质 - 蛋白质相互作用网络图表明蛋白质参与许多细胞过程，包括转录、蛋白质折叠和降解、突触传递、线粒体功能。在接下来的四个部分，我们将介绍 mHTT 存在下每个细胞过程中发生的功能异常。

Transcriptional Disruption

The effects of mHTT on the transcriptome are one of the key molecular events in HD. Expression of mHTT causes the formation of heterochromatin, effectively closing off transcription of genes located in affected chromosome regions. In addition, smaller soluble mHTT fragments interact with molecular components of transcription and obstruct the functions of the promoter-binding transcription factors and the proteins necessary

转录干扰

mHTT 对转录物组的影响是 HD 发生的关键分子事件之一。mHTT 表达会导致异染色质形成，有效关闭位于染色体受影响区内的基因转录。此外，较小的可溶性 mHTT 片段与转录的分子组成成分相互作用，阻碍了启动子结合转录因子和转录起始必需蛋白质的功能，最终导致整个基因组启动子亲和力降低和转录起始减少。

for transcription initiation. The net result is reduced promoter accessibility and transcription initiation across the genome.

Impaired Protein Folding and Degradation

The correct folding of proteins depends on the action of proteins called chaperones that mediate folding. In HD, transcriptional disruption inactivates several families of chaperones, which leads to the accumulation of incorrectly folded proteins in the cytoplasm. The result is disruption of normal folding and a slowdown in degradation of misfolded proteins.

In normal cells, misfolded proteins are tagged by a small protein (ubiquitin) and directed to a cellular structure called the proteasome where they are degraded. In brain cells of individuals with HD, ubiquitin binds to aggregated mHTT in the cytoplasm, which is then targeted for degradation, but the proteasome system is inhibited by an unknown mechanism. As a result, mHTT aggregates accumulate and impair cellular functions, triggering apoptosis and cell death. Together, the inhibition of chaperone function and proteasome function causes a collapse of normal protein function and turnover in brain cells of individuals with HD.

蛋白质折叠与降解异常

蛋白质的正确折叠依赖一类可引导折叠的蛋白质——分子伴侣的作用。在HD中，转录异常会导致多个分子伴侣蛋白质家族发生失活，从而造成错误折叠的蛋白质在细胞质中大量积累，最终导致蛋白质的正常折叠异常，且错误折叠的蛋白质降解速度减缓。

在正常细胞中，错误折叠的蛋白质会被小蛋白——泛素标记，并被指引到达细胞结构——蛋白酶体，在此处错误折叠的蛋白质会被降解。在 HD 患者个体的脑细胞中，泛素与细胞质中聚集的 mHTT 相结合，然后引导其进行降解，但是蛋白酶体系统功能受到抑制且目前机制尚不清楚。结果便是，mHTT 聚集体积累，损害细胞功能，触发细胞凋亡和细胞死亡。分子伴侣功能和蛋白酶体功能均被抑制一同导致 HD 患者正常蛋白质的功能崩溃和脑细胞更新失败。

Synaptic Dysfunction

In individuals with HD, subtle changes in motor function resulting from synaptic dysfunction can appear decades before the onset of neuronal death. To investigate synaptic defects, researchers constructed transgenic

突触功能障碍

在 HD 患者中，由于突触功能障碍引起的运动功能细微变化可能在神经元坏死发作前数十年即会出现。为了研究突触缺陷，研究人员构建了携带突变型人类 HD 等位基因的转基因果蝇系。在转基因果蝇

Drosophila strains carrying a mutant human HD allele. In transgenic flies, the synaptic vesicles carrying neurotransmitters were much smaller than normal. As a result, synaptic transmission was disrupted, causing behavioral changes in locomotion.

中，携带神经递质的突触小泡比正常突触小泡要小得多。因此，突触传递被破坏，引起了行进运动的行为改变。

Impaired Mitochondrial Function

mHTT binds to the outer mitochondrial membrane and impairs electron transport, reducing the amount of ATP available to the cell. Disruption of the electron transport chain also increases the levels of reactive oxygen species including free radicals, which cause widespread oxidative damage to cellular structures.

Within neurons, mitochondria migrate to synapses when rates of nerve impulse transmissions increase. Mitochondrial movement is inhibited by aggregation of N-terminal mHTT fragments that physically block migration along microtubules. This reduces the energy available for transmission of nerve impulses at synapses.

In sum, the damage to mitochondria caused by expression of mHTT includes the disruption of ATP production, promotion of oxidative damage within mitochondria and the cytoplasm, lowered synaptic transmission, and reduction of mitochondrial numbers to a level that can no longer support the core activities of cells, which eventually triggers apoptosis and cell death.

Although progress has been made in defining the defects that follow the expression and accumulation of mHTT, several important questions about the underlying mechanisms of

受损的线粒体功能

mHTT 与线粒体外膜相结合会阻碍电子传输，从而减少可供给细胞的 ATP 的量。电子传输链受损同时也会提高包括可对细胞结构造成广泛氧化损伤的自由基在内的活性氧的含量。

在神经元内，当神经冲动传递的速率增加时，线粒体会迁移至突触。N 末端 mHTT 片段的聚集在物理上阻止了线粒体沿微管的迁移，从而抑制了线粒体运动，并减少了突触传递神经冲动所需的能量供给。

总之，由 mHTT 表达引发的线粒体损害包括 ATP 合成中断、线粒体和细胞质内氧化损伤加剧、突触传递能力降低、线粒体数量减少以至于不再能够支持细胞的核心活力，从而最终触发细胞凋亡和细胞死亡。

尽管伴随 mHTT 表达和积累产生的功能缺陷研究已经取得了进展，但是 HD 相关潜在机制的几个重要问题仍未得到解答。例如，尚不清楚任意单一细胞功能受损是

HD remain unanswered. For example, it is not known whether any single disruption of cell function is sufficient to cause neurodegeneration or cell death, or whether one or more of these pathways must interact to bring about these results. To answer these and other questions, researchers turned to the use of transgenic animal models of HD.

否足以导致神经退行性病变或细胞死亡，或者这些通路中的一个或多个是否必须通过相互作用才能产生这些结果。为了回答这些问题和其他相关问题，研究人员转向使用 HD 转基因动物模型进行研究。

12.4 Transgenic Animal Models of Huntington Disease

12.4 亨廷顿病的转基因动物模型

Shortly after the *HTT* gene was cloned, researchers constructed transgenic model organisms to analyze the disease process at the molecular level.

在 *HTT* 基因被克隆后不久，研究人员便构建了转基因模式生物，以便在分子水平层面分析疾病进程。

Animal models of human behavioral disorders present an opportunity to separate behavioral phenotypes into their components. This makes it possible to study the developmental, structural, and functional neuronal mechanisms related to these behaviors that are difficult or impossible to do in humans. In addition, behavior in animal models can be studied in controlled conditions that limit the impact of environmental factors. Although it is possible to construct transgenic models of human HD in a wide range of organisms, including yeast, *Caenorhabditis elegans*, and *Drosophila melanogaster*, the mouse is the most widely used model organism for these studies. Researchers favor mice because humans and mice share about 90 percent of their genes and because a wide range of strains with specific behavioral phenotypes are available.

人类行为障碍动物模型为将行为表型分解成它们的组成部分提供了机会。这使得在人类中难以或不可能进行的与这些行为相关的发育、结构和功能神经元机制的研究成为可能。此外，研究人员可以在可控条件下研究动物模型的行为，从而减少环境因素的影响。尽管在多种生物中均可构建人类 HD 转基因疾病模型，包括酵母、秀丽隐杆线虫和果蝇，但是这些研究中使用最为广泛的模型生物还是小鼠。研究人员偏爱使用小鼠的原因在于人与小鼠大约存在 90% 的同源基因，同时也因为存在许多具有特定行为表型的小鼠品系可供选择。

Using Transgenic Mice to Study Huntington Disease

The first mouse model of HD was constructed using the promoter sequence and first exon of the human mutant *HTT* allele, which contains an expanded CAG repeat. Examination of transgenic mouse brains a year after gene transfer showed abnormalities in the levels of neurotransmitter receptors and the presence of protein aggregates, a significant finding that was later confirmed to exist in the brains of humans with HD.

Soon after, researchers began to examine the relationship between CAG repeat length and disease progression. Transgenic mice carrying human *HD* genes with 16, 48, or 89 CAG repeats were monitored from birth to death to determine the age of onset and stages of abnormal behavior. Mice carrying 48- or 89-repeat human *HD* genes showed behavioral abnormalities as early as 8 weeks, and by 20 weeks they showed problems with motor coordination.

At various ages, brains of wild-type and transgenic mice carrying these mutant alleles were examined for changes in structure. Degenerating neurons and cell loss were evident in mice carrying 48 and 89 repeats, but no changes were seen in brains of wild-type mice or of those carrying a 16-repeat transgene (Figure 12.2).

There are now more than 20 different mouse models of HD, which are used to examine changes in molecular and cellular processes or in brain structure that occur before or just after the

使用转基因小鼠研究亨廷顿病

首个HD小鼠模型的构建使用了人类突变*HTT*等位基因的启动子序列和第一个外显子区，其中含有扩增的CAG重复序列。基因转移一年后对转基因小鼠的大脑进行检查，结果显示神经递质受体水平异常并且存在蛋白质聚集体，这一重要发现后来证实在HD患者大脑中同样存在。

不久之后，研究人员开始研究CAG重复长度与疾病进展之间的关系。研究人员对分别携带16、48或89个CAG三核苷酸重复的人源*HD*基因转基因小鼠进行了从出生到死亡的连续监测，以确定发病年龄和出现异常行为的阶段。携带48或89个CAG序列重复的人源*HD*基因小鼠早在第8周时即出现了行为异常，到第20周时它们显示出了运动协调问题。

在不同年龄段，研究人员对野生型小鼠和携带突变型等位基因的转基因小鼠的大脑进行了结构变化方面的检查。在携带48和89个CAG序列重复的小鼠脑中出现了明显的神经元退化和脑细胞损失，但在野生型小鼠或携带16个CAG序列重复的转基因小鼠的大脑中未见变化（图12.2）。

现在已有20多种HD小鼠模型可用于考察症状发作之前或之后发生的分子和细胞过程变化或大脑结构变化，并开发了实验方法以减缓或逆转细胞损失。HD转基因

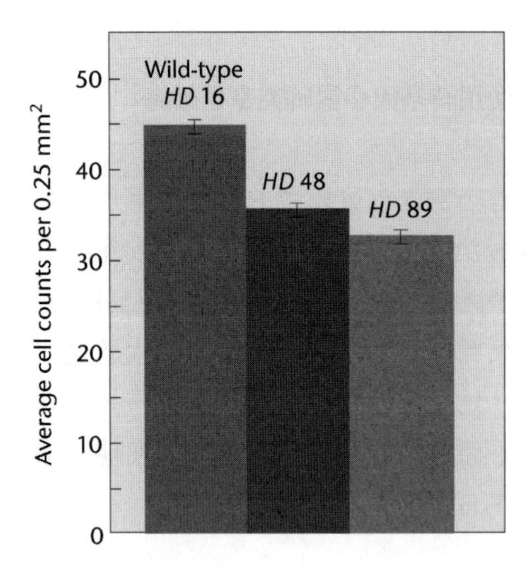

FIGURE 12.2 Relative levels of neuronal loss in HD transgenic mice. Cell counts show a significant reduction of neurons in the striatum in the brains of *HD* 48 mutants (middle column) and *HD* 89 mutants (right column). Cell loss in this brain region is also found in humans with HD, making these transgenic mice valuable models to study the course of this disease.

图 12.2 HD 转基因小鼠神经元退化的相对水平。细胞数量显示 *HD* 48 突变体（位于中间的柱形图）和 *HD* 89 突变体（位于右侧的柱形图）大脑纹状体内的神经元出现显著减少。大脑区域内细胞的减少在人类 HD 患者中同样存在，从而使得这些转基因小鼠成为研究该疾病进程十分有价值的模型。

onset of symptoms and to develop experimental treatments to slow or reverse cell loss. Transgenic models for HD allow researchers to administer treatment at specific times in disease progression and to evaluate the outcome of treatments in the presymptomatic stages of HD, something that is not possible in humans with HD.

模型使研究人员可以在疾病进展的特定时间实施治疗，并评估在 HD 症状发生前阶段的治疗效果，这些在 HD 人类患者中是不可能进行的。

In one important study, A. Yamamoto and colleagues constructed a transgenic mouse with inducible expression of the *HTT* gene. In this case, researchers constructed a human *HTT* exon 1 fragment containing 94 CAG repeats with an adjacent promoter that could be switched off when the antibiotic doxycycline was added to the drinking water. When the gene was switched off shortly after motor symptoms of HD developed, protein aggregates in the brain were rapidly degraded and disappeared, along with the abnormal motor symptoms. This provided the first clue that treatment in the early stages of the disease might be effective in controlling or reversing the symptoms in humans.

在一项重要研究中，山本及其同事构建了 *HTT* 基因可诱导表达的转基因小鼠。在该研究中，当饮用水中添加抗生素多西环素时，构建在含 94 个 CAG 重复序列的人源 *HTT* 外显子片段 1 相邻位置的启动子将被关闭。当 HD 运动症状发作后不久关闭该基因，则大脑中的蛋白质聚集体将迅速降解并消失，异常运动症状也会随之消失。这为在人类疾病的早期阶段介入治疗可能实现有效控制或逆转症状提供了首个线索。

Transgenic Large-Animal Models of Huntington Disease

Despite the important discoveries made in mice, the main problem with transgenic mouse models of HD is that mice do not show the same pattern of neurodegeneration seen in humans. In addition, mice have a smaller brain, a shorter lifespan (two to three years), and different physiology than humans. To overcome some of these problems, large-animal models, including sheep, mini-pigs, and a number of nonhuman primates are being developed to study the mechanisms of disease and the testing of drugs for human therapies.

Transgenic sheep were one of the first large-animal models developed to study HD. One transgenic sheep model carries a full-length human *HTT* gene with 73 CAG repeats. Given the relatively long life span of sheep (>10 years), this model is being used to study the long-term development of HD. In addition, because the size and structure of sheep brains are similar to those of humans, it is possible to do MRI and PET scans that can be directly compared to the results of scans in humans affected with HD.

A recently developed pig model carrying a human *HD* gene with an expanded set of CAG repeats shows a pattern of neurodegeneration that mirrors the brain changes seen in affected humans. These pigs also develop the characteristic abnormal movements seen in humans with HD. Researchers who developed this model anticipate that it will be valuable in testing treatments for HD. Continued development of large-animal models will

亨廷顿病的转基因大型动物模型

尽管在小鼠中取得了重要的发现，但HD转基因小鼠模型的主要问题在于小鼠并没有表现出与人类相同的神经退行性病变模式。此外，与人类相比，小鼠的脑较小、寿命较短（2—3年），并且生理学也不同。为了克服其中的一些问题，大型动物模型，包括绵羊、迷你猪和许多非人类灵长类动物等，正被开发用于研究疾病机理和人类疾病治疗的药物测试。

转基因绵羊是最早用于研究HD的大型动物模型之一。一种转基因绵羊模型携带含73个CAG重复序列的全长人源*HTT*基因。由于绵羊的寿命相对较长（>10年），该模型正被用于HD的长期发展研究。此外，由于绵羊脑的大小和结构与人类相似，因此可进行MRI（磁共振成像）和PET（正电子发射断层显像）扫描，并将其与人类HD患者的扫描结果直接进行比较。

最近开发的携带一组扩展CAG重复序列的人类*HD*基因的猪模型显示出神经退行性病变模式，该模式反映了人类患者的大脑变化。这些猪同样出现HD患者人群中特征性的运动异常特征。开发此模型的研究人员期望该模型能在HD治疗试验中发挥重要价值。大型动物模型的不断开发将有助于更为详细地了解HD的致病机制和开发更为有效的治疗方法。

hopefully lead to a more detailed understanding of the mechanism of HD and the development of effective therapies.

12.5 Cellular and Molecular Approaches to Therapy

12.5 治疗的细胞生物学和分子生物学方法

Because the mHTT protein affects a large number of cellular processes through protein-protein interactions and the accumulation of mHTT aggregates, researchers are using multiple approaches to investigate treatment strategies. We will now briefly review several of these approaches.

由于 mHTT 蛋白通过蛋白质 - 蛋白质相互作用和 mHTT 聚集体的积累影响大量细胞过程，因此研究人员正在使用多种方法研究治疗策略。我们将简要回顾其中的几种方法。

Stem Cells for Transplantation

Stem cells are undifferentiated somatic cells with two properties: (a) the ability to renew their numbers by mitosis, and (b) the ability to differentiate and form tissue-specific specialized cell types. Research on HD uses human stem cells for studies of HD mechanisms, drug screening, drug testing, genetic correction of mHTT accumulation, and as donor cells for transplantation.

干细胞移植

干细胞是未分化的体细胞，具有两个特性：（a）通过有丝分裂进行细胞更新的能力；（b）进行分化并形成组织特异性专门细胞类型的能力。HD 研究使用人类干细胞进行 HD 机制研究、药物筛选、药物测试、mHTT 积累的遗传校正，并以其作为移植的供体细胞。

HD is associated with transcriptional repression of many gene sets, including the gene that encodes brain-derived neurotrophic factor (BDNF), which is essential for the survival and function of cells in the striatum. Loss of this factor contributes to the death of striatal brain cells in HD. Human mesenchymal stem cells (MSCs) can be genetically programmed to overexpress BDNF. Injection of human MSCs modified to overexpress BDNF into the brains

HD 与许多基因集的转录抑制有关，包括编码脑源性神经营养因子（BDNF）的基因，该基因对于纹状体细胞的存活和功能至关重要。该因子的缺失会导致 HD 患者纹状体脑细胞死亡。人类间充质干细胞（MSCs）经遗传改造后可以过表达 BDNF。将经遗传修饰后可过表达 BDNF 的人类 MSCs 注射到携带含 128 个 CAG 重复序列全长人源 *HTT* 基因的转基因小鼠脑中，可以显著提高新生神经细胞的形成，延长

of a transgenic mouse strain carrying a full-length human *HTT* gene with 128 CAG repeats significantly increased the production of new nerve cells, increased life span, and reduced HD-associated behaviors. Further development of MSCs as a delivery system in other mouse models and large-animal models will be required before the system is ready for human clinical trials, which are expected to begin as soon as additional animal studies are completed.

小鼠寿命，并减少 HD 相关行为的出现。在用于人类临床试验之前，MSCs 传送系统在其他小鼠模型和大型动物模型中仍需进一步开发，并且有望在完成其他动物研究后立即开始人类临床试验。

Gene Silencing to Reduce mHTT Levels

Because mHTT initiates a chain of events that lead to cell death and the onset of HD symptoms, therapies that reduce or eliminate the expression or accumulation of mHTT, either alone or in combination with other therapies, might be especially effective in treating HD.

A therapy using synthetic zinc-finger nucleases (ZFNs) to repress transcription of m*HTT* alleles (see Chapter 11—Gene Therapy for a discussion of ZFNs) in mouse neuronal cell lines and a cell line from an individual with HD carrying a mutant allele with 45 CAG repeats showed that ZFN treatment significantly repressed m*HTT* expression with no effects on the expression of other genes containing CAG repeats.

The ZFN construct was then tested on a transgenic mouse strain carrying 115 to 160 CAG repeats by injection into the striatal region of one side of the brain. A control vector was injected into the other side of the brain. Two weeks later, levels of mRNA from the normal and mutant alleles were assayed in tissue from each side of

基因沉默降低 mHTT 水平

由于 mHTT 启动会引发细胞死亡和 HD 病症发作的链式事件，因此减少或消除 mHTT 表达或积累的治疗方法不论是单独使用还是联合其他疗法共同使用都可能对 HD 尤其有效。

一种使用合成锌指核酸酶（ZFNs）抑制 m*HTT* 等位基因转录的疗法（ZFNs 相关讨论请参阅第 11 章——基因治疗）被用于小鼠神经细胞系和携带含 45 个 CAG 重复序列突变等位基因的 HD 患者细胞系，结果显示 ZFN 治疗能显著抑制 m*HTT* 表达，同时对其他含 CAG 重复序列的基因表达没有影响。

随后，ZFN 构建系统被注射入携带 115—160 个 CAG 序列重复的转基因小鼠脑的一侧纹状体区域进行测试，对照载体则被注入小鼠脑的另一侧。两周后，对脑的每侧组织分别检测正常等位基因和突变等位基因的 mRNA 水平。总体而言，与对照侧相比，注射 ZFN 的一侧 m*HTT* mRNA 降

the brain. Overall, on the ZFN-injected side, there was a 40 to 60 percent reduction of m*HTT* mRNA compared with the control side, with no effect on expression of the normal allele on either side. Mice injected on both sides of the brain with the ZFN construct showed no differences in behavioral tests compared with normal mice. This was an important proof of principle that ZFN repression of m*HTT* transcription reduces m*HTT* mRNA and protein levels and may be an effective therapy for HD.

低40%—60%，而正常等位基因表达在任一侧均未受影响。在脑两侧均注射ZFN构建体系的小鼠在行为测试方面与正常小鼠没有差异。这是ZFN抑制m*HTT*转录、降低m*HTT* mRNA和蛋白质水平的重要证据，并且表明ZFN抑制m*HTT*转录可能是一种有效的HD治疗方法。

Instead of inhibiting transcription, gene-silencing techniques can be used to intervene in gene expression after transcription but before translation takes place. Two widely used methods of gene silencing use antisense oligonucleotides (ASOs) and RNA interference (RNAi). ASOs are short single-stranded DNAs (8 to 50 nucleotides long) that bind to target mRNAs by complementary base pairing (Figure 12.3). The mRNA strand of the resulting DNA-RNA hybrid is degraded by RNase H, a cytoplasmic enzyme. The intact ASO is released and can bind other copies of the target mRNA, marking them for degradation by RNase H.

除了转录抑制，基因沉默技术还可用于在转录后、翻译发生前干预基因表达。两种广泛使用的基因沉默方法是使用反义寡核苷酸（ASOs）和RNA干扰（RNAi）。ASOs是短的单链DNAs（长8—50个核苷酸），通过互补碱基配对与靶mRNAs相结合（图12.3）。产生的DNA-RNA杂交双链中的mRNA链被胞质中的核糖核酸酶H降解，完整的ASO则被释放并可继续与靶mRNA的其他拷贝相结合，对靶mRNA进行标记使之被核糖核酸酶H降解。

A ground-breaking study by Lu and Yang transfused an ASO complementary to a human m*HTT* allele into the cerebrospinal fluid of transgenic mice carrying a full-length human m*HTT* allele with 97 CAG repeats. m*HTT* mRNA and mutant HTT protein levels were selectively reduced for up to 12 weeks, with a rebound to near normal levels by 16 weeks. With two weeks of continuous infusion at the age of six months, treated mice still showed

Lu和Yang的一项开创性研究是将与人源m*HTT*等位基因互补的ASO序列输注至含携带97个CAG重复序列的全长人源m*HTT*等位基因的转基因小鼠脑脊液中。m*HTT* mRNA水平和突变型HTT蛋白水平发生选择性地降低长达12周，并在16周时反弹至接近正常水平。对6月龄小鼠连续输注2周，小鼠可表现出运动协调性和行为改善，直至输注后9个月，且输注后5个月m*HTT* mRNA水平和突变型HTT蛋白水平

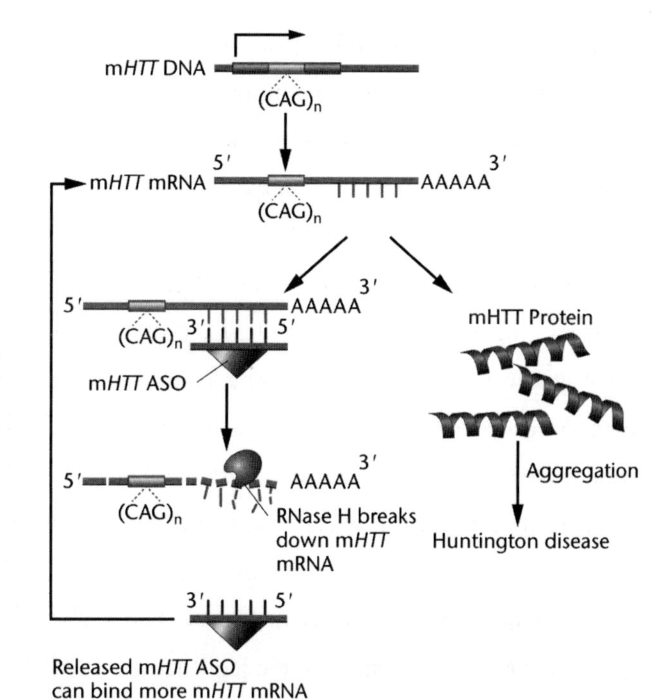

FIGURE 12.3 An ASO constructed to bind to m*HTT* mRNA forms a DNA-RNA hybrid, which then attracts RNase to degrade the m*HTT* mRNA, inhibiting translation of the *HTT* protein and halting disease progression. The released ASO is free to bind other m*HTT* mRNAs and continue the cycle of degradation.

图 12.3 合成的等位基因特异寡核苷酸（ASO）与 m*HTT* mRNA 形成 DNA-RNA 杂交双链，从而吸引核糖核酸酶降解 m*HTT* mRNA，抑制 HTT 蛋白翻译并延缓疾病进程。释放的 ASO 将继续与 m*HTT* mRNA 结合，循环往复地进行降解。

improved motor coordination and behavior nine months after infusion and five months after m*HTT* mRNA and protein levels rebounded to pretreatment levels. Similar ASO studies in nonhuman primates showed a 25 to 68 percent reduction in levels of *HTT* mRNA in brain regions involved in HD, with no adverse effects.

会反弹至治疗前水平。在非人类灵长类动物中开展的类似 ASO 研究表明与 HD 相关的脑区域中 *HTT* mRNA 水平会降低 25%—68%，并且没有不良影响。

Human Phase I clinical trials on ASO therapy for HD began in 2015 in Canada and Europe. The drug is being infused into the cerebrospinal fluid, which surrounds and bathes the brain. This trial is designed to evaluate the safety of ASO infusion. If this and subsequent clinical trials are successful, the first treatment to directly target the cause of HD will soon be available for the thousands of people affected with this devastating disease.

HD ASO 疗法的人类 I 期临床试验于 2015 年在加拿大和欧洲启动。药物被注入包裹并浸润大脑的脑脊液中。该试验旨在评估 ASO 输注的安全性。如果这项试验和随后的临床试验获得成功，那么首个直接靶向 HD 病因的治疗方法将很快面向成千上万罹患这种毁灭性疾病的患者。

Genome Editing in Huntington Disease

Over the last decade, there has been rapid

亨廷顿病的基因编辑

在过去 10 年中，基因组编辑方法已经

development of methods for genome editing. While details of genome editing techniques differ, conceptually they all work in a similar way. A nuclease is guided to cut a specific DNA sequence, which then allows for the replacement or deletion of all or part of a given gene. Because of its specificity and ease of use, the CRISPR-Cas9 system (see Chapter 11—Gene Therapy for a discussion of genome editing) is the most widely used method. The enzyme Cas9 can be directed to cut DNA at specific sequences at nearly any location in the genome by means of a single-stranded guide RNA (sgRNA). The guide RNA is complementary to the sequence to be cut and directs cutting by the Cas9 nuclease. Since HD is a dominant disorder, in theory, this disorder can be treated by using this technology to edit and silence the mutant allele.

获得了长足发展。尽管基因组编辑技术的细节存在不同，但是它们的工作原理都很相似。核酸酶在引导下切割特定的DNA序列，然后允许给定基因的全部或部分进行替换或删除。由于其特异性和使用便利性，CRISPR- Cas9系统（基因组编辑相关讨论请参阅第11章——基因治疗）已经成为最被广泛使用的方法。在单链指导RNA（sgRNA）的引导下，Cas9酶几乎可以在基因组中任何位置的特定序列处切割DNA。指导RNA与待切割的靶向序列互补，并由Cas9酶指导切割。由于HD是一种显性疾病，因此理论上讲，该疾病可通过使用该技术对突变等位基因进行编辑和沉默。

Using CRISPR-Cas9 technology and human cells carrying a mutant and a normal *HTT* allele, Jong-Min Lee and colleagues edited the disease-associated mutant *HTT* allele, removing a 44-kb DNA fragment, completely inactivating the mutant allele, while leaving the normal allele intact. This experiment shows that inactivation of a disease gene can be individually tailored to edit a mutant allele carried by an affected individual and, in theory, can be used to inactivate disease alleles of any gene by editing.

Jong-Min Lee及其同事使用CRISPR-Cas9技术以及携带*HTT*突变等位基因和*HTT*正常等位基因的人类细胞对与疾病相关的突变体*HTT*等位基因进行了编辑，移除了一段长44 kb的DNA片段，使突变体等位基因完全失活，同时保留了完整的正常等位基因。该试验表明，疾病基因失活可以通过编辑患者携带的突变型等位基因量身定制，并且理论上可以通过编辑来失活任何基因的疾病等位基因。

A research team led by Xiao-Jiang Li used CRISPR genome editing to reverse symptoms of HD in transgenic mice carrying exon 1 of a mutant human *HTT* allele containing 140 CAG repeats. They injected one vector carrying Cas9 into the striatum along with another vector that

由Xiao-Jiang Li领导的研究小组使用CRISPR基因组编辑逆转了携带含140个CAG重复序列的人源突变*HTT*等位基因外显子1的转基因小鼠的HD症状。他们将携带Cas9的载体与携带sgRNA的载体同时注射到纹状体中，sgRNA会引导Cas9靶向

carried an sgRNA to direct Cas9 to target the mutant *HTT* allele to edit out the expanded CAG repeat region. Three weeks after injection, analysis of striatal cells showed that production of the mutant human HTT protein was suppressed and the number of aggregated protein clumps was reduced.

突变 *HTT* 等位基因并编辑删除扩增的 CAG 重复区域。在注射 3 周后，纹状体细胞的分析结果显示，突变型人源 HTT 蛋白的合成受到抑制，蛋白聚集体数量减少。

In subsequent experiments, sgRNAs and Cas9 were injected into 9-month-old mice. Over the next three months, these mice showed improvements in motor skills including balance, mobility, and muscle coordination. In addition, increases in motor skills were correlated with the amount of aggregated protein cleared from striatal cells.

在随后的实验中，sgRNAs 和 Cas9 被注入 9 月龄小鼠体内。在接下来的 3 个月中，这些小鼠的运动技能，包括平衡能力、运动能力和肌肉协调性均得到改善。此外，运动技能的提升与纹状体细胞中蛋白聚集体的数量减少有关。

These results are encouraging, but before CRISPR-Cas9 genome editing can be used in humans, further work in model systems and the elimination of potential safety problems are needed. However, genome-editing technology offers the possibility that a cure for HD and many other genetic disorders is not far off.

这些结果十分令人鼓舞，但是在将 CRISPR-Cas9 基因组编辑技术用于人类之前，模式系统和消除潜在安全问题方面依然还有进一步的工作尚需开展。但是，基因组编辑技术为在不久的将来治愈 HD 和许多其他遗传疾病提供了可能性。

Looking back, the focus on researching the cause of HD began with a foundation established by a family with a history of HD. Efforts rapidly expanded into a large-scale international program that pioneered the use of genetics and molecular genetics to identify an expanded stretch of polyglutamines in the huntingtin protein as the cause of this disorder. Along the way, researchers developed an integrative strategy that combined old and new methods including pedigree analysis, RFLP markers, somatic cell genetics, Southern blots, new cloning vectors, predictive genetic testing, population genetics,

回顾过去，聚焦 HD 致病原因的研究起始于一个具有 HD 遗传家族史的家族建立的基金会。这一努力迅速扩展成为一项大规模的国际项目，该项目率先使用遗传学和分子遗传学鉴定出亨廷顿蛋白质中存在的扩增多聚谷氨酰胺是造成这种疾病的原因。在该过程中，研究人员开发了整合策略，将旧方法和新方法联合使用，包括系谱分析、RFLP 标记、体细胞遗传学、DNA 印迹法、新型克隆载体、预测性遗传检测、群体遗传学以及目前正在遗传学中普遍使用的其他方法。然而，尽管 30 多年的工作已经取得了巨大进步，但是目前仍

and other methods that are now universally used in genetic research. Nevertheless, in spite of the progress made in more than 30 years of work, there are still no therapies that can halt, reverse, or prevent the onset and progression of this devastating neurogenetic disorder. Looking forward, recent results from the combined use of human cells, animal models, genome editing, and clinical trials suggest that this last barrier may fall in the near future. If this is the case, the approach used in HD research will stand as a paradigm for understanding the structure and function of the brain and the development of therapeutic methods for a range of genetic disorders.

然缺乏任何治疗方法可以阻止、逆转或预防这种毁灭性的神经遗传性疾病的发作和发展。展望未来，联合使用人类细胞、动物模型、基因组编辑和临床试验的最新研究结果表明这一最后的障碍在不久的将来即可被攻克。如果是这样，HD 的研究历程将成为了解脑结构和功能、开发遗传疾病治疗方法的范例。

13 DNA Forensics

13 DNA 法医学

Forensic science (or forensics) uses technological and scientific approaches to answer questions about the facts of criminal or civil cases. Prior to 1986, forensic scientists had a limited array of tools with which to link evidence to specific individuals or suspects. These included some reliable methods such as blood typing and fingerprint analysis, but also many unreliable methods such as bite mark comparisons and hair microscopy.

法医学利用技术手段和科学方法回答与刑事案件或民事案件事实相关的问题。在 1986 年之前，法医科学家只有十分有限的技术手段将证据与特定个体或嫌犯相联系。这些技术包括一些可靠方法，如血型测定和指纹分析，但是也包含许多可靠性不高的方法，如咬痕对比和毛发显微检查。

Since the first forensic use of **DNA profiling** in 1986, **DNA forensics** (also called **forensic DNA fingerprinting** or **DNA typing**) has become an important method for police to identify sources of biological materials. DNA profiles can now be obtained from saliva left on cigarette butts or postage stamps, pet hairs found at crime scenes, or bloodspots the size of pinheads. Even biological samples that are degraded by fire or time are yielding DNA profiles that help the legal system determine identity, innocence, or guilt. Investigators now scan large databases of stored DNA profiles in order to match profiles generated from crime scene evidence. DNA profiling has proven the innocence of people who were convicted of serious crimes and even sentenced to death.

自 1986 年 **DNA 分析**首次用于法医鉴定以来，**DNA 法医学**（也被称为**法医 DNA 指纹分析**或 **DNA 分型**）已经成为一种帮助警察鉴定生物材料来源的重要方法。目前，DNA 图谱可以从香烟烟嘴、邮票上留下的唾液、犯罪现场发现的宠物毛发或针头大小的血液斑点中获得。即使生物样本由于火或时间的原因发生降解，我们依然能从这些样本中获得 DNA 图谱，从而有助于法律系统确定身份，判定无罪或有罪。现在，调查人员可以通过快速扫描大型数据库中已保存的 DNA 图谱文件，与犯罪现场证据中获得的 DNA 图谱进行比对。DNA 分析已经成功地用于证明被判重罪甚至被判死刑的人员无罪。法医科学家已利用 DNA 分析对巨大灾难，如 2004 年亚洲海啸和 2001 年纽约“9 · 11”恐怖袭击事件中的罹难者

Forensic scientists have used DNA profiling to identify victims of mass disasters such as the Asian Tsunami of 2004 and the September 11, 2001, terrorist attacks in New York. They have also used forensic DNA analysis to identify endangered species and animals trafficked in the illegal wildlife trade.

进行身份确定。他们还用法医 DNA 分析鉴定濒危物种和非法野生生物交易中所涉及的动物。

The applications of DNA profiling extend beyond forensic investigations. These include paternity and family relationship testing, identification of plant materials, verification of military casualties, and evolutionary studies.

DNA 分析的应用不仅限于法医调查，还可用于亲子关系和家庭关系鉴定、识别植物材料、核实军事伤亡和开展进化研究。

In this chapter, we will explore how DNA profiling works and how the results of profiles are interpreted. We will learn about DNA databases, the potential problems associated with DNA profiling, and the future of this powerful technology.

在本章中，我们将探索如何进行 DNA 分析以及如何诠释 DNA 指纹图谱结果。我们将了解 DNA 数据库、DNA 分析相关的潜在问题及这种功能强大技术的未来。

13.1 DNA Profiling Methods

13.1 DNA 分析方法

VNTR-Based DNA Fingerprinting

The era of DNA-based human identification began in 1985, with Dr. Alec Jeffreys's publication on DNA loci known as **minisatellites**, or **variable number tandem repeats** (**VNTRs**). VNTRs are located in noncoding regions of the genome and are made up of DNA sequences of between 15 and 100 bp long, with each unit repeated a number of times. The number of repeats found at each VNTR locus varies from person to person, and hence VNTRs can be from 1 to 20 kb in length, depending on the person. For example, the VNTR

基于 VNTR 的 DNA 指纹分析

基于 DNA 的人类识别时代始于 1985 年，亚历克 · 杰弗里斯博士发表了 DNA 位点**微卫星**或**可变数目串联重复**（**VNTRs**）相关的文章。VNTRs 位于基因组的非编码区，由 15—100 bp 长度的 DNA 序列单元组成，并且每个单元重复多次。每个 VNTR 基因座中的重复次数在个体间存在差异，因此 VNTRs 长度也因人而异，可从 1 kb 至 20 kb。比如：VNTR 序列

5'-GACTGCCTGCTAAGAT
GACTGCCTGCTAAGAT
GACTGCCTGCTAAGAT-3'

is composed of three tandem repeats of a 16-nucleotide sequence (highlighted in bold).

VNTRs are useful for DNA profiling because there are as many as 30 different possible alleles (repeat lengths) at any VNTR in a population. This creates a large number of possible genotypes. For example, if one examined four different VNTR loci within a population, and each locus had 20 possible alleles, there would be approximately 2 billion possible genotypes in this four-locus profile.

To create a VNTR profile (also known as a DNA fingerprint), scientists extract DNA from a tissue sample and digest it with a restriction enzyme that cleaves on either side of the VNTR repeat region (Figure 13.1). The digested DNA is separated by gel electrophoresis and subjected to Southern blot analysis. Briefly, separated DNA is transferred from the gel to a membrane and hybridized with a radioactive probe that recognizes DNA sequences within the VNTR region. After exposing the membrane to X-ray film, the pattern of bands is measured, with larger VNTR repeat alleles remaining near the top of the gel and smaller VNTRs, which migrate more rapidly through the gel, being closer to the bottom. The pattern of bands is the same for a given individual, no matter what tissue is used as the source of the DNA. If enough VNTRs are analyzed, each person's DNA profile will be unique (except, of course, for identical twins) because of the huge number of possible VNTRs

5'-GACTGCCTGCTAAGAT
GACTGCCTGCTAAGAT
GACTGCCTGCTAAGAT-3'

含一段 16 个核苷酸序列（加粗标示）的 3 次重复。

VNTRs 在 DNA 分析中具有应用价值，因为每个种群中任何一个 VNTR 基因座都存在多达 30 种不同的等位基因（重复序列的长度），从而产生了大量可能的基因型。比如，如果研究一个种群内 4 种不同的 VNTR 基因座，并且每个基因座含 20 种可能的等位基因，那么该 4 基因座图谱将存在大约 20 亿种可能的基因型。

为了绘制一张 VNTR 图谱（或 DNA 指纹图谱），科学家要先从组织样本中提取 DNA，然后利用限制性内切酶在 VNTR 重复序列两端进行酶切（图 13.1）。酶切后的 DNA 片段再利用凝胶电泳进行分离，最后使用 DNA 印迹法进行分析。简言之，分离后的 DNA 片段从电泳凝胶转移至膜上，再与可以特异性识别 VNTR 区域中 DNA 序列的放射性探针进行杂交。杂交后的膜暴露于 X 射线胶片中，DNA 片段分离的电泳区带模式即可被显示，分子量较大的 VNTR 重复等位基因位于距离电泳起点较近的位置，而分子量较小的 VNTR 重复等位基因由于在凝胶中迁移速度更快而位于距离电泳起点较远的位置。该电泳区带模式对于某个个体来说始终不变，不论 DNA 样本来自生物体的哪种组织。因为存在大量可能的 VNTRs 和等位基因，所以如果分析足够数量的 VNTRs，那么每个个体的 DNA 图谱都将是独一无二的（当然除了同卵双生的双胞胎）。事实上，科学家一般只需分析 5—

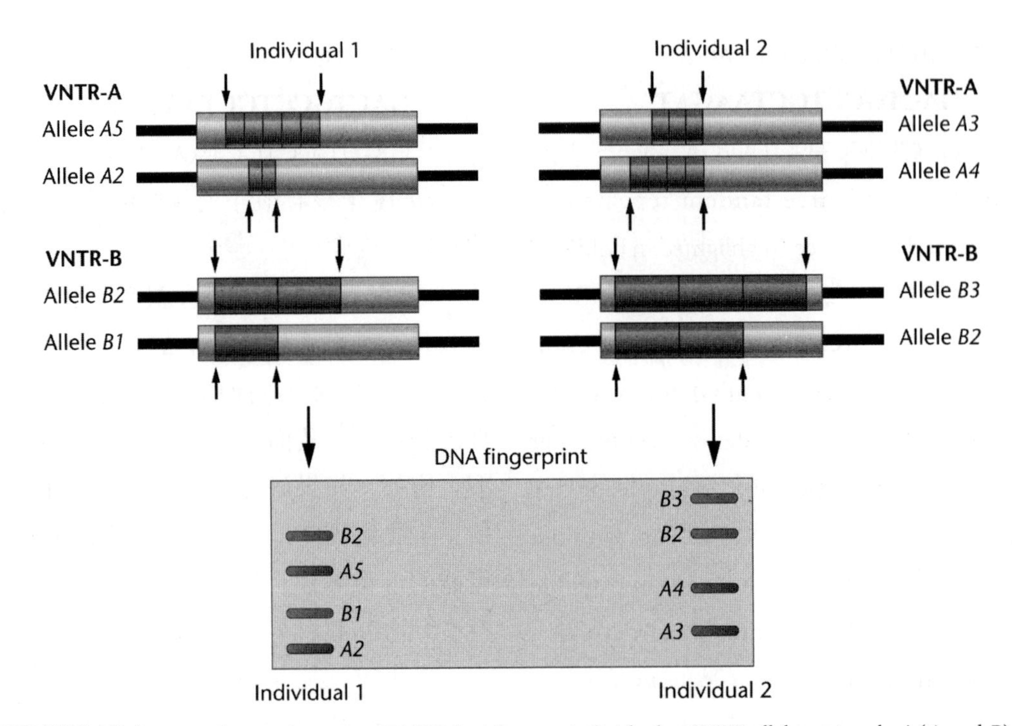

FIGURE 13.1 DNA fingerprint at two VNTR loci for two individuals. VNTR alleles at two loci (*A* and *B*) are shown for two different individuals. Arrows mark restriction-enzyme cutting sites that flank the VNTRs. Restriction-enzyme digestion produces a series of fragments that can be separated by gel electrophoresis and detected as bands on a Southern blot (bottom). The number of repeats at each locus is variable, so the overall pattern of bands is distinct for each individual. The DNA fingerprint profile shows that these individuals share one allele (*B2*).

图 13.1 对两个个体的两个 VNTR 位点进行 DNA 指纹图谱分析。两个个体中位点 *A* 和位点 *B* 的 VNTR 等位基因如图所示。箭头用于标记位于 VNTR 两侧的限制性内切酶酶切位点。限制性内切酶通过酶切会产生一系列片段，这些片段可利用电泳进行分离，再通过 DNA 印迹法进行检测，从而显现条带（图的底部）。每个位点内的重复次数是可变的，因此每个个体条带的整体图案彼此不同。DNA 指纹图谱显示这两个个体存在一个相同位点（*B2*）。

and alleles. In practice, scientists analyze about five or six loci to create a DNA profile.

A significant limitation of VNTR profiling is that it requires a relatively large sample of DNA (10,000 cells or about 50 μg of DNA) — more than is usually found at a typical crime scene. In addition, the DNA must be relatively intact (nondegraded). As a result, VNTR profiling has been used most frequently when large tissue samples are available— such as in paternity testing. Although VNTR profiling is still used in some cases, it has mostly been replaced by more sensitive methods, as described next.

6 个 VNTRs 基因座即可创建一张 DNA 图谱。

VNTR 分析受到的明显限制在于它需要相对大量的 DNA 样本（10 000 个细胞或者大约 50 μg DNA），该样本需求量通常超过典型犯罪现场可能提供的样本量。此外，DNA 必须相对完整，即未被降解。因此，VNTR 分析多用于能提供大量组织样本的情形，比如亲子鉴定。尽管 VNTR 分析仍用于一些案件的调查，但是大多已被将要介绍的灵敏度更高的方法替代。

Autosomal STR DNA Profiling

The development of the **polymerase chain reaction** (**PCR**) revolutionized DNA profiling. PCR is an *in vitro* method that uses specific primers and a heat-tolerant DNA polymerase to amplify specific regions of DNA. Within a few hours, this method can generate a million fold increase in the quantity of DNA within a specific sequence region. Using PCR-amplified DNA samples, scientists are able to generate DNA profiles from trace samples (e.g., the bulb of single hairs or a few cells from a bloodstain) and from samples that are old or degraded (such as a bone found in a field or an ancient Egyptian mummy).

常染色体 STR DNA 分析

聚合酶链式反应（**PCR**）的发展为DNA 分析带来了革新。PCR 是一种体外方法，它利用特定引物和耐热的 DNA 聚合酶来扩增 DNA 的特定区域。该方法在几小时内便可使特定区域的 DNA 片段在数量上增加百万倍。利用 PCR 扩增 DNA 样本，科学家能够从微量样本（例如单根毛发的毛囊或者血迹中的少量细胞）、陈旧或已降解样本（例如野外发现的骨骼或者古埃及木乃伊）中获得 DNA 图谱。

The majority of human forensic DNA profiling is now done by amplifying and analyzing regions of the genome known as **microsatellites**, or **short tandem repeats** (**STRs**). STRs are similar to VNTRs, but the repeated motif is shorter—between two and nine base pairs, repeated from 7 to 40 times. For example, one locus known as D8S1179 is made up of the four base-pair sequence TCTA, repeated 7 to 20 times, depending on the allele. There are 19 possible alleles of the locus that are found within a population. Although hundreds of STR loci are present in the human genome, only a subset is used for DNA profiling. At the present time, the FBI and other U.S. law enforcement agencies use 20 STR loci as a core set for forensic analysis. Most European countries now use 12 STR loci as a core set.

现如今大多数人类法医 DNA 分析均通过扩增和分析基因组中的**微卫星**或**短串联重复序列**（**STRs**）来进行。STRs 与 VNTRs 类似，但发生重复的序列更短，介于 2—9 个碱基对，重复次数介于 7—40 次。例如，D8S1179 基因座由 4 碱基对序列“TCTA”组成，根据等位基因不同重复 7—20 次不等，该基因座在种群中存在 19 种可能的等位基因。尽管人类基因组中存在上百种 STR 基因座，但是其中只有一部分可被用于 DNA 分析。目前，美国联邦调查局和其他美国法律执行机构会使用 20 种 STR 基因座作为法医分析的一套核心组合。大多数欧洲国家则使用 12 种 STR 基因座作为一套核心组合。

Several commercially available kits are

一些商品化试剂盒可用于 STR 基因座

used for forensic DNA analysis of STR loci. The methods vary slightly, but generally involve the following steps. Each primer set is tagged by one of four fluorescent dyes— represented as blue, green, yellow, or red. Each primer set is designed to amplify DNA fragments, the sizes of which vary depending on the number of repeats within the region amplified. For example, the primer sets that amplify the TH01, vWA, D21S11, D7S820, D5S818, TPOX, and DYS391 STR loci are all labeled with a fluorescent tag indicated as yellow. The sizes of the amplified DNA fragments allow scientists to differentiate between the yellow-labeled products. For example, the amplified products from the D21S11 locus range from about 200 to 260 bp in length, whereas those from the TPOX locus range from about 375 to 425 bp, and so on.

的法医 DNA 分析。分析方法存在细微差别，但总体上均包括以下步骤。每组引物分别用四种荧光染料进行标记，例如用蓝、绿、黄、红标识。每组引物均用于扩增 DNA 片段，片段大小随被扩增的 DNA 区域内串联重复次数的不同而不同。例如，用于扩增 TH01、vWA、D21S11、D7S820、D5S818、TPOX 和 DYS391 等 STR 基因座的引物组均用黄色荧光染料标记，扩增获得的 DNA 片段大小使科学家能够区分黄色标识的各种产物。比如，D21S11 基因座扩增产物长度在 200—260 bp 之间，而 TPOX 基因座扩增产物长度在 375—425 bp 之间，以此类推。

After amplification, the DNA sample will contain a small amount of the original template DNA sample and a large amount of fluorescently labeled amplification products (Figure 13.2). The sizes of the amplified fragments are measured by **capillary electrophoresis**. This method uses thin glass tubes that are filled with a polyacrylamide gel material similar to that used in slab gel electrophoresis. The amplified DNA sample is loaded onto the top of the capillary tube, and an electric current is passed through the tube. The negatively charged DNA fragments migrate through the gel toward the positive electrode, according to their sizes. Short fragments move more quickly through the gel, and larger ones more slowly. At the bottom of the tube, a laser detects each fluorescent fragment as it migrates

在 PCR 扩增后，DNA 样本将含少量原始模板 DNA 样本和大量荧光标记的扩增产物（图 13.2）。扩增的 DNA 片段大小通过**毛细管电泳**进行测定。该方法使用管内充满类似平板电泳中所用的聚丙烯酰胺凝胶材料的细玻璃管。扩增的 DNA 样品上样至毛细管的起始端部，电流则通过整个毛细管。带有负电荷的 DNA 片段在凝胶中向正极泳动，并且泳动速度根据片段大小不同各有差异。短 DNA 片段在凝胶中泳动速度较快，而长 DNA 片段则泳动速度较慢。在毛细管末端，由激光检测电泳经过毛细管的每种荧光标记 DNA 片段，然后通过软件分析计算 DNA 产物的片段大小及数量，这些结果以层析峰的形式呈现于图中。一般而言，自动化系统能在一小时内同时分析几十个样品。

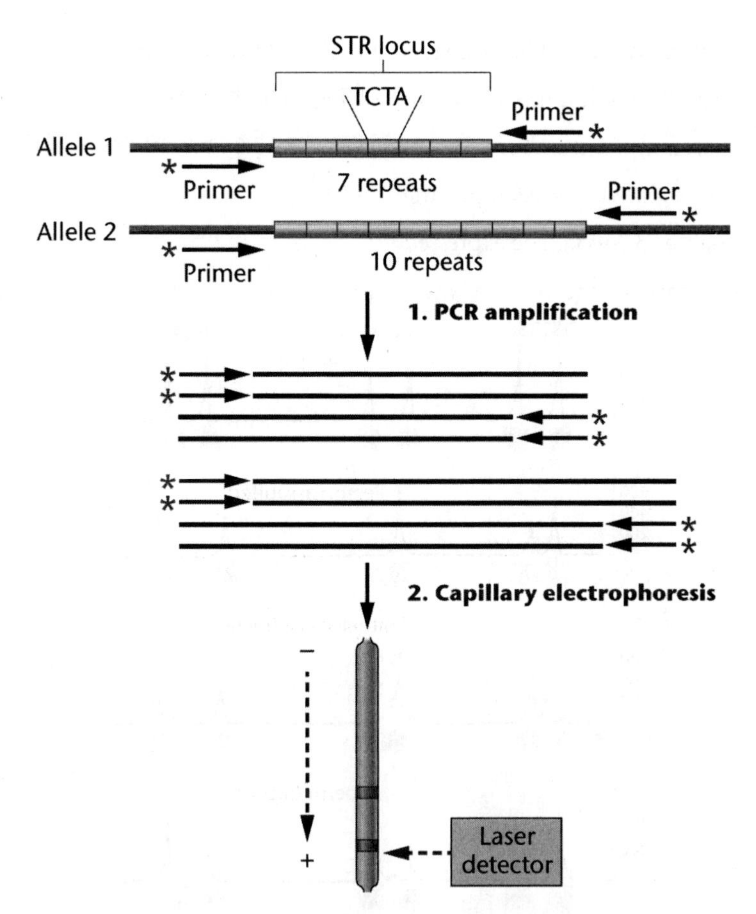

FIGURE 13.2 Steps in the PCR amplification and analysis of one STR locus (D8S1179). In this example, the person is heterozygous at the D8S1179 locus: One allele has 7 repeats and one has 10 repeats. Primers are specific for sequences flanking the STR locus and are labeled with a colored fluorescent dye. The double-stranded DNA is denatured, the primers are annealed, and each allele is amplified by PCR in the presence of all four dNTPs and Taq DNA polymerase. After amplification, the labeled products are separated according to size by capillary electrophoresis, followed by fluorescence detection.

图 13.2 PCR 扩增和分析 STR 位点 (D8S1179) 的步骤。在本例中，个体的 D8S1179 位点杂合：一个等位基因含 7 个重复，另一个等位基因含 10 个重复。STR 位点两侧序列特异的引物用彩色荧光染料进行标记。双链 DNA 变性，与引物复性，每个等位基因在 4 种 dNTP 和 Taq DNA 聚合酶的存在下进行 PCR 扩增。在扩增后，被标记的产物通过毛细管电泳依产物大小进行分离，随后使用荧光进行检测。

through the tube. The data are analyzed by software that calculates both the sizes of the fragments and their quantities, and these are represented as peaks on a graph. Typically, automated systems analyze dozens of samples at a time, and the analysis takes less than an hour.

After DNA profiling, the profile can be directly compared to a profile from another person, from crime scene evidence, or from

在 DNA 分析后，图谱可直接与来自另一个人的图谱或来自犯罪现场物证的图谱，抑或来自 DNA 图谱数据库中存储的档案

other profiles stored in DNA profile databases (Figure 13.3). The STR profile genotype of an individual is expressed as the number of times the STR sequence is repeated. For example, the profiles shown in Figure 13.3 would be expressed as shown in Table 13.1.

信息进行比对（图 13.3）。个体 STR 图谱的基因型表现为 STR 序列的重复次数。例如，图 13.3 所示的 DNA 图谱可以表达为表 13.1。

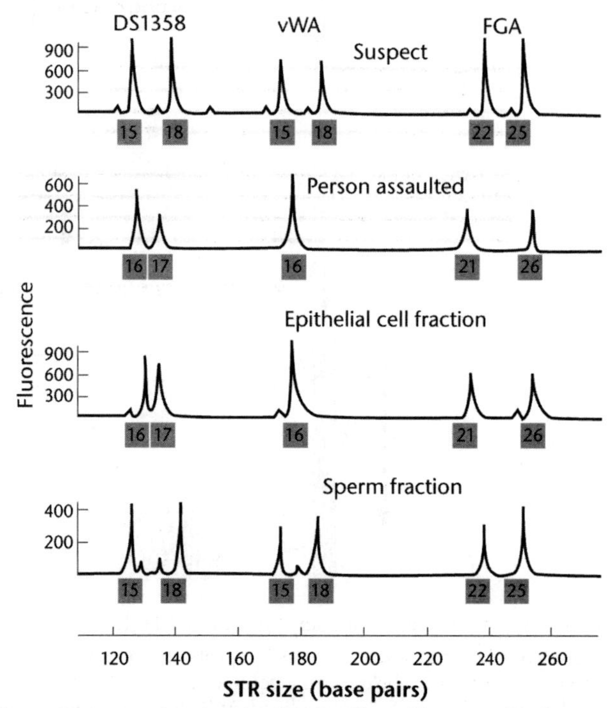

FIGURE 13.3 An electropherogram showing the STR profiles of four samples from a rape case. Three STR loci were examined from samples taken from a suspect (male), the person who was sexually assaulted (female), and two fractions from a vaginal swab taken from the female. The *x*-axis shows the DNA size ladder, and the *y*-axis indicates relative fluorescence intensity. The number below each allele indicates the number of repeats in each allele, as measured against the DNA size ladder. Notice that the STR profile of the sperm sample taken from the female matches that of the suspect.

图 13.3 来自一例强奸案的 4 个现场物证 STR 图谱的电泳图。在嫌疑人（男性）样本、被性侵个体（女性）样本、从女性个体采集的 2 个阴道拭子样本中分别检测 3 个 STR 位点。*x* 轴显示 DNA 分子大小梯度，*y* 轴显示相对荧光强度。每个等位基因下的数字表明依据 DNA 分子大小梯度的检测结果得到的该等位基因所含重复次数。注意从女性体内采集的精子样本与嫌疑人样本相吻合。

TABLE 13.1 STR Profile Genotypes from the Four Profiles Shown in Figure 13.3

表 13.1 图 13.3 所示四个图谱得出的 STR 图谱基因型

STR Locus	Profile Genotype from			
	Suspect	Person Assaulted	Epithelial Cells	Sperm Fraction
DS1358	15, 18	16, 17	16, 17	15, 18
vWA	15, 18	16, 16	16, 16	15, 18
FGA	22, 25	21, 26	21, 26	22, 25

Scientists interpret STR profiles using statistics, probability, and population genetics, and these methods will be discussed in the section Interpreting DNA Profiles.

科学家使用统计学、概率、群体遗传学来诠释 STR 图谱，这些方法将在 13.2 节“DNA 图谱诠释”中进行介绍。

Y-Chromosome STR Profiling

In many forensic applications, it is important to differentiate the DNA profiles of two or more people in a mixed sample. For example, vaginal swabs from rape cases usually contain a mixture of female somatic cells and male sperm cells. In addition, some crime samples may contain evidence material from a number of male suspects. In these types of cases, STR profiling of Y-chromosome DNA is useful. There are more than 200 STR loci on the Y chromosome that are useful for DNA profiling; however, fewer than 20 of these are used routinely for forensic analysis. PCR amplification of Y-chromosome STRs uses specific primers that do not amplify DNA on the X chromosome.

Y 染色体 STR 分析

在很多法医学应用中，从混合样品中区分两个或多个个体的 DNA 指纹十分重要。例如，强奸案中的阴道拭子通常是包含女性个体体细胞、男性精子细胞的混合样本。此外，一些犯罪样本可能含有源于多位男性嫌疑人的证据材料。在这类案件中，Y 染色体 DNA 的 STR 分析就有了用武之地。Y 染色体上存在 200 种以上的 STR 基因座可用于 DNA 分析，然而通常可用于法医分析的不足 20 种。Y 染色体 STRs 区 PCR 扩增所用的特异性引物不会扩增 X 染色体上的 DNA 片段。

One limitation of Y-chromosome DNA profiling is that it cannot differentiate between the DNA from fathers and sons or from male siblings. This is because the Y chromosome is directly inherited from the father to his sons, as a single unit. The Y chromosome does not undergo recombination, meaning that less genetic variability exists on the Y chromosome than on autosomal chromosomes. Therefore, all patrilineal relatives share the same Y-chromosome profile. Even two apparently unrelated males may share the same Y profile, if they also share a distant male ancestor.

Y 染色体 DNA 分析的局限性在于其不能区分父与子或者男性亲缘个体间的 DNA。这是因为 Y 染色体作为一个独立单元由父亲直接遗传给儿子。Y 染色体不参与重组，这也意味着 Y 染色体的遗传多样性比常染色体少。因此，所有父系亲属共享同样的 Y 染色体图谱。即使两个明显无亲缘关系的男性，如果他们拥有共同的、遥远的男性祖先，那么他们的 Y 染色体图谱也可能完全一样。

Although these features of Y-chromosome

尽管 Y 染色体图谱的这些特征在一些

profiles present limitations for some forensic applications, they are useful for identifying missing persons when a male relative's DNA is available for comparison. They also allow researchers to trace paternal lineages in genetic genealogy studies.

法医学应用中存在局限性，但是当存在男性亲属样本可比对时，它们对于确定失踪人口身份还是十分有用的，并且也可帮助研究人员在遗传系谱研究中追踪父系血统。

Mitochondrial DNA Profiling

Another important addition to DNA profiling methods is **mitochondrial DNA (mtDNA)** analysis. Between 200 and 1700 mitochondria are present in each human somatic cell. Each mitochondrion contains one or more 16-kb circular DNA chromosomes. Mitochondria are passed from the human egg cell to the zygote during fertilization; however, as sperm cells contribute few if any mitochondria to the zygote, they do not contribute these organelles to the next generation. Therefore, all cells in an individual contain multiple copies of specific mitochondrial variants derived from the mother. Like Y-chromosome DNA, mtDNA undergoes little if any recombination and is inherited as a single unit.

线粒体 DNA 分析

DNA 分析技术的一个重要补充是**线粒体 DNA（mtDNA）**分析。每个人类体细胞中都存在 200—1700 个线粒体。每个线粒体中含有一条或多条长达 16 kb 的环形 DNA 染色体。线粒体在受精过程中由卵细胞遗传至合子细胞，然而，精细胞对于合子几乎不贡献线粒体，所以它们对于后代的这种细胞器也没有任何贡献。因此，个体中所有细胞所含特定线粒体变体的多个拷贝均由其母本遗传而来。与 Y 染色体 DNA 类似，mtDNA 几乎不发生重组，并以独立单元进行遗传。

Scientists create mtDNA profiles by amplifying regions of mtDNA that show variability between unrelated individuals and populations. After PCR amplification, the DNA sequence within these regions is determined by automated DNA sequencing. Scientists then compare the sequence with sequences from other individuals or crime samples, to determine whether or not they match.

科学家通过扩增在无关联个体和群体间呈现多样性的 mtDNA 区域来创建 mtDNA 图谱。在 PCR 扩增后，这些区域内的 DNA 序列通过 DNA 自动测序进行确定。然后科学家将这些序列与其他个体或犯罪样本中获得的 DNA 序列进行比对，从而确定它们之间是否匹配。

The fact that mtDNA is present in high copy numbers in cells makes its analysis useful in

mtDNA 在细胞中的高拷贝存在特性使得其在样本量小、样本陈旧或存在降解

cases where samples are small, old, or degraded. mtDNA profiling is particularly useful for identifying victims of mass murders or disasters, such as the Srebrenica massacre of 1995 and the World Trade Center attacks of 2001, where reference samples from relatives are available. The main disadvantage of mtDNA profiling is that it is not possible to differentiate between the mtDNA from maternal relatives or from siblings. Like Y-chromosome profiles, mtDNA profiles may be shared by two apparently unrelated individuals who also share a distant ancestor—in this case a maternal ancestor. Researchers use mtDNA profiles in scientific studies of genealogy, evolution, and human population migrations.

的案件调查中十分有用。mtDNA 分析在大屠杀或大灾难中由亲属提供参照样本进行受害者身份确认时尤其有用，如 1995 年斯雷布雷尼察大屠杀、2001 年纽约世贸中心恐怖袭击。mtDNA 分析的主要不足之处在于其无法区分母系亲属或兄弟姐妹间的 mtDNA。与 Y 染色体图谱一样，两个明显无亲戚关系的个体同样可能具有相同的 mtDNA 图谱，如果他们拥有同一位遥远祖先——在这里是女性祖先。研究人员利用 mtDNA 图谱进行系谱、进化和人类种群迁移方面的科学研究。

Single-Nucleotide Polymorphism Profiling

Single-nucleotide polymorphisms (SNPs) are single-nucleotide differences between two DNA molecules. They may be base-pair changes or small insertions or deletions. SNPs occur randomly throughout the genome, approximately every 500 to 1000 nucleotides. This means that there are potentially millions of loci in the human genome that can be used for profiling. However, as SNPs usually have only two alleles, many SNPs (50 or more) must be used to create a DNA profile that can distinguish between two individuals as efficiently as STRs.

Scientists analyze SNPs by using specific primers to amplify the regions of interest. The amplified DNA regions are then analyzed by a number of different methods such as automated DNA sequencing or hybridization to

单核苷酸多态性分析

单核苷酸多态性（SNPs）是指两个 DNA 分子间存在单个核苷酸差异，它们可能是碱基变化、小的插入或缺失。SNPs 在整个基因组内随机发生，大约每 500—1000 个核苷酸中就含 1 个 SNP 基因座，这就意味着人类基因组中含有数以百万计的潜在 SNPs 基因座可用于图谱分析。然而，因为 SNPs 通常仅有两种等位基因形式，所以必须同时利用多个 SNPs（50 个或以上）才能绘制出与 STRs 图谱同样有效的、可区别两个个体差异的 DNA 图谱。

科学家通过使用特异引物扩增感兴趣的 DNA 区域来分析 SNPs 基因座。扩增的 DNA 片段可以利用各种可区别含单核苷酸差异的 DNA 分子方法进行分析，如 DNA 自动测序、与 DNA 芯片上固定化探针进行

immobilized probes on DNA microarrays that distinguish between DNA molecules with single-nucleotide differences.

杂交等。

Forensic SNP profiling has one major advantage over STR profiling. Because a SNP involves only one nucleotide of a DNA molecule, the theoretical size of DNA required for a PCR reaction is the size of the two primers and one more nucleotide (i.e., about 50 nucleotides). This feature makes SNP analysis suitable for analyzing DNA samples that are severely degraded. Despite this advantage, SNP profiling has not yet become routine in forensic applications. More frequently, researchers use SNP profiling of Y-chromosome and mtDNA loci for lineage and evolution studies.

法医SNP分析与STR分析相比具有一个巨大优势。因为DNA分子中1个SNP基因座仅涉及1个核苷酸，因此PCR反应所需的DNA模板理论大小仅为两条引物的大小再加上一个核苷酸的总和（也就是说，大约50个核苷酸）。这一特点使得SNP分析适用于对已发生严重降解的DNA样本进行分析。尽管具有这一优势，但是SNP分析依然尚未成为法医学应用中的常规方法。在多数情况下，研究人员会利用Y染色体和mtDNA的SNP分析进行系谱和进化研究。

DNA Phenotyping

An emerging and controversial method, known as DNA phenotyping, is gaining popularity as a new DNA forensics tool. Unlike DNA profiling, which is used to confirm or exclude sample identities, DNA phenotyping uses DNA sequence information to reveal a person's physical features and ancestral origins.

DNA表型预测技术

DNA表型预测技术作为一种新兴的、备受争议的法医取证工具正越来越受到世人关注。DNA分析用于证实或排除样本身份，而DNA表型预测与此不同，它利用DNA序列信息来揭示个体的物理特征及其祖先起源。

Currently, DNA phenotyping methods can predict a person's eye, hair, and skin colors based on their DNA SNP patterns. For example, scientists have found six SNPs in six genes that are related to blue and brown eye color. Using statistical models based on these six SNPs, it is possible to predict with 95 percent accuracy whether a person has brown or blue eyes. Using 22 SNPs associated with 11 genes, it is possible to predict with 90 percent accuracy whether a

当前，DNA表型预测可以基于个体DNA的SNP模式预测他们的眼睛、头发和皮肤颜色。例如，科学家已经在6种与蓝色、棕色眼睛颜色相关基因中找到6个SNPs基因座。基于这6个SNPs基因座建立统计学模型，可以预测个体眼睛是棕色还是蓝色，并且准确率可达95%。基于与头发颜色相关的11种基因中的22个SNPs基因座，可以预测个体的头发是不是黑色，准确率达90%；或者预测个体的头发是红色还是棕色，

person has black hair and 80 percent accuracy whether a person has red or brown hair. Skin color predictions involve 36 SNPs associated with 15 genes, with prediction accuracies similar to those for hair colors. Both biological sex and geographic ancestry can also be accurately determined from a person's DNA sequence.

准确率可达 80%。肤色预测则涉及 15 种基因中的 36 个 SNPs 基因座，准确率与头发颜色预测的准确率相当。此外，从个体的 DNA 序列中还可准确预测个体的生物学性别及地理学祖先。

Some researchers and private companies have taken DNA phenotyping well beyond prediction of these features. Their algorithms claim to predict three-dimensional facial structures which allow them to compile full-color photographic representations of a person's face, based only on their DNA sample.

一些研究人员和私人企业已将 DNA 表型预测技术应用于远超过这些特征的预测。他们的算法声称可以预测三维立体面部结构，即仅基于个体 DNA 样本来编译个体面部全彩照片图像。

At the present time, DNA phenotyping has not been validated sufficiently to be presented in court. However, police are using the method to help identify unknown missing persons and to provide leads in cold cases.

当前，DNA 表型预测技术在法庭上尚不具有充足的有效性。然而，警方正在使用这种方法帮助确定未知失踪人口并为一些悬案提供信息。

13.2 Interpreting DNA Profiles

13.2 DNA 图谱诠释

After a DNA profile is generated, its significance must be determined. In a typical forensic investigation, a profile derived from a suspect is compared to a profile from an evidence sample or to profiles already present in DNA databases. If the suspect's profile does not match that of the evidence profile or database entries, investigators can conclude that the suspect is not the source of the sample(s) that generated the other profile(s). However, if the suspect's profile matches the evidence profile or a database entry, the interpretation becomes more complicated. In this case, one could conclude that the two

DNA 图谱构建成功后，它的内在含义必须被一一解读。在一项典型的法医调查中，嫌疑人的图谱要与取证样本的图谱或 DNA 数据库中的已存图谱进行对比。如果嫌疑人的图谱与取证样本的图谱或数据库中的已存图谱信息不匹配，那么调查人员可以推断该嫌疑人并不是已获取证样本的图谱来源。然而，如果嫌疑人的图谱与取证样本的图谱或数据库中某已存图谱的信息相匹配，那么该图谱的诠释将变得更加复杂。在这种情况下，调查人员可以推断这两个图谱或者源于同一个体；或者来自两个不同但恰巧拥有同样 DNA 图谱的个

profiles either came from the same person—or they came from two different people who share the same DNA profile by chance. To determine the significance of any DNA profile match, it is necessary to estimate the probability that the two profiles are a random match.

体。为了确定任何一个 DNA 图谱匹配的显著性，人们有必要对两个图谱发生随机匹配的概率进行评估。

The *profile probability* or *random match probability* method gives a numerical probability that a person chosen at random from a population would share the same DNA profile as the evidence or suspect profiles. The following example demonstrates how to arrive at a profile probability.

图谱概率或随机匹配概率法用数值表示群体中随机选取的个体与取证所得图谱或嫌疑人图谱完全匹配的概率。以下例子将展示如何得到图谱概率。

The first locus examined in this DNA profile (D5S818) has two alleles: 11 and 13. Population studies show that the 11 allele of this locus appears at a frequency of 0.361 in this population and the 13 allele appears at a frequency of 0.141. In population genetics, the frequencies of two different alleles at a locus are given the designation p and q, following the Hardy-Weinberg law. We assume that the person having this DNA profile received the 11 and 13 alleles at random from each parent. Therefore, the probability that this person received allele 11 from the mother and allele 13 from the father is expressed as $p \times q = pq$. In addition, the probability that the person received allele 11 from the father and allele 13 from the mother is also pq. Hence, the total probability that this person would have the 11, 13 genotype at this locus, by chance, is $2pq$, $2pq$ is 0.102 or approximately 10 percent. It is obvious from this sample that using a DNA profile of only one locus would not be very informative, as about 10

DNA 图谱（D5S818）中考察的第一个基因座存在两种等位基因：11 和 13。群体研究表明，等位基因 11 在群体中的基因频率为 0.361，而等位基因 13 在群体中的基因频率为 0.141。在群体遗传学中，依据哈迪 - 温伯格定律，一个基因座上两种不同等位基因的基因频率分别用字母 p 和 q 来表示。我们假定拥有该 DNA 图谱的个体从其每个亲本中随机获得 11 和 13 两种等位基因。因此，该个体从其母亲遗传得到等位基因 11，同时从其父亲遗传得到等位基因 13 的概率可以表示为 $p \times q=pq$。此外，该个体从其父亲遗传得到等位基因 11，同时从其母亲遗传得到等位基因 13 的概率也是 pq。所以，在这个基因座上，该个体基因型为 11、13 的总概率为 $2pq$，而 $2pq$ 约为 0.102 或者大约为 10%。从这个例子中很显然可以看出仅含一个基因座的 DNA 图谱所能提供的有效信息十分有限，因为群体中大约 10% 的个体在 D5S818 基因座上同为 11、13 基因型。

percent of the population would also have the D5S818 11, 13 genotype.

The discrimination power of the DNA profile increases when we add more loci to the analysis. The next locus of this person's DNA profile (TPOX) has two identical alleles— the 11 allele. Allele 11 appears at a frequency of 0.243 in this population. The probability of inheriting the 11 allele from each parent is $p \times p = p^2$, the genotype frequency at this locus would be 0.059, which is about 6 percent of the population. If this DNA profile contained only the first two loci, we could calculate how frequently a person chosen at random from this population would have this genotype, by multiplying the two genotype probabilities together. This would be $0.102 \times 0.059 \approx 0.006$. This analysis would mean that about 6 persons in 1000 (or 1 person in 166) would have this genotype. The method of multiplying all frequencies of genotypes at each locus is known as the *product rule*. It is the most frequently used method of DNA profile interpretation and is widely accepted in U.S. courts.

随着分析的基因座数目增多，DNA 图谱的区分能力也随之提升。该个体 DNA 图谱的另一个基因座（TPOX）含两种相同的等位基因——11。并且等位基因 11 在群体中的基因频率为 0.243。所以从每个亲本遗传得到等位基因 11 的概率是 $p \times p = p^2$，那么该基因座的这种基因型频率将约为 0.059，即在群体中所占比例约为 6%。如果这张 DNA 图谱仅含前两种基因座，那么我们可以计算出群体中随机选取的个体具有这种基因型的概率是多少，通过将两种基因型概率相乘即可，即 $0.102 \times 0.059 \approx 0.006$。该分析结果表明在 1000 个个体中大约有 6 人（或 1/166）将含有该基因型。这种将每种基因座上的基因型频率进行相乘的方法被称为乘法定律。这是 DNA 图谱分析中最为频繁被使用的方法，并且在美国法庭中也被广泛接受。

By multiplying all the genotype probabilities at the five loci, we arrive at the genotype frequency for this DNA profile: 9×10^{-7}. This means that approximately 9 people in every 10 million (or about 1 person in a million), chosen at random from this population, would share this 5-locus DNA profile.

通过将五种基因座上的所有基因型概率相乘，我们将得到 DNA 图谱中的基因型频率：9×10^{-7}。这意味着如果在群体中进行随机选取，那么每 1000 万人中大约有 9 人（或 1×10^{-6} 的概率）含有相同的 5 基因座 DNA 图谱。

The Uniqueness of DNA Profiles

As we increase the number of loci analyzed in a DNA profile, we obtain smaller probabilities

DNA 图谱的唯一性

随着 DNA 图谱中被分析的基因座数目增加，随机匹配的概率将逐渐变小。理

of a random match. Theoretically, if a sufficient number of loci were analyzed, we could be *almost* certain that the DNA profile was unique. At the present time, law enforcement agencies in North America use a core set of 20 STR loci to generate DNA profiles. Using this 20-loci set, the probability that two people selected at random would have identical genotypes at these loci would be approximately 1×10^{-28}.

论上讲，如果分析的基因座数目足够多，我们几乎可以确定所得的 DNA 图谱是唯一的。当前，北美执法机构使用含 20 种 STR 基因座的成套核心组合来绘制 DNA 图谱。利用该 20 基因座组合，任何随机选取的两名个体在这些基因座上均拥有同样基因型的概率大约仅有 1×10^{-28}。

Although this would suggest that most DNA profiles generated by analysis of the 20 core STR loci would be unique on the planet, several situations can alter this interpretation. For example, identical twins share the same DNA, and their DNA profiles will be identical. Identical twins occur at a frequency of about 1 in 250 births. In addition, siblings can share one allele at any DNA locus in about 50 percent of cases and can share both alleles at a locus in about 25 percent of cases. Parents and children also share alleles, but are less likely than siblings to share both alleles at a locus. When DNA profiles come from two people who are closely related, the profile probabilities must be adjusted to take this into account. The allele frequencies and calculations that we describe here are based on assumptions that the population is large and has little relatedness or inbreeding. If a DNA profile is analyzed from a person in a small interrelated group, allele frequency and calculations may not apply.

尽管这表明大多数通过分析 20 种 STR 基因座创建的 DNA 图谱在地球上将是独一无二的，但是在有些情形下也不尽如此。例如，同卵双生的双胞胎拥有同样的 DNA，因此他们的 DNA 图谱也完全一样。同卵双生双胞胎的发生概率大约是 1/250。此外，兄弟姐妹之间任意 DNA 基因座上一个等位基因相同的概率大约是 50%，两个等位基因均相同的概率大约是 25%。父母与子女之间也拥有同样的等位基因，但是同一基因座上两个等位基因均相同的可能性比兄弟姐妹之间小。因此，当 DNA 图谱源自两个亲缘关系很近的个体时，图谱概率必须考虑这些因素并加以修正。这里所介绍的等位基因频率及其计算都是基于所研究群体很大并且个体间无相关性、无杂交的假设。如果 DNA 图谱来源于小且彼此间存在关联的群体中的个体，那么这里的基因频率及计算方法则不适用。

DNA Profile Databases

Many countries throughout the world maintain national DNA profile databases. The

DNA 图谱数据库

全球很多国家已经建立了国家 DNA 图谱数据库，其中第一个数据库是 1995 年在

first of these databases was established in the United Kingdom in 1995 and now contains more than 6 million profiles. In the United States, both state and federal governments have DNA profile databases. The entire system of databases along with tools to analyze the data is known as the **Combined DNA Index System (CODIS)** and is maintained by the FBI. As of August 2018, there were more than 17 million DNA profiles stored within the CODIS. These include the *convicted offender database*, which contains DNA profiles from individuals convicted of certain crimes, and the *forensic database*, which contains profiles generated from crime scene evidence. In addition, some states have DNA profile databases containing profiles from suspects and from unidentified human remains and missing persons.

英国建立的，目前已包含 600 万份以上的 DNA 图谱文件。在美国，州政府和联邦政府同时拥有 DNA 图谱数据库。整个数据库系统及其数据分析工具被称为 **DNA 联合索引系统 (CODIS)**，由美国联邦调查局维护。截至 2018 年 8 月，CODIS 中已存有 1700 万份以上的 DNA 图谱文件，其中包括已定罪罪犯数据库（其所存 DNA 图谱文件来自案件中已被宣判有罪的个体）和法医数据库（其所存 DNA 图谱文件来自犯罪现场的取证样本）。此外，一些州还建立了包含来自犯罪嫌疑人、身份未被确认的人体遗骸和失踪人口的 DNA 图谱文件数据库。

DNA profile databases have proven their value in many different situations. As of August 2018, use of CODIS databases had resulted in more than 400,000 profile matches that assisted criminal investigations and missing persons searches. Despite the value of DNA profile databases, they remain a concern for many people who question the privacy and civil liberties of individuals versus the needs of the state.

DNA 图谱数据库已经在很多情形下证明了它们的存在价值。截至 2018 年 8 月，CODIS 数据库中已有 400 000 份以上的图谱文件被用于辅助犯罪调查和失踪人口搜寻工作。尽管 DNA 图谱数据库具有应用价值，但是同时也引发了人们的担忧，即如何权衡个人隐私、公民自由与国家需求之间的关系。

13.3 Technical and Ethical Issues Surrounding DNA Profiling

13.3 围绕 DNA 分析的技术与伦理问题

Although DNA profiling is sensitive, accurate, and powerful, it is important to be aware of its limitations. One limitation is

尽管 DNA 分析灵敏、准确、功能强大，但是了解其局限性也十分重要。其中一个不足就是在大多数刑事案件中或者无 DNA

that most criminal cases have either no DNA evidence for analysis or DNA evidence that would not be informative to the case. In some cases, potentially valuable DNA evidence exists but remains unprocessed and backlogged. Another serious problem is that of human error. There are cases in which innocent people have been convicted of violent crimes based on DNA samples that had been inadvertently switched during processing. DNA evidence samples from crime scenes are often mixtures derived from any number of people present at the crime scene or even from people who were not present, but whose biological material (such as hair or saliva) was indirectly introduced to the site. Crime scene evidence is often degraded, yielding partial DNA profiles that are difficult to interpret.

证据可供分析；或者即使有 DNA 证据，但对于案件侦破却无法提供有用线索。在一些案件中，虽然存在有潜在价值的 DNA 证据，但是这些证据未得到充分利用并被积压。另一个严重问题是人为错误。在一些暴力犯罪案件中，由于 DNA 样本在处理过程中被不经意地替换，从而使得基于这些 DNA 样本的无辜人员被判有罪。犯罪现场的 DNA 取证样本通常是案发现场多名人员的混合样本，或者甚至会混入不在案发现场的人员样本，这些人员的生物材料（如头发或者唾液）经一些间接方式而被引入现场。犯罪现场的证据通常存在降解情况，由此得到的部分 DNA 图谱也很难进行分析解读。

One of the most disturbing problems with DNA profiling is its potential for deliberate tampering. DNA profile technologies are so sensitive that profiles can be generated from only a few cells—or even from fragments of synthetic DNA. There have been cases in which criminals have introduced biological material to crime scenes, in an attempt to affect forensic DNA profiles. It is also possible to manufacture artificial DNA fragments that match STR loci of a person's DNA profile. In 2010, a research paper reported methods for synthesizing DNA of a known STR profile, mixing the DNA with body fluids, and depositing the sample on crime scene items. When subjected to routine forensic analysis, these artificial samples generated perfect STR profiles. In the future, it may be necessary to develop methods to detect the presence of

DNA 分析中最严重的干扰问题之一是其易于被蓄意篡改。DNA 图谱技术十分灵敏，以至于图谱文件只需少量细胞，甚至一些合成 DNA 片段即可被创建。曾经在一些案件中，罪犯将一些生物材料引入案发现场，蓄意影响法医 DNA 图谱分析。还可能通过制造人造 DNA 片段来匹配一个人 DNA 图谱中的 STR 基因座。2010 年，一篇科研论文报道了根据已知 STR 图谱合成 DNA 片段，将 DNA 片段混入体液，再将该样本沉积至犯罪现场物品中的方法。如果依循常规的法医分析，那么这些人造样本将创建十分完美的 STR 图谱。未来，人们有必要开发对犯罪现场样本中的合成 DNA 或克隆 DNA 进行检测的方法。专家指出这些检测方法可基于天然 DNA 中含诸如甲基化等表观遗传学标记的事实来实现。

synthetic or cloned DNA in crime scene samples. It has been suggested that such detections could be done, based on the fact that natural DNA contains epigenetic markers such as methylation.

Many of the ethical questions related to DNA profiling involve the collection and storage of biological samples and DNA profiles. Such questions deal with who should have their DNA profiles stored on a database and whether police should be able to collect DNA samples without a suspect's knowledge or consent.

与 DNA 分析相关的很多伦理问题涉及生物样本和 DNA 图谱的采集和保存。这些问题包括哪些人应该将其 DNA 图谱保存至数据库中，警方是否有权在嫌疑人未知或未经嫌疑人允许的情况下采集 DNA 样本。

Another ethical question involves the use of DNA profiles that partially match those of a suspect. There are cases in which investigators search for partial matches between the suspect's DNA profile and other profiles in a DNA database. On the assumption that the two profiles arise from two genetically related individuals, law enforcement agencies pursue relatives of the person whose profile is stored in the DNA database. Testing in these cases is known as *familial DNA testing*. Should such searches be considered scientifically valid or even ethical?

另一个伦理相关问题则涉及与嫌疑人 DNA 图谱部分匹配的那些 DNA 图谱文件的使用。曾经在一些案件中，调查人员利用 DNA 数据库搜寻与嫌疑人 DNA 图谱文件部分匹配的相关文件。基于遗传上相关联的两个个体的 DNA 图谱存在部分匹配这一假设，执法机构追捕了 DNA 数据库中存有图谱文件的个体的亲属。这些案件所采用的方法即为家族性 DNA 检测技术。那么这种搜寻罪犯的方法在科学上是否有效或者是否合乎道德呢？

As described previously, it is now possible to predict some facial features and geographic ancestries of persons based on information in their DNA sample—a method known as DNA phenotyping. Should this type of information be used to identify or convict a suspect?

正如前面所述，现在基于人类 DNA 样本所提供的信息可以预测人类的面部特征及其地理学祖先，即 DNA 表型预测技术。那么这些信息是否可以用来确定或者判定嫌疑人有罪呢？

As DNA profiling becomes more sophisticated and prevalent, we should carefully consider both the technical and ethical questions that surround this powerful new technology.

随着 DNA 分析技术越来越成熟，使用越来越广泛，我们应该认真仔细地考虑这种新型有力的技术相关的技术问题和伦理问题。

14 Genetically Modified Foods

14 转基因食品

Throughout the ages, humans have used selective breeding techniques to create plants and animals with desirable genetic traits. By selecting organisms with naturally occurring or mutagen-induced variations and breeding them to establish the phenotype, we have evolved varieties that now feed our growing populations and support our complex civilizations.

古往今来，人类已经利用选育技术创造了很多具有理想遗传性状的植物物种和动物物种。人类通过自然选择生物，利用诱变剂诱导获得变异，并将其进行培育以建立表型，现在这些品种不仅养活着地球上日益增长的人口，而且支撑着我们复杂的人类文明。

Although we have had tremendous success shuffling genes through selective breeding, the process is a slow one. When recombinant DNA technologies emerged in the 1970s and 1980s, scientists realized that they could modify agriculturally significant organisms in a more precise and rapid way by identifying and cloning genes that confer desirable traits, then introducing these genes into organisms. Genetic engineering of animals and plants promised an exciting new phase in scientific agriculture, with increased productivity, reduced pesticide use, and enhanced flavor and nutrition.

尽管我们通过选育在改组基因方面取得了巨大成功，但是整个进程依然是缓慢的。当重组 DNA 技术在 20 世纪 70 年代和 80 年代应运而生时，科学家意识到他们能够通过一种更加精确、更加快速的方式来改良农业上的重要生物，该方法即鉴定并克隆理想性状的决定基因，然后将这些基因导入生物体体内。动物和植物基因工程为人类进入令人振奋的农产品产量提高、农药使用减少、食品风味和营养价值提升的科学农业新时期提供了保证。

Beginning in the 1990s, scientists created a large number of **genetically modified (GM) food** varieties. The first one, approved for sale in 1994, was the Flavr Savr tomato— a tomato that stayed firm and ripe longer than non-GM tomatoes. Soon afterward, other GM foods were

自 20 世纪 90 年代，科学家创造了大量**转基因（GM）食品**品种。首个转基因食品是1994年被批准上市的“Flavr Savr”番茄——一种不发生软化、摘下后保鲜时间比非转基因番茄更长的产品。此后不久，其他一些转基因食品也陆续被开发了出来：抗病

developed: papaya and zucchini with resistance to virus infection, canola containing the tropical oil laurate, corn and cotton plants with resistance to insects, and soybeans and sugar beets with tolerance to agricultural herbicides.

毒侵染的木瓜和西葫芦、含热带油月桂酸酯的油菜、抗虫玉米和棉花、抗除草剂大豆和甜菜。

Although many people see great potential for GM foods—to help address malnutrition in a world with a growing human population and climate change— others question the technology, oppose GM food development, and sometimes resort to violence to stop the introduction of GM varieties.

尽管很多人看到了转基因食品的巨大潜力——在全球人口不断增长并伴随气候变化的情况下有助于解决全世界范围内的营养不良问题——但是同时也有很多人质疑这项技术，反对转基因食品的发展，有时甚至采用暴力来阻止转基因品种的引进。

Some countries have outright bans on all GM foods, whereas others embrace the technologies. Opponents cite safety and environmental concerns, whereas some scientists and commercial interests extol the almost limitless virtues of GM foods. The topic of GM food attracts hyperbole and exaggerated rhetoric, information, and misinformation— on both sides of the debate.

一些国家全面禁止转基因食品，而另一些国家则十分欢迎相关技术。反对者认为存在食品安全和环境方面的隐患，但是一些科学家和商业利益团体则称赞转基因食品拥有几乎无限的优良品质。转基因食品话题在反对方和支持方均引来了夸大其词、华而不实的论述、信息以及虚假信息。

So, what are the truths about GM foods? In this chapter, we will introduce the science behind GM foods and examine the promises and problems of the new technologies. We will look at some of the controversies and present information to help us evaluate the complex questions that surround this topic.

因此，转基因食品的真相到底是什么？在本章中，我们将介绍转基因食品背后的科学知识，探讨这些新技术的应用前景及所存在的问题。我们还将了解一些争论和当前已有的信息，从而帮助我们评估该话题相关的一些复杂问题。

14.1 What Are GM Foods?

14.1 什么是转基因食品？

GM foods are derived from **genetically modified organisms** (**GMOs**), specifically plants and animals of agricultural importance. GMOs are defined as organisms whose genomes

转基因食品由**遗传修饰生物体**（**GMOs**）衍生而来，特别是农业上具有重要意义的植物和动物。GMOs 被定义为基因组以非自然方式被改变的生物体。尽管

have been altered in ways that do not occur naturally. Although the definition of GMOs sometimes includes organisms that have been genetically modified by selective breeding, the most commonly used definition refers to organisms modified through genetic engineering or recombinant DNA technologies. Genetic engineering allows one or more genes to be cloned and transferred from one organism to another— either between individuals of the same species or between those of unrelated species. It also allows an organism's endogenous genes to be altered in ways that lead to enhanced or reduced expression levels. When genes are transferred between unrelated species, the resulting organism is called **transgenic**. The term **cisgenic** is sometimes used to describe gene transfers within a species. In contrast, the term **biotechnology** is a general term, encompassing a wide range of methods that manipulate organisms or their components—such as isolating enzymes or producing wine, cheese, or yogurt. Genetic modification of plants or animals is one aspect of biotechnology.

GMOs 的定义有时也包括通过遗传选育方式进行遗传修饰而获得的生物，但是最为常用的定义是通过基因工程或重组 DNA 技术改造而获得的生物。基因工程将一种或多种基因进行克隆后，从一种生物转移至另一种生物——既可以是同一物种的不同个体之间，也可以是毫无亲缘关系的不同物种个体之间。基因工程也可以使生物体的内源基因发生变化，从而使其表达水平增强或降低。当基因在毫无亲缘关系的物种间进行转移时，被称为**异源转基因**，而**同源转基因**则被用于描述同一物种间进行的基因转移。与此不同的是，“**生物技术**”是一个通用术语，涵盖了操作生物体及其组分的多种方法——如分离生物酶和生产葡萄酒、奶酪或酸奶。对植物或动物进行遗传修饰只是生物技术的一个方面。

It is estimated that GM crops are grown in approximately 30 countries on 11 percent of the arable land on Earth. The majority of these GM crops (almost 90 percent) are grown in five countries—the United States, Brazil, Argentina, Canada, and India. Of these five, the United States accounts for approximately half of the acreage devoted to GM crops. According to the U.S. Department of Agriculture, 93 percent of soybeans and 90 percent of maize grown in the United States are from GM crops. In the United

据估计，转基因作物（GM 作物）目前在大约 30 个国家进行耕种，并占据全球可耕种面积的 11%。这些 GM 作物中绝大部分（几乎 90%）耕种于以下 5 个国家——美国、巴西、阿根廷、加拿大和印度。其中，美国的 GM 作物耕种面积大约占到全部面积的 50%。根据美国农业部的统计，美国种植的 93% 的大豆和 90% 的玉米为 GM 作物，70% 以上的加工食品中含有源自 GM 作物的原料成分。

States, more than 70 percent of processed foods contain ingredients derived from GM crops.

Approximately 200 different GM crop varieties are approved for use as food or livestock feed in the United States. However, only about two dozen are widely planted. Table 14.1 lists some of the common GM food crops available for planting in the United States. Only one GM food animal, the AquAdvantage salmon, has been approved for consumption.

在美国，大约有200种GM作物品种被批准用于食品或家畜饲料。然而，其中仅有20多种被广泛种植。表14.1列出了美国种植的常见转基因食品农作物。此外，仅有一种具食品用途的转基因食品动物——AquAdvantage转基因三文鱼被批准上市。

TABLE 14.1 Some GM Crops Approved for Food, Feed, or Cultivation in the United States*

表14.1 在美国可用于食品、饲料和种植的部分GM作物

Crop	Number of Varieties	GM Characteristics
Soybeans	19	Tolerance to glyphosate herbicide Tolerance to glufosinate herbicide Reduced saturated fats Enhanced oleic acid Enhanced omega-3 fatty acid
Maize	68	Tolerance to glyphosate herbicide Tolerance to glufosinate herbicide Bt insect resistance Enhanced ethanol production
Cotton	30	Tolerance to glyphosate herbicide Bt insect resistance
Potatoes	28	Bt insect resistance
Canola	23	Tolerance to glyphosate herbicide Tolerance to glufosinate herbicide Enhanced lauric acid
Papaya	4	Resistance to papaya ringspot virus
Sugar beets	3	Tolerance to glyphosate herbicide
Rice	3	Tolerance to glufosinate herbicide
Zucchini squash	2	Resistance to zucchini, watermelon, and cucumber mosaic viruses
Alfalfa	2	Tolerance to glyphosate herbicide
Plum	1	Resistance to plum pox virus

* Information from the International Service for the Acquisition of Agri-Biotech Applications, www.isaaa.org.

Herbicide-Resistant GM Crops

Herbicide-tolerant varieties are the most widely planted of GM crops, making up approximately 70 percent of all GM crops. The majority of these varieties contain a bacterial gene

抗除草剂GM作物

GM作物中种植面积最广的是除草剂耐受品种，大约占据所有GM作物的70%。这些品种中的绝大多数含有一种源自细菌的基因，从而可以提供对广谱性除草剂——

that confers tolerance to the broad-spectrum herbicide **glyphosate**—the active ingredient in commercial herbicides such as Roundup®.

草甘膦的耐受能力。草甘膦是商用除草剂如Roundup®的活性成分。

Farmers who plant glyphosate-tolerant crops can treat their fields with glyphosate, even while the GM crop is growing. This approach is more efficient and economical than mechanical weeding and reduces soil damage caused by repeated tillage. It is suggested that there is less environmental impact when using glyphosate, compared with having to apply higher levels of other, more toxic, herbicides.

种植草甘膦耐受型农作物的农民可以使用草甘膦进行农田除草，即使GM作物正处于生长期。这一方法比机械除草更加经济有效，同时也减少了由重复耕作对土壤造成的损耗。人们认为相比于使用浓度更高、毒性更强的其他类型除草剂，草甘膦对于环境的影响更小。

Insect-Resistant GM Crops

抗虫GM作物

The second most prevalent GM modifications are those that make plants resistant to agricultural pests.

第二类种植面积最广的转基因改造则是让植物获得抵抗农业害虫的能力。

The most widely used GM insect-resistant crops are the **Bt crops**. ***Bacillus thuringiensis*** (Bt) are soil-dwelling bacterial strains that produce crystal (Cry) proteins that are toxic to certain species of insects. These Cry proteins are encoded by the bacterial *cry* genes and form crystal structures during sporulation. The Cry proteins are toxic to Lepidoptera (moths and butterflies), Diptera (mosquitoes and flies), Coleoptera (beetles), and Hymenoptera (wasps and ants). Insects must ingest the bacterial spores or Cry proteins in order for the toxins to act. Within the high pH of the insect gut, the crystals dissolve and are cleaved by insect protease enzymes. The Cry proteins bind to receptors on the gut wall, leading to breakdown of the gut membranes and death of the insect.

抗虫GM作物中应用最为广泛的是**苏云金杆菌（Bt）作物**。Bt是土壤细菌，能够合成对特定种类昆虫有毒害作用的结晶（Cry）蛋白。这些Cry蛋白由细菌的结晶蛋白基因（*cry*）编码，并且在孢子形成过程中形成晶体结构。这些Cry蛋白对于鳞翅目(飞蛾和蝴蝶)、双翅目(蚊子和苍蝇)、鞘翅目（甲虫）和膜翅目（黄蜂和蚂蚁）昆虫等均有毒害作用。昆虫必须摄取细菌孢子或Cry蛋白才能引发毒性。在昆虫肠道内的高pH环境中，蛋白质晶体溶解，并被昆虫体内的蛋白酶降解。Cry蛋白与位于肠道壁上的受体相结合，从而导致肠膜瓦解和昆虫死亡。

Each insect species has specific types of gut

每种昆虫含特定类型的肠道受体并仅

receptors that will match only a few types of Bt Cry toxins. As there are more than 200 different Cry proteins, it is possible to select a Bt strain that will be specific to one pest type.

与少数几种 Bt Cry 毒素相作用。由于存在 200 种以上的 Cry 蛋白，所以有可能从中挑选出一种 Bt 菌株特异性地针对某一种害虫类型。

Bt spores have been used for decades as insecticides in both conventional and organic gardening, usually applied in liquid sprays. Sunlight and soil rapidly break down the Bt insecticides, which have not shown any adverse effects on groundwater, mammals, fish, or birds. Toxicity tests on humans and animals have shown that Bt causes few negative effects.

Bt 孢子在传统农业和有机农业中作为杀虫剂已有几十年的历史，通常情况下采用液态喷洒方式使用。阳光和土壤会快速降解 Bt 杀虫剂，因此尚未发现对地下水、哺乳动物、鱼类或鸟类存在任何不良效应。对于人类和动物的毒性检测试验也显示 Bt 几乎不会引发负面效应。

To create Bt crops, scientists introduce one or more cloned *cry* genes into plant cells using methods described in the next section. The GM crop plants will then manufacture their own Bt Cry proteins, which will kill the target pest species when it eats the plant tissues.

为了创建 Bt 作物，科学家将一种或多种 *cry* 基因利用接下来将要介绍的方法克隆入植物细胞中，然后 GM 作物将自行合成 Bt Cry 蛋白，从而有针对性地杀死以这些作物组织为食的害虫种类。

GM Crops for Direct Consumption

To date, most GM crops have been designed to help farmers increase yields. Also, most GM food crops are not consumed directly by humans, but are used as animal feed or as sources of processed food ingredients such as oils, starches, syrups, and sugars. For example, 98 percent of the U.S. soybean crop is used as livestock feed. The remainder is processed into a variety of food ingredients, such as lecithin, textured soy proteins, soybean oil, and soy flours. However, a few GM foods have been developed for direct consumption. Examples are rice, squash, and papaya.

直接消费的 GM 作物

至今，大多数 GM 作物改造均致力于帮助农民提高作物产量。而且，大多数 GM 食品作物并非直接供人类消费，而是用作动物饲料或加工食品组分的原料，如食用油、淀粉、糖浆和糖。例如，美国 98% 的大豆作物用于牲畜饲料，剩下的则用于加工系列食品添加剂，如卵磷脂、大豆组织化蛋白、大豆油、大豆粉。然而，一些 GM 食品则用于直接消费，如大米、南瓜和木瓜。

14.2 Methods Used to Create GM Plants

Most GM plants have been created using one of two approaches: the **biolistic method** or ***Agrobacterium tumefaciens*-mediated transformation** technology. Both methods target plant cells that are growing *in vitro*. Scientists can generate plant tissue cultures from various types of plant tissues, and these cultured cells will grow either in liquid cultures or on the surface of solid growth media. When grown in the presence of specific nutrients and hormones, these cultured cells will form clumps of cells called calluses, which, when transferred to other types of media, will form roots. When the rooted plantlets are mature, they are transferred to soil medium in greenhouses where they develop into normal plants.

The biolistic method is a physical method of introducing DNA into cells. Particles of heavy metals such as gold are coated with the DNA that will transform the cells; these particles are then fired at high speed into plant cells *in vitro*, using a device called a **gene gun**. Cells that survive the bombardment may take up the DNA-coated particles, and the DNA may migrate into the cell nucleus and integrate into a plant chromosome. Plants that grow from the bombarded cells are then selected for the desired phenotype.

Although biolistic methods are successful for a wide range of plant types, a much improved transformation rate is achieved using *Agrobacterium-mediated technology*. *Agrobacterium tumefaciens* (also called

14.2 构建 GM 植物的方法

绝大多数 GM 植物是通过两种途径进行构建的：**基因枪法**或**土壤根癌农杆菌转化法**。这两种方法均以体外生长的植物细胞为靶标。科学家利用不同类型的植物组织进行植物组织培养，这些培养细胞既可以生长在液体培养基中，也可以生长在固体培养基的基质表面。在培养过程中提供特定的营养物质和激素，这些培养细胞将形成被称为愈伤组织的细胞团，随后将愈伤组织转移至其他类型培养基中形成植物的根。当这些生根的苗成熟后，再将它们转移至温室土壤中进行生长，最终发育成正常植株。

基因枪法是一种将 DNA 导入细胞的物理方法。将待转入细胞内的 DNA 包裹在重金属颗粒如金颗粒的表面，然后使用名为基因枪的设备将这些颗粒在体外高速射入植物细胞中。经基因枪轰击后幸存的细胞将接纳带有 DNA 的颗粒，且 DNA 会进入细胞核内并整合入植物染色体中。由基因枪轰击的细胞发育而成的植株随后将根据所需表型进行筛选。

尽管基因枪法在许多植物种类中均获得了成功，但是土壤根癌农杆菌转化法的转化效率则更高。土壤根癌农杆菌（也被称为放射根瘤菌）是一种土壤微生物，能够感染植物细胞产生肿瘤。这些特性源自

Rhizobium radiobacter) is a soil microbe that can infect plant cells and cause tumors. These characteristics are conferred by a 200-kb tumor-inducing plasmid called a **Ti plasmid**. After infection with *Agrobacterium*, the Ti plasmid integrates a segment of its DNA known as transfer DNA (T-DNA) into random locations within the plant genome (Figure 14.1). To use the Ti plasmid as a transformation vector, scientists remove the T-DNA segment and replace it with cloned DNA of the genes to be introduced into the plant cells.

大小为 200 kb 的肿瘤诱导质粒——**Ti 质粒**。在农杆菌侵染植物细胞后，Ti 质粒会将其所含的一段 DNA 片段——转移 DNA（T-DNA）随机整合入植物基因组内（图 14.1）。为了利用 Ti 质粒作为转化载体，科学家将其所含的 T-DNA 片段移除，并用克隆的需转入植物细胞中的基因 DNA 序列来替代 T-DNA。

In order to have the newly introduced gene expressed in the plant, the gene must be cloned next to an appropriate promoter sequence that will direct transcription in the required plant tissue. For example, the β -carotene pathway genes were cloned next to a promoter that directs transcription of the genes in the rice endosperm. In addition, the transformed gene requires appropriate transcription termination signals and signal sequences that allow insertion of the encoded protein into the correct cell compartment.

为了使新导入植物细胞内的基因能够表达，该基因必须克隆至合适的启动子序列附近，从而由启动子指导其在指定植物组织中进行转录。例如，β - 胡萝卜素代谢途径系列基因应被克隆至可以引导它们在水稻籽粒胚乳中进行基因转录的启动子附近。此外，被转化的基因还需要合适的转录终止信号以及能够引导编码蛋白在细胞内进行正确定位的信号序列。

Selectable Markers

The rates at which T-DNA successfully integrates into the plant genome and becomes appropriately expressed are low. Often, only one cell in 1000 or more will be successfully transformed. Before growing cultured plant cells into mature plants to test their phenotypes, it is important to eliminate the background of nontransformed cells. This can be done using either positive or negative selection techniques.

选择标记

T-DNA 成功整合入植物基因组内并能进行准确表达的概率不高，通常转化成功率在 1/1000 或以上。在培养植物细胞至成熟植株以检测其表型之前，清除未成功转化的细胞背景十分重要，这既可以采用正向选择技术，也可以采用反向选择技术来完成。

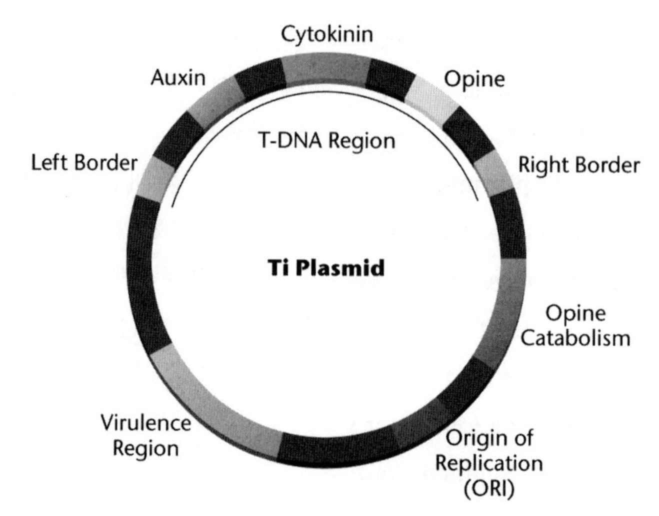

FIGURE 14.1 Structure of the Ti plasmid. The 250-kb Ti plasmid from *Agrobacterium tumefaciens* inserts the T-DNA portion of the plasmid into the host cell's nuclear genome and induces tumors. Genes within the virulence region code for enzymes responsible for transfer of T-DNA into the plant genome. The T-DNA region contains auxin and cytokinin genes that encode hormones responsible for cell growth and tumor formation. The opine genes encode compounds used as energy sources for the bacterium. The T-DNA region of the Ti plasmid is replaced with the gene of interest when the plasmid is used as a transformation vector.

图 14.1 Ti 质粒结构。来自土壤根癌农杆菌的 Ti 质粒（250 kb）将质粒 T-DNA 插入宿主细胞核基因组内诱发肿瘤形成。毒性区的基因负责编码将 T-DNA 转移至植物基因组所需的相关酶类。T-DNA 区含植物生长素基因和细胞分裂素基因，分别编码细胞生长和肿瘤形成相关的激素。冠瘿碱基因编码为细菌提供能量源泉的化合物。当 Ti 质粒被用作转化载体时，质粒 T-DNA 区将被目的基因替换。

An example of negative selection involves use of a **marker gene** such as the hygromycin-resistance gene. This gene, together with an appropriate promoter, can be introduced into plant cells along with the gene of interest. The cells are then incubated in culture medium containing hygromycin—an antibiotic that also inhibits the growth of eukaryotic cells. Only cells that express the hygromycin resistance gene will survive. It is then necessary to verify that the resistant cells also express the cotransformed gene. This is often done by techniques such as PCR amplification using gene-specific primers. Plants that express the gene of interest are then tested for other characteristics, including the phenotype conferred by the introduced gene of interest.

一个反向选择的例子涉及使用潮霉素抗性基因等**标记基因**。潮霉素抗性基因随同合适的启动子与目的基因一同被转化入植物细胞中，随后这些细胞将在含有潮霉素的培养基中进行培养。这种抗生素具有抑制真核生物细胞生长的作用，所以仅有能够表达潮霉素抗性基因的细胞才能够存活。接下来还要继续验证具有抗性的细胞同时也表达了共转化基因，这通常通过利用基因特异性引物的 PCR 扩增等技术来完成。最后，能够表达目的基因的植株还需采用其他特性进行检验，包括被导入的目的基因所决定的表型。

An example of positive selection involves the use of a selectable marker gene such as that encoding **phosphomannose isomerase (PMI)**. This enzyme is common in animals but is not found in most plants. It catalyzes the interconversion of mannose 6-phosphate and fructose 6-phosphate. Plant cells that express the *pmi* gene can survive on synthetic culture medium that contains only mannose as a carbon source. Cells that are cotransformed with the *pmi* gene under control of an appropriate promoter and the gene of interest can be positively selected by growing the plant cells on a mannose-containing medium. Studies have shown that purified PMI protein is easily digested, nonallergenic, and nontoxic in mouse oral toxicity tests. A variation in positive selection involves use of a marker gene whose expression results in a visible phenotype, such as deposition of a colored pigment.

一个正向选择的例子涉及使用编码**磷酸甘露糖异构酶**（PMI）基因等选择标记基因。磷酸甘露糖异构酶在动物中普遍存在，而在大多数植物中尚未被发现，它催化 6- 磷酸甘露糖与 6- 磷酸果糖间的可逆转化。表达 *pmi* 基因的植物细胞能够在仅提供甘露糖作为碳源的合成培养基上存活。位于合适启动子下的 *pmi* 基因与目的基因进行了共转化的细胞能够在含有甘露糖的培养基上被正向选择。小鼠口服毒性试验研究表明纯 PMI 蛋白易被消化、不引起过敏、没有毒性。另一种正向选择方法则是使用能表达可见表型的标记基因，如有色色素的降解。

Genome Editing and GM Foods

The previously described methods are those that have been used to create the majority of GM plants. In the last few years, several new and revolutionary methods of genome modification have entered the field of GM foods. Collectively, these are known as **genome-editing methods**. They include zinc-finger nuclease (ZFN), transcription activator-like effector nuclease (TALEN), and CRISPR-Cas techniques.

基因组编辑与 GM 食品

以上所介绍的方法已被用于绝大多数 GM 植物的构建中。在过去数年里，一些用于基因组修饰的新型、具有革命性的方法也已进入 GM 食品领域，这些方法被称为**基因组编辑法**，包括锌指核酸酶（ZFN）、转录激活物样效应核酸酶（TALEN）和 CRISPR-Cas 技术。

Genome-editing methods have had significant effects on the speed at which scientists can induce genetic changes in plants and animals as well as on the types of changes that

基因组编辑法对科学家诱导植物和动物遗传改变的速度以及可能的遗传改变类型产生了重大影响。基因组编辑使科学家能够进行精确的核苷酸或单基因突变或缺

are possible. Genome editing allows researchers to create precise nucleotide or single-gene mutations or deletions without introducing foreign DNA into the organism. To create genome-edited plants or animals, scientists typically mutate or inactivate only one or two of the organism's endogenous genes.

失，而不需将外源DNA引入生物体内。科学家通常通过突变或失活生物体内仅一个或两个内源性基因来创建基因组编辑植物或动物。

One example of a genome-edited food is a potato developed by the biotechnology company Calyxt, Inc. Using TALEN methods, they inactivated the *vacuolar invertase* gene that encodes an enzyme responsible for degrading sugars in cold-stored potatoes. This gene inactivation resulted in a potato with an increased storage life as well as one that does not produce harmful acrylamides when the potato is fried. The genome-edited potato is currently in field trials.

一个有关基因组编辑食品的例子是由美国农业生物技术公司Calyxt研发的土豆。他们利用TALEN方法失活液泡转化酶基因，该基因编码的酶负责在土豆低温储存过程中降解糖类。该基因的失活可延长土豆储存期，并且使土豆在煎炸时不再产生有害的丙烯酰胺。基因编辑土豆目前正在田间试验阶段。

Because genome-edited organisms contain no transgene material, they have not been considered genetically modified by most regulatory agencies and therefore do not have the same oversight as other GM foods. For example, in March of 2018 the United States Department of Agriculture (USDA) announced that they will not regulate crops "that could otherwise have been developed through traditional breeding techniques as long as they are not plant pests or developed using plant pests." Without the need to go through a lengthy USDA approval process, genome-edited foods may quickly move from development to commercialization. Futhermore, the development of genome-edited crops has become faster, cheaper, and more efficient with CRISPR-Cas technology.

因为基因编辑生物不含转基因成分，因此多数监管机构并不认为它们被遗传改造过，也不认为它们需要类似其他转基因食品的监督机制。例如，2018年3月，美国农业部（USDA）宣布他们将不再监管"原本已通过传统育种技术培育的作物，只要它们不是植物有害生物或利用植物有害生物培育获得的"。无须经过美国农业部漫长的审批程序，基因组编辑食品可能将迅速从开发走向商业化。此外，CRISPR-Cas技术也使基因组编辑作物的开发速度更快、成本更低、效率更高。

Scientists are also using genome-editing technologies to introduce gene alterations into farm animals. For example, the Roslin Institute in Scotland is developing a strain of pigs that is immune to the African swine fever virus. Using ZFN and CRISPR-Cas9 methods, they have introduced small changes to one of the pigs' immune system genes (*RELA*) so that the gene has the same DNA sequence as the *RELA* gene from warthogs which are resistant to the virus. The pigs are now in infection trials.

科学家也会利用基因组编辑技术将基因改变引入家畜中。例如，苏格兰的罗斯林研究所正在研制一种对非洲猪瘟病毒具有免疫力的猪。他们使用 ZFN 和 CRISPR-Cas9 法将一些小的基因改变引入猪的免疫系统基因（*RELA*）中，从而使该基因的 DNA 序列与能够抵抗非洲猪瘟病毒的疣猪 *RELA* 基因相同。这些基因编辑猪目前正处于感染测试阶段。

Another example is that of the double-muscled pig, developed at Seoul National University. Using TALEN methods, researchers introduced mutations into both copies of the myostatin (*MSTN*) gene, inactivating it. Myostatin is responsible for inhibiting the growth of muscle cells. When the gene is inactivated, muscle tissue grows to produce muscle-enhanced animals. Such animals produce higher yields of lean meat.

另一个例子是由首尔大学研发的双肌性状改良猪。研究人员使用 TALEN 法在肌生长抑制蛋白基因（*MSTN*）的两个拷贝中均引入突变，使其失活。肌生长抑制蛋白可以抑制肌肉细胞生长，当该基因发生失活时，肌肉组织将持续生长并产生肌肉发达的动物，这样的动物瘦肉产量更高。

Although genome-edited crops and animals are currently not regulated in the same way as GM foods, some countries are reviewing their guidelines and new regulations may be introduced in the near future.

尽管基因编辑作物和基因编辑动物目前的监管方式与 GM 食品不同，但是有些国家正在审查相关指南，并且在不久的将来可能引入新法规。

14.3 GM Foods Controversies

14.3 GM 食品之争

GM foods may be the most contentious of all products of modern biotechnology. Advocates of GM foods state that the technologies have increased farm productivity, reduced pesticide use, preserved soils, and have the potential to feed growing human

GM 食品可能是所有现代生物技术产品中最饱受争议的。GM 食品的拥护者指出该技术可以提升农场生产力、减少杀虫剂的使用、保护土壤，并且有潜力满足日益增长的人口所需。反对者则称 GM 食品对于人类和环境均不安全，因此他们向监管机

populations. Critics claim that GM foods are unsafe for both humans and the environment; accordingly, they are applying pressure on regulatory agencies to ban or severely limit the extent of GM food use. These campaigns have affected regulators and politicians, resulting in a patchwork of regulations throughout the world. Often the debates surrounding GM foods are highly polarized and emotional, with both sides in the debate exaggerating their points of view and selectively presenting the data. So, what are the truths behind these controversies?

构施压要求禁止或严格限制 GM 食品的使用范围。这些争议影响了监管机构和政治家，并在全球范围内建立了错综复杂的法规。通常，围绕 GM 食品展开的辩论高度两极分化并伴随着情绪化，辩论双方均会夸大他们各自的观点并选择性地展示数据。那么，这些争议背后的真相到底是什么呢？

One point that is important to make as we try to answer this question is that it is not possible to make general statements about all "GM foods." Each GM crop or organism contains different genes from different sources, attached to different expression sequences, accompanied by different marker or selection genes, and inserted into the genome in different ways and in different locations. In addition, the recent proliferation of genome-edited food organisms further complicates the situation. Many advocates and regulatory agencies state that genome-edited foods are not GM foods, as they do not fit the previous definitions of GMOs. GM foods are created for different purposes and are used in ways that are both planned and unplanned. Each construction is unique and therefore needs to be assessed separately.

当尝试回答这一问题时，有一点很重要，那就是对所有“GM 食品”做出统一的概括性陈述是不可能的。每一种 GM 作物或 GM 生物所含外源基因的来源不同、外源基因相邻的表达基因序列不同、伴随外源基因导入的标记基因或选择基因不同、插入基因组的方式及插入位点也不同。此外，近期基因组编辑食品生物的增多使这种情况进一步变得更加复杂。许多支持者和监管机构认为基因组编辑食品不是 GM 食品，因为它们并不符合原先对于 GMOs 的定义。GM 食品的构建目的各有不同，并且在具体使用过程中既有计划内的也有计划外的不同使用方式。每一种 GM 生物的构建都是独特的，因此需要单独进行评估。

We will now examine two of the main GM foods controversies: those involving human health and safety, and environmental effects.

下面，我们将分析 GM 食品相关的两个主要争议：涉及人类健康与安全以及环境效应。

Health and Safety

GM food advocates often state that there

健康与安全

GM 食品的拥护者常说目前尚无证据表

is no evidence that GM foods currently on the market have any adverse health effects, either from the presence of toxins or from potential allergens. These conclusions are based on two observations. First, humans have consumed several types of GM foods for more than 20 years, and no reliable reports of adverse effects have emerged. Second, the vast majority of toxicity tests in animals, which are required by government regulators prior to approval, have shown no negative effects. A few negative studies have been published, but these have been criticized as poorly executed or nonreproducible.

明已上市的GM食品对于健康有不利影响，无论是毒素存在情况还是潜在过敏原。这些结论基于两个观察结果：第一，20多年来，人类已经消费了多种类型的GM食品，但尚未出现不良影响相关的可靠报道；第二，政府监管部门在食品获批之前要求进行的绝大多数动物毒性试验均未显示负面效应。虽然有一些显示负面效应的研究结果已发表，但是这些研究由于操作不当或不可重复而受到了批评。

Critics of GM foods counter the first observation in several ways. First, as described previously, few GM foods are eaten directly by consumers. Instead, most are used as livestock feed, and the remainder form the basis of purified food ingredients. To date, no adverse effects of GM foods in livestock have been detected. In addition, the processing of many food ingredients removes most, if not all, plant proteins and DNA. Hence, ingestion of GM food-derived ingredients may not be a sufficient test for health and safety. Second, GM food critics argue that there have been few human clinical trials to directly examine the health effects of most GM foods. They also say that the toxicity studies that have been completed are performed in animals—primarily rats and mice—and most of these are short-term toxicity studies.

GM食品反对者则从多个角度反驳了上述的第一个观察结果。第一，如前所述，消费者几乎不直接食用GM食品，取而代之的是，大多数GM食品用于牲畜饲料，其余的则用作制备食品配料纯品的基础原料。迄今为止，尚未在家畜中检测出GM食品的不利影响。此外，许多食品配料的加工过程去除了大部分（即使不是全部）植物蛋白质和DNA，因此，GM食品来源成分的摄入可能不足以检验其对健康的影响和食品安全。第二，GM食品反对者还指出，目前几乎还没有人类临床试验直接检验大多数GM食品的健康影响。他们还指出，已经完成的毒性研究是在动物（主要是大鼠和小鼠）中进行的，并且其中大多数是短期毒性研究。

Supporters of GM foods answer these criticisms with several other arguments. The first argument is that short-term toxicity studies in animals are well-established methods for

GM食品支持者使用其他一些论据回应了这些批评。第一个论据是，利用动物进行的短期毒性研究是检测毒素和过敏原的行之有效的方法。任何GM食品在获批之

detecting toxins and allergens. The regulatory processes required prior to approval of any GM food demand data from animal toxicity studies. If any negative effects are detected, approval would not be given. Supporters also note that several dozen long-term toxicity studies have been published that deal with GM crops such as glyphosate-resistant soybeans and Bt corn, and none of these has shown long-term negative effects on test animals. Those few studies reporting negative effects have been shown to have serious design flaws and their conclusions are considered unreliable. GM food advocates note that human clinical trials are not required for any other food derived from other genetic modification methods such as selective breeding. During standard breeding of plants and animals, genomes may be mutagenized with radiation or chemicals to enhance the possibilities of obtaining a desired phenotype. This type of manipulation has the potential to introduce mutations into genes other than the ones that are directly selected. Also, plants and animals naturally exchange and shuffle DNA in ways that cannot be anticipated. These include interspecies DNA transfers, transposon integrations, and chromosome modifications. These events may result in unintended changes to the physiology of organisms—changes that could potentially be as great as those arising in GM foods.

前所需进行的监管程序均要求提供来自动物毒性研究的数据。如果发现任何负面影响，GM 食品将无法获批。支持者还指出，人们已经发表了几十项有关 GM 作物的长期毒性研究，如抗草甘膦大豆和抗虫 Bt 玉米，结果均显示对试验动物不存在长期负面影响。那些少数报道 GM 食品有负面影响的研究均显示出存在严重的实验设计缺陷，因此这些研究结论并不可靠。GM 食品拥护者指出，通过其他遗传修饰方法如选择育种衍生而来的任何其他食品并不需要进行人类临床试验。在动植物的标准育种过程中，可能通过使用射线或化学试剂进行基因组诱变，从而提高获得理想表型的可能性。这种操作可能将突变引入任何基因，而不仅是人类直接选择的基因。而且在自然条件下，动物和植物也以多种无法预期的方式对自身 DNA 进行交换和重组，包括种间 DNA 转移、转座子整合和染色体修饰。这些事件可能导致生物体的生理发生意想不到的变化——这些变化可能与 GM 食品中发生的变化一样大。

Environmental Effects

Critics of GM foods point out that GMOs that are released into the environment have both documented and potential consequences for the

环境效应

GM 食品反对者指出释放到环境中的 GMOs 对于环境既存在已有文字记载的影响，也存在潜在的影响，因此可能间接影

environment—and hence may indirectly affect human health and safety. GM food advocates argue that these potential environmental consequences can be identified and managed. Here, we will describe two different aspects of GM foods as they may affect the natural environment and agriculture.

响人类健康和安全。GM 食品支持者则认为这些潜在的环境影响是可确定并可控的。我们在这里将讲述 GM 食品可能影响自然环境和农业的两个方面。

1. Emerging herbicide and insecticide resistance. Many published studies report that the planting of herbicide-tolerant and insect-resistant GM crops has reduced the quantities of herbicides and insecticides that are broadly applied to agricultural crops. As a result, the effects of GM crops on the environment have been assumed to be positive. However, these positive effects may be transient, as herbicide and insecticide resistance is beginning to emerge.

1. 日益涌现的除草剂抗性和杀虫剂抗性。许多已发表的研究报告指出，耐除草剂和抗虫的 GM 作物的种植已经减少了广泛应用于农作物种植的除草剂、杀虫剂的使用量。因此，人们认为 GM 作物对于环境是有利的。但是，随着除草剂和杀虫剂耐药性的出现，这些积极效应可能只是短暂的。

Since glyphosate-tolerant crops were introduced in the mid-1990s, more than 24 glyphosate-resistant weed species have appeared in the United States. Resistant weeds have been found in 18 other countries, and in some cases, the presence of these weeds is affecting crop yields. One reason for the rapid rise of resistant weeds is that farmers have abandoned other weed-management practices in favor of using a single broad-spectrum herbicide. This strong selection pressure has brought the rapid evolution of weed species bearing gene variants that confer herbicide resistance. Scientists point out that herbicide resistance is not limited to the use of GM crops. Weed populations will evolve resistance to any herbicide used to control them, and the speed of evolution will be affected by the extent to which the herbicide is used.

自 20 世纪 90 年代中期美国开始种植耐草甘膦作物以来，美国已经出现了 24 种以上的抗草甘膦农田杂草。在其他 18 个国家中，同样也发现了抗性杂草，并且在某些情况下，这些杂草的存在正在影响着农作物的产量。抗性杂草迅速增长的原因之一是农民已经放弃了使用其他杂草管理措施，转而仅使用单一的广谱除草剂。这种强大的选择压力使得带有除草剂抗性基因突变的杂草物种得以迅速进化。科学家指出除草剂抗性并不仅限于 GM 作物的应用，不论使用哪种除草剂控制杂草生长，杂草种群都将通过进化获得除草剂抗性，并且进化速度受除草剂使用程度的影响。

Since 1996, more than eight different species of insect pests have evolved some level of resistance to Bt insecticidal proteins. In order to slow down the development of Bt resistance, several strategies are being followed. The first is to develop varieties of GM crops that express two Bt toxins simultaneously. Several of these varieties are already on the market and are replacing varieties that express only one Bt *cry* gene. The second strategy involves the use of "refuges" surrounding fields that grow Bt crops. These refuges contain non-GM crops. Insect pests grow easily within the refuges, which place no evolutionary pressure on the insects for resistance to Bt toxins. The idea is for these nonselected insects to mate with any resistant insects that appear in the Bt crop region of the field. The resulting hybrid offspring will be heterozygous for any resistance gene variant. As long as the resistance gene variant is recessive, the hybrids will be killed by eating the Bt crop. In fields that use refuges and plant GM crops containing two Bt genes, resistance to Bt toxins has been delayed or is absent. As with emerging herbicide resistance, farmers are also encouraged to combine the use of Bt crops with conventional pest control methods.

1996 年以来，已经有 8 种以上不同种类的害虫通过进化获得了一定程度的 Bt 杀虫蛋白抗性。为了减缓 Bt 抗性的发展，人们正在采取一些不同的策略。第一种策略是开发可以同时表达两种 Bt 毒素的 GM 作物品种，这些品种有些已经上市，并且正在替代仅表达一种 Bt *cry* 基因的品种。第二种策略则是在种植 Bt 作物的田地周围建立“避难所”，在这些避难所内种植非 GM 作物。害虫在避难所内容易生存，从而不会对害虫 Bt 毒素抗性能力的形成产生进化压力。这一策略的理论基础在于未被选择的害虫与 Bt 作物种植区域内任何具抗性能力的昆虫交配后，所得的杂交后代对于任何抗性基因变异均是杂合的。只要该抗性基因变异是隐性的，那么杂交后代通过食用 Bt 作物即可被杀死。在使用避难所和种植含两种 Bt 基因的 GM 作物田地中，害虫对于Bt毒素的抗性效应均发生延迟或消失。与日益显现的除草剂抗性一样，农民同样被建议采用联合抗虫的方法，即将 Bt 作物的使用与常规害虫防治方法相结合。

2. The spread of GM crops into non-GM crops. There have been several documented cases of GM crop plants appearing in uncultivated areas in the United States, Canada, Australia, Japan, and Europe. For example, GM sugar beet plants have been found growing in commercial top soils. GM canola plants have been found growing in ditches and along roadways, railway

2. GM 作物向非 GM 作物的传播。在美国、加拿大、澳大利亚、日本和欧洲各国，已出现多例 GM 作物出现在非种植区的报道。例如，在商业表层土壤中发现正在生长中的 GM 甜菜植株；在远离种植区域的沟渠中、道路旁、铁轨沿线和填土中发现正在生长中的 GM 油菜植株。

tracks, and in fill soils, far from the fields in which they were grown.

One of the major concerns about the escape of GM crop plants from cultivation is the possibility of **outcrossing** or **gene flow**—the transfer of transgenes from GM crops into sexually compatible non-GM crops or wild plants, conferring undesired phenotypes to the other plants. Gene flow between GM crops and adjacent non-GM crops is of particular concern for farmers who want to market their crops as "GM-free" or "organic" and for farmers who grow seed for planting.

GM 作物植株从栽培区中逃逸的主要令人担忧的问题之一是**异交**或**基因流**的可能性——外源基因从 GM 作物转移至具有性亲和的非 GM 作物或野生物种中，从而赋予其他植物人类并不希望其具有的表型。GM 作物和与之相邻的非 GM 作物之间的基因流对于那些希望销售"非转基因"或"有机"作物的农户以及以收获种子为目的的农户而言尤为担忧。

Gene flow of GM transgenes has been documented in GM and non-GM canola as well as sugar beets, and in experiments using rice, wheat, and maize.

在 GM 油菜和非 GM 油菜、GM 甜菜和非 GM 甜菜中，以及在使用水稻、小麦和玉米的试验中，均存在 GM 转基因基因流的报道。

It is thought that the presence of glyphosate-resistant transgenes in wild plant populations is not likely to be an environmental risk and would confer no positive fitness benefits to the hybrids. The presence of glyphosate-resistant genes in wild populations would, however, make it more difficult to eradicate the plants. This is illustrated in a case of escaped GM bentgrass in Oregon, where it has been difficult to get rid of the plants because it is no longer possible to use the relatively safe herbicide glyphosate. The potential for environmental damage may be greater if the GM transgenes did confer an advantage—such as insect resistance or tolerance to drought or flooding.

人们认为野生植物种群中存在草甘膦抗性基因不可能给环境带来风险，并且也不会给杂交个体生长带来积极的适应性生长益处。然而，野生种群中草甘膦抗性基因的存在使得根除这些植物更加困难。在俄勒冈州发生的 GM 剪股颖逃逸事件正说明了这一点，因为不可能继续使用相对安全的除草剂草甘膦，所以摆脱这些植物已经变得十分困难。如果 GM 转基因确实赋予了植物一些其他优势，如抗虫能力、耐寒能力或耐涝能力，那么其对于环境造成破坏的潜力可能更大。

In an attempt to limit the spread of transgenes from GM crops to non-GM crops, regulators are considering a requirement to

为了限制转基因从 GM 作物向非 GM 作物的传播，监管机构正在考虑将作物种植相互分开，从而降低花粉在它们之间进

separate the crops so that pollen would be less likely to travel between them. Each crop plant would require different isolation distances to take into account the dynamics of pollen spreading. Several other methods are being considered. For example, one proposal is to make all GM plants sterile using RNAi technology. Another is to introduce the transgenes into chloroplasts. As chloroplasts are inherited maternally, their genomes would not be transferred via pollen. All of these containment methods are in development stages and may take years to reach the market.

行传播的可能性。考虑到花粉扩散的动力学特性，每种农作物所需的隔离距离并不相同。此外，还有一些其他方法可供考虑。例如，一种方案是使用RNAi技术使所有GM植株不育；另一种方案则是将转基因引入叶绿体中，由于叶绿体是母性遗传，因此其基因组不会通过花粉发生转移。所有这些遏制方法均处于开发阶段，可能需要经历数年方能投入市场。

14.4 The Future of GM Foods

Over the last 20 years, GM foods have revealed both promise and problems. GM advocates are confident that the next generation of GM food, especially those created using genome-editing technologies, will show even more promising prospects—and may also address many of the problems.

Research is continuing on ways to fortify staple crops with nutrients. For example, Australian scientists are adding genes to bananas that will not only provide resistance to Panama disease—a serious fungal disease that can destroy crops—but also increase the levels of β -carotene and other nutrients, including iron. Other GM crops in the pipeline include plants engineered to resist drought, high salinity, nitrogen starvation, and low temperatures.

Researchers are also devising more creative ways to protect plants from insects and

14.4 GM食品的未来

在过去的20年里，GM食品既展现出了它的应用前景，也显示出了它所带来的问题。GM倡导者对下一代GM食品很有信心，尤其是使用基因组编辑技术构建的GM食品，这种食品将呈现更为广阔的应用前景，并且也可能解决许多问题。

提高主要粮食作物营养价值的研究仍在持续进行中。例如，澳大利亚科学家将特定基因转入香蕉中，不仅使香蕉能够抵抗一种非常严重的、足以摧毁作物的真菌病——巴拿马病，而且能够提升香蕉中β-胡萝卜素、铁元素及其他营养物质的含量。设计构建中的其他GM作物还包括抗干旱、耐高盐、耐低氮和耐低温等特性。

研究人员也在策划更多有创意的途径，用于帮助植物抵御虫害和病害。一个项目

diseases. One project introduces into wheat a gene that encodes a pheromone that acts as a chemical alarm signal to aphids. If successful, this approach could protect the wheat plants from aphids without using toxins. Another project involves cassava, which is a staple crop for many Africans and is afflicted by two viral diseases—cassava mosaic virus and brown streak virus—that stunt growth and cause root rot. Although some varieties of cassava are resistant to these viruses, the life cycle of cassava is so long that it would be difficult to introduce resistance into other varieties using conventional breeding techniques. Scientists plan to transform plants with genes from resistant cassava. This type of cisgenic gene transfer is more comparable to traditional breeding than transgenic techniques.

是将编码信息素的基因导入小麦中作为蚜虫的化学警报信号。如果该项目取得成功，那么这种方法可以在不使用毒素的情况下保护小麦免受蚜虫侵害。另一个项目涉及木薯，木薯是许多非洲人的主要粮食作物，它同时遭受两种病毒病的困扰——木薯花叶病毒和褐色条纹病病毒——从而阻碍作物生长，并导致根部腐烂。尽管一些木薯品种能够抵抗这些病毒，但是由于木薯生长周期很长，因此很难采用常规杂交技术将这种抗病毒能力引入其他木薯品种中。科学家计划将源自抗病毒木薯的基因转入其他品种植株中。与转基因技术相比，这种同源基因转移和传统育种更为相似。

In the future, GM foods will likely include additional GM animals. In one project, scientists have introduced a DNA sequence into chickens that protects the birds from spreading avian influenza. The sequence encodes a hairpin RNA molecule with similarity to a normal viral RNA that binds to the viral polymerase. The presence of the hairpin RNA inhibits the activity of the viral polymerase and interferes with viral propagation. If this strategy proves useful *in vivo*, the use of these GM chickens would not only reduce the incidence of avian influenza in poultry production, but also reduce the transmissibility of avian influenza viruses to humans.

未来，GM 食品可能还将包括 GM 动物。在一个科研项目中，科学家将一段 DNA 序列导入鸡中从而保护禽类免于禽流感的传播。该序列编码一种具有发夹结构的 RNA 分子，其与正常病毒 RNA 分子类似，并能与病毒聚合酶相结合。该发夹 RNA 的存在抑制了病毒聚合酶的活性，并干扰了病毒扩增。如果该策略在体内证明有效，那么这些 GM 鸡的应用将不仅会降低禽流感在禽类生产中的发病率，而且会降低禽流感病毒对人类的传播能力。

Although these and other GM foods show promise for increasing agricultural productivity and decreasing disease, the political pressure from anti-GM critics remains a powerful force.

尽管这些案例和其他 GM 食品在提高农业生产力与降低病害方面展现出了广阔前景，但是来自反 GM 批评人士的政治压力依然保持强劲。深入理解这些技术背后

An understanding of the science behind these technologies will help us all to evaluate the future of GM foods.

的科学知识将有助于我们更好地评估 GM 食品的未来。

15 Genomics and Precision Medicine

15 基因组学与精准医疗

Over the last decade, the terms *precision medicine* and *personalized medicine* have entered public consciousness as emerging, and likely revolutionary, approaches to disease prevention, diagnosis, and treatment. In 2015, the United States announced the Precision Medicine Initiative—a $215 million investment into the molecular tools required to bring precision medicine into routine clinical use. In the same year, the United Kingdom announced its Precision Medicine Catapult—aimed to accelerate the development and application of precision medicine technologies.

在过去 10 年中，精准医疗、个性化医疗作为新兴的、有望革新的疾病预防、疾病诊断和疾病治疗手段已经进入公众视野。2015 年，美国宣布实施“精准医疗计划”，该计划拟投资 2.15 亿美元，用于将精准医疗引入常规临床应用所需的分子工具中。同年，英国宣布开展“精准医疗工程”，旨在加速精准医疗相关技术的开发与应用。

So, what are precision medicine and personalized medicine? **Precision medicine** can be defined as an individualized, molecular approach to disease diagnosis and treatment—one that examines a patient's individual genomic, proteomic, gene expression, and other molecular profiles and applies that information to select precise disease treatments and to develop new treatments and drugs. Precision medicine classifies patients into subpopulations based on their molecular profiles, and then directs each group into a treatment regimen that will bring about maximum benefit. Although often used interchangeably with precision medicine,

那么，什么是精准医疗和个性化医疗呢？**精准医疗**可以定义为针对疾病诊断和治疗采用个性化的分子技术——检查患病个体的基因组、蛋白质组、基因表达和其他分子概况，将这些信息应用于选择疾病的精准治疗方案、开发新的治疗方法和新型药物中。精准医疗依据病人自身的分子概况分析将他们归于不同的亚群，然后对每个亚群制定不同的治疗方案以期获得最好的治疗效果。尽管通常被作为精准医疗同义词进行替换使用，但是**个性化医疗**则指基于个体独一无二的分子概况分析为每位患者设计特定的，甚至是独一无二的治疗方案。个性化医疗被认为是精准医疗的

personalized medicine is defined as a way to design specific, even unique, treatments for each individual, also based on their unique molecular profiles. Personalized medicine can be considered a part of precision medicine.

一部分。

In this chapter, we will examine some of the new developments in precision medicine, with an emphasis on pharmacogenomics and precision oncology. The types of genetic tests currently used in precision medicine are described in Chapter 10—Genetic Testing.

在本章中，我们将介绍精准医疗的一些新进展，尤其是药物基因组学和精准肿瘤学。精准医疗中目前应用的各种遗传检验技术在第 10 章——遗传检测中已进行了详细介绍。

15.1 Pharmacogenomics

15.1　药物基因组学

Perhaps the most developed area in precision medicine is pharmacogenomics. **Pharmacogenomics** is the study of how an individual's genetic makeup determines the body's response to drugs. It also involves the development and use of drugs that are specifically targeted to a patient's genetic profile. The term *pharmacogenetics* is often used interchangeably with pharmacogenomics but refers to the study of how sequence variation within specific genes affects an individual's drug responses.

精准医疗中最长足发展的领域也许是药物基因组学。**药物基因组学**研究生物个体的遗传组成如何决定其机体对于药物的响应，同时还包括针对病人遗传信息进行的特异性药物开发和应用。“药物遗传学”通常与“药物基因组学”相互替代使用，但药物遗传学主要研究特定基因的序列变异如何影响个体对药物的响应。

In this section, we examine two ways in which precision medicine is changing the development and use of drugs: by optimizing drug responses and by developing molecularly targeted drugs.

在这一部分，我们将细述精准医疗改变药物开发和药物应用的两种途径：优化药物响应、开发分子靶向性药物。

Optimizing Drug Responses

Every year, approximately 2 million people in the United States have serious side effects from pharmaceutical drugs and of these,

优化药物响应

美国每年大约有 200 万人在药物使用过程中会产生严重副作用，其中约 10 万人因此丧命。此外，还有很多病人在药物治疗

approximately 100,000 will die. In addition, many patients do not respond to drug treatment as well as expected, due in part to their genetic makeup and the genomic variants that are associated with their diseases.

过程中并未取得预期效果，而这一现象的部分原因在于个体的遗传信息组成及其疾病相关的基因组变体。

Sequence variations in dozens of genes affect a person's reactions to drugs. The proteins encoded by these gene variants control many aspects of drug metabolism, such as the interactions of drugs with carriers, cell-surface receptors, and transporters; with enzymes that degrade or modify drugs; and with proteins that affect a drug's storage or excretion.

很多基因的序列变化会影响个体对药物的响应。这些基因变体编码的蛋白质控制着药物代谢的诸多方面，如药物与载体、细胞表面受体或药物转运体之间的相互作用，药物与药物降解酶、药物修饰酶之间的相互作用，药物与影响药物储存或分泌的蛋白质之间的相互作用。

Examples of genes that are involved in drug metabolism are members of the cytochrome P450 gene family. People with some cytochrome P450 gene variants metabolize and eliminate drugs slowly, which can lead to accumulations of the drug and overdose side effects. In contrast, other people have variants that cause drugs to be eliminated quickly, leading to reduced effectiveness. An example is the *CYP2D6* gene, which encodes debrisoquine hydroxylase. This enzyme is involved in the metabolism of approximately 25 percent of all pharmaceutical drugs, including acetaminophen, clozapine, β -blockers, tamoxifen, and codeine. There are more than 70 variant alleles of this gene. Some variants reduce the activity of the encoded enzyme, and others can increase it. Approximately 80 percent of people are homozygous or heterozygous for the wild-type *CYP2D6* gene and are known as extensive metabolizers. Approximately 10 to 15 percent of people are homozygous for alleles that decrease

细胞色素 P450 基因家族成员是一类参与药物代谢的基因。体内含使药物代谢和药物清除速度减慢的细胞色素 P450 基因变体的个体，将产生药物在体内积累和药物过量的副作用。与此相反，另一些个体体内所含的基因变体可引发药物迅速被清除从而导致药效降低。例如 *CYP2D6* 基因，该基因编码异喹胍羟化酶，该酶参与药品总量中约 25% 的药物代谢，所涉药品包括对乙酰氨基酚、氯氮平、β - 受体阻滞药、他莫昔芬、可待因。该基因有 70 种以上的等位基因变体，有些基因变体降低编码蛋白的酶活性，有些则提升酶活性。大约 80% 的人类个体是 *CYP2D6* 野生型基因纯合子或杂合子，即为众所周知的强代谢型；10%—15% 的人类个体是酶活性降低型等位基因的纯合子（弱代谢型）；剩下的个体则含基因重复（超级快速代谢型）。表 15.1 中列出了影响药物效率的其他基因变体的例子。

activity (poor metabolizers), and the remainder of the population have duplicated genes (ultra-rapid metabolizers). Examples of other gene variants that influence the effectiveness of drugs are presented in Table 15.1.

TABLE 15.1 Examples of Drug Responses Affected by Gene Variants*
表 15.1 基因变体影响药物响应的例子 *

Gene	Drug Affected	Description
TPMT	Mercaptopurine, thioguanine, azathioprine	People with low levels of TPMT enzyme develop toxic side effects after taking thiopurine drugs for the treatment of leukemia or inflammatory conditions.
HLA-B	Allopurinol, carbamazepine, abacavir	Alleles of *HLA-B* are associated with allergic reactions to these drugs used to treat gout, epilepsy, and HIV, respectively.
CYP2D6	Codeine, tramadol, tricyclic antidepressants	Numerous alleles in the population affect metabolism of many drugs leading to underdoses and overdoses.
VKORC1	Warfarin	Warfarin anticoagulant inactivates VKORC1 protein. Variants in the *VKORC1* gene produce less protein, resulting in overdose effects at normal warfarin dosages.
CYP2C9	Warfarin	This gene encodes a liver enzyme that oxidizes warfarin. Variants metabolize warfarin less efficiently, leading to overdoses.
CYP2C19	Tricyclic antidepressants, clopidogrel, voriconazole	CYP2C19 protein is a liver enzyme that metabolizes 10–15% of drugs. Alleles result in poor metabolizers to ultra-metabolizers.
SLCO1B1	Simvastatin	This gene encodes a liver transporter. Variants are less efficient at removal of statins, which are used to control cholesterol levels.

*For more information, visit the PharmGKB Web site (https://www.pharmgkb.org/index.jsp).

One of the primary goals of precision medicine is to provide screening to patients prior to treatment so that the choice of drug and its dosage can be tailored to the patient's genomic profile. Normally, physicians order a single-gene test only when a specific drug needs

精准医疗的首要目标之一是在治疗之前对病人的基因组进行分析，从而对药物选择及其剂量进行量体裁衣。通常，医生仅在开特定药物处方或处方药物的疗效不及预期时申请单基因检测。目前，可提供基因变体检测服务的基因大约有 20 种。这

to be prescribed or when a prescribed drug is not performing as expected. Currently, tests for genetic variants in about 20 genes are available. These tests predict reactions to approximately 100 drugs representing about 18 percent of all prescriptions in the United States. Several research hospitals have initiated programs to bring extensive genomic screening to all patients prior to treatment, and prior to development of future diseases— an approach called preemptive screening.

些检测可以预测大约 100 种药物的反应，占全美所有处方药物的约 18%。一些研究型医院已经启动一些项目旨在疾病治疗之前和疾病尚未发展之前对所有病人进行大规模的基因组筛选——该方法被称为抢占筛选。

Developing Targeted Drugs

Another goal of pharmacogenomics is to develop drugs that are targeted to the genetic profiles of specific subpopulations of patients. The most advanced applications are in the treatment of cancers. Large-scale sequencing studies show that each tumor is genetically unique. This genomic variability has been exploited to develop new drugs that specifically target cancer cells that may express mutant proteins or overexpress others.

开发靶向性药物

药物基因组学的另一个目标则是针对病人中特定亚群的遗传分析开发药物。靶向性药物最为前沿的应用是肿瘤治疗。大规模测序研究表明每种肿瘤在遗传上都是独特的。这种基因组多元化已被用于开发新型药物，以特异性靶向表达突变蛋白或过量表达其他蛋白质的肿瘤细胞。

One of the first success stories in precision targeted therapeutics was that of the ***HER-2*** gene and the drug **Herceptin®** in breast cancer. The *HER-2* gene codes for a transmembrane tyrosine kinase receptor protein. These receptors are located within the cell membranes of normal breast epithelial cells and, when bound to other growth factor receptors and ligands on the cell surface, they send signals to the cell nucleus that result in the transcription of genes whose products stimulate cell growth and division.

首批精准靶向治疗的成功案例之一是 ***HER-2*** 基因及用于乳腺癌治疗的靶向药物**赫赛汀**®。*HER-2* 基因编码跨膜酪氨酸激酶受体蛋白。这些受体定位于正常乳腺上皮细胞的细胞膜内，当其与细胞表面的其他生长因子受体和配基相结合时，它们会将信号传递至细胞核内，最终导致刺激细胞生长和分裂的相关基因进行转录。

In about 25 percent of invasive breast

在大约 25% 的浸润性乳腺癌中，

cancers, the *HER-2* gene is amplified and the protein is overexpressed on the cell surface. The presence of *HER-2* overexpression is associated with increased tumor invasiveness, metastasis, and cell proliferation, as well as a poorer patient prognosis.

HER-2 基因均存在扩增情况，并且该蛋白质在细胞表面过表达。*HER-2* 基因的过表达与肿瘤侵袭性增强、肿瘤转移、细胞增殖及患者的预后较差均相关。

Based on this knowledge, Genentech Corporation in California developed a monoclonal antibody known as trastuzumab (or Herceptin) that binds to the extracellular region of the HER-2 receptor, inhibiting HER-2 signaling, triggering cell-cycle arrest, and leading to destruction of the cancer cell.

基于这些了解，美国加利福尼亚州基因泰克公司开发了一种单克隆抗体药物——曲妥珠单抗（或赫赛汀），该药可与HER-2 受体的胞外区相结合，抑制 HER-2 信号转导，引发细胞周期终止，从而引导肿瘤细胞毁灭。

Because Herceptin acts only on cancer cells that have amplified *HER-2* genes, it is important to know the HER-2 status of each tumor. A number of molecular assays have been developed to determine the gene and protein status of breast cancer cells. These include immunohistochemistry (IHC) and fluorescence *in situ* hybridization (FISH) assays. In IHC assays, an antibody that binds to the HER-2 protein is added to fixed tissue on a slide. The presence of bound antibody is then detected with a stain and observed under the microscope. The FISH assay assesses the number of *HER-2* genes by comparing the fluorescence signal from a HER-2 probe with a control signal from another gene that is not amplified in the cancer cells.

由于赫赛汀仅作用于 *HER-2* 基因发生扩增的肿瘤细胞，因此了解每种肿瘤中 HER-2 的状况十分重要。人们已经开发了大量用于确定乳腺癌细胞内基因和蛋白质状况的分子检测方法，包括免疫组织化学（IHC）和荧光原位杂交（FISH）。在IHC 检测中，与 HER-2 蛋白结合的抗体被添加至切片上被固定的组织中。随后通过染色检验抗体的结合状况，也可在显微镜下进行观察。在 FISH 检测中，以肿瘤细胞中不发生扩增的其他基因作为对照，比较HER-2 探针的荧光信号，从而评估 *HER-2* 基因的拷贝数。

Herceptin has had a major effect on the treatment of HER-2 positive breast cancers. When Herceptin is used in combination with chemotherapy, there is a 25 to 50 percent increase in survival, compared with the use of chemotherapy alone. Herceptin has also been

赫赛汀在 HER-2 呈阳性的乳腺癌治疗中起到了重要作用。当赫赛汀与化学治疗进行联合治疗时，病人存活率比单纯使用化学治疗提升了 25%—50%。赫赛汀还被发现在其他 HER-2 过表达的恶性肿瘤治疗中也是有效的，如胃癌、胃食管连接部癌症。

found effective in the treatment of other HER-2 overexpressing cancer cells, including those of the stomach and gastroesophageal junction.

There are now dozens of drugs that are targeted to the genetic status of the cancer cells (Table 15.2). For example, about 40 percent of colon cancer patients respond to the drugs **Erbitux®** (cetuximab) and **Vectibix®** (panitumumab). These two drugs are monoclonal antibodies that bind to **epidermal growth factor receptors** (**EGFRs**) on the surface of cells and inhibit the EGFR signal transduction pathway. To work, cancer cells must express EGFR on their surfaces and must also have a wild-type *K-ras* gene. The presence of EGFR protein can be assayed using a staining test and observation of cancer cells under a microscope. Mutations in the *K-ras* gene can be

现在已有很多药物能够针对肿瘤细胞的遗传状况进行治疗（表 15.2）。如**爱必妥**®（西妥昔单抗）和**维克替比**®（帕尼单抗）对于大约 40% 的结肠癌病人有效。这两种药物均为单克隆抗体，可以与细胞表面的**表皮生长因子受体**（EGFRs）相结合，从而抑制 EGFR 信号转导通路。肿瘤细胞必须在其表面表达 EGFR，同时还必须拥有野生型 *K-ras* 基因才能发挥其肿瘤细胞特性。EGFR 蛋白的存在状况可利用染色实验和显微镜下观察肿瘤细胞来进行分析。*K-ras* 基因的突变也可通过之前介绍的 PCR 方法进行检测分析。

TABLE 15.2 Examples of Drugs That Specifically Target Proteins Mutated or Abnormally Expressed in Cancer Cells

表 15.2 针对肿瘤细胞中特异靶蛋白的突变表达或异常表达进行治疗的药物示例

Drug	Cancer Types	Target	Description
Imatinib (Gleevec)	Ph+ CML and ALL	BCR-ABL kinase	Imatinib binds to and inhibits BCR-ABL, which is encoded by the *bcl-abl* fusion gene located on the Philadelphia chromosome.
Olaparib (Lynparza)	BRCA1/2 mutated ovarian cancer	Poly ADP ribose polymerase (PARP)	BRCA1/2-defective cancers rely on PARP for DNA repair. Olaparib inhibits PARP repair, blocking cell division.
Trametinib (Mekinist)	Melanoma	Mitogen-activated protein kinase (MEK)	Trametinib inhibits mutated constitutively active MEK pathways, resulting in cell cycle arrest and increased apoptosis.
Crizotinib (Xalkori)	NSCLC	EML4-ALK fusion kinase	Crizotinib inhibits fusion kinase activity, reducing cancer cell growth and invasion.
Vismodegib (Erivedge)	Basal-cell carcinoma	Smoothen receptor	Inhibits transcription factors that are necessary for expression of tumor genes.

Ph+ CML = Philadelphia chromosome-positive chronic myelogenous leukemia
Ph+ ALL = Philadelphia chromosome-positive acute lymphoblastic leukemia
NSCLC = non-small-cell lung carcinoma

detected using assays based on the polymerase chain reaction (PCR) method.

15.2 Precision Oncology

One of the promises of precision medicine is to treat cancer patients with therapies that target specific gene mutations and gene expression defects in their tumors, leading to effective remissions and even cures. To support these promises, advances in exomic and whole-genome sequencing methods are making these technologies more cost effective for the diagnosis of many diseases including cancers. Large research programs, such as The Cancer Genome Atlas project, are mapping the genomes of thousands of tumor types to identify mutations and expression profiles for which targeted drugs can be developed. Targeted therapies and diagnostics also benefit from high-throughput proteomic and metabolomic assays.

As described in the previous section, many cancer drugs targeted to specific genetic and gene expression profiles are already being used, sometimes with dramatic effects. So far, the percentage of patients that can be successfully treated with precision cancer drugs is small. One clinical trial showed that only 6.4 percent of enrolled patients could be matched to a targeted drug based on their tumor's genomic profile. Another challenge is to deal with tumor resistance. To circumvent resistance, it will be necessary to use multiple treatment approaches simultaneously— both targeted and generalized.

Beyond the use of targeted drugs,

15.2 精准肿瘤学

精准医疗的应用前景之一是针对肿瘤特异性基因突变和基因表达缺陷对肿瘤患者进行靶向治疗，使其病情得到有效缓解甚至治愈。外显子测序方法和全基因组测序方法的改进使得这些技术在许多疾病包括肿瘤的诊断中成本更低，为精准医疗的实现提供了有力支持。一些大型研究计划，如肿瘤基因组图谱计划，对上千种类型的肿瘤进行基因组测序，确定突变和表达图谱，从而为靶向性药物的开发提供理论依据。靶向性治疗和诊断也受益于高通量的蛋白质组学和代谢组学分析。

正如前一部分所述，很多以特异性遗传图谱或基因表达图谱为靶向的肿瘤药物已得到应用，并且有时疗效还十分显著。迄今为止，能够使用精准肿瘤药物获得成功治疗的病人比例依然很小。一项临床试验表明被试病人中仅有6.4%的患者能够基于他们的肿瘤遗传图谱与靶向性药物进行匹配。另一个挑战则是如何应对肿瘤耐药性。为了抵御肿瘤耐药性，同时采用多种治疗方法十分必要——包括靶向性治疗和广谱性治疗。

除了使用靶向性药物外，研究人员也

researchers are also making progress in the use of other targeted modalities, including targeted cancer immunotherapies, which are described next.

在其他靶向性治疗应用中取得了一些进展，包括将要介绍的靶向性肿瘤免疫治疗。

Targeted Tumor Immunotherapies

Some of the most promising new developments in precision medicine are in the field of cancer **immunotherapy**. These therapies harness the patient's own immune system to kill tumors, and some have brought remarkable therapeutic effects in clinical trials and triggered billions of dollars of investment into their development. In this section, we will describe two of the most promising precision cancer immunotherapies—adoptive cell transfer and engineered T-cell methods.

靶向性肿瘤免疫治疗

精准医疗中最具应用前景的一些新进展属于肿瘤**免疫治疗**领域。这些治疗方法利用病人自身免疫系统杀灭肿瘤，有些在临床试验中已经取得了不同寻常的治疗效果，并且吸引了数十亿美元用以投资它们的发展。在这一部分，我们将详细介绍两种最具应用前景的精准肿瘤免疫治疗——过继细胞移植法和工程 T 细胞法。

To understand how these therapies work, we need to briefly review how the immune system, particularly **T cells**, defends against the development of cancer. As summarized in Figure 15.1, the immune system consists of cell types

为了理解这些治疗方法的工作机理，我们有必要简要回顾一下免疫系统，尤其是 **T 细胞**如何抵御肿瘤的发展。正如图 15.1 的总结所示，免疫系统所含的细胞类型和化学信号共同组成了先天性免疫系统和

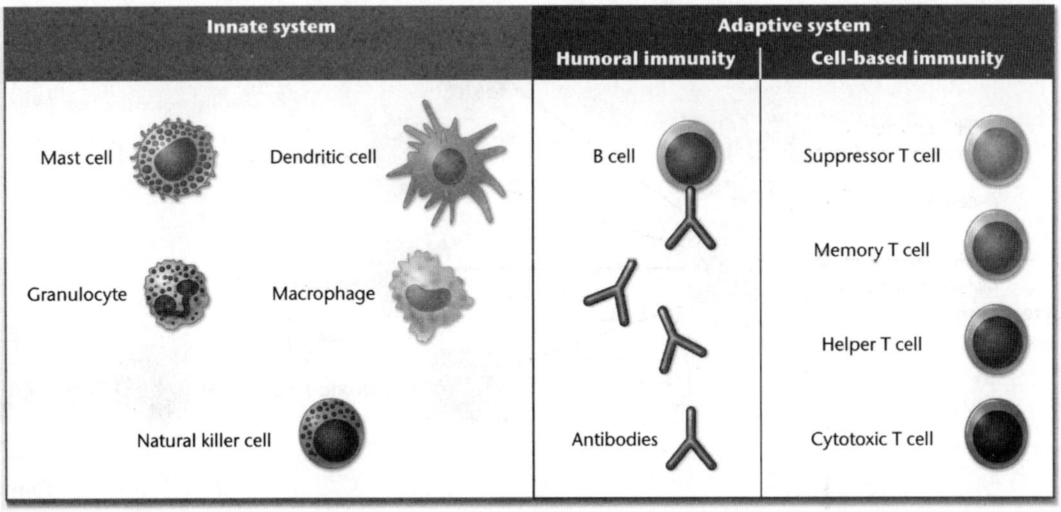

FIGURE 15.1 Cell types of the innate and adaptive immune systems.

图 15.1 先天性免疫系统和适应性免疫系统中的细胞类型。

and chemical signals constituting the innate and the adaptive systems.

适应性免疫系统。

Both adoptive cell transfer and engineered T-cell methods exploit **cytotoxic T lymphocytes (CTLs)** to recognize specific antigens on the surface of cancer cells, bind to the cells, and destroy them. Figure 15.2 summarizes the steps involved in normal T-cell recognition and destruction of cancer cells.

过继细胞移植法和工程 T 细胞法两种方法利用**细胞毒性 T 淋巴细胞**（CTLs）来识别肿瘤细胞表面的特异性抗原，与肿瘤细胞相结合，然后摧毁肿瘤细胞。图 15.2 总结了正常 T 细胞识别和消灭肿瘤细胞的步骤。

Cancer cells express many proteins that are specific to the tumor and have the capacity to be recognized by the patient's immune system as nonself antigens. These nonself antigens result from abnormal gene expression and mutations in the coding regions of both cancer driver and passenger genes. For example, 30 percent of human cancers contain mutated *ras*-family genes (such as *K-ras* and *H-ras*), which act as cancer driver genes. Many different point mutations can occur in these genes, each encoding an altered protein that is not found in normal cells. Cancer cells also contain up to hundreds of

肿瘤细胞表达很多肿瘤特异性蛋白，这些特异性蛋白能被病人免疫系统识别且认定为异己抗原。这些异己抗原源自肿瘤驱动基因和肿瘤乘客基因的异常基因表达和基因编码区突变。例如，30% 的人类肿瘤含突变的 *ras* 家族基因（比如 *K-ras* 和 *H-ras*），这类基因为肿瘤驱动基因，它们可发生多种点突变，每种突变均会编码一种在正常细胞中不存在的突变蛋白。肿瘤细胞中也含有上百种乘客基因突变，这类基因的基因产物不仅涉及肿瘤表型，而且编码突变的异己蛋白。总而言之，这些蛋白质中所含的新生异己抗原被称为**新生抗原**。

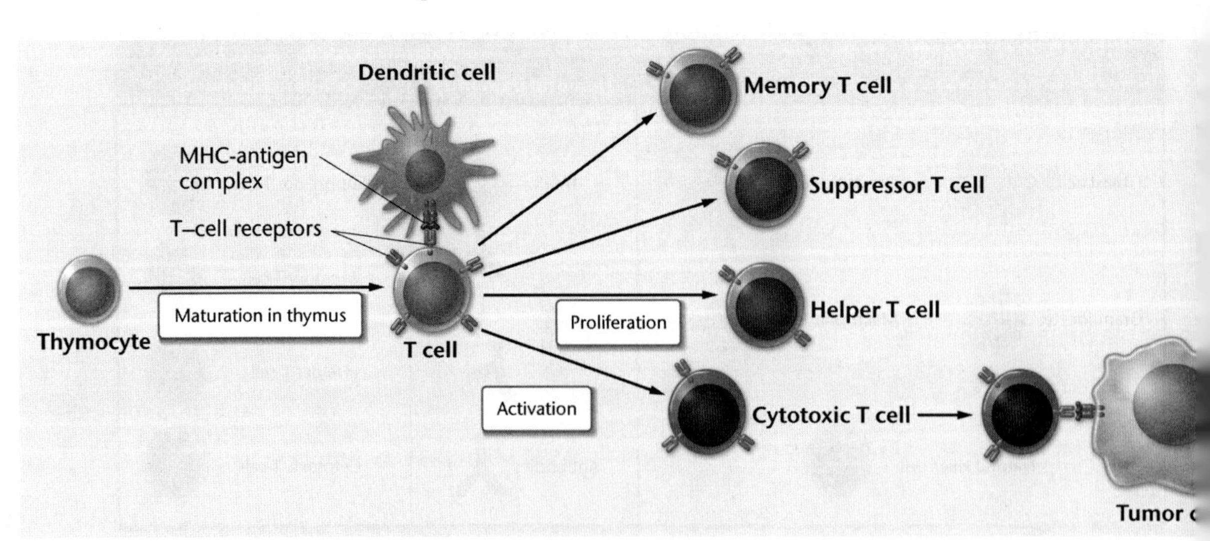

FIGURE 15.2 Steps in the maturation and activation of T cells.

图 15.2 T 细胞成熟和激活的步骤。

mutations in passenger genes whose products are not involved in the cancer phenotype, but also encode mutated, and hence nonself, proteins. Collectively, the novel, nonself antigens that are contained within their proteins are known as **neoantigens**.

Although T cells are known to associate with tumors and are able to recognize tumor neoantigens, they are often not able to destroy tumor cells. These tumor-associated T cells are also known as **tumor-infiltrating lymphocytes (TILs)**.

尽管T细胞已知与肿瘤相关，并且能够识别肿瘤新生抗原，但是它们通常并不能摧毁肿瘤细胞。这些肿瘤相关T细胞也被称为**肿瘤浸润淋巴细胞（TILs）**。

Cancers use many different strategies to suppress T-cell responses. These strategies include the synthesis of molecules that bind to T cells and repress their activity. Interestingly, some effective new drugs called *checkpoint inhibitors* help T cells avoid these checkpoint molecules, thereby enhancing the tumor-killing ability of TILs. Another way that tumors avoid immune system activity is that they are often abnormal in their expression of cell-surface major histocompatibility complex (MHC) molecules, which are essential to stimulate antigen-presenting cells, which in turn are necessary to stimulate T cells to recognize and kill cells that bear nonself antigens. A third way that tumors avoid immune responses is through the presence of tumor-associated **regulatory T cells** called T-regs (including **suppressor T cells**), whose role is to repress the activities of activated T cells. The presence of other tumor-infiltrating cells such as **macrophages** and **monocytes** also repress the activities of T cells.

肿瘤使用多种策略抑制T细胞应答，这些策略包括合成能结合T细胞并抑制其活性的分子。有趣的是，一些被称为检查点抑制剂的新型有效药物能够帮助T细胞躲避这些检查点分子，从而加强TILs的肿瘤杀伤能力。肿瘤逃避免疫系统活性的另一种策略是它们细胞表面的主要组织相容性复合体分子（MHC）通常会表达异常，这些分子对于刺激抗原呈递细胞十分重要，从而对于刺激T细胞识别和杀死含异己抗原的细胞必不可少。肿瘤逃避免疫系统的第三种途径是通过肿瘤相关**调节性T细胞**（T-regs，包括**抑制性T细胞**）抑制已被激活的T细胞活性。其他肿瘤浸润细胞如**巨噬细胞**、**单核细胞**的存在同样可以抑制T细胞活性。

To circumvent and overwhelm the

为了克服和打败肿瘤用于抑制抗肿瘤

mechanisms that cancers use to repress anticancer immune responses, scientists have developed the following personalized T-cell-based therapies.

免疫应答的机制，科学家们已经开发了以下几种个性化的基于 T 细胞的治疗方法。

Adoptive cell transfer (ACT) involves removing TILs from a patient's tumor, selecting those that specifically recognize tumor antigens, amplifying these specific TILs *in vitro*, and reintroducing them back into the patient.

过继细胞移植法（ACT）涉及从病人肿瘤中移除 TILs、遴选其中能特异性识别肿瘤抗原的 TILs、在体外扩增这些特异性 TILs，以及把它们重新植入病人体内。

The steps in ACT are summarized in Figure 15.3. In the first step, tumor specimens that contain TILs are removed from the patient and digested into small samples containing one or several cells. Each sample is grown in a culture dish in the presence of tumor material and IL-2, a growth factor for T cells. As the T cells grow in the dish, those with reactivity to the tumor cells destroy the tumor cells within two to three weeks. These T cells are selected and retested for their tumor-destroying activity in coculture assays. Positive T cells are then grown to high numbers (10^{11} cells) in the lab, in the presence of several growth-stimulatory factors. The process requires about six weeks from obtaining the tumor specimen to harvesting the amplified reactive T-cell preparation. At this point, the patient is treated with chemotherapy to rid the body of immune system cells such as T-regs and macrophages that repress the activity of activated T cells. Then, the patient is reinfused with the amplified T cells and IL-2. The adoptive T cells can continue to expand up to 1000-fold after reinfusion. In some patients, the tumor-reactive T cells can be found in the circulation months after the initial infusion, where they make up as

ACT 的所有步骤总结如图 15.3 所示。将含 TILs 的肿瘤样本从病人体内取出，并消解成含有一个或多个细胞的微样品。每个样品在培养皿中与肿瘤物质、T 细胞生长因子 IL-2 在共存条件下进行生长。当 T 细胞在培养皿中生长时，那些对肿瘤细胞有响应的 T 细胞将在 2—3 周内摧毁肿瘤细胞。这些 T 细胞将被选出，并通过共培养分析再次检测它们的肿瘤细胞杀伤能力。分析结果呈阳性的 T 细胞在实验室通过与多种生长刺激因子共存培养直至高达 10^{11} 个细胞。整个过程从获得肿瘤样本到收获经增殖的活性 T 细胞大约需要 6 周时间。在这一阶段，病人采用化学疗法来消除机体内调节性 T 细胞、巨噬细胞等免疫系统细胞对被激活 T 细胞活力的抑制作用。然后，病人被重新输注增殖后的 T 细胞和 IL-2，过继移植 T 细胞还将在输注后继续增殖扩大 1000 倍。在一些病人体内，初次移植后数月即能在病人循环系统中发现肿瘤反应性 T 细胞，并占整个 T 细胞总量的 80%。过继移植 T 细胞的持续存在与阳性抗肿瘤效应直接相关。

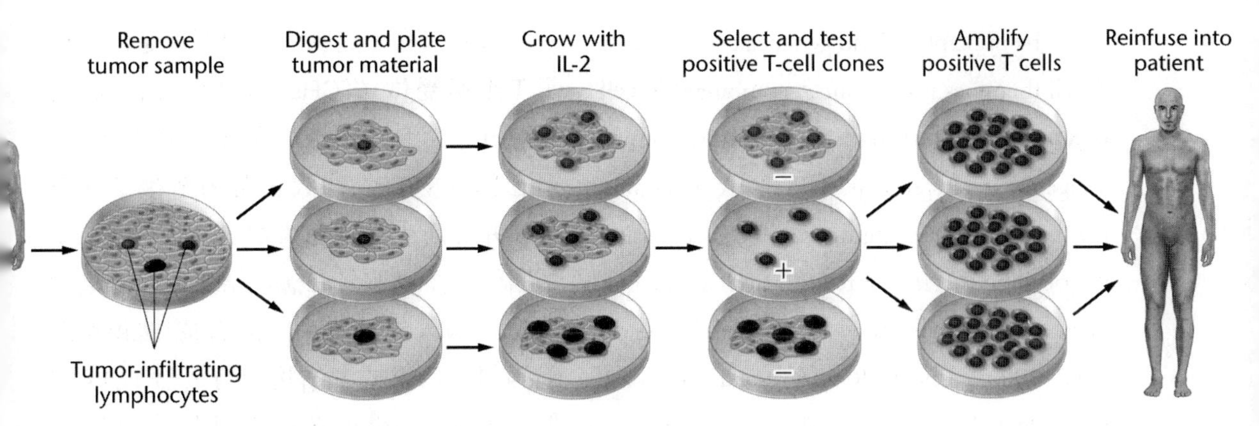

FIGURE 15.3 Summary of adoptive cell transfer method.

图 15.3 过继细胞移植法。

much as 80 percent of the T-cell population. The persistence of the adoptive T cells correlates with a positive antitumor effect.

The results of some ACT clinical trials have produced positive results, particularly in patients with metastatic melanoma—a cancer that normally has a poor outcome. For example, in trials conducted by the National Cancer Institute, after only one treatment, the outcome was remarkable. Between 53 and 72 percent of patients showed positive responses to the treatment, 22 percent showing complete regressions and 20 percent having no recurrence of their cancer up to 10 years later.

一些 ACT 临床试验已经产生了积极的效果，尤其是对转移性黑色素瘤患者，转移性黑色素瘤是一种通常治疗效果较差的肿瘤。例如，在美国癌症研究所进行的试验中，仅经过一次治疗即获得非同凡响的治疗效果。53%—72% 的病人在治疗中显示出积极的治疗效果，22% 的个体肿瘤完全消退，20% 的个体甚至在此后 10 年内都未出现肿瘤的复发。

The promising results of ACT clinical trials on metastatic melanoma have encouraged attempts to use this method to treat other malignancies, such as cervical cancer and some blood cancers. To extend ACT therapies to patients who may not have activated TILs within their tumors that recognize unique neoantigens on those tumors, scientists are developing other ways to target the immune system to cancer. One of these methods is described next.

ACT 在转移性黑色素瘤临床试验中获得的这些令人充满希望的结果已经鼓舞人们开始尝试将该方法应用于其他恶性肿瘤如宫颈癌和一些血液癌症的治疗中。为了将 ACT 治疗拓展应用至那些自身肿瘤组织中不含可识别肿瘤独特新生抗原的激活 TILs 的病人，科学家正在开发其他途径使免疫系统靶向肿瘤，接下来介绍的即为这些方法中的一种。

The principle behind genetically engineered T-cell therapies is to create recombinant **T-cell receptors (TCRs)** that specifically recognize antigens on cancer cells. The DNA sequences that encode these engineered TCRs are then introduced *in vitro* into a patient's normal, naïve T cells which then express these TCRs on their surfaces. The TCR-transduced T cells are then selected, amplified, and reinfused into the patient. The synthetic TCR genes encode either TCRs that are structurally similar to natural TCRs or **chimeric antigen receptors (CARs)** that can directly recognize antigens on the tumor cell without requiring T-cell activation by antigen-presenting cells. In this subsection, we will describe CAR T-cell therapies.

基因工程 T 细胞疗法的原理是创建重组 **T 细胞受体（TCRs）**特异性识别肿瘤细胞抗原。编码这些工程 TCRs 的 DNA 序列在体外被转入病人正常的新生 T 细胞中，并在新生 T 细胞表面进行表达。TCR 基因转导 T 细胞经筛选、增殖后，再重新输注入病人机体内。人工合成 TCR 基因编码的蛋白质或者结构上与天然 TCRs 类似，或者是不需抗原呈递细胞激活 T 细胞即可直接识别肿瘤细胞抗原的**嵌合抗原受体（CARs）**。在这一部分，我们将介绍 CAR-T 细胞治疗。

CAR proteins are fusions of several proteins derived from a variety of sources. (The structures of normal TCRs and CARs are shown in Figure 15.4.) The extracellular portion of a CAR consists of the variable regions of immunoglobulin heavy and light chains, separated by a linker sequence. These variable regions fold in such a way that they mimic the specificity of an antibody that recognizes a specific antigen—such as a tumor neoantigen. The variable antibody regions are preceded by a signal peptide to direct the CAR to the surface of the T cell in which it is expressed. A spacer region allows the variable regions to orient themselves to bind to antigens on the cancer cell. A transmembrane region anchors the CAR in the T-cell membrane, and the intracellular region is responsible for sending various activation signals to the T cell, after the variable antibody regions have made contact

CAR 蛋白是多种来源蛋白质的融合蛋白（标准 TCRs 和 CARs 的结构如图 15.4 所示）。CAR 的细胞外部分由免疫球蛋白的重链和轻链的可变区组成，二者之间是一段连接序列。这些可变区经折叠模拟抗体识别特定抗原的特异性——恰如肿瘤的新生抗原。可变抗体区由信号肽引导将 CAR 定位至其蛋白质表达所在的 T 细胞表面。间隔区使可变区能够自我调整，从而与肿瘤细胞表面抗原相结合。跨膜区将 CAR 锚定在 T 细胞膜上，CAR 的胞内区则负责在可变抗体区与抗原识别结合后将不同的激活信号传递给 T 细胞。这些激活信号包括指示细胞增殖、指令细胞分化、合成细胞因子以及杀灭靶细胞。

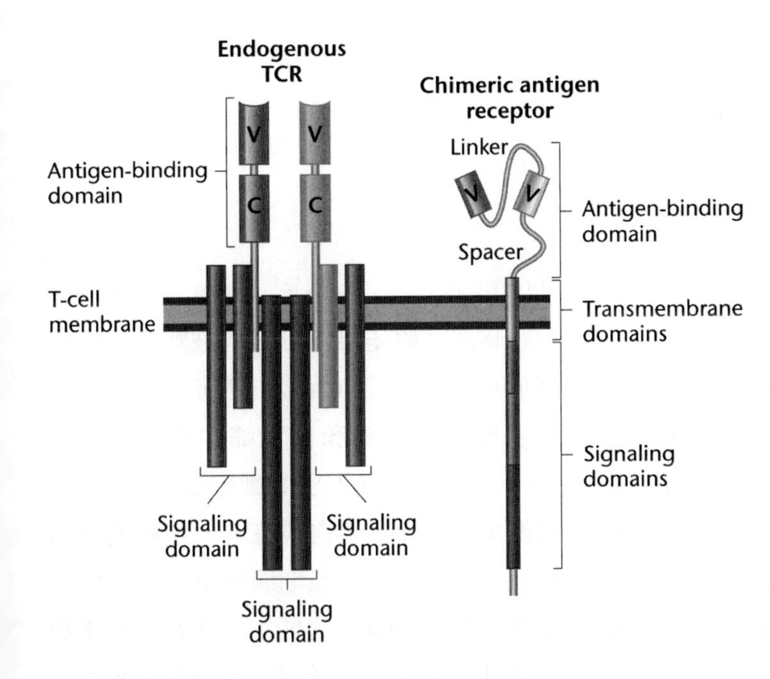

FIGURE 15.4 Structures of an endogenous T-cell receptor (TCR) and a recombinant chimeric antigen receptor (CAR). Variable regions of antigen-binding domains are labelled V and constant regions as C.

图 15.4 内源 T 细胞受体（TCR）和重组嵌合抗原受体（CAR）的结构。抗原结合域的可变区标记为 V，恒定区标记为 C。

with an antigen. The activation signals include instructions to proliferate, differentiate, produce cytokines, and kill the target cell.

To create CARs, scientists clone DNA fragments encoding each of the regions described above into a single linear recombinant DNA molecule, which encodes the entire CAR fusion protein. The DNA fragment encoding immunoglobulin variable regions is usually cloned from cells that secrete monoclonal antibodies. The cells that produce these monoclonal antibodies have previously been screened and selected for their reactivity against the desired neoantigens found on the surface of cancer cells. Once the chimeric DNA molecules have been cloned, they are introduced into normal T cells that have been purified from the patient's peripheral blood. The types of vectors and methods used to introduce therapeutic constructs into human cells are described in more detail in Chapter 11—Gene Therapy. The

为了制备 CARs，科学家将编码上述每一区域的 DNA 片段克隆整合至单一线性重组 DNA 分子中，从而编码整个 CAR 融合蛋白。编码免疫球蛋白可变区的 DNA 片段通常源自分泌单克隆抗体的细胞，这些生产单克隆抗体的细胞在此之前已根据它们对于肿瘤细胞表面新生抗原的反应能力而进行了筛选。一旦这些嵌合 DNA 分子克隆成功，它们就会被导入从病人外周血中纯化而来的正常 T 细胞中。医学治疗的基因构建所需载体类型和导入人类细胞的方法在第 11 章——基因治疗中有详细介绍。接下来的步骤与过继细胞移植法类似——筛选和增殖 T 细胞，然后将这些 T 细胞重新输注入已通过化学治疗或放射治疗减少内源性免疫系统细胞数量的病人体内。

remainder of the procedure is similar to adoptive cell transfer—screening and amplifying the T cells, and reinfusing them into patients who have been treated with chemotherapy or radiation therapies to reduce the numbers of endogenous immune system cells.

Results of clinical trials have been encouraging. The majority of trials have tested CAR T cells for effectiveness in treating B-cell cancers such as leukemias and lymphomas. The CAR T cells recognize the CD19 surface proteins that are expressed on B cells but not on other cells. The response rates have varied between 70 and 100 percent with reports of long-term remissions of up to several years. Several clinical trials have tested CAR T-cell therapies against solid tumors, with less encouraging results.

临床试验结果十分令人鼓舞。大部分试验已经验证了 CAR-T 细胞在治疗 B 细胞肿瘤如白血病和淋巴瘤中的有效性。CAR-T 细胞可以识别 B 细胞表达的 CD19 表面蛋白，而该蛋白质在其他细胞表面不存在。长达数年的长期缓解报告显示应答率介于 70%—100% 之间。一些针对实体瘤的 CAR-T 细胞疗法临床试验结果则相对不太理想。

As of October 2017, two CAR T-cell therapies have been approved by the US Food and Drug Administration (FDA). In August 2017, a CAR T-cell therapy called Kymriah™ was approved for treating children with acute lymphoblastic leukemia (ALL) who had relapsed twice or did not respond to earlier treatments. In clinical trials of these patients, Kymriah™ treatment produced complete remissions in 83 percent of patients. In October 2017, the FDA approved a CAR T-cell therapy called Yescarta™ for treating adults with some types of large B-cell lymphomas.

2017 年 10 月，两种 CAR-T 细胞疗法通过美国食品药品监督管理局（FDA）批准正式上市。2017 年 8 月，CAR-T 细胞疗法 Kymriah™ 被批准用于治疗急性淋巴细胞白血病（ALL）的儿童患者，这些患者的病情已两次复发并对之前的治疗响应无效。在对这些病人的临床试验中，Kymriah™ 疗法使 83% 的病人病情得到了完全缓解。2017 年 10 月，FDA 批准了另一种 CAR-T 细胞疗法—— Yescarta™，用于治疗某些类型大 B 细胞淋巴瘤的成人患者。

Despite promising results with CAR T-cell therapies, these treatments have several serious side effects in many patients. These include systemic inflammatory responses, neurotoxicity, and eventual tumor resistance. Researchers

尽管 CAR-T 细胞疗法取得了令人满意的结果，但是这些治疗在很多病人身上也引发了多种严重副作用，包括系统性炎症反应、神经毒性，最终引起肿瘤耐药性。研究人员期待随着这些新型疗法的应用经

expect that these side effects will become manageable as more experience is gained with these new therapies.

验越来越丰富，这些副作用能够变得可控。

15.3 Precision Medicine and Disease Diagnostics

15.3 精准医疗与疾病诊断

The ultimate goal of precision medicine is to apply information from a patient's full genome to help physicians diagnose disease and select treatments tailored to that particular patient. Not only will this information be gleaned from genome sequencing, but it will also be informed by gene-expression information derived from transcriptomic, proteomic, metabolomic, and epigenetic tests.

精准医疗的终极目标是提供病人的全基因组信息，帮助医生诊断疾病并针对特定病人量身选择治疗方案。这些信息不仅需要从基因组测序中收集，而且还要从转录组、蛋白质组、代谢组、表观遗传检验衍生出的基因表达信息中获得。

Presently, the most prevalent use of genomic information for disease diagnostics is genetic testing that examines specific disease-related genes and gene variants. Most existing genetic tests detect the presence of mutations in single genes that are known to be linked to a disease. Currently, more than 45,000 genetic tests are available. A comprehensive list of genetic tests can be viewed on the NIH Genetic Testing Registry at www.ncbi.nlm.nih.gov/gtr/. The technologies used in many of these genetic tests are presented elsewhere in the text.

现如今，基因组信息在疾病诊断中最为广泛的应用是遗传检测，即检测特定疾病相关的基因及基因变体。大多数现有的遗传检测技术可以检测已知与疾病相关的单基因突变的存在。目前，市面上提供的遗传检测项目多达 45 000 种以上，在 NIH 遗传检测注册中心的官方网站（www.ncbi.nlm.nih.gov/gtr/）上可以查阅遗传检测项目的详细清单。许多遗传检测所涉及的技术在本书中均有介绍。

Over the last decade, genome sequencing methods have progressed rapidly in speed, accuracy, and cost effectiveness. In addition, other "omics" technologies such as transcriptomics and proteomics are providing major insights into how DNA sequences lead to gene expression and, ultimately, to phenotype.

过去 10 年中，基因组测序法在测序速度、精确度、成本效益各方面均有长足发展。此外，其他“组学”技术如转录组学和蛋白质组学也为 DNA 序列如何引导基因表达、最终表现为表型提供了重大发现。当这些技术的速度越来越快、成本效益越来越好时，它们将开始为精准医疗做出重

As these technologies become more rapid and cost effective, they will begin to make important contributions to precision medicine.

要贡献。

Although the application of omics technologies to precision diagnostics has not yet entered routine medical care, several proof-of-principle cases have illustrated the ways in which whole-genome analysis may be applied in the future. They also reveal some of the limitations that must be overcome before genome-based medicine becomes commonplace and practical.

尽管组学技术在精准诊断中的应用尚未纳入常规医疗中，但是多个原理验证案例已经表明全基因组分析未来可被应用的途径，同时它们还揭示了基于基因组的医学在变得普遍和实用之前必须克服的一些局限性。

One such case study is described here. This study combined data from whole-genome sequencing, transcriptomics, proteomics, and metabolomics profiles from a single patient at multiple time points over a 14-month period. This in-depth, multilevel personal profiling followed the patient through both healthy and diseased states, as he contracted two virus infections and a period of Type 2 diabetes. This research points out how complex changes in gene expression may affect phenotype and shows the importance of looking beyond the raw sequence of DNA. It also indicates that gene-expression profiles can be monitored by current technologies and may be applied in the future as part of personalized medical testing.

这里描述其中一例案例研究。该研究整合了来自单一病人14个月间多个时间点的全基因组测序、转录组、蛋白质组和代谢组信息。这些深入、多层次的个人信息分析追踪了该病人经历两次病毒感染和一段2型糖尿病发病过程中的健康状态和患病状态。该研究指出基因表达的复杂变化可能影响表型，展示了读懂DNA原始序列所蕴含信息的重要性，同时也表明利用当前技术可以检测基因表达图谱，并且作为个性化医学检测的组成部分可能可以应用于未来。

15.4 Technical, Social, and Ethical Challenges

15.4 技术、社会和伦理的挑战

There are still many technical hurdles to overcome before precision medicine will become a standard part of medical care. The technologies of genome sequencing, omics profiling,

精准医疗在成为医疗的标准组成部分之前依然存在很多亟待解决的技术障碍。基因组测序、组学分析、芯片分析和SNP检测等技术还需更快、更准确、更价廉。

microarray analysis, and SNP detection need to be faster, more accurate, and cheaper.

另一个挑战则是海量基因组数据和其他组学数据的储存与诠释。例如，每个个体基因组产生的字符数与200本英文大型电话簿相当，这些数据必须储存在数据库中，并用于挖掘相关的序列突变体。随后，每一种序列突变体必须被赋予意义。为了从事此类分析，科学家需要从研究序列突变体与表型、疾病易感性或药物反应间相互关联的大规模人群基因分型中采集数据。专家推测这些研究在公共和私人研究团队共同努力下需要10年以上才能完成。科学家还需要开发处理这些海量信息的高效自动化系统和算法。精准医疗还需整合环境、个人生活方式和表观遗传因素等信息。

Another challenge will be the storage and interpretation of vast amounts of genomic and other omics data. For example, each personal genome generates the letter-equivalent of 200 large phone books, which must be stored in databases and mined for relevant sequence variants. Then, meaning must be assigned to each sequence variant. To undertake these kinds of analyses, scientists need to gather data from large-scale population genotyping studies that will link sequence variants to phenotype, disease susceptibility, or drug responses. Experts suggest that such studies will take the coordinated efforts of public and private research teams more than a decade to complete. Scientists will also need to develop efficient automated systems and algorithms to deal with this massive amount of information. Precision medicine will also need to integrate information about environmental, personal lifestyle, and epigenetic factors.

精准医疗的第三个挑战则是自动化健康信息技术的发展。卫生保健机构需要利用电子健康文档储存、检索和分析每位患者的基因组信息，并且将这些信息与不断取得进展的基因与疾病相关知识进行比对。目前在美国，大约有10%的医院和医生使用了这些信息技术。

Another technical challenge for precision medicine is the development of automated health information technologies. Health-care providers will need to use electronic health records to store, retrieve, and analyze each patient's genomic profile, as well as to compare this information with constantly advancing knowledge about genes and disease. Currently, approximately 10 percent of hospitals and physicians in the United States have access to these types of information technologies.

精准医疗引起了巨大的社会关注。为了让精准医疗惠及大众，遗传检测费用以

Precision medicine raises a number of societal concerns. To make precision medicine available to everyone, the costs of genetic tests, as well as the genetic counselling that accompanies them, must be reimbursed by insurance companies, even in cases where there are no prior diseases or symptoms. Regulatory changes are required to ensure that genetic tests and genomic sequencing are accurate and that the data generated are reliably stored in databases that guarantee the patient's privacy.

及与之相伴的遗传咨询费用必须由保险公司承担，甚至在没有疾病或症状先兆的情况下。同时还需要监管变化，以确保遗传检测和基因组测序的准确性以及生成的数据可靠储存于能保障病人隐私的数据库中。

Precision medicine also requires changes to medical education. In the future, physicians will be expected to use genomics information as part of their patient management. For this to be possible, medical schools will need to train future physicians to interpret and explain genetic data. In addition, more genetic counsellors and genomics specialists will be required. These specialists will need to understand genomics and disease, as well as to manipulate bioinformatic data. As of 2017, there were only about 4000 licensed genetic counsellors in the United States.

精准医疗还需对医学教育进行改革。未来，基因组信息的使用能力将成为医生管理病人工作的一部分。为了实现这一可能，医学院需要培训未来医生阐明和解读遗传数据。此外，社会需要更多的遗传咨询师和基因组学专家，这些专家能够理解基因组学和疾病，同时熟练操作生物信息学数据。截至2017年，美国仅有大约4000名已获资质的遗传咨询师。

The ethical aspects of precision medicine are also diverse and challenging. For example, it is sometimes argued that the costs involved in the development of genomics and precision medicine are a misallocation of limited resources. Some argue that science should solve larger problems facing humanity, such as the distribution of food and clean water, before allocating resources on precision medicine. Similarly, some critics argue that such highly specialized and expensive medical care will not be available to everyone and represents a worsening of economic inequality. There are also concerns about how we will protect the privacy

精准医疗的伦理学方面同样多元化并富有挑战性。例如，时有争论认为发展基因组学和精准医疗的成本存在有限资源分配不当问题。还有人主张科学应该解决人类正在面临的更大问题，如食物和清洁水的供给，而不是将资源分配至精准医疗中。类似地，一些批评人士指出如此高特异性、昂贵的医疗不可能面向每个人，且会使得经济不平等现象日趋恶化。对于如何保护储存在数据库及个人医疗服务记录中的基因组信息隐私，人们也存在担忧。

of genome information that is contained in databases and private health-care records.

Most experts agree that we are at the beginning of a precision medicine revolution. Information from genetics and genomics research is already increasing the effectiveness of drugs and enabling health-care providers to predict diseases prior to their occurrence. In the future, precision and personalized medicine will touch almost every aspect of medical care. By addressing the upcoming challenges of precision medicine, we can guide its use for the maximum benefit to the greatest number of people.

大多数专家认为我们当前正处于精准医疗革命的起始阶段。从遗传学和基因组学研究中获得的信息已经提升了药物的有效性，也已赋予卫生保健机构在疾病发生之前对其进行预测的能力。未来，精准医疗和个性化医疗将可能涉及医疗的方方面面。通过设法解决精准医疗中不断出现的挑战，我们将引导精准医疗最大限度地发挥其优势，为尽可能多的人服务。